THE EUROPEAN OIL AND GAS CONFERENCE

A Multidisciplinary Approach in Exploration and Production R&D

This conference was organised by the Commission of the European Communities, Directorate-General for Science, Research and Development, in cooperation with AGIP, Italy

THE EUROPEAN OIL AND GAS CONFERENCE

A Multidisciplinary Approach in Exploration and Production R&D

Proceedings of the conference held in
Altavilla Milicia (Palermo, Sicily, Italy), 9-12 October 1990

Edited by

G. IMARISIO, M. FRIAS AND J.M. BEMTGEN
Commission of the European Communities
Directorate-General for Science, Research and Development

Published by
Graham & Trotman
for the Commission of the European Communities

Graham & Trotman Ltd
Sterling House
66 Wilton Road
London SW1V 1DE

Kluwer Academic Publishers Group
101 Philip Drive
Assinippi Park
Norwell, MA 02061 USA

Publication arrangements:
Commission of the European Communities
Directorate-General Telecommunications, Information Industries and Innovation, Luxembourg

ISBN 1 85333 622 X

EUR 13793

First published 1991

British Library Cataloguing in Publication Data

European Oil and Gas Conference (1990 : Palermo, Italy)
The European Oil and Gas Conference : a multidisciplinary approach in exploration and production R&D.
I. Imarisio, G. II. Frias, M. III. Bemtgen, J.M.
333.823

Library of Congress Cataloguing in Publication Data is available.

Printed and bound in Great Britain by
Hartnolls Ltd, Bodmin, Cornwall

ORGANIZING COMMITTEE

Chairman:

G.Imarisio
Commission of the European Communities
Directorate General for Science, Research and Development
Joint Research Centre

Members:

H.L.Beckers, Shell (NL)
E. Bertocco, AGIP (I)
J. Bosio, ELF Aquitaine (F)
G.L. Chierici, University of Bologna (I)
H.J. de Haan, University of Delft (NL)
D.L. Ducate, SPE (USA)
B. Garcia-Siñeriz, Repsol (E)
D.M. Grist, BP (UK)
J.P. Joulia, CEC, DG XVII (B)
J. Makris, University of Hamburg (D)
E. Millich, CEC, DG XVII (B)
J.M. Øverli, Statoil (N)
A.D. Pinto, Partex (P)
F. Rocca, Polytechnic of Milan (I)
P. Simandoux, IFP (F)
W.F. Steenken , EAPG (NL)
D.H. Welte, KFA (D)

CONFERENCE SECRETARIAT

J.M . Bemtgen, CEC, DG XII (B)
M. Frias, CEC, DG XII (B)
L. Nola, AGIP (I)

CONTENTS

INTRODUCTION 1

OPENING SESSION

Opening Address 4
Dr. PH. BOURDEAU, Director Environment and Non Nuclear Energy Research, Directorate General for Science, Research and Development, Commission of the European Communities, Brussels

Opening Address 7
G.M. SFLIGIOTTI, Deputy Director for Planning and Development, AGIP SpA, Rome (I)

Opening Address 8
R. DE BAUW, Director for Energy Technology, Commission of the European Communities

WELL PRODUCTION AND DRILLING

Horizontal wells and reservoir management strategy 14
A. SPREUX and A. JOURDAN, Elf Aquitaine, Pau (F)

Developments in drilling and production 23
L.M.J. VINCKEN, Dietsmann (International), Antwerp (B)

NATURAL GAS

Synthetic fuels from natural gas 34
Å. SOLBAKKEN, Statoil Research Center, Trondheim (N)

More natural gas – A multi-disciplinary challenge 44
P. LEPRINCE and M. VALAIS, Institut Français du Pétrole, Mission Information-Documentation, Rueil-Malmaison (F)

OFFSHORE CHALLENGES IN EUROPE

Offshore challenges in Europe – Progress and future developments 54
G. M. BOZZO, M. BENETTI, Tecnomare S.p.A. Italy, Venezia (I)

GEOSCIENCES

Reservoir development in a complex deposit environment: Germigny-sous-Coulombs aquifer storage 70

M. LEBLANC, J. ROUGE and E. TALLEC, Gaz de France, Research and Development Division, Underground Storage Department, La Plaine Saint Denis Cedex (F)

El Borma field - SITEP - Tunesia - An integrated geological, petrophysical and numerical approach aimed at an infill drilling project 80

C. ROSSINI, P. TERDICH, B. VOLPI, C. DESCALZI, F. GENOVESI and P.P. PIRAS, AGIP - Production Services - Reservoir Engineering Dpt., San Donato Milanese (I), K. KADDOUR, R. GRIBAA, H. GAAYA, M. AZOUZ and C. ATTAYA, SITEP - Direction des Etudes et Developpement, Tunis (TU)

Geological modelling of turbidite reservoirs 90

J.D. SCHUPPERS, M.E. DONSELAAR and C.R. GEEL, Faculty of Mining and Petroleum Engineering, Delft University of Technology, Delft (NL)

The Cleeton reservoir test: Reducing uncertainty 97

I.W. WRIGHT, BP Exploration, Southern North Sea Field Group, Easington, North Humberside (UK)

Geological modelling in HC exploration: Reconstruction of geometries, temperatures, pore-pressures - Principles and applications 106

J. BURRUS, F. SCHNEIDER, S. WOLF, W. SASSI, R. ZOETEMEIJER, Institut Français du Pétrole, Rueil-Malmaison (F)

Further development of hydrocarbon potential forecast system is a basis for efficient oil and gas prospecting and exploratory works 116

N. KULAKHMETOV, A. BREKHUNTSOV, Ministry of Geology of U.S.S.R., Tyumengeologiya Concern, Tyumen (U.S.S.R.)
I. NESTEROV, A. RYLKOV, V. SHPILMAN, Ministry of Geology of U.S.S.R., West Siberian Research Institute of Petroleum Geology, Tyumen (U.S.S.R.)

Sequence architecture and modelling in the western province of the Troll reservoir, Norway - Part 1: sequences, systems tract and sedimentary facies development 121

S.D. NIO, C.S. YANG, Intergeos, Leiderdorp (NL), K. GIBBONS, K. VEBENSTAD, Statoil, Stavanger (N), T. HELLEM, A. KJEMPERUD, Read Production Geology Services A/S, Sandvika (N)

Pore-pressure-induced fracturing of petroleum source rocks. Implications for primary migration 131

F.K. LEHNER, Koninklijke/Shell Exploratie en Produktie Laboratorium, Rijswijk Z.H. (NL)

A multidisciplinary approach to primary migration pathways of petroleum – the Lower Toarcian of NW–Germany as a model source rock – 142
U. MANN, Institute of Petroleum and Organic Geochemistry at the Research Centre (KFA) Jülich, Jülich (FRG)

Ultrasonic velocity: A breakthrough for oil and gas characterization 155
S. YE, B. LAGOURETTE, J. ALLIEZ, H. SAINT–GUIRONS, P. XANS, L.P.M.I. Université de Pau et des Pays de l'Adour (F), F. MONTEL, Elf Aquitaine, CSTJF, Pau (F)

Gravity–assisted inert gas injection: Micromodel experiments and model based on fractal roughness 166
R. LENORMAND, Institut Français du Pétrole, Rueil–Malmaison (F)

Prediction of hydrocarbon phase state at great dephts for oil & gas provinces and areas of the USSR 176
N. NEMCHENKO, All Union Petroleum Exploration Research Institute (VNIGNI), Moscow (USSR), A. ROVENSKAYA, Institute of Geology and Exploration of Fossil Fuel (IGIRGI), Moscow (USSR)

Fines migration in argillaceous sandstones 182
D.C. BULLER & T.R. HARPER, BP Research Centre, Sunbury–on–Thames, Middlesex (UK)

A multidisciplinary approach in reservoir description and production planning 192
A. COSTA E SILVA, J. CARVALHO, Partex, CPS, Lisboa (P)

Facies estimation using wireline logs (electrofacies): A case history from South–East Sicily 204
P. BALOSSINO & A. VALDISTURLO, AGIP SpA, Milano (I), M. LOMBARDINI, Western Atlas International, Ravenna (I)

Reservoir rock characterization using computed tomography 205
B. GORIČNIK, Z. KRILOV, INA–Naftaplin, Zagreb (YY), M. MAROTTI, University Hospital "Dr. M. Stojanović", Zagreb (YU)

Scleroglucan rheological properties and coreflood performances at high temperature 206
C. NOÏK and J. LECOURTIER, Institut Français du Pétrole, R. RIVENQ and A. DONCHE, Elf Aquitaine (F)

Characterization of source rock and reservoir qualities using infrared analysis 207
SH. N. GANZ, F. ÖNER, TU Berlin, SFB 69, Berlin (FRG), W. KALKREUTH, ISPG Calgary, Calg. Alberta (CDN), M.J. PEARSON, Marishal College, The University, Aberdeen (UK) and H. WEHNER, BGR Hannover, Hannover (FRG)

A numerical study on the effect of clay distribution on shaly sand conductivity 208
X.D. JING, J.S. ARCHER and T.S. DALTABAN, Imperial College of Science, Technology and Medicine, London (UK)

Gamma–ray tomography inspection technique for flooding experiments in porous medium 209
J.–R. URSIN, Rogaland University Centre, Stavanger (N)

Characterization of oil components sorbed onto clay minerals and quartz grains in oil sand 210
A. FENDEL and K. SCHWOCHAU, Institute of Petroleum and Organic Geochemistry at the Research Centre (KFA) Jülich, Jülich (FRG)

The effects of brine migration and salt cementation on hydrocarbon reservoirs 211
H. DRONKERT, Faculty of Mining and Petroleum Engineering, Delft University of Technology, Delft (NL) and International Geoservices B.V., Leiderdorp (NL)

Measurement of gas tracer retention under simulated reservoir conditions 212
Ø. DUGSTAD and T. BJØRNSTAD, Institutt for energiteknikk, Kjeller (N)

OPERATIONS

Environmental protection in AGIP hydrocarbons E & P 216
L. CEFFA, G. DI LUISE, G. DOSSENA, Geodynamics & Environment, AGIP SpA, San Donato Milanese (I)

An integrated emergency management system for spills of chemical substances 228
J.B. NIELSEN and T. GUDMUNDSSON, Danish Hydraulic Institute, Hørsholm (DK), H. BACH, Water Quality Institute, Hørsholm (DK)

A multi–disciplinary European approach to developing technologies for the removal of offshore platforms 237
R.W. EBDON, Advanced Mechanics & Engineering Ltd. (UK), R. SURLE, Comex Services, (F), P. MINARDI, Tecnomare S.p.A. (I)

The need of rational criteria for fire safe design of offshore structures 250
R. CAZZULO, C. MURGIA, F. ZILIOTTO, Registro Italiano Navale, Genova (I)

In situ measurement and numerical modelling of the reservoir compaction and of the induced surface subsidence 260
M.J. BOUTECA, Y. MEIMON, Institut Français du Pétrole (F), D. FOURMAINTRAUX, Elf–Aquitaine (Production)

The application of physics in exploration and production of oil and gas 272
T.M. QUIGLEY, A.P. FOAKES, S. SIMMONDS, V.H.Y. TAM, BP Research, Sunbury Research Centre, Sunbury on Thames, Middlesex (UK)

Abandonment and removal of steel platforms 282
J.M. MARTIN BOURGON, Repsol Exploracion S.A., M. MOREU, A. MORON, Seaplace Iberia S.A. (E)

MECHANICS AND NEW MATERIALS

POSEIDON: The multiphase production theoretical approach becomes reality 286
A. LAFAILLE, TOTAL Compagnie Française des Pétroles, Paris La Défense (F)

The unmanned platforms management in North Adriatic Sea 296
M. GUIDA and S. TONELLI, AGIP S.p.A., Production Dpt., San Donato Milanese (I)

PLATINE – Research project on unmanned offshore field 307
P. LEFEVRE, Elf Aquitaine (Production)

Column stabilized production platforms 317
A. REY-GRANGE, Seamet International, Paris (F), D. ARMENIS, ALFAPI, Athens (GR), R.L. JACK, Noble Denton, London (UK)

Determination of in-situ stresses for the design of hydraulic fractures: A three-years study 330
J.P. SARDA, P.J. PERREAU and M. BOUTECA, Institut Français du Pétrole, Rueil-Malmaison (F), P. CHARLEZ, Total-CFP, Paris (F), J.L. DETIENNE, Elf Aquitaine, Pau (F)

The forged nodes application for off-shore steel jackets 340
F. NICOLUSSI, L. PELLIZZARI, Tecnomare S.p.A., Venezia (I), W. STORESUND, Veritec A/S, Hovik (N)

Eureka EU-191 advanced underwater robots: The WIR technology development program 349
W. PRENDIN, D. MADDALENA, A. TERRIBILE, Tecnomare S.p.A., Venezia (I), T. HEDGE, Ferranti ORE, Great Yarmouth (UK)

The development of a subsea three phase metering system 359
T.L. DEAN, Texaco Limited, London (UK), E.L. DOWTY, Texaco EPTD, Bellaire, Texas (USA) and R.J.J. JISKOOT, Jiskoot Autocontrol Limited, Tunbridge Wells (UK)

An advanced system to prepare and to follow drilling and completion operations 370
J.-M. COURTEILLE and P. BOUTROLLE, Elf Aquitaine, CSTJF, Pau (F)

Monitoring of flexible risers by acoustic emission 376
A. SUGIER, P. MARCHAND, Institut Français du Pétrole, Rueil-Malmaison (F), J. MALLEN, A. MARION, COFLEXIP, Paris (F)

Injection of glycol into CO_2-containing natural gas in pipelines for corrosion mitigation. Glycol/water phase behaviour 387
J.N.J.J. LAMMERS, Koninklijke/Shell-Laboratorium, Shell Research B.V., Amsterdam (NL)

A new instrument for measuring gross heating value of natural gas 397
D. INGRAIN, J.H. ZORIO, Direction des Etudes et Techniques Nouvelles, Gaz de France (F), G. DESERT and L. MONDEIL, Division Recherche et Développement en Production, Société Nat. Elf Aquitaine Production (F)

New selective water control processes: Main characteristics and field testing 398
N. KOHLER and A. ZAITOUN, Institut Français du Pétrole (F), J.F. COSTE and J.P. ZUNDEL, Elf Aquitaine (F)

MARELAB – An advanced offshore laboratory platform for oceanological research and field testing of new underwater technologies 399
G. SEBASTIANI, CEOM S.C.p. A., Palermo (I)

Mass balance evaluation in gas treating and processing plants by PC-oriented programme package 400
F. SIMON, Z. CSERMELY, Hungarian Oil and Gas Research Institute, Veszprém (H), J. SIKLOS, Trust of Hungarian Oil and Gas Industry, Budapest, (H)

A flow loop to test drilling fluids under bottomhole conditions 401
D. DEGOUY and J. LECOURTIER, Institut Français du Pétrole, Rueil-Malmaison (F)

403
Applications of high performance composite tubes to deepwater risers
C.P. SPARKS and P. ODRU, Institut Français du Pétrole, Rueil-Malmaison (F)

THOR 2 – Tig Hyperbaric Orbital Robot. Second generation automatic hyperbaric welding system 404
J. BLIGHT & R. ROUGIER, COMEX, Marseille (F)

Computational tools for the analysis and design of TLP tethers 405
J.F. McNAMARA and M. LANE, Marine Computation Services International, Galway (IRL)

Ultrasonic sensor for measuring free- and dissolved gas in drilling fluids 406
O.M. VESTAVIK, B. AAS and G.W. HALSEY, Rogaland Research Institute, Stavanger (N)

String failure and remedial actions in ultradeep drilling 407
G. BORRIELLO and M. PRAZZOLI, AGIP Drilling Technologies R&D Department

Dimensional verification to be carried out at open sea on offshore structures by means of photogrammetry 409
M. RAMPOLLI, Agip SpA, Offshore Department, Industrial Photogrammetry Group, Cologno Monzese (Mi) (I)

Development and tests of a supervisory controlled telemanipulator system for industrial underwater applications 410
P. MINARDI and D. MADDALENA, Tecnomare S.p.A., Venezia (I)

INFORMATION PROCESSING

The use of seismic velocity anomalies in predicting gas saturated layers 412
A. CARLINI, S. CORNINI, Agip S.p.A., Geophysical Research and Development Department, Milan (I)

Drill bit noise as a seismic source in geophysical surveys 422
G.P. ANGELERI, AGIP, Milano (I), S. PERSOGLIA, F. POLETTO, OGS, Trieste (I)

Horizon processing in 3-D seismic interpretation 432
J.C. MONDT, Koninklijke/Shell Exploratie en Produktie Laboratorium, Rijswijk ZH (NL)

Monte Carlo simulation of lithology and porosity from seismic data 445
P.M. DOYEN and T.M. GUIDISH, Western Geophysical, a Division of Western Atlas International, Isleworth, Middlesex (UK)

Transmission and reflection tomographic inversion of offset VSP-data and the use of the results for a target oriented processing of reflection seismic data 455
B. LEHMANN, C. GELBKE, Deutsche Montan Technologie, Institut für Angewandte Geophysik, Bochum (FRG), D. KROLLPFEIFER, L. DRESEN, Ruhr-Universität Bochum, Institut für Geophysik, Bochum (FRG)

Neural networks applications within IFP 465
B. BRAUNSCHWEIG, J.-M. LAMBERT, Institut Français du Pétrole, Rueil-Malmaison (F), P. NAIM, Auralog, Palaiseau (F)

LOBSTER: An expert system for the interpretation of sedimentary environment from core analysis 476
L. MATTEINI, F. FONNESU and G. DI DIO, AGIP S.p.A., Milan (I)

Exchange format for optimal transfer of geological and geophysical 3D subsurface model data 483
M.H. MULDER, S. PEN, I.L. RITSEMA, TNO Institute of Applied Geoscience, Geoenergy Department, Delft (NL)

XPS–FROCKI – An expert system for fluid rock interaction problems in oil production 489
M. ALBERTSEN, DGMK Deutsche Wissenschaftliche Gesellschaft für Erdöl, Erdgas und Kohle e.V., Hamburg (FRG), H. KNOKE, D. SCHENK, Geologisch–Paläontologisches Institut und Museum der Universität Kiel, Kiel (FRG), A. PEKDEGER, L. THOMAS, Institut für Angewandte Geologie, Fachbereich Geowissenschaften der Freien Universität Berlin, Berlin (FRG)

Outcrop studies and geostatistical modelling of a Middle Jurassic brent analogue 497
C. RAVENNE, R. ESCHARD, D. GUERILLOT, HERESIM GROUP, Institut Français du Pétrole, Rueil–Malmaison (F), A. GALLI, H. BEUCHER, HERESIM GROUP, Centre de Géostatistique, Ecole des Mines de Paris, Fontainebleau (F)

Analytic non–linear forward and inverse seismic scattering: A perturbation approach 521
B. KUMMER, University of Hamburg, Institute of Geophysics, Hamburg (FRG)

Petrophysical characterization of carbonatic and clastic reservoirs using acoustic waveforms logs 522
M. GONFALINI, M. PIANA and H. ANXIONNAZ

Damage diagnosis using knowledge processing techniques 523
M. TAMBINI and G. COSENZA, AGIP S.p.A., Milan (I)

Polarization: A new tool 524
CH. CLIET and M. DUBESSET, Institut Français du Pétrole, Rueil–Malmaison Cedex (F)

COMPLEX: An expert system for well completion design 525
J. MONNET, C. PRESLES, A. RICORDEAU, D. EHRET and J. VANDEVELDE, Société Nationale Elf Aquitaine (Production), CSTJF EP/S/PRO/FIP, Pau (F)

CONVERSION OF NATURAL GAS TO LIQUIDS

Natural gas exploitation: Snamprogetti technologies network 528
D. SANFILIPPO and A. PAGGINI, Snamprogetti S.p.A., San Donato Milanese (I)

The oxidative coupling of methane in a recirculating fast fluid bed reactor 537
S.J. KORF, J.A. ROOS, P.T. COOLEN, J.G. VAN OMMEN and J.R.H. ROSS, Faculty of Chemical Technology, University of Twente, Enschede (NL)

IFP processes for the direct conversion of methane into higher hydrocarbons 544
J. WEILL, Vernaison (F), C.J. CAMERON and C. RAIMBAULT, Institut Français du Pétrole, Rueil–Malmaison (F)

The shell middle distillate synthesis process 553
H.M.H. VAN WECHEM, Shell Internationale Petroleum Maatschappij, The Hague (NL), P.L. ZUIDEVELD, Shell International Gas Limited, London (UK), M.M.G. SENDEN, Koninklijke/Shell Laboratorium Amsterdam, Amsterdam (NL)

Computer–aided thermodynamic analysis of light olefin conversion 560
M. MOLINARI and M. BRUNELLI, Eniricerche S.p.A., San Donato Milanese (I), M. LUNELLI, Dipartimento di Matematica, Università degli Studi di Milano (I)

Conversion of synthesis gas to liquid oxygenated products over promoted rhodium–based catalysts 561
P. COMOTTI, S. MARENGO, Stazione sperimentale per i Combustibili, San Donato Milanese (I), S. MARTINENGO, Dipartimento di Chimica Inorganica e Metallorganica, Università di Milano, Milan (I), L. ZANDERIGHI, Dipartimento di Chimica Fisica ed Elettrochimica, Università di Milano, Milan (I)

Methane conversion by reaction with O_2/H_2, O_2/Cl_2, O_2/Metal Oxide 562
I. VEDRENNE, J. SAINT–JUST, A. BENHADID, Gaz de France (DETN), La Plaine Saint Denis Cedex (F), G.M. CÔME, R. LE BEC, P.M. MARQUAIRE, P. BARBE, F. BARONNET, CNRS (URA 328), INPL (ENSIC) et Université de Nancy I, Nancy Cedex (F)

Single step catalytic oxidation of methane to formaldehyde 563
E. MAC GIOLLA CODA, M. KENNEDY, J.B. MCMONAGLE, B.K. HODNETT, Dept of Materials and Industrial Chemistry, University of Limerick, (IRL), J. VAN OMMEN and J.R.H. ROSS, Dept of Chemical Technology, University of Twente, (NL), J.W.M.H. GEERTS and K. VAN DER WIELE, Eindhoven University of Technology, (NL)

FINAL DISCUSSION AND CLOSURE SESSION 564

LIST OF PARTICIPANTS 570

INDEX OF AUTHORS 600

INTRODUCTION

Although a number of regular European Conferences are dedicated to the research and technical fields in hydrocarbon discovery and exploitation, most are specialized in particular disciplines or aspects either by specific planning or by long association. The European Oil and Gas Conference held in 1990 for the first time was fully designed to cover multidisciplinary aspects of a number of disciplines in the above techniques and scientific fields. This approach was fully supported by the Commission of the European Communities in connection with its R&D and Technological programmes in the same fields, which deliberately aim at grouping together a wide variety of disciplines in collaboration with the oil and gas industry.

The Commission of the European Communities sponsored the Conference and organized it in collaboration with AGIP. The following companies gave their support towards its organization : BP, Elf Aquitaine, Gaz de France, IFP, Repsol, Shell, Teconomare and Veba Oil and are thanked for their contributions.

The advent of modern information processing and transfer, the clear perception that further significant advances could be obtained only by means of a broader view, even in very specialized domains, and the more integrated nature of science education have all shown that boundaries between areas of science and engineering are artificial and should be crossed with real benefit. Frequently we now seek solutions to problems or search for new approaches by using the know-how developed in another discipline. The multidisciplinary character which was the main motivation of this Conference responds to this specific necessity.

The areas covered by the first European Oil and Gas Conference were: Information Processing, Geosciences, Mechanics and New Materials, Gas Conversion and Operations.

The Organizing Committee selected, from the large number of abstracts received, a set of papers that convey technical advances and a variety of approaches within each session, while maintaining the multidisciplinary approach of the Conference. Some of the papers presented in each session result from novel multinational projects part-sponsored by the European Community.

Seven plenary papers emphasized the theme of discipline integration. Major poster sessions also successfully encouraged free discussion around a further presentation of concepts. The lay out of the Torre Normanna conference centre promoted communication and discussion. The principal sessions of the Conference may be summarized as follows:

- Information Processing is fundamental to exploration technology and featured as one of the larger sessions, with papers extending from the use of seismic velocity anomalies to the development of a hydrocarbon potential forecast system.
- The Geosciences session ranged from the numerical modelling of sedimentary basins to uncertainty reduction in gas reservoir sizing.
- The Mechanics and New Materials session ranged in scope from drilling systems

through hostile well completions to unmanned platforms.

- The rise in resource importance of remote gas accumulations was acknowledged with a session on the Conversion of Natural Gas to Liquids.
- The Operations session primarily featured environmental and safety issues.

In order to accommodate an adequate selection of the papers offered while maintaining a certain homogeneity, two parallel session streams were organized. This was done in such a way, however, that cross-disciplinary exchanges were easy, since disciplines likely to interact were never treated contemporarily in the parallel sessions. Therefore participants were able to move from one to the other to attend presentations of their direct or indirect interest. The final plenary session was organized as a synthesis of the technical sessions and finished with an open debate on the main problems treated and on the potential future evolution of petroleum related technologies.

With 300 inscriptions, participation in the Conference was very satisfactory and nearly all participants were present until the last session. The site chosen, far from any major town, favoured exchanges and discussions in and outside the meeting rooms. The multidisciplinary approach of the Conference and the high quality of the papers presented were no doubt the major factors which raised the interest in the event and promoted the inter-disciplinary exchange observed during the Conference.

OPENING SESSION

OPENING ADDRESS

Dr. Ph. Bourdeau
Director, Environment and Non Nuclear Energy Research
Directorate General
for Science, Research and Development
Commission of the European Communities
Brussels

Ladies and Gentlemen,

It is a great pleasure for me to address you this morning and to tell you about the role of EC in research and also how the Commission views this conference.

Its originality lies in the fact that it attempts to bring together experts from several disciplines all of which are important in oil and gas exploration and production, in the hope that there will be an opportunity for mutual benefits, for instance in terms of methodology transfer. We hope that there will be synergistic effects between the sciences or disciplines which are represented in this conference : information processing which is of paramount importance and is broadly treated in the programme; earth sciences which are obviously fundamental for the sector; mechanics and new (advanced) materials; and operations with emphasis on the safety and environmental aspects.

The programme includes many papers, to be presented orally or as posters. These originate from large oil companies, small high technology firms, universities and public research centres. We are particularly glad to see that some come from countries beyond the boundaries of the European Community, not only from EFTA countries but also from Hungary, USSR and Algeria. This multidisciplinary and multi-national approach augurs well for the success of this conference. We hope that this formula will prove to be useful and that perhaps a second conference could be held along the same lines in future.

The interest of the EC in research and development was significantly strengthened in 1987 when the Treaty establishing the European Economic Community was amended by the Single European Act. This stipulates inter alia, that the promotion of research and technological development was one of the objectives of the EC, aimed at increasing the competitiveness of European industry. The Act also prescribes that this should be done through the implementation of "Framework Programmes for Research and Technical Development" which would determine for periods of 4 to 5 years the various specific programmes to be undertaken as well as their level of funding. We are now initiating the third Framework Programme (1990-1994), which has superseded the 2nd one (1987 to 1991). This formula of rolling or sliding programmes allows for the regular updating of research priorities in order to cover newly emerging needs.

In the energy sector the on-going JOULE programme is concerned with the supply and use of all forms of energy , with the exception of nuclear. Research on fossil fuels, which represent 80% of the energy supply in our modern societies, includes hydrocarbons exploration and exploitation. This is complemented by an important action on technological development in the same field which is managed by the Directorate General for Energy of the Commission and about which you will hear later on from my colleague Mr. Robert de Bauw.

The rules for R&D support by the Community favour medium to long-term, pre-competitive projects. Pre-competitiveness is a concept that may be interpreted in various ways, but essentially it means that the projects are at such a stage of development that companies which are competing on the market, can still cooperate because the results of these projects do not attain the degree of marketable systems or equipments.

In the past most projects submitted in response to public calls for proposals and eventually funded originated from a single proposer. Now this has changed: proposals must be multipartner and multinational, involving at least two member states. In some programmes, countries belonging to EFTA are allowed to participate under special conditions. Nowadays the Community is also looking forward to cooperation with countries of Eastern and Central Europe which have become closer and it is thinking of formulae to involve them in its programmes.

In the field of non-nuclear energy, the JOULE programme which is in full course, provides a test for the feasibility of those multinational projects.
We have tried to encourage applicants to prepare large projects - projects which have a well identified objective and bring together people from various disciplines and from various countries, ideally mixing scientists and engineers from industry, universities and public research centres. With regard to exploration and production of hydrocarbons, the Commission has recently launched a project which is perhaps a model of this style. The "Geosciences Project" involves 13 universities, 3 public research centres and 3 oil companies from 9 EEC countries. The total cost is 14 MECU from which 7 MECU are provided by the Community budget (as a rule, the intervention of the Community budget is limited to 50%; however, contracts with universities may cover all of marginal costs, that is all additional costs that are needed to implement the project).

Besides favouring large multi-partner projects, the EC also allows the possiblity of implementing exploratory research, so that innovative research proposal by a single person or group may still be considered.

The new Framework Programme (1990-1994) which was approved in April 1990 is now being implemented. It provides for a specific R&D programme in non-nuclear energy with a budget of 157 MECU, which is currently under discussion in the European Parliament and the Council of Ministers. It is hoped that a final decision will be obtained within the next few months.

With regard more specifically to hydrocarbon exploration and production, an area for which particularly close cooperation with the THERMIE programme is necessary, R&D will concentrate on the more basic aspects in the geosciences such as advanced prospection methods, high resolution techniques, 3D methods, etc., as well as on integrated advanced tools and methods in reservoir characterization and reservoir management and fluid mechanics in reservoirs. Additionally, work on natural gas will be supported as well.

In the longer term, it is likely that EC resources for R&D in general may well increase. In addition there is clearly a renewed interest in energy R&D. Consequently one may reasonably expect a reinforcement of EC support for research in the subject of this conference, which I hope will be successful. Thank you for your attention.

OPENING ADDRESS

G.M. SFLIGIOTTI

Deputy Director for Planning and Development, AGIP SpA, Rome

Mr. Chairman, Distinguished Participants, Ladies and Gentlemen,

I will be very brief and confine myself to only a few remarks.

When the Organizing Committee of this Conference started working on it, in November 1988, the oil scenario was characterized by a situation of

- Abundance of supply
- Low Prices: The average price of Brent crude for the month of November 1988 was 12,90 $/b.

Now, the situation has reversed:

- We have, or may have, a situation of actual or potential shortage of oil;
- Prices have reached the level of 40 $/b and nobody can predict the level they may reach should the present situation of uncertainty last for long, or should it break out into war.

I know that present oil prices of about 40 $/b are lower in today's $ than the prices reached in 1981–82. Nevertheless, we cannot but be concerned about the consequences on our economies – and those of developing countries – of such high level of prices, should the present situation last for a long period of time.

This Conference, which had to take place with a calm background scenario, takes place instead in a situation of worries and concern. This is certainly good for the organizers because it is going to attract more attention from the specialists, the media and the laymen in general! But this is a meager consolation indeed for a situation of great concern for the lives of so many and for the risks of a world catastrophe.

I know quite well that the present difficult situation of the oil industry is not due to problems of technical nature; it is due to political and military reasons.

Nevertheless, present difficulties show that they can be eased, if not in the short term certainly in the medium and long term, by further improving the technical sides of the oil industry.

I have in mind the still existing margins for further reduction in the costs of the upstream phases of our industry and for new technical breakthrough. Thanks to these improvements, we are going to have a better exploitation of existing reservoirs and we could afford to invest for exploration and production in more hostile and difficult areas. All this will bring about more oil and gas, and achieve a wider geopolitical diversification, which recent events prove to be of paramount importance.

At this point, going into more technical details it would mean for me "to bring coal to Newcastle". Something one should carefully avoid doing. I therefore think I should stop now.

One thing, however, I would like to underline: tackling the problems of the oil industry with a global, multidisciplinary approach, as you have done in this Conference, is very important, indeed it is mandatory.

I am sure your work will bring new light to the difficult, complex problems of our industry and help oil companies to take the right decisions and make wise investments.

Let me end by whishing you all a profitable attendance to the Conference and a pleasant stay here in Sicily.

OPENING ADDRESS

R. DE BAUW
Director for Energy Technology
Commission of the European Communities

Ladies and Gentlemen,

It is my pleasure and honour to address this European Oil and Gas conference.

As Mr Sfligiotti said, the present geopolitical situation, due to the Gulf crisis, underlines the importance of Oil and Gas for Community energy supply : this increases the timeliness of this conference. For the European consumers, oil and gas supply is not a game where the price of the barrel goes up and down by several dollars per day. Supply requires growth under stable conditions. The availability of new cost efficient technologies with minimal impact on the environment is an important factor in achieving such stability and therefore a key element for the future.

This conference, organized by the Directorates-General XII (research) and XVII (energy), with the support of AGIP, will significantly contribute to the debate on progress in the development and promotion of energy technologies. In fact, the need for multidisciplinarity along the whole chain, from science to engineering, has become more and more evident during the past few years. I want to thank the organizers, and especially AGIP, represented here by my old friend Mr Sfligiotti, for having focused the objectives of this first conference on this notion of multidisciplinarity. Nobody could expect that all of the issues linked to it will be solved by the end of this week, but your proceedings will undoubtedly have the merit of highlighting the complexity of the problems and of opening up new avenues for intersectorial cooperation.

1 - PAST ACTION

As we all know, Community involvement in the development of oil technologies is not recent. I do not wish to give the impression that Community action is wholly responsible for all that the Oil industry has achieved during the past 15 years, particularly in the development of the North Sea. Nevertheless, I would like to express my firm belief that the Community has made a significant and positive contribution to it. The process started in 1973, in a "crisis" atmosphere, similar to the one in which we are living now. At that time Member states recognized the need to develop sound and efficient techniques to improve Community security of supply.

Over a period of 16 years more than half a billion Ecus have been allocated to some 500 projects. This substancial effort has been successful in the sense that new cost-effective technologies have been introduced to the market. Being in Sicily, I am happy to quote as an example the laying of pipelines in the Strait of Messina, whose studies and tests established the feasibility of deep sea transport of gas. Based on this success, the Transmed was constructed, which by bringing Algerian gas to Europe, represents a major contribution to our natural gas supply.

Later on, other major innovatory techniques have benefited from Community support : horizontal drilling, tension leg platform design, dynamic positioned drilling, etc.

More recently, high-tech in the field of seismic investigation has been supported; also subsea operations with a great variety of techniques from well-heads with automatic intervention means to subsea systems for production treatment or pumping.

Multidisciplinarity has been an emphasis of Community action

The Commission has always viewed technological progress in the hydrocarbons sector as dependent upon advances on a number of fronts, ranging from basic and fundamental research to development, demonstration and finally market promotion. Progress in reservoir management cannot be achieved without significant improvements in the knowledge of the relationships between fluid and rocks. Another example is multiphase technology, a subject of primary importance for the future exploitation of deep sea or marginal fields. Advancing in this field implies studying erosion-corrosion phenomenae as well as solving all the difficulties associated with the pumping of the effluent.

In summary, I would like to say that the future success of the scientific, technical and industrial community is highly dependent upon the circulation of information, the exchange of experience and the removal of barriers between disciplines and fields of activity.

2 - THERMIE to promote energy technologies

The removal of barriers is a requirement not only for research and development but also for the promotion of energy technologies, including those in the oil and gas sector. After an evaluation of the programme undertaken since 1973 for the development of new technologies in the hydrocarbons sector, the Commission concluded last year that this action had to be extended and set in a new, wider framework of support for energy technologies.
This new programme, under the name THERMIE, was finally approved on 29 June 1990 and came into force on 18 July.

It aims at the promotion of energy technologies in order to improve their market penetration. This is a multiannual, multidisciplinary energy programme covering the fields of :

- rational use of energy,
- renewable energy sources,
- solid fuels,
- hydrocarbons.

THERMIE offers Community industries and the Commission far greater possibilities than those which were offered by the previous so-called "hydrocarbons" and "demonstration" programmes.

Basically THERMIE will grant financial support for the promotion of 2 types of projects.

- Innovatory projects whose purpose is to implement innovatory techniques, or processes, or products. This type of project should prove the viability of new technologies by applying them on a sufficiently large scale for the first time;

- Dissemination projects which are designed to encourage broader utilization, either under different economic or geographical conditions, or with technical modifications of tested technologies where residual risks still impede market penetration.

A major innovation compared with the past programmes is the possibility given by THERMIE to encourage specific projects, the so-called "targeted projects". When a need is not being met or where significant technological advance could be achieved through cooperation, the Commission can encourage or coordinate the setting-up of such projects. Investigating which technological fields could benefit from such targeted projects is one of the tasks, in which the Commission services are currently involved.

Multiphase technology has often been mentioned as a suitable candidate for targeted projects. The results of an EEC workshop held at Copenhagen in September 1989 have emphasised that problems in this field are complex and require cooperation, which can only be achieved only by means of a multidisciplinary approach.

THERMIE provides financial support of up to 40% of the eligible project costs, with no requirement for repayment of the money in the case of commercial success. A total financial envelope of 350 Million Ecus has been allocated for the period from 1990 to 1992, with a provision for the same amount to be available for 1993-1994. However, for this first year, 1990, only 45 Mecus have been made available. This means that the budget for each of the next two years will be somewhere around 150 Mecus. Later, it could hopefully be increased to 175 Mecus/year.

Since I mentioned that several energy sectors are covered by THERMIE, the question arises of the division of the financial envelope between them. 75% of the total budget is to be shared equally between each of the 4 sectors of application. The remaining quarter is at the disposal of the Commission for orientating the programme towards priority areas and the most innovative technologies.

3 - COOPERATION AND MULTIDISCIPLINARITY

European "industries" in the Oil and Gas sector have shown their readiness to cooperate in the development of technologies under Community programmes. More and more cooperative projects are being presented, to such an extent that last year more than 50% of the funds were allocated to these. Obviously, this trend will continue, in view of the multidisciplinarity which is required to solve future problems in E & P. This is why THERMIE puts an emphasis on cooperative projects and requires that, for large projects - greater than 6 Mecus - such cooperation is compulsory.

Let me underline that this requirement is not only a dream of civil servants or Ministers, producing clever intellectual constructions. Cooperation is a very efficient way of achieving progress by selecting the most promising resources and allocating them in the best possible way, both financially and in terms of manpower. This, however, has to be mitigated by a reasonable degree of competition, stimulating innovation and effectiveness.

Multidisciplinarity - the general purpose of this conference - frequently requires a cooperative approach to innovation and THERMIE will encourage it.

Dissemination : a new approach to the promotion of energy technologies

Since it is more market orientated than its predecessors, the THERMIE programme provides the possibility for **associated measures** such as :

- **evaluation** of the market potential together with sectoral diagnosis and feasibility studies,
- **monitoring** and **evaluation** of projects
- **dissemination** of information and results. A wide variety of media are available for this.

One of the most novel features of THERMIE is the stress which it puts on the dissemination of energy technologies. Dissemination may appear relatively easy to achieve in the Oil and Gas sectors, and also in the Coal sector, due to the structure of the industries. However, the obstacles to the penetration of energy technologies are much more difficult to overcome in other sectors covered by the programme. Despite this structural differentiation, there is a strong case for improving dissemination in all of the fields of application of THERMIE.

THERMIE, therefore, pursues an integrated approach which will provide European industries with Community financial support for achieving greater efficiency of their development and promotion efforts. In the sector covered by this conference - Oil and Gas - this means a stronger presence of the oil and its related industries on export markets, resulting in a diversification of resources and an increased security of supply.

THERMIE a unique opportunity for European industries

It is now clear that THERMIE opens up new opportunities to European industry. The onus is now on industry to seize these opportunities. The implementation of THERMIE should result in a new relationship between the Community, industry and all of the national bodies involved. Events such as this conference facilitate the circulation of information in order to achieve an increased efficiency of Community action. If, as I said earlier, problems are growing more complex and require investigation in a number of fields in order to achieve adequate solutions, it is obvious that the circulation of information is more and more important. This is particularly the case for the Commission, which could not manage the THERMIE programme efficiently without close contacts with industrialists.

However, civil servants and politicians are no substitute for industrial leaders. It is their responsibility to coordinate their research, development and marketing approaches, whilst respecting the competion rules so as to maintain the leading role which they have already taken in many of the most up-to-date E. & P. technologies.

I am sure many of you will share my view that this is one of the primary objectives of this conference, if not the prime. I hope that these 3 days will reinforce the perception in the industry that Community action can help in this approach towards multidisciplinarity, and I am sure that the intellectual climate of this region - Sicily, a crossroads of civilisations and a link between continents - will help in achieving this.

WELL PRODUCTION AND DRILLING

Session Chairman: J. BOSIO – ELF AQUITAINE (F)

"HORIZONTAL WELLS AND RESERVOIR MANAGEMENT STRATEGY"

Alain SPREUX and André JOURDAN

Elf Aquitaine
Avenue Larribau,
64018 Pau, France

Summary

During the past ten years, horizontal drilling has assumed ever greater importance in the oil industry. Initially reserved for reservoirs of a specific nature, today the horizontal well is no longer the exception and should be regarded in exactly the same light as a conventional well, i.e. as an option to be considered for the best economical and technical reservoir development scheme.

After a reminder of the main characterisics of horizontal wells and the related advantages, this paper reviews the concept of the horizontal well in the different phases of the life of a field, from initial appraisal to final recovery.

1. INTRODUCTION

The spectacular increase in the number of horizontal wells drilled in recent years is an obvious sign of the interest that is being focused on horizontal drilling technology. In response to the growing evidence of successful field applications, the oil industry has been obliged to review its approach to reservoir management. Horizontal well technology, initially applied only to marginal and uneconomic reservoirs, has subsequentely been utilized for other types of applications where it has demonstrated its potential for increasing production and cost saving. Today, horizontal wells are attractive in most situations and many are convinced that this technology could completely transform the exploration, development and production of petroleum resources.

With this rapid growth, both industry and U.S. regulatory agencies have begun to address the concerns of horizontal wells. New regulations are appearing in order to overcome problems such as the disproportion between the length of the wells and the production block size, or the fact that the rig site is situated a long way away from the pay zone (Figure 1). An possible change concerns prevention of oil and gas wastage as a horizontal well may drastically modify the rate of reservoir depletion.

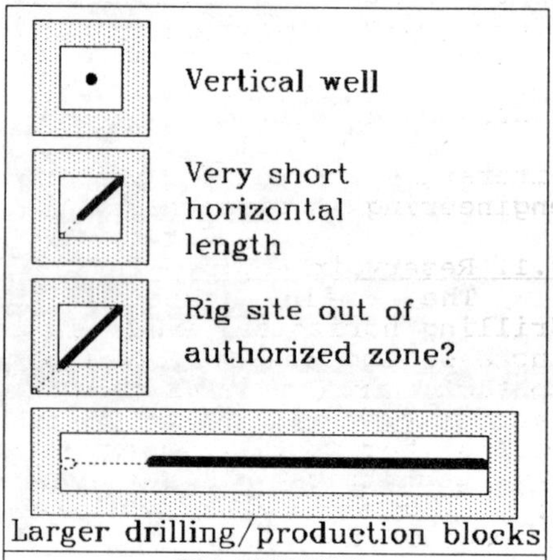

Fig.1: REGULATION

2. HORIZONTAL WELL HISTORY

The concept of the horizontal well has tantalized the oil industry since the late 1920's. But it is only in the last decade that horizontal wells have gained a permanent foothold.

The literature describes a first attempt to drill a horizontal well in Texas in 1929. It is mentioned that two lateral drains 23 to 24 foot long, drilled from an existing vertical well, resulted in a production forty times greater than that of the initial well. A good start!

Subsequently, a number of successes followed: in 1936 (daily production multiplied by four), in 1939/40 (an increase in production of 667%!), in 1946 (a well shut down for 18 years as it was producing only 1 BOPD with substantial sand entries (35%) was side-tracked with eight lateral drains; this resulted in production stabilizing at around 20 BOPD with sand entries reduced to 8%). In all these cases, short-radius laterals were drilled from existing vertical wells. In the 1950's, horizontal wells are mentioned in the U.S.S.R..

The ardour for horizontal wells faded in the 1950's. It was during this period that the first convincing results were obtained with hydraulic fracturing which continually won out as a more economical means of improving production. Horizontal drilling was still in its infancy, lateral extension remaining fairly limited due to the inability to accurately control the well trajectory.

Therefore horizontal drilling was used only in a workover situation, i.e. for remedying the insufficient production of a well. A turning point came around ten years ago with a new approach which considered the horizontal well as a conventional means of development with a greater production potential and/or enabled specific problems to be solved. Several field experiments were conducted including Fort Mc Murray (Texaco Canada), Normal Wells and Cold Lake (Esso Canada), Lacq Superieur and Rospo Mare (Elf Aquitaine)...

In the mean time, spectacular technology advances considerably increased the drilling and completion capability and reliability. Furthermore, conventional production improvement techniques, such as multifracturation, water or steam injection, etc, were modified to maximize the advantages offered by the horizontal hole configuration.

3. HORIZONTAL WELL TECHNOLOGY

These points have been extensively discussed in the literature so only a brief reminder of drilling and reservoir engineering is given here.

3.1. Reservoir engineering

The main reason for drilling horizontal wells is to increase the reservoir/wellbore contact area. The two key factors governing this are the well's horizontal length and its geometrical configuration in relation to the discontinuities of the reservoir.

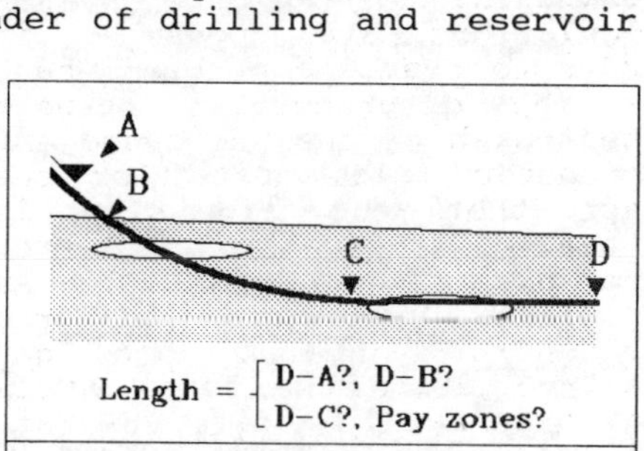

Fig.2: HORIZONTAL LENGTH DEFINITION

$$J = \frac{q}{P_e - P_w} = \frac{2\pi h k_h}{\mu B} \bullet \frac{1}{f\,(\text{geometry})}$$

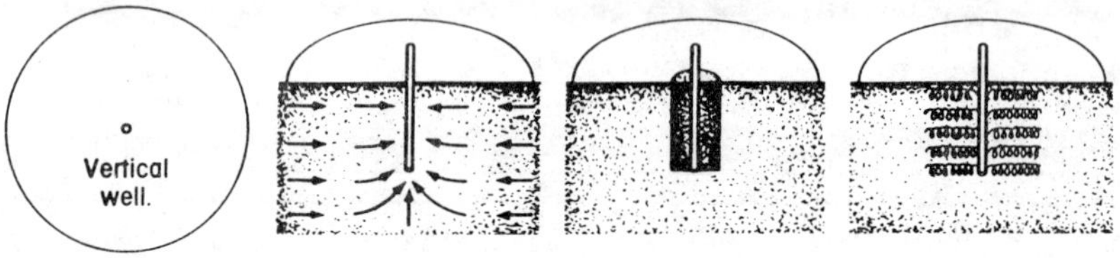

$$f\,(\text{geometry}) = \text{Ln}\left(\frac{r_e}{r_w}\right) + S_g + \frac{h}{h_p} S_c + \frac{h}{h_p^2}\,\frac{u}{2\pi r_w}\,\frac{\rho}{\mu}\,q$$

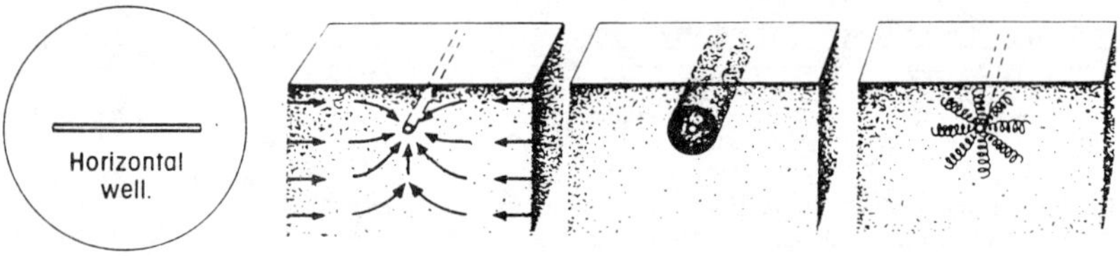

$$f\,(\text{geometry}) = \text{Ln}\left(\frac{r_e}{L/4}\right) + a\frac{h}{L}\,\text{Ln}\left(\frac{h}{2\pi r_w}\;\frac{1}{\cos\left(\pi \frac{e}{h}\right)}\;\frac{2a}{1+a}\right) + a\frac{h}{L} S_c + \frac{2a^2}{1+a}\,\frac{h}{L^2}\,\frac{u}{2\pi r_w}\,\frac{\rho}{\mu}\,q$$

$L < r_e \quad ah < r_e$

This figure shows the different terms of the pressure drop equation between a cylindrical isopotential with radius r_e and the central well when flow is permanent.

Merkulov's formula corresponds to the case of a horizontal well. It should not be used when the conditions mentioned are not complied with ($L < r_e$, $ah < r_e$).

- r_e = External radius
- r_w = Well radius
- h = Reservoir thickness
- h_p = Perforated interval (vertical well)
- L = Perforated length (horizontal well)
- e = Vertical distance between the horizontal well and the middle of the reservoir
- k_h = Horizontal permeability
- k_v = Vertical permeability
- u = Turbulence parameter
- a = $\sqrt{k_h/k_v}$
- u = βk_h (u and β are assumed here to be isotropic)
- q = Production rate
- β = Usual turbulence factor (see Katz)
- μ = Fluid viscosity
- ρ = Fluid gravity
- B = Fluid volume factor
- P_e = External pressure
- P_w = Well pressure
- S_c = Skin effect
- S_g = Geometrical skin effect (convergence effect)

Figure 3 - Generalization of Merkulov's formula

Firstly the horizontal length: today, it is possible to drill several hundred metres, even over a thousand metres in a reservoir. The advantage of this length is easily seen in the general expression of Merkulov's law, taking into account the permeability anisotropy, the convergence effects, the formation damage and the turbulence (Figure 3).

Secondly the geometrical configuration: the possibility of staying well away from unwanted fluids, of optimizing the orientation of the drain-hole according to fractures, of taking advantage of the variations of reservoir properties related to the stratification or to discontinuities are just a few of the factors in favour of the horizontal well (Figure 4).

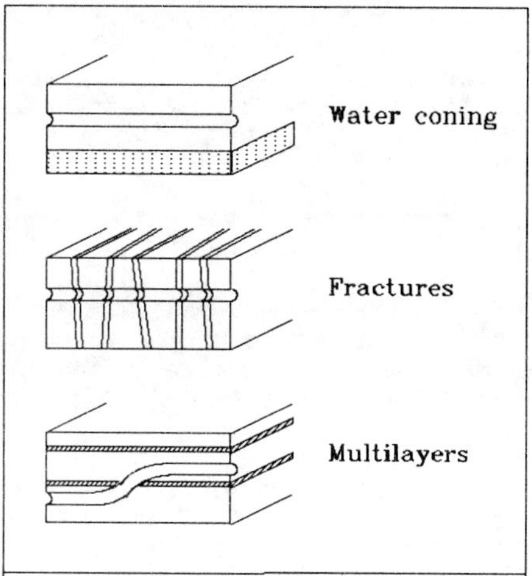

Fig.4: Drainhole vs reservoir geometry

The combination of these two points enables improved well productivity, a higher critical flow rate when there is coning, an increase in recovery, a better sweep efficiency, an increment of drainage area, and, in the case of sand production, a reduction of the rate of sand influx into the well.

It should not be forgotten that we have the same methods at our disposal for characterizing the reservoir as for conventional wells, e.g. logging, well testing, geostatics... as well as analytical and numerical tools for forecasting well performance.

3.2. Reminder of drilling techniques

Horizontal drilling techniques are widely discussed in the literature. Here, let us simply recall that we generally classify the wells in three main categories according to their radius, : long-radius, medium-radius and short-radius wells. The geometrical characteristics of these different radius wells are shown in Figure 5.

	Well classification		
	Long radius	Medium radius	Short radius
Radius of curvature (m)	300–750	100–200	6–30
Build up rate (°/10m) (°/m)	2–0.75	6–3	10–2
Standard hor. length (m)	400–1500	300–600	100–200

Fig.5: WELL CHARACTERISTICS

In a long-radius well, the build-up of the angle is achieved using conventional deviation equipment, and the inclination gradient is limited to 2 degrees per 10 meters. The medium-radius method requires modified drilling assemblies, generally called "steerable systems", which are designed to match with standard oil field hardware.

The maximum practical dogleg corresponds to the semi-standard API tubular stresses limitations for bending and torsion.

Short-radius systems are defined by use of specific flexible or articulated tools.

A particular concept to be borne in mind is that of the target. With a conventional well, the target can usually be sufficiently well defined by an area at the top of the reservoir. With a horizontal well, the target must be defined by the area of impact at the top of the reservoir and by a "horizontal" volume as shown in Figure 6.

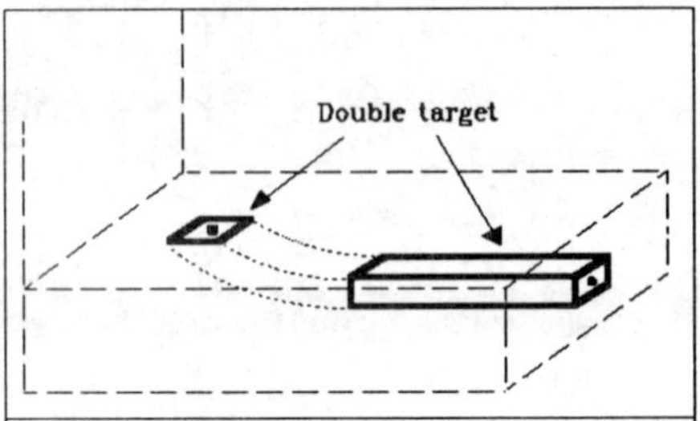

Fig.6: DRILLING TARGET

Finally, it is interesting to note that the greatest vertical depth reached up to now is 3500 metres and the longest horizontal reservoir drilled is 1750 metres.

4. APPLICATION OF A HORIZONTAL WELL

Initially, most horizontal wells were considered as a remedial or ultimate solution, i.e. as a means of prolonging the life of the well. Often, such a well was drilled as a last resort, after all other aids to increasing production had failed.

Technology advances have now made the horizontal well a "high performance" and reliable tool with potential application at any stage of a reservoir life.

4.1. The horizontal well as an exploration/appraisal tool

A simplified description of how the elements making up a reservoir are organized could be given as a vertical distribution of fluids and a distribution which is both vertical (stratification) and horizontal (heterogeneities) for the rock. It is obvious that under such conditions a horizontal exploration well (or horizontal wildcat) cannot be envisaged: due to the lack of crucial information, it is not possible to decide at what depth the drainhole should be positioned and, consequently, deeper levels would not be reached.

In the case of intensive exploration, i.e. when a minimum amount of data is already available, the problem is not the same. Figure 7 shows two examples where a horizontal well could be very useful.

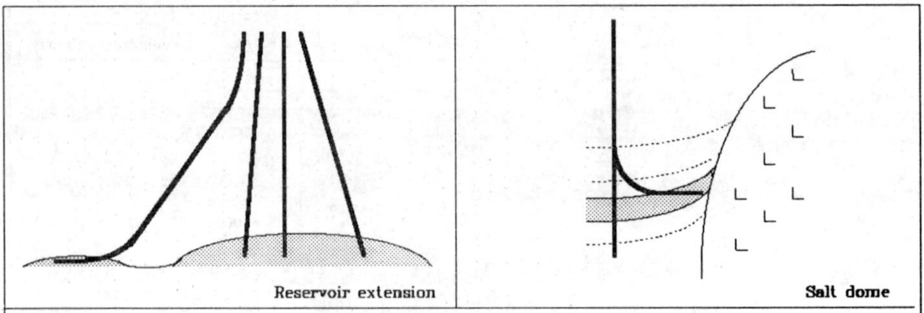

Fig.7: EXPLORATION/APPRAISAL WELL

In the first case, after having explored the levels on the flank of a diapir, it would be possible to look for the extensions of these levels and for a possible stratigraphic trap by means of a lateral drain.

The second case shows the existence of a reservoir; extrapolation of the seismic data indicates the possibility of a secondary reservoir. What is more, we would have a good idea of its depth. It would be possible to envisage drilling a horizontal well in this reservoir which, after confirmation of a discovery, could be directly converted into a production well. Naturally, the decision to drill a horizontal well will depend on the corresponding technical and economical factors.

Reservoir appraisal concerns both the exploration and the development phases. From the exploration point of view, what differentiates them from the above wells is the fact that we are looking for specific parameters.

A horizontal well may be a useful tool for investigating the geological characterization as it can provide new data concerning the evolution of facies, the distribution of heterogeneities, the presence of fractures or the lateral limits of the reservoir...

Among the exploration techniques used, logging is currently employed, either MWD in real time, or conventional drill pipe conveyed logging, as well as horizontal coring. Experiments with VSP have also been carried out with a triaxial geophone displaced along a horizontal drain recording the signals from vibrators placed vertically to the well (Figure 8).

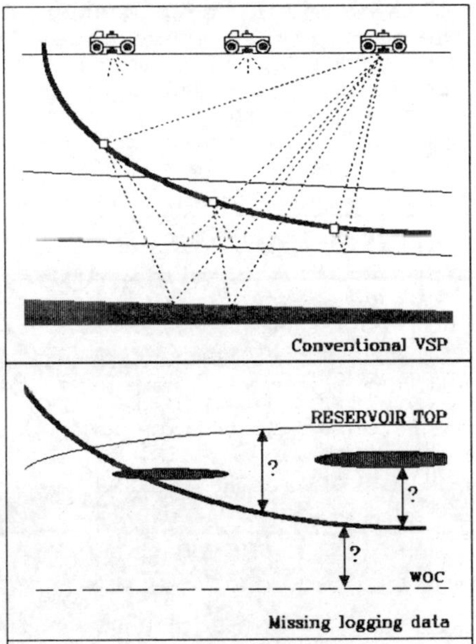

Fig.8: EXPLORATION DATA

However, geological knowledge of the reservoir provided by a horizontal well must be treated with caution as considerable uncertainties remain with regard to its vertical environment. Deep investigation logging tools are not yet available which would tell us either the distance of the well from an interface or from the top at any point along the drainhole, or the existence of barriers (Figure 8).

There are several examples of geological data and/or surprises provided by horizontal wells (even development wells) drilled in supposedly well-known reservoirs, e.g. reservoir limit, presence of fractures, shaly layers, etc.

4.2. Appraisal/long term well test

In the case of a discovery, appraisal of the production potential of the reservoir provides valuable information to aid in the development decision. For this purpose, it might be useful to consider the advantages of a horizontal well. For partial development or as a production pilot, the drilling of a horizontal well could be justified. Due to its high production potential, it can be used in the detailed evaluation phase as a self-paying early production system. This well could be either a new well, or the result of sidetracking an earlier discovery well, depending on the best technical and economical conditions.

4.3. Development well

It is in the development of oil fields that the horizontal well is most widely used. By associating horizontal wells and conventional wells, it is possible to increase the size of the area drained from a platform or a cluster and to reduce the total number of wells while obtaining optimum recovery.

In the past few years, we have witnessed a number of horizontal pilot wells which, once their potential was confirmed, subsequently replaced programmed conventional development wells. Examples are Prudhoe Bay (BP), Zuidwal (Elf Petroland), Dan Field (Maersk). In these cases, the horizontal wells were introduced when development was already underway. The majority of these cases confirmed that the horizontal well was well-adapted to deal with the problems arising and it is possible to imagine that these fields could have been developed solely by means of horizontal wells. This is certainly the case of Rospo Mare (Elf Aquitaine) in Italy (Figure 9).

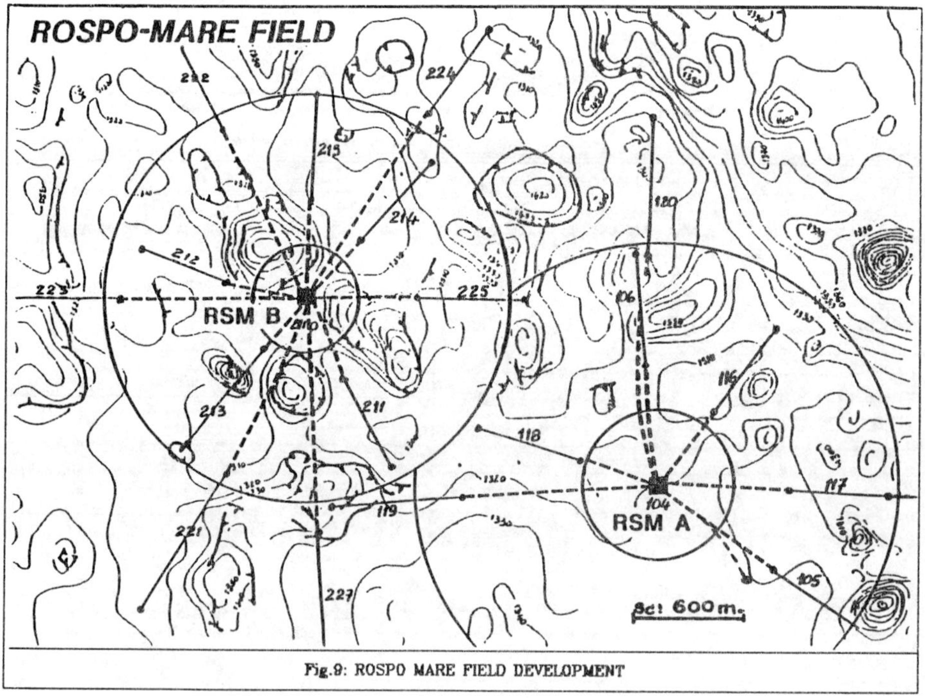

Fig.9: ROSPO MARE FIELD DEVELOPMENT

It is also necessary to consider the case of a complementary development. This generally poses additional technological problems, for example, due to distances to be reached which are a long way from the zone already developed. Extending the drainage area to distant zones is well within the capacity of horizontal wells.

Looking at existing horizontal wells, it can be noted that know-how regarding completion techniques is rapidly progressing. So horizontal wells can be envisaged in many different types of reservoirs: unconsolidated sand (Chateaurenard / Elf Aquitaine, Helder Field / Unocal, Rabi / Shell), compact reservoir requiring fracturing (Dan Field / Maersk), multi-level reservoir (Zuidwal / Elf Petroland)...

We cannot leave this chapter without mentioning the current boom in the production of the Austin Chalks reservoirs in Texas ("The drilling activity in the Pearsall Field ... has been described as "hysterical"). These reservoirs consist of thin fractured layers located inside large compact zones and containing enormous reserves. Low producer conventional wells have been replaced by horizontal wells, some of which produce up to a thousand barrels per day.

A development case worth a particular mention is that of underground gas storage. Although to our knowledge no horizontal well has yet been used for gas storage, it is easy to imagine a considerable reduction of the volume of the gas cushion remaining in the reservoir by positioning a horizontal well at the top of the structure, resulting in a substantial saving.

4.4. Enhanced recovery

The horizontal well can also be used as a tool for enhanced recovery. At the end of the production life of a reservoir, it is sometimes necessary to displace the drainage points, for instance, by raising the perforation depth in an attempt to reduce water production.

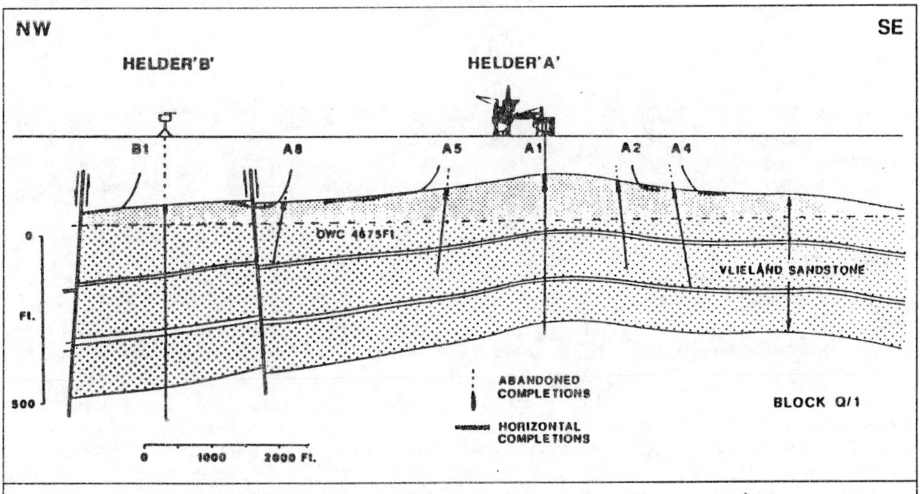

Fig.10: HELDER FIELD CROSS SECTION (from P.J.Murphy, SPE., JPT June 1990)

A sidetrack drain can be the right option to maintain its level of production under very attractive economic conditions. One of the most successful examples is Helder Field (Unocal/Netherlands) which "was virtually redeveloped by drilling eight horizontal wells", all of which were drilled off existing wells (Figure 10).

The horizontal well is also used in association with enhanced recovery techniques, typically as a water injection well. In this instance, high rates of injection over long distances can be achieved and be an efficient solution to the problems of maintaining pressure in the reservoir or oil sweeping... Other uses are for injecting steam (Lacq Supérieur / Elf Aquitaine) or for draining the oil displaced by steam injection in a conventional well (North Tangleflags / Sceptre Reservoir).

5. CONCLUSIONS

The horizontal well represents a significant technological development with positive implications for both exploration and production. This does not mean that horizontal wells can be successfully employed everywhere. The successful developments are in fact the result of very careful selection of suitable wells, based on good reservoir knowledge, efficient use of well performance prediction tools and being in full control of the conditions under which the well is drilled.

Through this review of horizontal well applications, our wish has been to show that the technical and economical advantages of horizontal wells deserve consideration at all the stages in the life of a field. Today the horizontal well is a well like any other. As part of field development and for an optimum reservoir management strategy, horizontal boreholes should not be neglected.

ACKNOWLEDGEMENTS

The Authors thank Elf Aquitaine for its authorization to publish this paper

REFERENCES

. C.R. HYLAND - "Drain Hole Drilling - An Old Idea Whose Time is Now" - paper SPE 12792, Long Beach CA, April 11-13, 1984

. O. de MONTIGNY and al. - "Hole benefits, reservoir types key to profit" - *Oil & Gas Journal*, July 11, 1988

. O. de MONTIGNY and al. - "Horizontal-well drilling data enhance reservoir appraisal" - *Oil & Gas Journal*, July 4,1988

. A. JOURDAN and al. - "How to Build an Hold 90 Degree Angle Hole" - paper SPE/IADC 18707, New Orleans, Feb.28-Mar.3, 1989

. D. GUST - "Horizontal drilling evolving from art of science" - *Oil & Gas Journal*, July 24, 1989

. B.J. MAHONY - "Horizontal Technology as Applied to Marginal Production" - paper NMT 890007, Socorro NM, October 16-19, 1989

. P.C. CROUSE - "Reserve potential due to horizontal drilling is substantial" - *World Oil*, October 1989

. "Horizontal drilling scores more successes" - *Oil & Gas Journal*, February 26, 1990

DEVELOPMENTS IN DRILLING AND PRODUCTION

L.M.J. Vincken
Technical Director
Dietsmann (International)
Noorderlaan 133, Box 23,
B - 2030 Antwerp

Summary:

1) Operator - The oil company
2) Contractor - A service rendering equipment owner directly responsible to the oil company
3) Service company - A third party with or without ownership of equipment working either directly or as a subcontractor for the oil company

Developments in Drilling and Production

Mr Chairman, Ladies and Gentlemen, if this conference would have taken place in January 1990, an oil and gas industry analyst would probably have opened this session with an optimistic note indicating that there was evidence that the fundamental turnaround in the industry was underway. The same analyst would probably have changed his mind, maybe even more than one time during the course of the year.

I am not an analyst and whilst I do not envy them with the difficult task of predicting the future of the oil and gas industry I must confess that although they were right in their views of a turnaround, they missed the date. The fundamental turnaround started some four years ago and has so far not come to an end. The price per barrel dropped, did not recover to previous levels before the Gulf crisis and the expectation of such lasting is low. The oil companies embarked on a cost cutting exercise and this has not abated. New and stricter demands on safety and protection of the environment place additional financial burdens on the industry. Contractors and service companies, with some exceptions, are still struggling with low margins and one can validly ask the question "What's new"?

It is my intention to let you share with me some facts and views, not new, to some of you, but definitely put in a different context. I have taken the liberty of presenting the views of the oilproducer, the contractor and of other people whose views are often not published.

DRILLING

Let us first have a look at the drilling industry. We all know the immediate effect of the 1986 collapse of the oilprice. Most operating companies made large cuts in their budgets and the number of active drilling rigs fell sharply. Consequently day rates dropped accordingly but not enough to entice operators into low cost drilling.

The following years drilling companies aimed for cash flow rather than profit and a number of them did not survive. Staff numbers were reduced and many good "hands" decided to try their luck elsewhere. What we did not know at the time is that many would not come back in better times. I will later on address the staff problem.

Some operators and drilling contractors who had weathered through earlier ups and downs took this time a fundamentally different look at drilling, more particularly they decided to investigate the possibilities of different approaches in drilling leading to reductions in footage costs and acceptable margins for the drilling contractor. I would like to discuss the three most important outcomes of this approach.

Research and Development

Firstly Research and Development. It is clearly understood that R&D is essential to further ones business. One only has to look at Japan to understand the significance they attach to R&D programmes. Japanese companies will spend 150% more on factories, equipment and research this year than they did in 1985. In the US such spending rose only 23% over the same period. At the 1987 World Petroleum Congress I had the honour to present a paper on R&D for the drilling industry. At the time there was much debate on the necessity and usefulness of R&D for what was considered a mature and somewhat declining industry. Moreover, drilling contractors were of the opinion that this was not the time to invest in fancy new equipment and even if new technology would become available financial constraints would limit development and application thereof. Examples are the fabrication of fourth generation semi-submersibles, further development and application of MWD, the mechanisation (automation) of the total drilling process and the development of alternative methods and equipment for horizontal drilling.

Notwithstanding these reservations the forward vision and the preparedness to invest of some operators and contractors albeit in most cases companies with a strong cash position, allows us to note significant progress today.

The development of horizontal drilling and completion is definitely of great importance to the industry and most rewarding in yield. The next speaker will amplify this technology.

Another significant development is the new generation drilling rig. Although this new piece of equipment has so far no track record expectations are high.

Not many operators have taken the bold step to introduce a novel drilling machine but the trend towards alternative drilling methods can best be illustrated by the number of top drives installed or on order.

Measurement while drilling (MWD), not so long ago considered to be the luxury horse for rich operators is rapidly becoming commonplace and offering significant advantages for instance in horizontal drilling. MWD and its newly emerging sister, logging while drilling (LWD), will be commodities in the decade ahead. Most systems offer reservoir evaluation tools.(Table I)

Table I

MWD SYSTEMS

* DRILLING MEASUREMENTS
    + INCLINATION (ANGLE)
    + DIRECTION (AZIMUTH)
    + MAGNETIC AND GRAVITY TOOLFACE
    + WEIGHT ON BIT
    + DOWNHOLE TORQUE
    + ANNULAR TEMPERATURE AND PRESSURE
* GEOLOGICAL DETAILS
    + HIGH RESOLUTION GAMMA RAY
    + RESISTIVITY
* TRUE LOGGING WHILE STEERING
    + DRILLING MEASUREMENTS
    + GEOLOGICAL DETAIL
    + QUICK DATA TRANSMISSION TO SURFACE
    + QUICK UPDATE AND INTERPRETATION
* SUBSURFACE/SURFACE INTERFACE
    + PRESSURE TRANSDUCER FOR SIGNAL DETECTION
    + ANALOG PRESSURE RECORDER
    + ELECTRONIC SIGNAL DECODING EQUIPMENT
    + DIGITAL AND ANALOG READOUTS AND PLOTTERS
    + SOFTWARE

Drilling contracts

Secondly I would like to mention the way in which drilling contracts were put together and some recently reintroduced, although modified, approaches. For many years most operators outside the US used day rate drilling contracts. In an up market it is time saving and the most familiar way of putting together a contract. Easy comparisons between drilling contractors can be made in a tender-board and once a contract is awarded everybody is happy as long as actual drilling times are not too far off budget. The operator has almost total responsibility, calls the shots and provided no major disasters occur has a reasonable grip on the total costs. There is no particular encouragement to improve as long as the equipment in use is operational and the crews are competent. Moreover the expertise of the drilling contractor is far from fully exploited, he executes a given programme and as a matter of fact longer drilling times, within limits, work to his advantage. The system is rather rigid and does not offer cost incentives.

Although commonplace for many decades in the United States, some operators and drilling contractors outside the US have lately had the courage to introduce incentive contracts with a more intense participation of the drilling contractor in the planning and supervision part of the well programme.

Staffing

Thirdly there is the problem of staffing, not so much staffing for office support of a drilling campaign, but staffing at the wellsite. The ups and downs in the drilling activity and particularly the rather drastic reduction of activity in 1986, leading to many lay off's of experienced professionals, made many engineers and drilling hands decide to turn away from the rig floor and try their luck somewhere else. Several studies conducted in the US and Europe have confirmed this trend. Moreover one may assume that with the experience of a cyclical industry, the untidiness of fieldwork, the many transfers of field-personnel and the availability of alternative jobs, few experienced "hands" will come back and new recruits will not last.

Some drilling contractors and operators subcontract rigcrewing to service contractors providing technical personnel and complete drilling crews. This enables flexibility in pay roll numbers and quick build up for rig crews. With new demands on the enhancement of safety and care for the environment the service contractors will have to train their personnel to the same technical and safety standards as their peers be it drilling contractor or operator. In the past, profit margins enabled all personnel to be trained properly but with present minimal margins the service contractor can only maintain training up to a point beyond which he will have to choose between two primary objectives: securing profit or training at par with the operator. The choice is obvious unless he finds a willing ear with the operator and training and the cost thereof can be shared.

I believe that parity in skills between operators, contractors and service company personnel deserves serious attention and should be of more importance in tender-board decisions. It could be dealt with as a special item and remunerated separately.

Table II

DRILLING

* RESEARCH AND DEVELOPMENT
* DRILLING CONTRACTS
* STAFFING

RECOMMENDATIONS:

* CAPITALISE ON THE RESULTS OF R&D
* MAKE USE OF INCENTIVE CONTRACTS
* ACCOMMODATE THE CONTRACTOR AND SERVICE COMPANY

In conclusion the drilling industry will in my opinion have to:

capitalise on the results of R&D,
make use of incentive contracts and
accommodate the contractor and the service company
(Table II)

PRODUCTION AND MAINTENANCE

Let us now turn to Production Operations.
Production operations are up to a point less dramatically influenced by relatively short-term price fluctuations. Nevertheless, also here events in the mid eighties led, as we shall shortly see, to new approaches.

It is extremely difficult to give a definition of production operations as the responsibilities vary from company to company. I take the liberty to include maintenance of equipment in the tasks of the production operator. Again I will discuss three areas of importance:
plant and equipment, the actual operations and the need to combine operators and maintainers.

Plant and equipment.

At first glance the essentials of an oil and gas processing plant have not changed too much. Individual wellstreams come together in a manifold, the well effluents are separated into two or three phases and intercooling or heating may be present at several stages in this process. Vessels, valves, coolers/heaters and regeneration units are bulky and demand a lot of space. Many design bureaus still use these basic building blocks in order to arrive at a process plant with ample redundancy. The same philosophy is used for most of the auxiliary equipment. For many years the trend was very much towards heavier and bulky platforms.

The collapse of the oil price in the mid-eighties prompted a new line of thinking: how to design, construct, install and operate cheaper. Where it concerned cutting costs of newly built plant, first reaction of the operators and the contractors was rather slow. As of late this picture is changing. The lead was taken by the process engineers who contributed to substantial reductions in deck space and deck load requirements. Several North Sea examples have been amply discussed in the available trade literature but if one examines them carefully what often transpires is that the mode of operation has changed and not the equipment proper. Let me try to illustrate this with some examples.

The need for two or three parallel separating trains and 100% redundancy was challenged and as a result we now see one train in some cases and no redundancy. The benefits of the elimination of a train of separators can easily be guessed. The vessel design, however has not changed significantly.

Turbine driven power generation for the drilling process and for other platform purposes, separated in the past, has been combined but the turbines are still the same.

Space and weight reductions have been secured by tender or jack-up drilling without introducing novel drilling units.

It is only now that the first fruits of many years of research come to the market, one of the most important being the introduction of the multi-phase pump. Tests conducted in laboratories are very promising and the concept of multi-phase pumping will no doubt have a profound impact on conventional and underwater oil and gas production.

Other areas of plant and equipment development worth mentioning are:

a) the coming of age of underwater oil and gas wells resulting in a line of standard wellhead and well completion equipment rather than custom made one-off solutions.
b) the successful economical application of unconventional platforms such as the tension leg platform, the single point moored production system and the floating production platform. All these applications incorporate in one way or another the functions of production, injection, storage and off-loading. Two significant details are the switch in design from different components to incorporated systems and the application in hostile environments not for trial reasons but based on solid economic grounds.
c) the application of large diameter, high pressure, flexible hoses in a dynamic mode. Whereas in the past the creation of a reliable S/N curve was a designers nightmare, laboratory and long duration field tests have now installed sufficient confidence in the dynamic behaviour of multi layer hoses.
d) the introduction and further development of alternative separators based on centrifugal forces amongst which some applications underwater. Furthermore, at least two operators have prototype flowline separators on the drawing board.

Operations.

Some 35 years ago a major operator produced 350,000 barrels from 32 unmanned offshore (lake Maracaibo) production platforms. In the mid eighties all that seemed to have been forgotten and the emphasis was on more and more people on the platform and huge demands for platform hotels and therefor on deckspace. There is of course no direct comparison to be made between a relatively simple production and injection platform in a moderate weather area at short distance from a shore base and a Northern North Sea sophisticated platform in the Brent or Statfjord field. It appears now, no doubt under pressure of necessary cost reductions and to a certain degree as a result of the North Sea learning curve, that vast savings can be made by reducing and in some cases eliminating the offshore requirements for platform personnel. Could it be that upmanning was a result of a wrong interpretation of the difference in sophistication between old and new?

The operator and the maintainer.

Whereas the need for continuous presence of the facility operator is much debated, events of the last two years have clearly re-emphasised the important role of maintenance on the platform. It is the least glamorous part of the entire life cycle of an offshore platform but the more important one in budgetary and manpower requirements terms. It can be stated that Northern North Sea platforms have a continuous need for maintenance and minor modifications.

Table III

PRODUCTION

* PLANT AND EQUIPMENT
* OPERATIONS
* OPERATORS AND MAINTAINERS

RECOMMENDATIONS:

* IMPLEMENT RESULTS OF R&D STEP BY STEP
* USE PROVEN EQUIPMENT IN SYSTEM ENGINEERING
* COMBINE OPERATORS AND MAINTAINERS
* USE CONTRACTORS WHERE POSSIBLE
* REDUCE NUMBER OF CONTRACTORS

On numerous occasions it has led to shortages of available beds and the deployment of flotels alongside the platform.
The introduction of multi-skilled teams (rather than multi-skilled individuals), also used in the late fifties, consisting of operators and maintainers with multi-skilled capabilities was almost born out of a need in order to reduce mass transportation and lodging of offshore workers. It requires training but above all convincing of people proud of their particular trade. I am a firm believer of the benefits of multi-skilfulness in teams provided it is properly prepared with due respect for the players.

The parallel development of maintenance management systems and preventive and predictive maintenance tools and techniques is to be applauded but should not lead to desk bound execution of maintenance or diverging courses between planners and doers.

Finally the role of the contractors in operations and maintenance. It serves no purpose to have more than 20 different contractors on a platform, which is a fairly regular feature on some Northern North Sea platforms. It complicates command lines and lowers the overall efficiency. It may at first glance seem to be cost effective to tender for short duration contracts for particular jobs but operators are discovering that their purpose is better served with longer contracts and a bundling of activities under the same supervisor. Variations of work load can be spread out over several platforms and in this manner peak shaving can be achieved.
(Table III)

In conclusion my recommendations to the operator would be to:

continue with R&D and apply the results in a step by step fashion

concentrate on system engineering making use of proven equipment

combine operations and maintenance staff into multi skilled crews

make efforts to reduce manpower and use contractors for peak shaving activities

reduce the number of different contractors to an absolute minimum and aim for longer duration contracts

NATURAL GAS

Session Chairman: D.M. GRIST, BP (UK)

SYNTHETIC FUELS FROM NATURAL GAS

Å. SOLBAKKEN

Statoil Research Center, Trondheim

1.INTRODUCTION

When discussing synthetic fuels ("syn fuels")from natural gas, there is a good reason to ask ourself why we want to convert the gas to other fuels and even suffering an energi loss by doing so. Natural gas is a very clean, convenient and efficient fuel as we see from Table 1. It is only hydro electric power and nuclear energy which have less emissions.

TABLE 1
ENERGY SOURCES

CONSUMPTION AND EMISSIONS
FOR 1 MWH

FUEL	Kg	Kg CO_2	Kg S
NAT. GAS. 80%EFF.	88	243	.0002
NAT.GAS. 50% EFF.	142	390	.0003
COAL. 40% EFF.	240	830	.45 1)
HYDROELECTRISITY	-	0	0
ATOMIC ENERGY	-	0	0

1) 100 PPM IN FLUE GAS. 'STATE OF THE ART'

STATOIL RESEARCH CENTRE. 8/8-90 AAS

However, in spite of this, there are important reasons for the conversion of natural gas to what we call <u>Synthetic Fuels.</u> We will mention some of these reasons and also some of the methods which are being used or have a potential to be used for such conversion.

2.WHY CONVERT NATURAL GAS?

There are both political,technical and environmental reasons for the conversion of natural gas to liquid fuels.

One technical/economical reason is **<u>transportability</u>**. This factor is important for "shut-in" gas fields far away from markets. The alternative value of the gas at the field is very low. Conversion to a liquid fuel is an alternative, and could in many cases well compete with LNG.

A **<u>political</u>** reason, which is very important, is crude oil pricing.In many countries it is an expressed policy to develope synfuels which can put a ceiling on oil prices when within the next 10-15 years a few nations virtually will have a monopoly on oil production (See ref. 1).

A third reason, and one which have increased in importance is the **<u>environmental aspect.</u>** Most of the syn fuels based on conversion of natural gas are high quality clean burning fuels.

In the following we will not include **methanol**, allthough it probably is the most convenient syn fuel product,with its good properties as a clean burning fuel.

3.THE ORIGIN OF MODERN SYNTHETIC FUELS.

A number of papers was printed in"Brennstoff Chemie" in the beginning of the 1920's. They were written by Franz Fischer and Hans Tropsch under a general title "How to produce mineral oil from carbon monoxide and hydrogen."

Their research led to an extensive activity in Germany in the 30'ies and during the second world war. It was all aimed at making Germany less vulnerable to its shortage of oil.

In addition to the now famous Fischer-Tropsch (F.T.) process, Germany also applied high pressure hydrocracking of coal. However,most important was the F.T process with the standardized boiling water cooled reactor, with more than 2200 double tubes filled with cobolt catalyst which converted synthesis gas into liquid hydrocarbons.

FIG.1 SYNFUELS

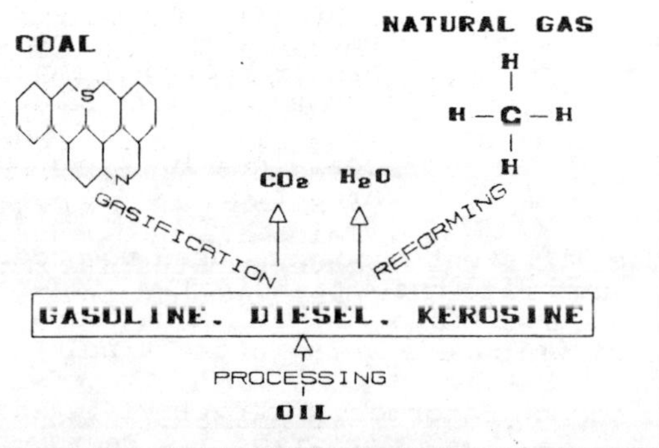

Fig 1 gives an illustration of the conversion of coal and natural gas to known fuels, and also the difference between those feedstocks.

Natural gas, with methane as a main constituent, is rich in hydrogen with a 4 to 1 ratio between hydrogen and carbon. This extra hydrogen ends up as water.

Coal however, has a low content of hydrogen, and to make products as gasoline, diesel and kerosine, carbon dioxid has to be discarded from the process.

One of the most important intermediates in making the products above, is **synthesis gas,** which is a mixture of mainly hydrogen and carbon monoxide.

4. REFORMING TO SYNTHESIS GAS

Methane is a very stable and unreactive molecule and conversion to other products requires very severe conditions.
There are two main reactions which converts natural gas to synthesis gas:

1. **Steam reforming**, which is a highly endothermic reaction between steam and natural gas following reaction 1:

$$CH_4 + H_2O \rightarrow CO + 3H_2 \qquad (1)$$

2. **Partial oxidation** which is an exothermic reaction between natural gas and oxygen:

$$CH_4 + 1/2\ O_2 \rightarrow CO + 2H_2 \qquad (2)$$

Separately or in combination these 2 reactions are the basis for modern **reforming plants** of which a few will be presented below.

Reaction (1) and (2) represents a highly ideal description of the processes. In addition we have important side reactions like the shift reaction which forms hydrogen and carbon dioxide from carbon monoxide and water.

FIG.2 NATURAL GAS CONVERSION

ENERGY YIELD [1]

PRODUCT	THEORETICAL	PRACTICAL
PARAFFINS	78-84%	55-65%
GASOLINE	70-73%	50-55%

1) HEAT OF COMBUSTION OF PRODUCTS/
HEAT OF COMBUSTION OF NATURAL GAS USED

Before we go into the question of synthesis gas reformers, we will have a look at fig. 2 which gives theoretical and practical energy yields in conversion processes.

When making paraffins from methane both the theoretical and practical energy yields are higher than for gasoline. The reason is of course the higher hydrogen content of the paraffins compared to gasoline which contains low hydrogen aromatics.

Ref.(3) gives a survey of the different processes used to produce synthesis gas. We will only summarize the 4 principally different types of commercial reformers available. In addition there are 2 processes under development which are very interesting.

STEAM REFORMING is the most common reformer. A flowchart is shown in fig 3. Energy is suplied by external heating of numerous tubes filled with catalyst.The pressure is max 20-24 bar, and the outlet temperature 850-900 °C. The process makes surplus hydrogen for most uses except for ammonia.

FIG. 3 STEAM REFORMING

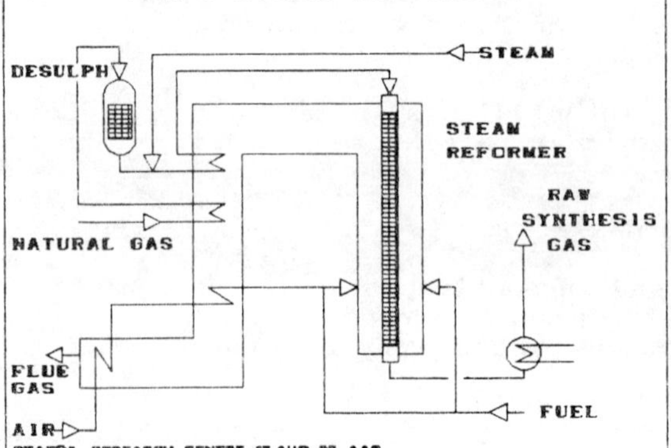

STATOIL RESEARCH CENTRE 17.AUG-90 AAS

UNCATALYZED PARTIAL OXIDATION is a process in which oxygen and natural gas is ignited in a combustion section. The peak temperature is is high and some soot is formed. In principle it is a simple process, but it uses a high amount of oxygen, and the composition of the product gas need to be corrected for the low hydrogen/carbon monoxide ratio for most uses.

In **AUTOTHERMIC REFORMING**, oxygen, steam and natural gas is mixed in a burner. Then the gas flows across a high temperature catalyst bed where the endothermic steam reforming reactions are taking place. The oxygen consumption is considerably less than in partial oxidation and the H_2/CO ratio can be made very near 2.

COMBINED REFORMING is an interesting process in many ways. It consists of a small steam reformer followed by an autothermic reformer. The dimensions are much less than for a conventional steam reformer,the pressure can be higher, up to 40 bar, due to lower outlet temperature from the tubes. The H_2/CO ratio can be controlled within a considerable range.Oxygen consumption is also low.Combined reforming could therefore be important for the production of synthetic fuels.

FIG.4 ICI GAS HEATED REFORMER

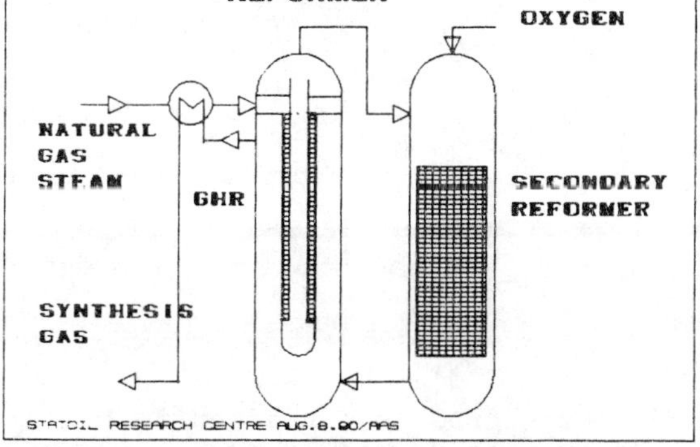

In the **ICI GHR** or Gas Heated Reformer, which is shown in Fig.4, the hot gases from the secondary reformer is directly used to furnish the energy for the steam reformer. There are a lot of savings possible by such a system: Reformer tubes are thin and light with only a small differential pressure across the walls. Volumes of the reactors are small and the oxygen consumption is comparatively low. (Ref.4)

The **UHDE's CAR**, or Combined Autothermic Reformer is similar to ICI's GHR in principle, as the heat from the secondary, oxygen blown reformer section is used to heat up the primary steam reforming tubes.The primary and secondary zones are however within the same vessel.

5. GASOLINE FROM NATURAL GAS.

The best known new synfuel technology in the 80's is probably MOBILs MTG process, or Methanol To Gasoline. This process has been commercialized in a 600.000 tonn pr.year unit in New Zealand. Technically it has been a success.

The process is divided in several steps:

1. Reforming of natural gas to synthesis gas:

$$CH_4 + H_2O\ (O_2) \dashrightarrow CO + 2H_2\ (CO_2,\ H_2O)$$

2. Methanol Synthesis:

$$CO + 2H_2 \text{ --> } CH_3OH$$

3. Gasoline from Methanol:

$$n\ CH_3OH \text{ --> } (CH_2)_n + nH_2O$$

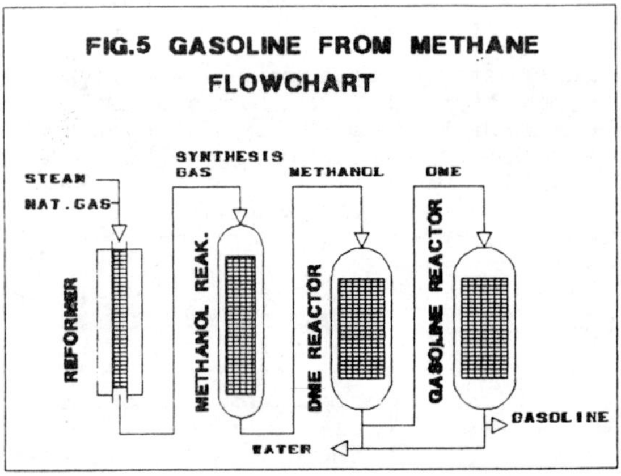

The last reaction is performed over a MOBIL ZSM5 catalyst.

From equation 3 we see that more than 50 % of the synthesis gas ends up as water. This is common for all processes based on synthesis gas from natural gas.

It is necessary in the commercial application to divide the last step in 2. In the first one makes DiMethyl Ether DME:

$$2\ CH_3OH \text{ --> } CH_3OCH_3 + H_2O$$

thereby removing 50% of the total water before the final conversion to higher hydrocarbons.

Fig.5 gives a simplified flowchart of the Mobil MTG Process.

6. DISTILLATES FROM NATURAL GAS

When we have synthesis gas we can produce paraffins in the wax region by the Fischer-Tropsch reaction as follows:

1. Synthesis gas from natural gas:

$$CH_4 + H_2O\ (O_2) \text{ --> } CO + 2H_2$$

2. The Fischer-Tropsch Reaction:

$$nCO + 2nH_2 \text{ --> } (CH_2)_n + nH_2O \text{ (Wax and water)}$$

3. Hydrocracking of the wax:

$$(CH_2)_n\ (+H_2) \text{ --> } (CH_2)_{kerosin} + (CH_2)_{diesel}$$

The Fischer- Tropsch reaction is not selective towards spesific products. The composition of the products follows the Schultz-Flory distribution. We see this in Fig.6.

If we want high selectivities to f.inst. middle distillates, we have to hydrocrack the wax to the products, according to step 3

FIG.6 F.T.PRODUCT DISTRIBUTION

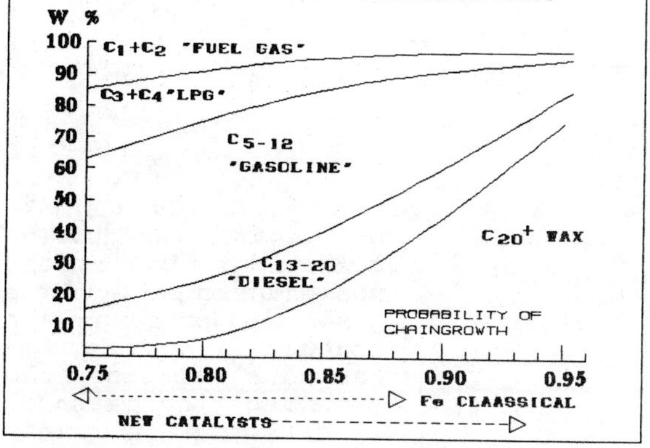

FIG.7 DISTILLATES FROM NATURAL GAS VIA FISCHER-TROPSCH

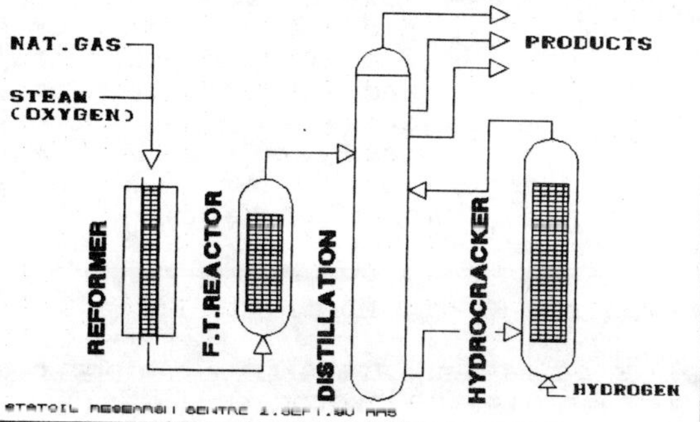

STATOIL RESEARCH CENTRE 2.SEPT.90 AAS

FIG.8 PROPERTIES OF F.T.DISTILLATES GMD PROCESS*

DIESEL	GMD QUALITY	CAR QUALITY
CETANE	72	45-48
CLOUD POINT	-13	-5 TO -10
DENSITY	0.78	0.84
SMOKEPOINT	> 60MM	

NO SULPHUR, NITROGEN OR AROMATICS

* STATOIL PROCESS

STATOIL RESEARCH CENTRE 2.SEPT.90 AAS

above.

This is just what is being done in modern Fischer-Tropsch technology. The catalysts are designed to give a high α, or high probability of chaingrowth, to make wax which can be hydrocracked back to middle distillates.

The fuel properties of the products are exceptionally good. This is shown in Fig 8 which gives the properties of the finished diesel from the GMD (STATOIL) process.(Ref.5)

The same qualities are reported by SHELL for their process.

This very good fuel quality can be utilized in several ways:

1. As a blendstock for other products, f.inst.light cycle oil.The extra value for this use will probably be in the order of 20-30% compared to regular diesel.

2. As a premium fuel for cars and trucks. There might be a large market for such "super diesel" or "super kerosin" with smoke points above 100 mm and no sulfur.

Fig 7. shows a flowchart for a modern Fischer-Tropsch plant for the production of distillates. The synthesis gas can be produced by several processes.The consumption ratio of H_2/CO is aproximately 2.15.

Also the Fischer-Tropsch reactor might change from process to process. Shell has published the use of tube-reactors. STATOIL are using three phase slurry reactors.

7.DISTILLATES VIA METHANOL

Fig.9 is a flowchart of a process developed by Mobil to make middle distillates. The conditions under the condensation phase of the gasoline process are changed in such a way that a considerable amount of the methanol is transformed into light olefins. These can then be oligomerized to higher paraffins which can be used as diesel.

The quality of this product is of course excellent as to sulfur and other contaminants. The Cetane number has been reported at 54-58. This is lower than the F.T. diesel, due to more extensive branching of the molecules.

FIG.9 DISTILLATES FROM NATURAL GAS VIA METHANOL

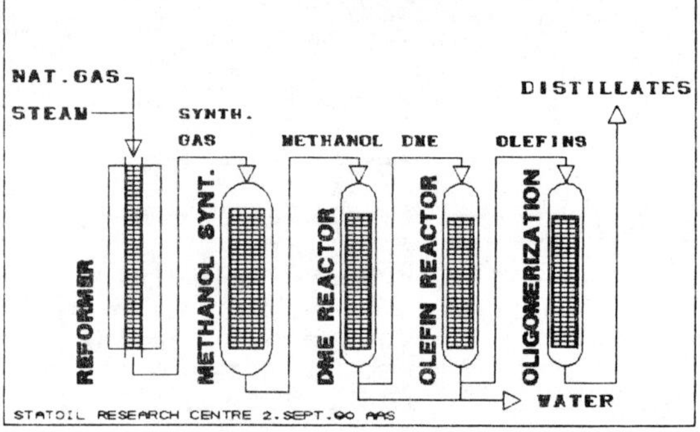

8. DIRECT COUPLING OF METHANE

The idea behind direct coupling of methane stems from the obvious advantages it would mean to maintain the hydrocarbon structure of the methane molecule:

$$nCH_4 \; \text{-->} \; (CH_2)_n + nH_2$$

This equation is the basis for <u>**thermal coupling**</u>. We understand immediately from the large production of H_2 that this reaction is highly endothermic and requires high temperature to be performed.

Fig.10 shows thermal coupling to acetylene and ethylene. The HÜLS electric arc process is still in use. Research data from SINTEF show a very high selectivity

FIG.10 THERMAL COUPLING OF METHANE

$$2CH_4 \rightarrow CH \equiv CH + 3H_2$$

ACETYLENE

	T °C	YIELD C_2	KWH /KG	REACT. SEK
HULS EL. ARC	?	75-80	12-14	.1-.01
SINTEF.DIR.*	1750	90+	8-9	.01-.001

* 1966-70 HOLMEN, ROTERUD, ROKSTAD, SOLBAKKEN

STATOIL RESEARCH CENTRE 1.SEPT.90 AAS

when using ultra short reaction times and high temperatures. However, it is difficult to find satisfactory reactor materials at this elevated temperatures. (Ref.6)
The consumption of electric energy is also high.

In order to avoid the high endothermic heat of formation, the hydrogen can be reacted with oxygen in situ:

$$CH_4 + O_2 \rightarrow CH_3CH_3 \quad \text{or} \quad CH_2{=}CH_2 + nH_2O$$

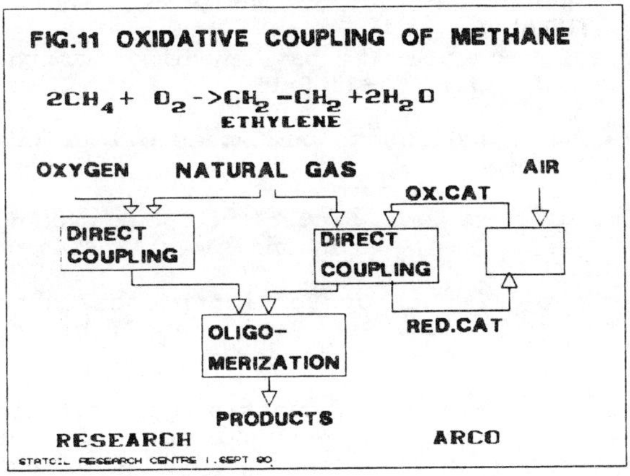

This is called **oxidative coupling** and has been the basis for a very extensive research and development activity for the last 10 years. Fig 11 shows 2 principles which are used for oxidative coupling. One is direct coupling by mixing methane and oxygen over a catalyst at 700-900°C. The products formed are a mixture of ethane and ethylene in a ratio of approximately 1:2. Ethane of course has to be dehydrogenated by recycling.
Selectivity is very low at interesting conversion levels, and a large portion of the oxygen is used for the combustion of methane, forming carbon oxides.

Fig.12 shows the poor selectivities being reached by oxidative coupling. The best data,(the curve in fig 12.), are data reported by ARCO which used a solid multivalent carrier for the oxygen, the carrier being oxidized by air and moved into the reaction zone where a stoichiometric reaction between methane and oxygen from the carrier took place.

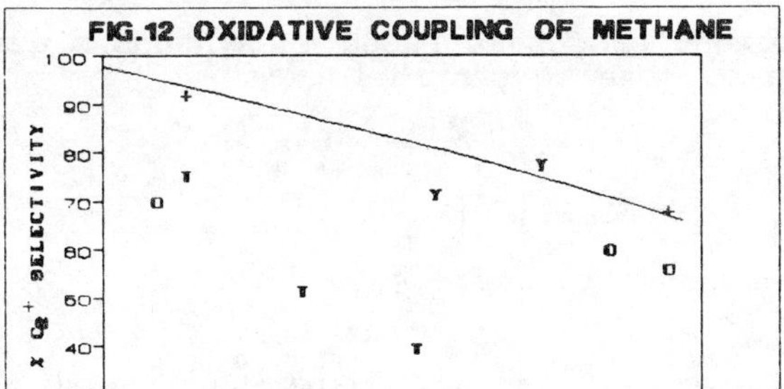

ARCO has declared their process as commercial and will license it to interested parties. It is a complicated process from a heat transfer and solid circulation point of view.

In the authors opinion direct oxidative coupling will not be economical as long as high temperatures have to be used. Uncontrolled gas reactions are taking place and will set severe limits to selectivity.

The low pressures used and the small conversion one finds at satisfactory selectivities leading to extensive resirculation,means very high area heat exchangers.

9. ECONOMICS AND CONCLUSIONS

Several authors (ref.7,8 and 9) have presented reviews of the cost of synthetic fuels . They can be summarized as follows:

The best new processes wil not be able to pay anything for the natural gas at oil prices lower than 20-25 $/bbl.

The first generation conversion processes would need oil prices of 25-40 $/bbl.

It is always hazardous to accept such figures without making engineering studies for spesific sites and conditions.

It is clear that the viability of such conversion processes is highly dependent on the oil price and the future availability of crude oil.

As we have seen, processes based on synthesis gas are in principle well developed. There is of course some concern over the cost of the reformer, which could amount to 50-65 % of the total cost of a plant. However, the selectivity and energy yields are excellent using this route.

Another factor which is important for synfuels is that there is being put a considerable effort into developing new and cheaper reformers.

Direct conversion via oxidative coupling does not seem to give any hope of commercial success. The selectivities and heat cycling in the present concepts are far away from any economic break through.

At the present time efficient processes based on synthesis gas seems to the most promising route to syn fuels.

11.REFERENCES

1. L.F.Ivanhoe, OGJ July 25. 1988 111-112

2. F.Fischer,H.Tropsch, Brennstoff-chemie 4,193, 1923
   ibid. 5, 201, 1924

3. Å.Solbakken, Nat.Gas Conversion Seminar, Oslo August 1990
   In print.

4. K. Mansfield,P.E.J.Abbott,M.R.Conduit, 1989 World Methanol Conference, Dec. 1989, p.XIV 1-10

5. K.Kinnari, P. Rokstad,E. Rytter, Å. Solbakken, AICHE, 1990 Spring Meeting,Orlando,Fl. In Press

6. O.A.Rokstad, A. Holmen, Å. Solbakken,IE&C,15,439 (1976)

7. J.M.Fox, Tan-Ping Chen,B.D.Degen:Direct Methane Conversion Process Evaluation, AICHE 1989, Houston April 2-5 1989.

8. S.C.Nirula: Economic Prospects. AICHE, Houston April 2-6 1989.

9. R.Moreno, Jr.,D.G.Fallen Bailey:Alternative Transport Fuels from Natural Gas.World Bank Techn.Paper 98.

MORE NATURAL GAS
A MULTI-DISCIPLINARY CHALLENGE

P. LEPRINCE and M. VALAIS

Institut Français du Pétrole
Mission Information-Documentation
1 et 4, avenue de Bois Préau
92 Rueil-Malmaison, France

Summary

Large resources and growing markets are the salient prospects of natural gas for the coming decades. The greater impact of natural gas on the worldwide energy market can become a reality if several scientific disciplines can be mobilized in order to succeed in cutting production costs. Modelling, mechanics of complex fluids and physical chemistry of interfaces are basic disciplines for understanding and mastering the gas processing technologies.

1. INTRODUCTION

Will natural gas be the leading energy in the post oil era? This is a legitimate question, if one simply observes that the proven gas reserves are growing one and a half times faster than consumption, while, for oil, new reserves between 1975 and 1988 have nearly compensated for annual consumption.

In recent years, in fact, new exploration techniques have led to the discovery of new fields for both oil and gas. At the same time, a closer knowledge of the reservoirs has helped to re-assess the reserves in place. Counted in years of consumption, reserves of oil as well as gas have never been as high: 44 years for oil, 63 for gas.

This situation of relative abundance obscures - as pointed out by many petroleum analysts - anomalies which are causes of crisis: the reserves distribution is extremely favorable to the Middle East for oil, and to the USSR and Iran for gas; due to the depletion of its oil reserves, the United States is becoming the biggest oil importer before Western Europe.

This problem, which relates to geopolitics, cannot conceal a long-term trend: the progressive depletion of low-cost resources. The situation is well known for oil, but, as emphasized at the World Energy Conference, it also affects gas resources.

Simultaneously, the prospects for gas demand are increasingly favorable. It is a clean fuel, distributed sulphur-free; it contributes less than any other fossil fuel to the greenhouse effect. Its markets, consisting mainly today of industrial and home heating, and to a smaller extend, of raw material for petrochemicals (fertilizer and methanol), could open up markets for fuels and for basic petrochemicals, when the new conversion technologies are successfully developed.

Large resources and growing markets are the salient characteristics of natural gas for the coming decades. This situation, apparently problem-free, conceals a far more complex reality that this paper proposes to analyze.

2. WORLD RESERVES OF NATURAL GAS - LOCATION AND PRODUCTION CONSTRAINTS

Two-thirds of the major discoveries of the past twenty years have occurred in zones where the production conditions are difficult (severe weather conditions, the deep offshore environment, or the distance from consumer zones). This pertains in particular to the Siberian Arctic and offshore fields. Accordingly, the share of reserves in difficult zones has grown from 31% in 1970 to 53% in 1990 (Table 1).

	1960	1970	1990
Easy onshore zones	15.8	27.5	60
Offshore	1.6	4.5	25
Arctic and Siberia	0.1	7.5	42
Other difficult onshore zones	-	0.5	2
World	17.5	40.0	129
Share of difficult zones (%)	10	31	53

Table 1 : Location of natural gas reserves (estimate Tcm)

Changes can also be observed in the size of the fields discovered. There are about 26,000 gas fields today (Table 2), most of which (nearly 20,000) are located in North America. But a closer analysis reveals that they include 24 supergiant fields, each containing over 1000 billion m^3 of recoverable gas, and corresponding to 36% of world reserves. Eleven of them are located in the USSR, and nine in the Middle East. In the other geographic zones, many fields are smaller, (10 to 100 billion m^3) and even more are often smaller than 10 billion m^3. Their exploitation is only feasible if they are easy to produce and located in zones near the consumer countries.

Size of gas fields (Bcm)	Small < 10	Medium 10 to 100	Giant 100 to 1000	Supergiant > 1000	Total (2)
North America	19,500	325	25	1	19,850
Latin America	1,260	70	7	-	1,340
Western Europe	1,450	150	12	2	1,610
USSR and Eastern Europe	940	210	50	11	1,210
Africa	580	90	7	1	680
Middle East	180	60	51	9	300
Asia and Oceania	985	90	15	-	1,090
World total (2)	24,900	995	167	24	26,100
% of world reserves (1)	27.5	15.5	21	36	100

(1) Proven reserves on 1 January 1990 + cumulative production.
(2) Values rounded off.

Table 2: Natural gas reserves (1) and fields
(Number of fields on 1 January 1990)

To have a more reliable picture of the situation of natural gas reserves, let us examine two offshore zones, the North Sea and the Gulf of Mexico, which are especially interesting in view of their relative proximity to consumer zones (Table 3). In these two zones, apart from the giant and medium reservoirs, whose exploitation raises no technical or economic problems, one finds many small reservoirs with reserves of 1 to 10 billion m^3, and even many reservoirs smaller than 1 billion m^3. Conventional production techniques are generally inapplicable to these reservoirs for economic reasons. There are 174 in the North Sea, but more than 6000 in the Gulf of Mexico, where they account for 65% of total reserves.

		North Sea number	North Sea reserves (Bcm)	Gulf of Mexico number	Gulf of Mexico reserves (Bcm)
Supergiant reservoirs	> 1000 Bcm	1	1,260	-	-
Giant reservoirs	100 to 1000 Bcm	12	1,490	-	-
Medium reservoirs	10 to 100 Bcm	110	2,745	35	500(e)
Small reservoirs	1 to 10 Bcm	164	520	660	1,000(e)
Marginal reservoirs	< 1 Bcm	174	80	6,340	1,600(e)
Total		461	6,095	7,035	3,100

(e) Estimated (non associated gas reservoirs).
(1) Proven reserves on 1 January 1990 + cumulative production.

Table 3 : Distribution of reserves in two 'mature' offshore zones

For gas fields, the economic conditions of production are related, through the investments to be made, to the physical processing of raw gas, before it can be sent to the consumer areas (Table 4). 50% of the reserves consist of dry and clean gas, for which processing is extremely simple (dehydration/compression), and 20% consist of wet and clean gas, for which the processing must include a separation of the C_2+ hydrocarbons before the gas can be shipped. 30% of the reserves consist of acid gas, dry or wet, which involves a number of field processing units, and may be considered too costly, so that production on this type of field could be delayed, especially if the reserves are small and the possibilities of shipment limited (distance, offshore location).

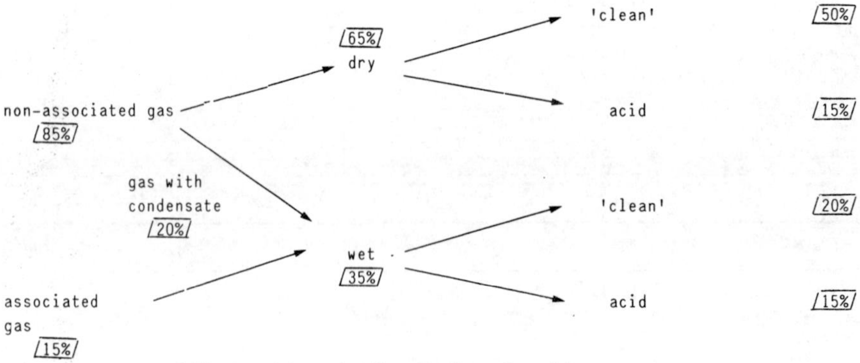

Table 4 : Schematic distribution of world gas reserves
(Estimated % of proven reserves)

From all the characteristics of each field a production cost can be estimated on the basis of current available techniques. This cost serves to classify world reserves roughly into four categories (Table 5).

Share of world reserves (%)	40	45	10	5
Technical cost:				
. $/MBtu	< 0.7	0.7 to 2.1	2.1 to 3.5	3.5 to 5.2
. $/boe	< 4	4 to 12	12 to 20	20 to 30

Table 5 : Classification of reserves by level of technical cost

Today, only the fields of the first two categories can be produced without economic difficulty in present conditions of international natural gas prices. On the other hand, the third category already falls into a cost bracket that is relatively incompatible with a number of European markets, and especially with the current range of supply prices in the United States.

3. NATURAL GAS IN EUROPE: TOWARDS STABILIZED PRODUCTION

Western Europe is the third-ranking natural gas consuming region, after the United States and the USSR, with 13.5% of world consumption. But it only possesses 4.2% of world reserves. Production, slightly under 200 billion m^3 in 1989, only covers 70.5% of requirements, and the remainder is supplied by imports from the USSR (19.1%), Algeria (9.7%) and Libya (0.6%), either by pipeline, or by shipment in the form of liquefied natural gas (LNG).

Within Western Europe, 64.6 billion m^3 is exported by two countries, Norway and the Netherlands. The United Kingdom, Western Europe's second largest producer (44.75 10^9 m^3), must import additional amounts of gas from fields in the Norwegian North Sea zone (10.53 10^9 m^3).

To complete this panorama, note that offshore production accounts for 58% of total Western European production today.

Production prospects for the coming years should not reveal any major discontinuity. Production should stabilize at a level close to 200 billion m^3. The decline of production in a number of traditional producing countries, including the Netherlands, should be offset by the development of the giant Norwegian Troll and Sleipner fields.

4. FUTURE DEMAND FOR NATURAL GAS IN EUROPE: NEW MARKETS EXPANDING

Before the oil crisis, natural gas acquired a significant share of the Western Europe energy market, with slightly over 10%. In the post crisis years, together with nuclear power, it was the winner in energy substitution. This share was 15.5% in 1985. Since that time it has only been growing slowly, so that, around the end of the century, it should be about 17%, whereas, in some countries like the United Kingdom, the USSR and the Netherlands, it will be 25 to 45%.

This evolution is due to a combination of several factors:

. Natural gas is a rather recent energy for Europe. Thirty years ago, annual consumption was 10 billion m^3, twenty times less than today, in other words a few per cent of total energy consumption.

- In the competition with other energies, natural gas has largely penetrated the space heating market (residential and commercial sector) (47%) which places Europe ahead of the other geographic zones for this sector. In the other sectors, however, Europe is not so advanced. Only 12.5% of natural gas is used for electricity generation, whereas the world average is 24%, and the figure is 73% for Japan.

This situation should gradually be reversed, because natural gas is a remarkable fuel and ideal for meeting the new environmental protection requirements:

- it contains no sulphur,
- emissions display low nitrogen oxide and particulate contents,
- gas emissions having a greenhouse effect are the lowest of all fossil fuels: only half of coal and two-thirds (0.71) of oil (per unit of electricity generated).

It should be added that the efficiency of gas turbines is being continuously improved, and the combined cycle power plants now under development are expected to reach 60% efficiency, which is much higher than conventional steam turbine power plants. Total energy efficiencies of up to 80% can also be achieved in co-generation (electricity/steam). To conclude, in this energy sector, new market segments are open in Europe to natural gas, chiefly as a substitute for other fossil fuels (fuel oil and coal) in electricity generation.

5. OBSTACLES TO THE PENETRATION OF NATURAL GAS

There are several types of constraints for a larger penetration of natural gas in the world energy market.

Even if we assume the extension of the existing fields and a few new discoveries, the next twenty years will probably see a reduction in Western Europe's natural gas reserves. In 2010, only Norway and the Netherlands will still have significant reserves (Table 6).

France	7	(for comparison)	
Italy	19	Algeria	54
Norway	68.5	USSR	63
Netherlands	24	Nigeria	112
West Germany	11	Iran	664
United Kingdom	11.5		

Table 6: Reserves of European countries (years of production)

Sources of natural gas that are available for the European market are far away and require for considerable investment. Four countries are generally suggested as capable of providing additional resources, Algeria, the USSR, Nigeria, and Iran. They have very large reserves, but for all of them, the distances are such that the investments for building the complete chain (processing/transportation) are very high if related to the quantity of energy. Compared with the oil chain, they are 2.5 to 5 times higher, $1.5 to 3 billion for 6 to 7 million tons oil equivalent.

The time between discovery and production of a gasfield may be longer than ten years, and even over fifteen, due to the lengthy technical and commercial contractual negotiations with the numerous partners and clients, who have to commit themselves to technical and commercial decisions for a period of at least twenty years.

6. THE GAS TECHNOLOGY CHAINS

The upstream part of the gas technologies chain is composed of the following main steps.

- The production installation, which is especially extensive for an offshore field (platform, subsea wellhead).
- The processing installation required for shipment of the natural gas. It differs according to the end use of the gas and its mode of transportation to the consumer centres. Figure 1 (see end of the paper) shows three alternatives which correspond to the main situations found today.

The operations involved represent a complex combination of methods and scientific disciplines which are determining factors for the efficiency and cost of the overall operation. These are (Table 7):

- multi-phase flow of the well effluent (gas, liquid hydrocarbons, water) between the well and the processing plant,
- the mechanisms of hydrate formation, which is undesirable in gas pipelines,
- the thermodynamic liquid/vapor equilibria of the complex mixture, which lead to gas/liquid separation,
- solvent extraction of the acid components (H_2S and CO_2),
- low-temperature heat exchanges for the liquefaction of natural gas.

Operation	Purpose	Data required
Injection of methanol or glycol	Control of hydrate formation	Solid-liquid equilibrium and crystallization kinetics
Solvent extraction	Removal of CO_2 and H_2S	Solubility of gases
Condensation	Extraction of NGL*, LPG and other condensates	Liquid/vapor equilibrium
Adsorption	Drying, removal of traces of water	Adsorption coefficients
Liquefaction	Production of LNG: long-distance transportation	Liquid/vapor equilibrium Equation of state Heat exchange coefficient

*NGL, natural gas liquids (ethane, LPG)

Table 7: Typical operations in the processing of natural gas

7. TECHNICAL AND TECHNOLOGICAL ADVANCES - OBJECTIVES AND METHODS

As we have shown, gas techniques are complex and capital intensive. This situation is especially unfavorable to offshore fields, small fields, and with complex chemical composition (acid gas with condensate, for example). In this context, the development of new techniques to cut production and transportation costs helps to boost the volume of exploitable reserves at a sufficiently low technical cost to allow their exploitation. Simultaneously, reducing utilization costs helps to expand markets. Table 8 offers an illustration of the objectives to be achieved in terms of the methods and techniques to be implemented.

	Objectives	Methods and techniques
Exploration/ production	Increase recovery in tight formations	Hydraulic fracturing
Field processing	Offshore fields: Reduction of platform weight and cost	Simplification/integration of gas processing
	Increase NGL recovery	Membrane separation
LNG chain	Reduce self-consumption and investments in plants and methane tankers	New heat exchanger design
		Axial compressors
		Modular construction
		Better insulation
		Diesel engine instead of steam turbines in LNG tanker
Pipelining	Reduction of construction time, investments and costs (maintenance)	Optimization and automation of pipeline laying
		Improving steel and welding quality
Use as transportation fuel	Improve performance of gas storage cylinders	New materials and equipment, improved loading stations
	Chemical conversion to liquid fuels: higher efficiencies, lower costs	Research on new routes to convert methane to fuels (and petrochemicals)
Energy uses	Greater efficiency, safety and flexibility	Optimization/adaptation of technology to specific end uses
	Lower investment	Direct heat transfer Recovery of lost heat Automation, safety control

Table 8: Reduction of technical costs
Technical advances and objectives

8. THE NEED FOR MULTI-DISCIPLINARY ACTION AT THE SCIENTIFIC LEVEL

It is obvious that an ambitious natural gas development programme must mobilize many scientific disciplines. Table 9 attempts a review on the basis of the objectives identified in Table 8.

Objectives	Basic techniques: Modelling and expert systems	Basic techniques: New Materials	Fluid mechanics and heat transfer	Physico-chemistry of interfaces	Rock and soil mechanics	Thermo-dynamics
Enhance recovery (tight formations)	++	+	+		++	
Simplify offshore gas processing	++	+	++	++		++
Improve the efficiency of the LNG chain	+	+	++	+		++
Cut pipelining costs	+	++	+		+	+
Improve storage cylinders	+	++	+			+
Develop direct routes for converting gas to liquid fuels	+	+		++		+
Improve the energy efficiency of end use	+	+	+			++

Table 9: Natural gas
Scientific developments and impact on the solution of technical problems

Modelling occurs at all steps of the gas chain, and is especially important in production. New materials with new properties are also likely to offer significant improvements in transportation, and especially in storage tank. This could actually be the key to the penetration of gas in the automotive sector. Note also that the mechanics of complex fluids and the physical chemistry of interfaces are basic disciplines for understanding and mastering processing.

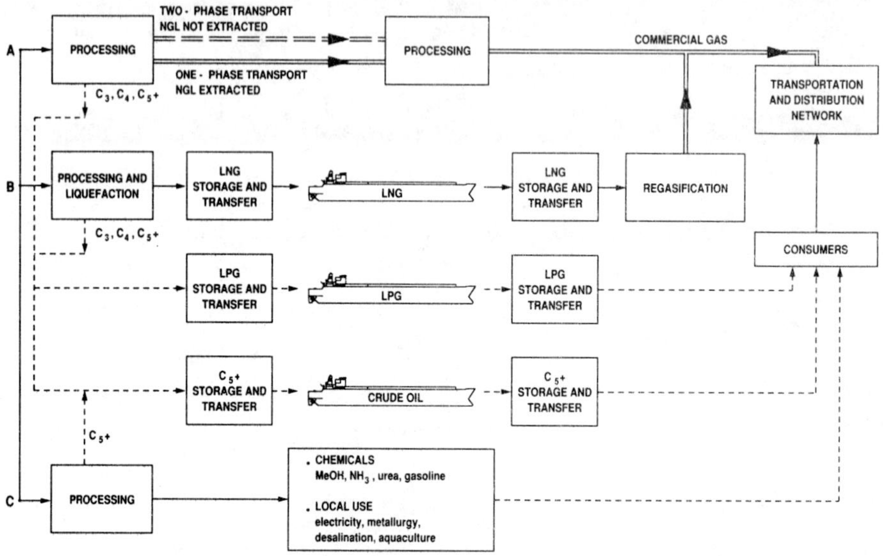

Figure 1: Processing of gas well

9. CONCLUSIONS

Will gas be the post-oil energy? To reply affirmatively, would indeed be foolhardy. The post-oil era is not just around the corner, and the new oil discoveries and new production methods will certainly prolong the oil era to the middle of the next century. Yet already today, natural gas must expand its role to satisfy hydrocarbon requirements.

. Plunged in a new Gulf crisis, it is clear that natural gas, through a different geographic distribution of the zones of production than that of oil, offers a strategy for substitution. This was already a factor from 1973 to 1988, and will continue to be one if the instability of the Middle East is prolonged.
. New gas fields, recently discovered or better known, are ready for production.
. The environmental constraints imposed on fossil fuels (greenhouse effect, sulphur emissions) favor natural gas.

The greater impact of natural gas can only become a reality if the obstacles to its broad penetration are overcome. These are the cost of the processing installations, and the high investment for the overall gas chains, from production to end use.

To achieve this is one of the greatest challenges of the hydrocarbon industry. To meet the challenge, the mobilization of the necessary scientific knowledge and of all the resources of technology is a top priority for the forthcoming decades.

OFFSHORE CHALLENGES IN EUROPE

Session Chairman: J.P. JOULIA, CEC

OFFSHORE CHALLENGES IN EUROPE
PROGRESS AND FUTURE DEVELOPMENTS

Gian Mario Bozzo, Massimo Benetti
Tecnomare S.p.A. - Italy
S. Marco 3584 - Venezia
Italy

Summary

The major developments in offshore technologies are briefly outlined focusing the important role of Europe on the enormous progress made by offshore activities. This process involved all sectors, from production systems to servicing equipment and operations. In particular, the main milestones achieved in the technology of fixed and floating platforms, pipelines and subsea production systems are considered.

For the future challenges in offshore technology, Europe will continue to be the leader in the most important developments and applications.

The European characteristic of high dependency on hydrocarbon supply strategies together with the important offshore discoveries in harsh environments, should force technological developments to achieve higher competitivity levels.

Within this framework an even more important role will be played by natural gas resources and the technological transfer from Europe to other world's most prospective hydrocarbon areas, including Eastern Countries.

1. INTRODUCTION

In the last twenty-five years Europe has been involved in one of the most significant evolutions of the exploration and development of offshore resources.

Offshore passed from a new technological area of development and innovation, to one that has reached an almost mature level.

The first offshore drilling in Europe was done by Agip in Gela (Offshore Sicily) in 1959 and until 1971 only 8 piled platforms had been installed in the Mediterranean Sea and 25 in the North Sea. At that time the role played by Europe in offshore hydrocarbon supply was of small influence.

European offshore production, particularly in the North Sea, became more significant from 1975-78, increasing to its present level of about 3.1 MMbopd and 280 MMscfd (figs.1,2).

In the last 10 years in Europe, Communist countries excluded, offshore oil production has risen from about 15% of oil consumption in 1980, to about 25% in 1989, with a peak value of 30% in 1985 (fig.3) [1].

In this scenario a significant role has been played by the European Companies and by the Commission of European Communities in securing hydrocarbon supply.

Particularly in the period 1973-1990 EEC, through the Directorate General for Energy, supported almost 700 projects for offshore oil related activities with an overall financial support of about half billion ECU. The EEC has been highly successful in stimulating technological advances and enhancing the capability of European hydrocarbon industries, starting with the Oil Companies and arriving at the Contractors, Engineeering Firms and Equipment Supply Services [2].

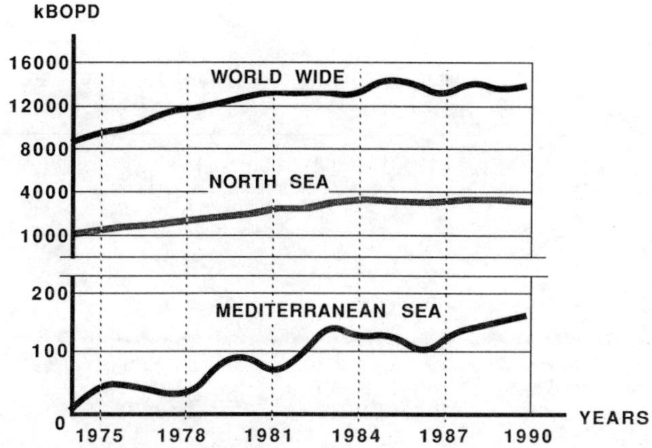

Fig 1 - *OFFSHORE OIL PRODUCTION*

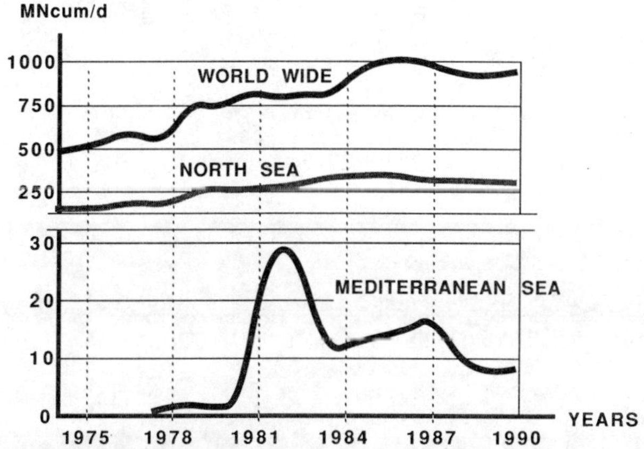

Fig 2 - *OFFSHORE GAS PRODUCTION*

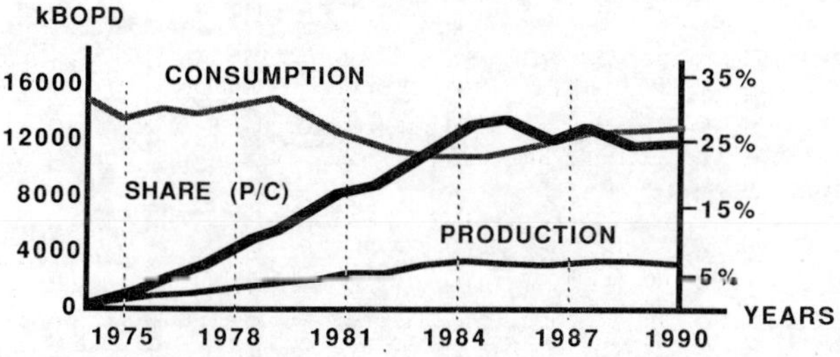

Fig 3 - *OIL PRODUCTION and CONSUMPTION IN EUROPE*

Table 1 - **Most significant milestones achieved in offshore technology**

Production system	Year	field	water	area	Operator
Piled platform in the North sea	1966	West Sole	27	North Sea	BP
Concrete storage	1973	Ekofisk	70	North Sea	Phillips
Concrete Gravity Platform	1975	Beryl	118	North Sea	Mobil
Steel Gravity Platform	1976	Loango	90	Off.Congo	Agip
Biggest Jacket	1982	Magnus	186	North Sea	Mobil
Guyed Tower	1983	Lena	300	Gulf of Mexico	Exxon
Steel Gravity Platform with integrated storage	1984	Maureen	96	North Sea	Phillips
Subsea X-mas Tree	1961	Molino		Off. California	Shell
Subsea Production System	1974	Experimental Station		Gulf of mexico	Exxon
Remote control systems for SPS	1976	Experimental station		Gabon	Elf
Underwater Maniford Center	1983	Cormorant	155	North Sea	Shell
Subsea Well. Acoustic Cont. Syst.	1987	Luna	186	Ionic Sea	Agip
Converted semisubmersible for floating production (FPS)	1975	Argyll	80	North Sea	Hammilton Brothers
Floating Production and Storage System (FPSO)	1977	Castellon	117	Mediterranean Sea	Eniepsa
Tension Leg Platform	1984	Hutton	147	North Sea	Conoco
FPSO with turret mooring for extended well testing-Petroljarl I	1986	Oseberg (norsk Hydro)		North Sea	Golat Nor

2. TECHNOLOGICAL DEVELOPMENTS

The most important milestones achieved in the offshore sector in the last twenty five years show the role of Europe, as listed in tab.1. The following main technological areas are considered below:

- fixed and floating platforms
- pipeline technology
- subsea production and operations

These are not the only areas where Europe has shown its capability.

Drilling and servicing technologies, such as IMR technology, are areas where significant improvements have been achieved.

FIXED AND FLOATING PLATFORMS

In the sixties the first installations in the European Offshore were relevant to gas fields developed by small pile platforms.

In the North Sea small platforms were installed one close to the other to obtain a large complex (e.g. West Sole,Leman, etc).

In 1972 the first giant oil-gas field (Ekofisk) was developed, other giant fields were discovered, with average recoverable reserves in oil bigger than 1,000 MMbbl, which required innovative production systems and new strategies for exploitation (figs. 4,5).

Large Jacket platforms

The severe environmental conditions governing North Sea exploitation in addition to large hydrocarbon discoveries dictated the development of new jacket platform configurations.

If compared to the design conditions of other traditional offshore areas, such as the Gulf of Mexico, waters deeper than 100 m, topside weights higher than 20,000 t and extremely high environmental loads led to the definition of a new structural configuration, foundation layout, topside arrangements,fabrication and installation procedures.

Important innovations were the development of the underwater pile driving hammer and the self-upending launching procedure [3].

Subsequently in the last years an additional improvement was achieved in crane barge technology which led to the fabrication of heavy lift vessels, able to handle more than 10,000 t. Their availability forced conventional installation and assembling procedures to be reconsidered, reducing the structural impact on the platform configuration and weight. The splitting of the topside plants was drastically modified reducing the number of modules with significant advantages over fabrication and offshore operations costs. The experience gained with giant platforms and the development of advanced drilling technologies allowed the Operators to simplify offshore plants reducing the living quarters and achieving low cost solutions.

As a result of the above mentioned developments and innovations the European offshore technology is a reference for other offshore areas, such as Brazilian and Australian offshore and the Gulf of Mexico itself.

Gravity platforms

They were introduced in the seventies to allow for the installation of platforms already equipped with the topside facilities and with a built-in integrated storage capacity. Gravity platforms permitted the exploitation of oil fields also in remote areas where an investment for a dedicated pipeline was not economic. The first application was the concrete storage tank for the Ekofisk field, while the first concrete gravity platform was installed in 1975 at the Beryl field in the North Sea.

The first steel gravity platform was installed at the Loango field in

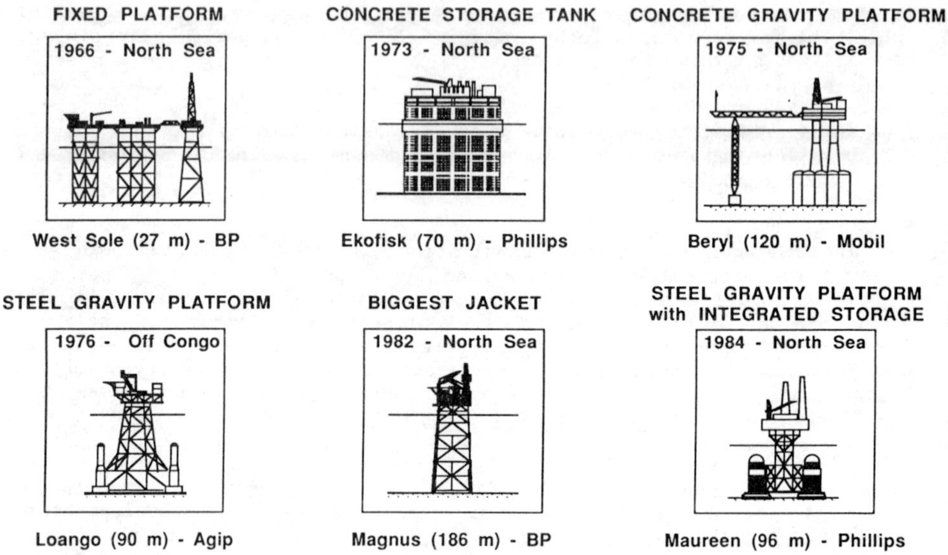

Fig 4 – Milestones on fixed platform technology

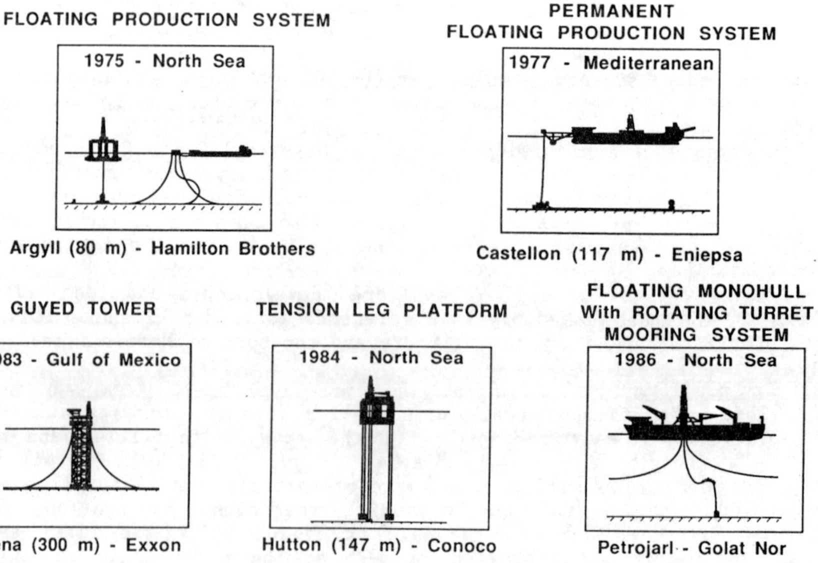

Fig 5 – Milestones on compliant and floating production system

1976 (offshore Angola)[4], followed in 1983 by the Maureen gravity platform, in the UK sector of the North Sea. This platform, the heaviest steel structure in the world (40,000 t) with an integrated storage capacity of 100,000 m3, was installed over a sea bed template for early production[5].

The early production strategy was also introduced in other development schemes, using both fixed or floating platforms, to improve field economics. This strategy was made possible by technological improvements achieved in drilling vessel capabilities and in platform design and installation procedures.

Compliant structures

The hydrocarbon discoveries in very deep water and the positive experience gained with the fixed platforms, forced the identification of a completely new platform concept: the compliant structure, capable of withstanding severe operational and environmental conditions by means of its ability to comply with loading conditions through its built-in flexibility. Two major categories have been designed and installed: the Compliant Towers and the Tension Leg Platforms [6]. The Tension leg platform, installed in 1984 at the Hutton field (147m wd), in the North Sea, involved new design approach and innovative fabrication and installation procedures.

At present the only compliant structure is Exxon's Lena Guided Tower in the Gulf of Mexico, but other applications are expected in the future as natural extensions of fixed substructures in very deep waters [7].

Floating production systems

The need to economically exploit offshore fields, with a reduced amount of recoverable reserves, in particular if located in deep waters, the so called "marginal fields", led to the consideration of innovative production systems.

Even in this case European technology was the leader as the first installation was a converted drilling semisubmersible TW-58 with subsea completion, in 1975 at the Argyll field.

The first floating production and storage system was installed in the Castellon field in 1977 in the Mediterranean Sea.

While these former innovations are the result of the conversion and evolution of existing systems (drilling rig for Argyll and tanker for Castellon), a completely new concept was developed.

The Petrojarl I, a dedicated floating production system, has been fabricated and operated during the last few years in the Oseberg field in the North Sea. It was the first FPSO (Floating Production, Storage and Offloading) system installed in severe environmental conditions. It is used for extended well testing operations thanks to the possibility of reducing the investment risk relevant to the uncertainties of field reservoir evaluation.

At present Offshore Brazil provides the major opportunities for applications of floating production systems, due to their very deep water discoveries.

PIPELINE TECHNOLOGY

One of the most important results achieved in this area is the Transmed pipeline, transporting 12 billion cubic meters per year of Algerian gas to Italy, 150 Km across the Mediterranean, at water depths of 610 m, which is the deepest achieved in the world by an operational offshore pipeline [2].

Another significant step was the installation of about 11,000 km of subsea pipeline network in the North Sea (fig. 6).

These results were obtained also by the utilization of:
- new advanced lay vessels;
- improvements in laying technology;
- trenching vehicles for burying pipelines and cables.

The European network, which facilitates oil and gas transportation to inland refineries, gives additional opportunities for the development of new fields, with the subsea production technology.

The opportunity became greater for marginal fields located close to operating production facilities which, due to the production profile decline in mature fields, provide a spare processing capacity viable for the satellite reservoir exploitation.

SUBSEA PRODUCTION

Up to the end of the seventies the subsea completions were considered as complementary technology for field exploitation based on fixed and/or floating platforms.

In this sector Europe, together mainly with Brazil, was among the innovators (fig.7).

Even if the first subsea completion was in 1961 at the Molino field, the installation of the Underwater Manifold Center (1983) represents one of the most significant steps in the subsea technology.

Another important technological achievement was the installation and operation of the Single Well Acoustic Control System at the Luna field by Agip in 1987 [8]. It represents the first subsea system autonomously controlled, and provides the technology for future issues such as the Emergency Shut Down Valve.

These results were made possible by:
- improvements to reliability levels of subsea components and systems
- underwater robotics
- utilization of existing pipeline networks to platform facilities

In the recent past multiphase production and transportation were the leading-edge technology of research and development activities in Europe.

The installation of the Subsea Slug Catcher in 1985 and of the first Underwater Multiphase Metering system on the Highlander field together with the Subsea Separator in the Argyll field in 1989, constitute a breakthrough in subsea production technology [9].

3. FUTURE OFFSHORE CHALLENGES

In the last few years Offshore activities and technologies were interested in large transformations, mostly directed towards improving safety and reducing costs for the exploitation.

At the same time the European offshore itself changed. Other new giant oilfields are very unlikely, while significant gas fields have been discoverd (fig.8).

Future production will deal with smaller oil fields, often located close to existing platforms or in deep waters.
New remote areas in very harsh environments, such as Sub-Arctic and Arctic regions (e.g. Barents Sea and Russian areas) will be explored and developed.

Before the present oil crisis it was expected that the near future would be characterized not so much by the development of completely new technologies, but by efforts directed towards the innovation and optimization of present achievements.

In the new scenario the following aspects will characterize future

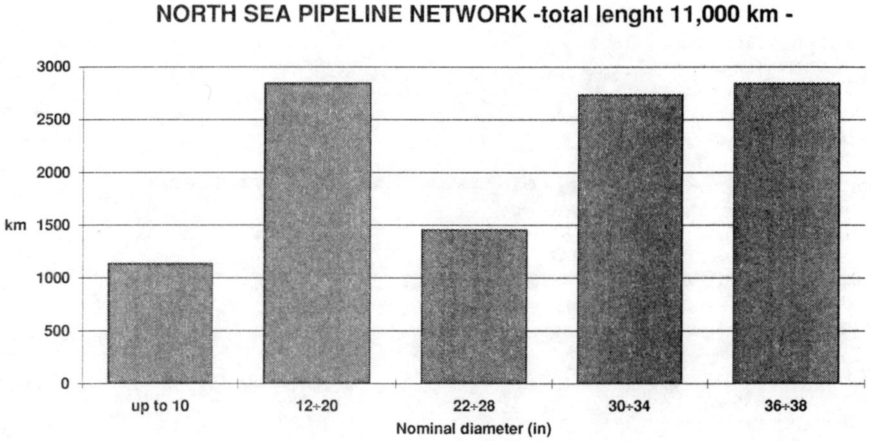

Fig 6 – North sea pipeline network – total length 11,000 km

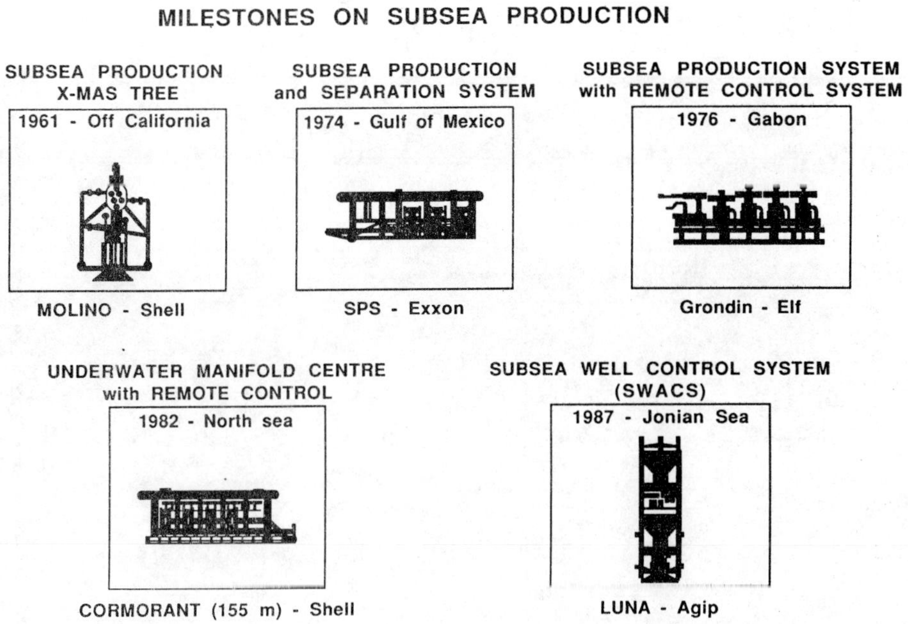

Fig 7 – Milestones on subsea production

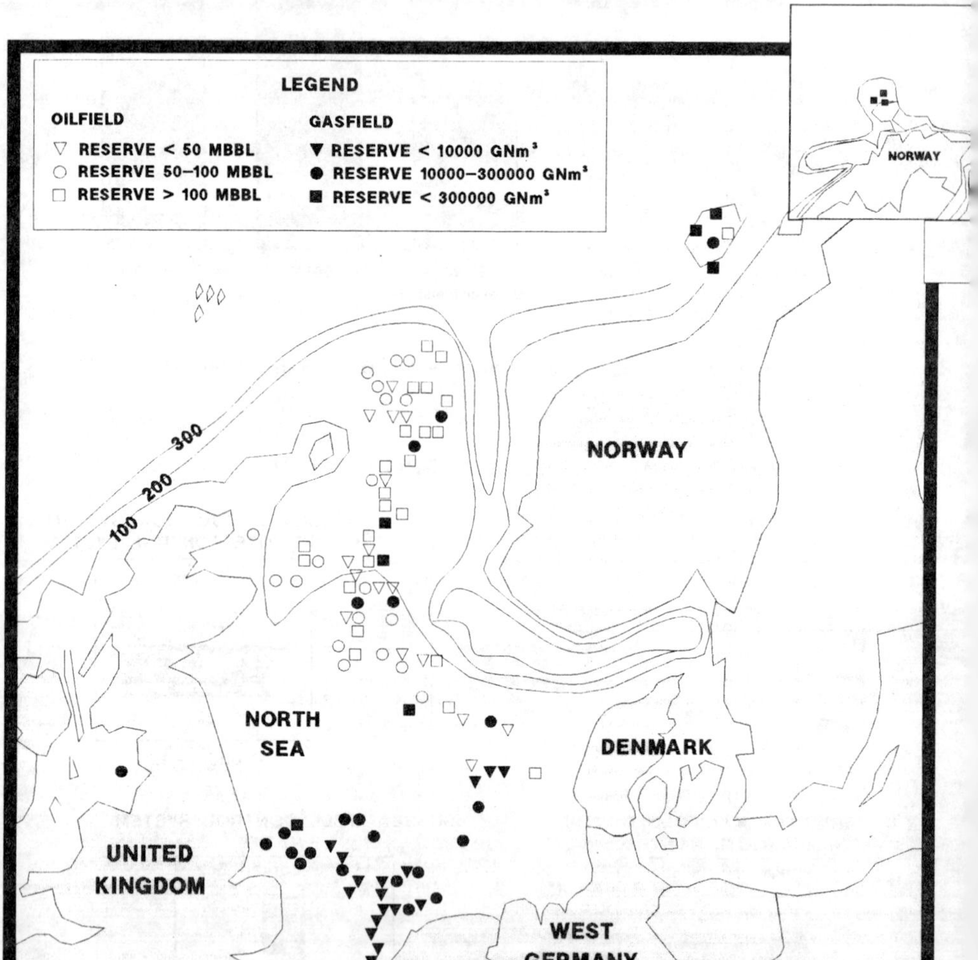

Fig 8 - North Sea discovered fields at 1990

offshore challenges:

- safety
- environmental protection
- exploitation of marginals deep water, subarctic and arctic fields
- natural gas as new energy source
- subsea and multiphase production
- platform demanning
- innovative materials and components

<u>Safety</u>

Safety continues to be the most important issue in the offshore industry.

Up to now large steps have been taken to improve safety levels, procedures, codes, regulations, development of new systems and components. The recent application of Emergency Shut Down Valves is an example of real time response to unexpected problems.

Even if the present situation could be considered satisfactory, further efforts must be dedicated to achieve higher safety levels and to reduce human exposure on risky operations such as diving and platform operations. This will be a result of both technlogical developments and new international regulations.

Significant contributions will be made by additional efforts in areas such as:

- platform demanning: to reduce platform operation costs and to increase the safety level;
- innovative production systems, such as multiphase subsea production and long distance transportation systems, to reduce the number of persons directly involved;
- innovative service equipment (e.g. pipeline repair systems, subsea wireline, etc.) to reduce the complexity of offshore operations and to facilitate human activities;
- subsea robotics: to improve the capabilities and effectiveness of remotely operated systems, not only for deep water operations.

<u>Environmental protection</u>

New rules and recommendations for offshore production and transportation are expected to prevent oil spills. Advanced technologies and methods for recovery at sea should be encouraged.

At present oil spills represent a catastrofic event for the environment. The efficiency and operability of existing systems is quite low and dependent on the environmental conditions.

A significant contribution to the environmental protection could be given by Environmental Impact Statment activities, not only for new systems, but also for the existing installations.

In this area human causes still remain the major ones, therefore, as for the safety of offshore operations, a great contribution could be given by new technologies such as advanced data acquisition systems and operation management systems for decision making processes.

<u>Platform removal</u>

The removal of obsolete offshore installations is required to guarantee the safety of navigation, to protect fishing and the environment.

It is a recent problem in Europe while greater experience has been gained in the Gulf of Mexico, where about 500 installations have been removed since 1975.

At present the legal constraints are decided by each coastal State but, in deciding what must be entirely or partially removed, Governments and Industry shall be guided in the future by common guidelines resulting from International co-operation.

In Europe an attempt at achieving a common regulation was made by the International Maritime Organization (IMO) in 1988 [10].

Exploitaion of marginal fields, and Arctic and Subarctic fields

At present about a thousand worldwide undeveloped marginal fields have been estimated. About half of them are likely to remain uncommercial, at least for a number of years. Another 10% remain undeveloped for other reasons.

In conclusion, about 300 unexploited hydrocarbon accumulations are considered of probable commercial value , and about 80 are located in the North Sea area, with about 44 of those considered as Satellite fields [11].

One important governing parameter is the distance from existing platforms having a surplus process capacity (fig. 9).

For long distance, floating production and FPSO systems [12], will be the most probable applications while for short distance subsea production systems show the largest potential.

The near future will probably be characterized by the increased application of monohull production and storage vessels, not only for extended well testing operations, but also as production systems, expecially in remote areas.

Exploitation of Arctic and Subarctic fields is another challenging opportunity. In Europe high hydrocarbon potential areas are in the Barents Sea and in the Russian regions.

Some of them might be exploited by FPSO with seasonal production strategy, others with fixed and removable platforms. Both need innovative systems and technologies.

Ice contingency plans and ice management procedures, new subsea completion and production systems, innovative materials and structural configurations, are the major areas of development.

Natural gas

In the past the gas market was considered as a sub-market of oil supply. The recent energy crisis and environmental considerations led Europe to a growing interest in gas as an alternative energy source.

Gas condensated fields, some of those giant, have been discovered in the Norwegian Sea and particularly in the Barents Sea.

These future developments need new technologies for the conversion of natural gas to liquid hydrocarbons.

Subsea production

Subsea production will continue to be an area of continuous technological developments mainly relevant to:

- higher reliability, safety levels and standardization by improving components and systems;
- very deep water installation;
- robotics for maintenance and diver substitution, also in shallower water;
- multiphase production technologies.

The latter represent the most important challenge, having the largest potential for cost saving and safety improvement.

A lot of research activities are now under way worldwide. Most of these activities are directed towards developing multiphase pumping units and metering systems (fig.10).

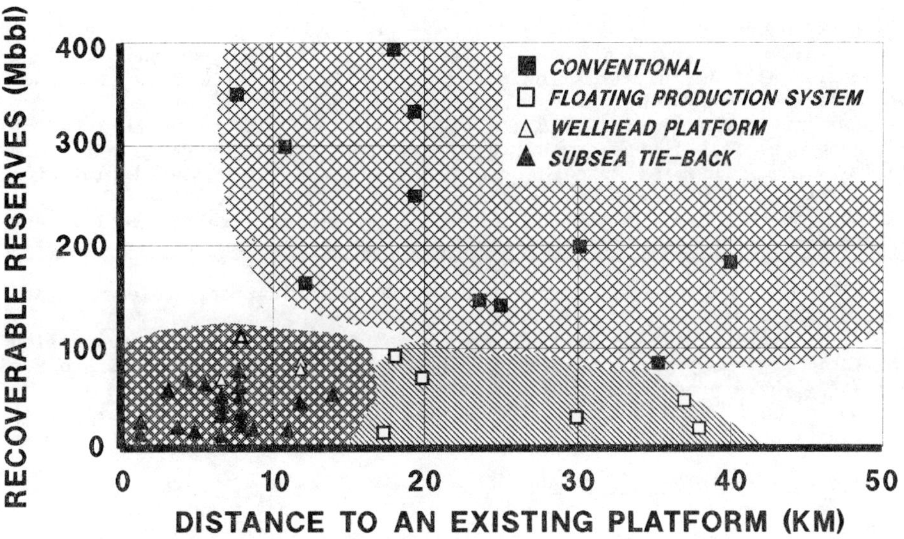

SOURCE;IFP/ECONOMICS DEPARTIMENT - 1989

Fig 9 - Planned probable oil fields in British sector

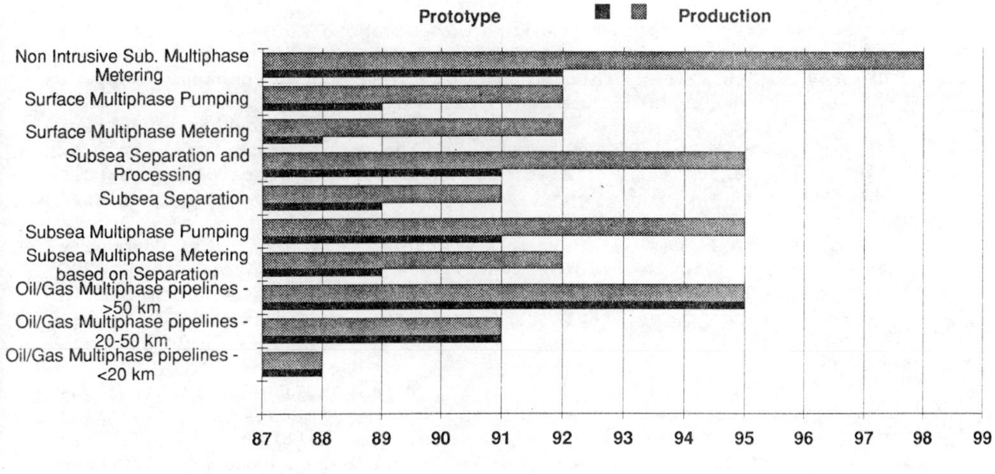

Fig 10 - Multiphase production technology

Extensive industrial applications will be put on stream before 2000 with the long term objective of developing offshore fields with reduced investment cost and minimal surface facilities [13].

Platform demanning

It is an example of how new technologies in process and drilling facilities allowed the reduction of the high costs for housing personnel offshore and the increase of safety levels in operations.

Several unmanned platforms are operated by Agip on gas fields in the Adriatic Sea. This technology has been also applied in the Perla oil field in the Sicily Channel. In the North Sea the number of unmanned platforms did not reach significant levels until the mid 1980s.

All these platforms are characterized by minimal process facilities, sufficient to send the produced gas to inland facilities in the Adriatic sea, or to the main platform in the North Sea .

This was achieved by improvements in communication, telemetry and control systems.

The future challenge is the development of fully remotely controlled platforms with higher automation levels applicable also on large process platforms where production operations are more complex.

Additional contribution will be made by new technologies such as fiber optics and robotics.

Innovative materials and components

Large expectations for the utilization of innovative materials and components in the offshore activities are mainly directed towards the reduction of the overall platform weight and the increase of safety levels.

Composite materials are foreseen for ancillaries of substructures and for process components and structural elements of topside facilities.

Innovative components such as forged nodes show high potential for jacket and compliant towers in mid to high water depth allowing for a better fatigue behaviour and reduced inspections during the operative life.[14]

4. CONCLUSIONS

In the last twenty-five years European Offshore has been the testing ground for much innovative technologies. In that period the Commission of the European Communities, through the hydrocarbon technology support programme, gave a European dimension to offshore activities .

EEC contributes to the Community security of hydrocarbon supplies, stimulating technological developments in the oil and natural gas by Industries, encouraging the efficient development of hydrocarbon resources and enhancing the technical capabilities of the European Oil related Companies.

During the same period the Oil Companies, secured the Community hydrocarbons supply by developing hydrocarbon fields outside the OPEC charter, also in harsh and remote environments, such as the North Sea in the past and Sub-Arctic or arctic regions in the near future.

They also established a large European network for oil and gas transportation.

Together with Contractors, Engineering Firms and R & D institutions the Oil Companies reduced investment and operating costs by developing innovative systems and technologies for higher competitiveness and safety levels and transferred European offshore technologies to other Offshore Areas (US included). For the future the same cooperative attitude among these figures is expected.

Before the present oil crisis it was expected that the near future would be characterized not so much by the development of completely new technologies, but by efforts directed towards the innovation and optimization of present achievements.

The present oil crisis focuses the weakness of hydrocarbon supply based on low cost resources. To overcome the problem European Oil Companies and the European Communities should allocate additional finance for further technological developments .

It is reasonable to foresee that the main technological areas should be:

- safety and environmental protection
- exploitation of marginal, deep water, Sub-Arctic and Arctic fields
- subsea and multiphase production
- platform demanning

As regard the hydrocarbon market, many forecasts have been proven wrong by events, but certainly an important role in energy supply will be played by natural gas.

Europe has already shown its capability on technological transfer to other world hydrocarbon areas. Within this framework the new political situation in Eastern Countries provides new opportunities for common developments and new strategies to secure energy supplies.

REFERENCES

[1] EUROIL (1990), Vol.1 - Noroil Publishing House Ltd. A/S

[2] EEC-Directorate General for Energy (1988), Evaluation of the Community Programme of Support for Technological Development in the Hydrocarbons Sectors 1974 to 1987.

[3] Bozzo G.M., Marazza R. (1984), Bouri DP-4 and DP-3 Jackets. The largest offshore platforms in the Mediterranean sea, Offshore Seminar, Tripoli

[4] Lalli D. (1976), Design, construction and installation of the Loango steel gravity Platforms - Istitution of Civil Engineers London

[5] Agostoni A., A.A. (1980), TSG - Integrated Storage Platform for Early Production in the North Sea, O.T.C. 3880, Houston

[6] Brandi R., Sala C. (1988), Floating Production System for the exploitation of deep water Mediterranean Oil Fields, 3rd Hydrocarbon Symposium EEC, Luxemburg.

[7] Nicolussi F., A.A. (1989), Compliant Steel Tower for Deep Water Platforms, D.O.T: Marbella.

[8] Galletti R., Franceschini G. (1989), The role of Autonomous Subsea Control Systems in Subsea Production, O.T.C. 5947, Houston.

[9] CEC-Directorate General for Energy (1989), Multiphase flow worshop, Copenhagen.

[10] "Guidelines and Standards for the Removal of Offshore Installations and Structures on the Continental Shelf and in the Exclusive Economic Zone" (1988) - Maritime Safety Committee of the International Maritime Organization.

[11] Le Prince P., Badour D. (1989), Multiphase Flow Production a perspective in Offshore Exploitation - Institut Francaise du petrole

[12] Forti P. (1990), An advaced turret floating production system for marginal fields, 6th International Conference on Floating Production System, London.

[13] Offshore Engineer , June 1989, Dealing with multiphase flow: a state of the art review.

[14] Michielon F., A. A. (1990), The application of the Forged Nodes to Offshore Light Weight Jackets, OMAE, Houston.

GEOSCIENCES

Session Chairmen: P. SIMANDOUX, IFP (F)

A.D. PINTO, Partex (P)

W.F. STEENKEN, EAPG (NL)

D.M. GRIST, BP (UK)

RESERVOIR DEVELOPMENT IN A COMPLEX DEPOSIT ENVIRONMENT :
GERMIGNY-SOUS-COULOMBS AQUIFER STORAGE

M. LEBLANC, J. ROUGE and E. TALLEC

GAZ DE FRANCE
Research and Development Division,
Underground Storage Department
361, Av. du Président Wilson
93211 - LA PLAINE SAINT DENIS CEDEX

Summary :

The GERMIGNY-SOUS-COULOMBS aquifer storage reservoir is a poorly structured anticline formed of alternating layers of fine sandstone and clay, deposited in a marginal littoral environment. As soon as exploitation began, the reservoir proved to be very heterogeneous and it soon became necessary to establish the detailed architecture of the sedimentary bodies. The complex image of lateral and vertical heterogeneities obtained had to be processed by a three-dimensional diphase numerical model. Operational information and improvements in the geological image continually enrich this model, which is thus able to provide new indications for the operator and geologist to work upon. On the basis of forecasts produced by the model, it was decided to inject gas into the deep layers, to pursue equipment with selective completions, to measure pressure gradients in-situ while seeking an optimal compromise between long term development and withdrawals over an annual operating cycle.

1. INTRODUCTION

Very early on, the gas industry was faced with the problem of modulating gas supply to meet variable demand while gas production remained pratically constant. Over the years, the problem amplified, finally settling into a pattern of summer-winter contrast. Today this rigidity remains, due to the type of supply contract binding gas companies to gas producers. As early as 1954, the possibility of storing gas in the subsoil was examined in France. Since then, 13 underground storage sites have been developed on French territory and have provided ample proof of their considerable potential and their excellent technical reliability. They provide around one quarter of national consumption in an average climatic year and more than half of the gas circulating in the network on very cold winter days. GAZ DE FRANCE operates 11 sites, of which 9 are comprised of aquifer reservoirs and 2 are salt cavities.

The GERMIGNY-SOUS-COULOMBS aquifer reservoir, located 60 km East of PARIS, is the last to be set into operation and provides a good illustration of the efforts deployed to minimize costs, optimize operation and improve development and equipment techniques or tools.

The phase of geological exploration by drilling, lasting from 1980 to 1984, followed on from previous oil prospection work. It revealed a very flat anticline structural trap with a closure of 35 metres, measuring 20 km in length and 10 km in width (see Fig. 1). The Wealdien facies reservoir level (Valanginian-Hauterivian), located at a depth of 900 m, has a thickness of 80 m with a mean porosity of 20 % and a permeability of around one darcy.

These characteristics indicate that total ultimate reservoir capacity is greater than 2 billion m^3 (n) of gas.

However, one of the essential characteristics of an aquifer reservoir is that over an annual cycle it is neither technically nor economically feasible to withdraw the entire recoverable stock. The working volume is therefore the quantity of gas that can be withdrawn during a winter season and reinjected the next summer. Based on our experience with different types of reservoir, we conclude that a poorly structured anticline of this type should lead to a working volume which is lower than on good sites (where it represents around half the total stock). The objective of studies and operation is therefore to increase this working volume.

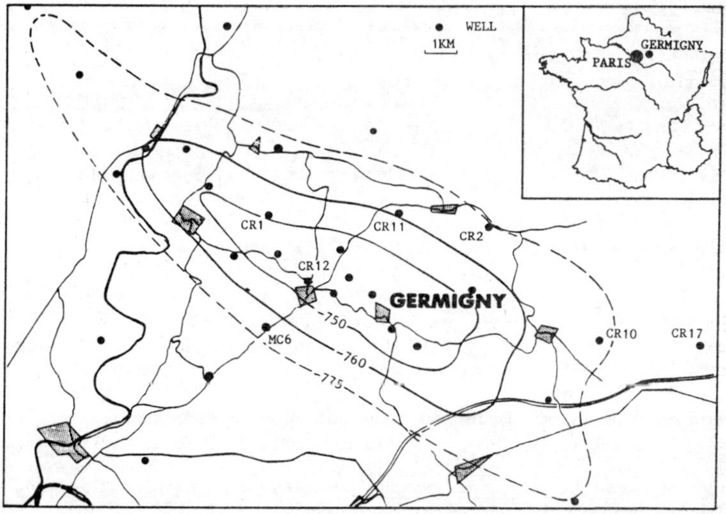

Fig. 1 : Map of isobaths on the Wealdian reservoir roof (m/sea level).

2. POINTING OUT HETEROGENEITIES AND COMPLEXITIES OF THE STORAGE

The geologist's knowledge

The clayous-sandstone reservoir was subdivided into 3 units already clearly delimited on all the exploration wells : R1, R2, R3 (see Fig. 2). The R1 unit constitutes the main reservoir, with a thickness of 30 m as the closure is only 35 m. This level is made up of 3 units R1-1, R1-2, R1-3, each measuring a few metres. The clayous caprock of the storage, with a thickness of 3.5 to 5 m, has a permeability measured on core samples and validated by a hydraulic interference test between wells, in the order of the microdarcy. This is sufficient for good gastightness.

Above the overburden, the Barremian reservoir, with a mean thickness of 5 m, sometimes absent, has excellent petrophysical characteristics (2 darcys) and thus constitutes the reservoir observation level, permanently monitored by 5 wells. Finally, 60 m of Barremian clayous series forms another very effective screen.

In 1984, the well distribution made possible to plot accurate maps of variations in thickness and petrophysical characteristics of each of the units R1-1, R1-2, R1-3. We noted :

- a significant degradation of all units in the East, indicating that it would doubtless be difficult to inject gas into the East of the anticline,
- a preferential North-South orientation of all drains, perpendicular to the tectonic axis. Gas injection was therefore likely to be disturbed by this severe anisotropy, aggravated by the flatness of the closure which reduces the gravitational effect on the gas.

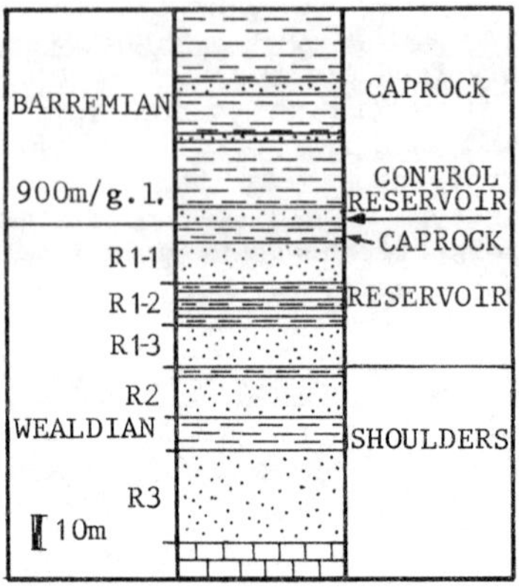

Fig. 2 : Lithostratigraphic cross section of the reservoir and caprock.

Events marking the first injections and withdrawals

Development began in 1982 with the injection of inert gas through a top well (see Fig. 3). Very soon, after 290 hm^3 (n) of natural gas has been injected, the gas reached an observation well at the North of the structure, CR 11, confirming the first geological analyses. Moreover, the gas reached another observation well, CR 01, much later than expected. These two facts confirmed that the geology was very heterogeneous and interfered seriously with the development of the gas bubble.

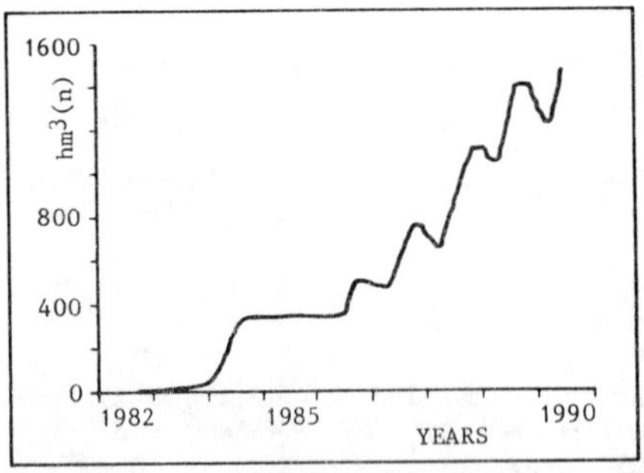

Fig. 3 : History of the stock from the outset.

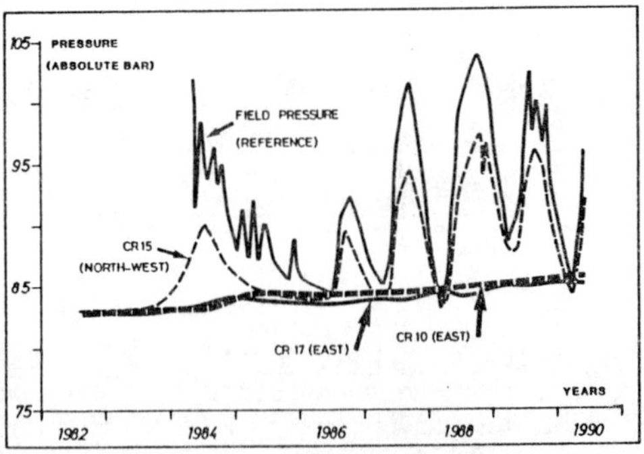

Fig. 4 : Pressure variation in observation well, and field pressure.

Furthermore, the water-gas interfaces measurements in the reservoir could not be explained and revealed pressure gradients between different layers, underlining the heterogeneous, multilayer characteristics of the reservoir. Finally, disparities in pressure variations in the observation wells (see Fig. 4) confirmed the heterogeneities of the spatial permeability field, particularly at the east of the reservoir (wells CR 10 and CR 17).

In view of the reservoir complexity and its lack of structure we conducted in-depth studies of its geology, involving the creation of numerical models, to gain a more detailed understanding of its behaviour.

3. A MULTI-DISCIPLINARY APPROACH

Detailed geological studies

Each of the sedimentary bodies with a thickness of a meter or more were marked out in space. Core samples were taken from several key wells representative of the site (eg. CR 12) and detailed sedimentological observations were made (see Fig. 5) to obtain detailed knowledge of the deposit environment. The various stratofacies range between two clearcut marine environments dominated by storms (or swell) and continental palustral environments dominated to a varying extent by rivers. The interfingering of the bodies is characteristic of a "deltaic" system in which marsh influenced by the sea develops between the lobes modelled by the swell.

Discontinuities, marking the limits of the deposit sequences, some with volcanic tuffs, are still being studied and should make it possible to build a definitive geological model, if numerical modelling so demands. High resolution sequential stratigraphy will form its basis.

R1-3 is dominated by the sea. R1-2 is somewhat fluviatile and the different clay banks are sufficiently continuous to isolate small reservoirs. As a result, the pressures of R1-1 and R1-3 are different and injections must be made selectively in this zone to reduce water influx during withdrawal. R1-1 is a wave dominated littoral bar (R1-1b) which links finally to the shore forming a beach (R1-1a) with excellent characteristics. It is the preferential drain through which the gas bubble spreads widely.

In the East of the anticline, all these reservoirs change to a more marine and more clayous environment, significantly deteriorating the reservoir character.

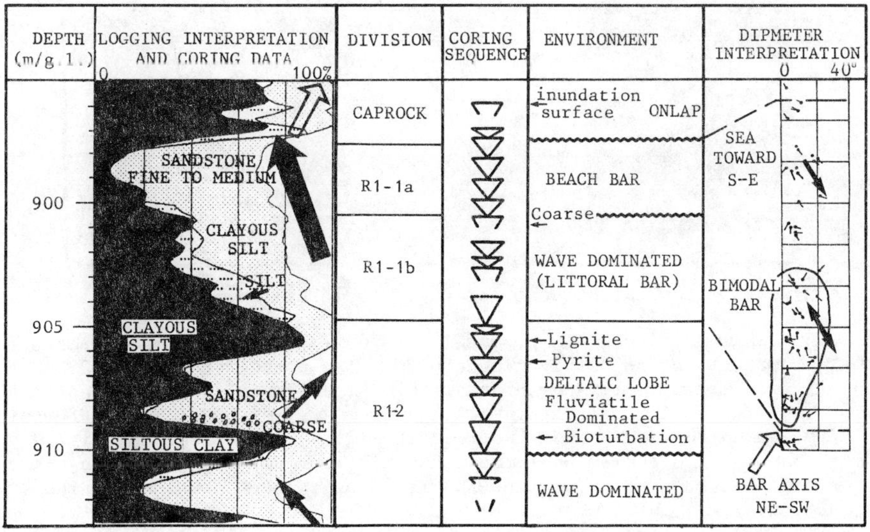

Fig. 5 : Geological analysis of well CR 12.

On the basis of the key wells, the different bodies were marked out on the other wells by sequential analysis of logging measurements. GAZ DE FRANCE took continuous cores from around 15 wells on this field and logged heavily and systematically all wells (double laterolog, micro-resistivity, acoustic, neutron, density, spectral radioactivity, dipmeter).

In this way, each body was oriented in space, like for example the North-East/South West oriented beach R1-1a. After matching the logging interpretation results with core sample measurements, isoporosity maps were plotted to identify the lateral variability of each body. Test during drilling, selective of each R1-1, R1-2, R1-3 zone, and core sample measurements were used to establish porosity/permeability laws useful for numerical modelling.

Numerical modelling

GAZ DE FRANCE has a large number of models and in the case of GERMIGNY-SOUS-COULOMBS, 2 three-dimensional models were built to meet very different objectives (see Fig. 6) :

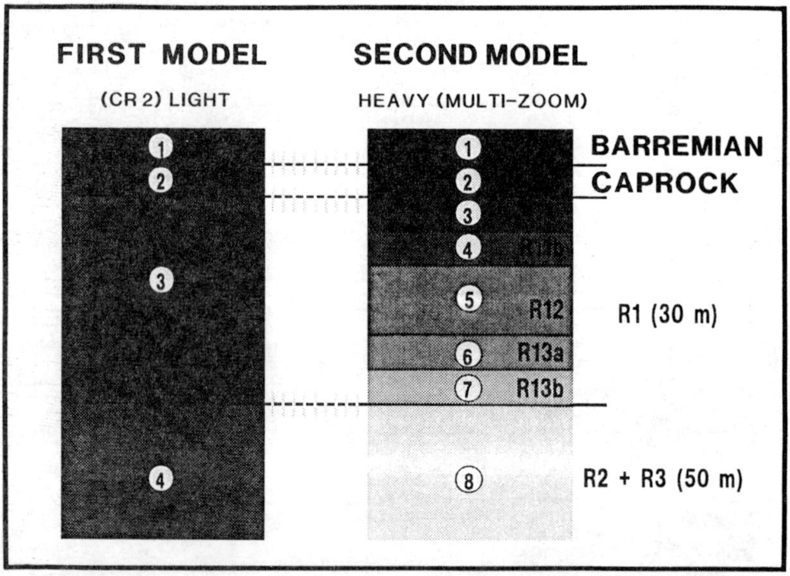

Fig. 6 : Schematic description of the numerical models.

The purpose of the first model was to explain rapidly a pressure anomaly observed in the observation aquifer at a perimeter well CR 2.

It was observed that a pressure rise in the main reservoir led to a significant pressure increase in the observation aquifer, very different to those measured on the other wells (see Fig. 7). The model comprise 4 layers, one for the entire Wealdian reservoir, one for the level subjacent to the Wealdian reservoir, one for the Wealdian caprock and one for the Barremien observation reservoir.

The model contains around 2 000 meshes and is able to restitute 5 years of operating history in 15 minutes of computer time (IBM 3090).

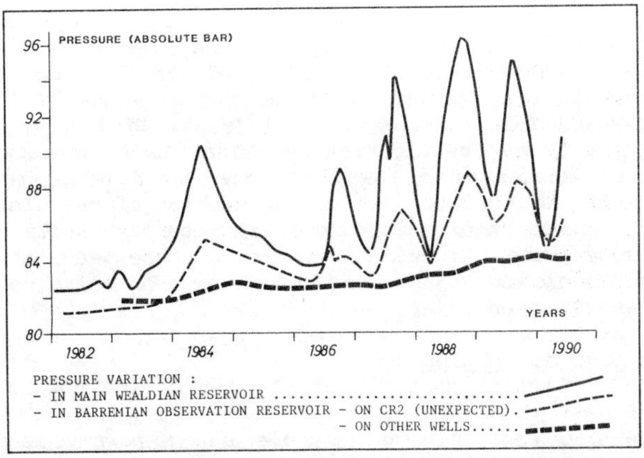

Fig. 7 : Pressure anomalies in the observation level.

It is thus possible to examine a large number of hypotheses (around 100), concerning caprock quality in the vicinity of the well where the anomaly was observed, anisotropies and heterogeneities of the observation aquifer, modifications in intrinsic caprock permeability, etc...

It was thus determined that pressure anomalies are probably due to significant and localized degradation of the caprock in the vicinity of the well or at the well itself which do not compromise overall reservoir development.

The purpose of the second model was to provide a prediction tool taking into account all production and detailed geological data concerning the reservoir. To this end, we used the ZOOM-MULTI model with which local grid refinements can be used (see Fig. 8) while remaining compositional for gas. The gas-filled reservoir is described in detail but, from a distance, the description is simple, using a small number of large meshes. It is thus possible to take account of most geological information and to describe the aquifer, even far from the reservoir, while keeping the number of meshes within the technical limits of the computing tools. The number of meshes is 13,000 spread over 8 level.

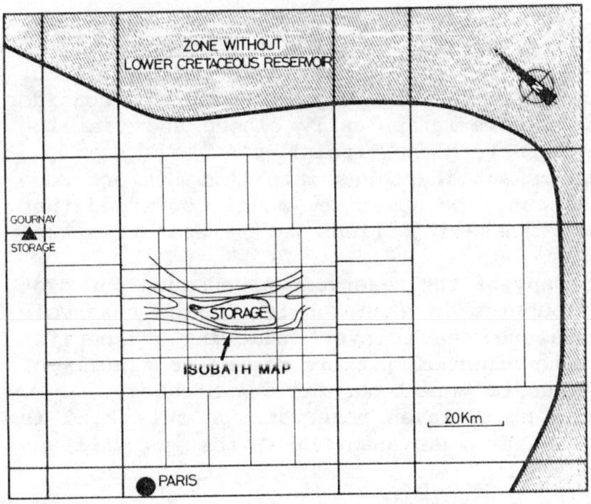

Fig. 8 : Structure of the zoom-multi model with local grid refinements

The subjacent levels, the Wealdien caprock and the Barremian reservoir are each represented by one layer, as in the previous model. The Wealdian reservoir is sub-divided into 5 layers taking account of geological units or groups of units corresponding to different deposit environments (R1-1a, R1-1b, R1-2, R1-3a, R1-3b).

The operating history is detailed for each well by introducing flow rate stages of a few days. Six years of history are thus handled in 5 hours of computer time. The matching of calculated pressures with measured pressures required around sixty attempts. As we have a reliable geological image, the values of porosity and permeability are modified only very slightly and adjustments concentrate on relative permeabilities and transfers between layers.

Reservoir development

Studies showed that the gas moving towards CR 11 followed a preferential drain and that this abnormal gas movement was not representative of the reservoir as a whole but a purely local phenomenon. Experience confirmed that the significant spatial anisotropy of permeability affect gas injection only at the very beginning and that the gas very quickly followed the structural form as predicted in the numerical models.

However, the pressure observations revealed a sharp degradation in permeability towards the East of the reservoir (well CR 10).

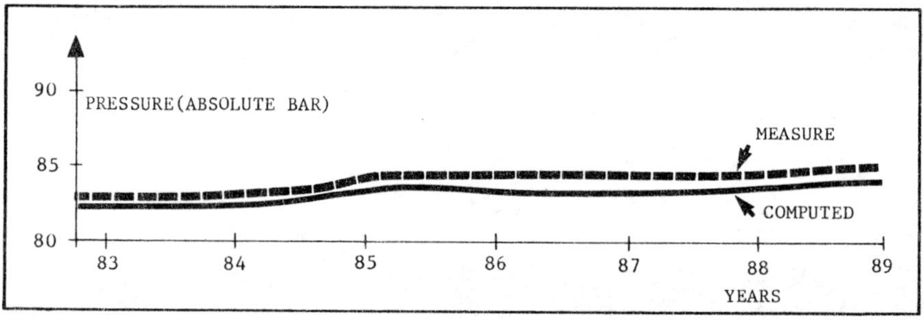

Fig. 9 : Pressure matching at observation well CR 10

Numerical modelling can easily account for this phenomenon and the field of permeabilities provided by the geological image leads, without modification, to a correct adjustment (see Fig. 9). We henceforth consider that the East of the reservoir is stratigraphically closed and that the gas will be unable to cross this structural spillpoint.

However, it was planned to inject inert gas into this Eastern zone to replace a part of the cushion gas. The numerical model confirmed that it would be preferable to transfer these injections to the Western structural spillpoint.

The extreme vertical anisotropy of the reservoir posed numerous problems for the monitoring of gas bubbles in each sub-level. The reservoir behaved like a multi-layer structure, each layer following a specific pressure pattern. This gave a good geological picture of the continuity of intermediate caprocks and enables us to orient our filling strategy.

A significant problem for a multi-layer reservoir of this kind is that of determining from where, and in what quantities, the gas will appear.

This can be very difficult if the observation wells are non-selective (of intermediate reservoir levels). The following theoretical scenario (see Fig. 10) illustrates the complexity of pressure gradient fields in two directions and three layers. If gas is injected into a given well, it meets 3 permeability fields in the x direction and 3 in the y direction. It it settles in place normally, and when K1x is smaller than K3y, the formation pressure loss is higher in lever 1 than in level 3 and the water pressure P3 is greater than P1. The opposite situation may well occur in a different direction (P'3 lower than P'1).

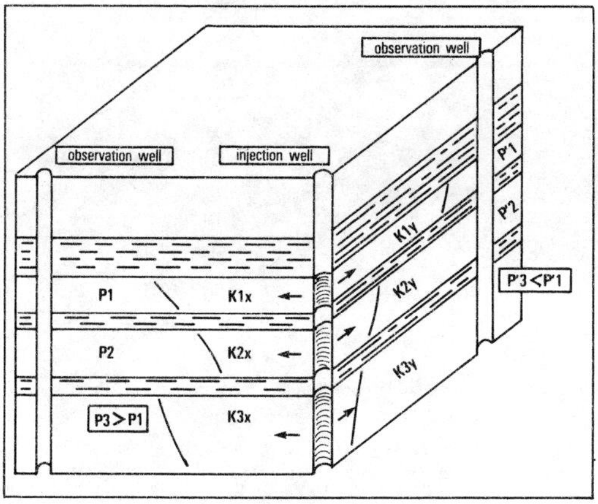

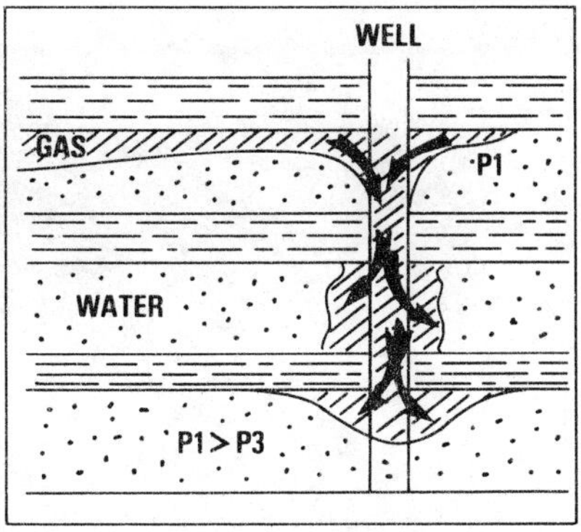

Figure 10 : Complexity of pressure gradients fields.

Figure 11 : Gas transfer encountered at well CR 11 and MC 06.

These complex pressure gradients fields may cause "abnormal" behaviour of the observation wells (and all non-selective wells) (see Fig. 11). The gas arrives via the upper level (observed on wells CR 11 and MC 06) at a pressure P1 greater than the pressure of the lower level P3 and will therefore invade the lower levels. The interfaces measured are therefore 20 to 30 m below "real" values in the case where gas replaces water to cancel the pressure gradient causing fluid transfer (measured by radioactive tracer or flowmetering or downhole pressure gage).

The opposite case has even been envisaged, in which water transfers from the lower level to the upper level could delay the arrival of gas at the well, the transferred water holding back locally the arrival of gas in the upper level.

The monitoring of these observation wells thus confirmed that the clayous levels between layers are practically impermeable and continuous within the scale of the structure. The use of selective completions, the presence of which was inferred from previous studies, was thereby confirmed. Similarly, to monitor the gas bubble, priority was given to neutron logging on the numerous wells opposite the reservoir which were specially cased and cemented for this purpose (GAZ DE FRANCE logging units).

The complete set of interactions between geological studies, numerical simulation studies and observation of gas injection into the reservoir serve to define the GERMIGNY-SOUS-COULOMBS reservoir as a multilayer heterogeneous reservoir (see Fig. 12).

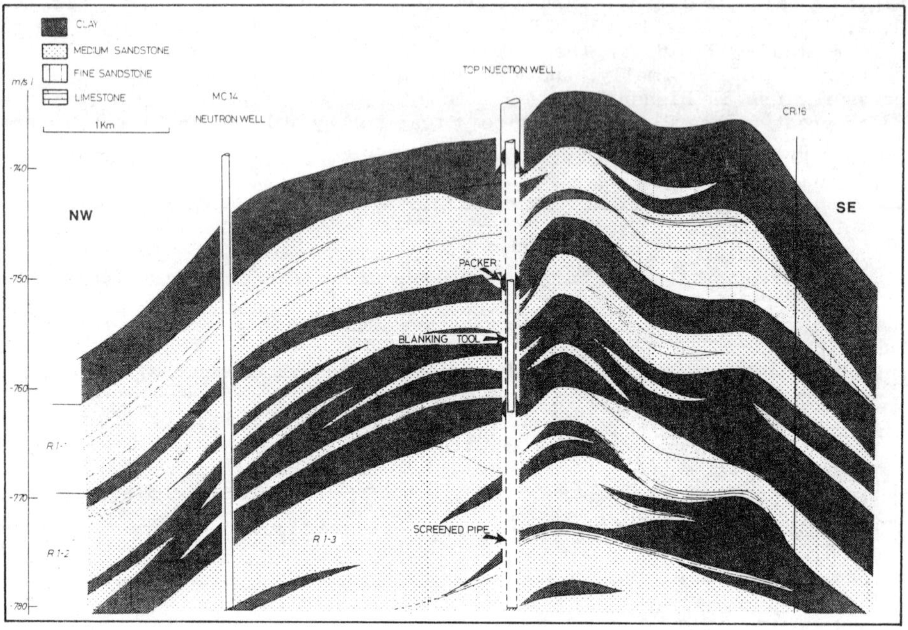

Fig. 12 : Axial NW-SE cross section with choice of completions

4. THE SEARCH FOR OPTIMUM OPERATION

As soon as the complex character of the reservoir was confirmed by the first gas injection, prospective filling simulations were performed to avoid overflow in critical zones and to determine optimum strategies. The optimum strategy is a compromise between the desire to create an immediately recoverable working volume by giving priority to gas injection to the point of saturation in the upper layers and the desire to inject gas deep down and not take the risk of overflowing out of the structure. To make selective injections, we have a suitable completion with two formation packers and three screened pipe zones. Each reservoir zone is accessible individually by the use of plugs or blanking tools for total selectivity. In addition, flow metering by the GAZ DE FRANCE logging teams ensures that the volumes really injected or withdrawn on each level correspond to optimum volumes.

Flow metering is used to measure or quantify transfers via production wells, whereas radioactive tracers are used to follow water transfers on the observation wells. Neutronic measurements serve to judge when saturation levels have been reached in each reservoir and to determine the extent of the gas bubble. It is thus possible to obtain an overall picture of the gas distribution per level and thereby to adjust the values of vertical permeability in the model.

The choice of deep injections in the upper levels has an important effect on the water/gas ratio at withdrawal, which is measured daily on each well. As the reservoir has a low gas saturation level, this ratio is still very large, at around 20 g of water per m^3 (n) of gas and causes corresponding sand ingress.

The problem of sand inflow is monitored using piezometric surface and well-bottom sensors and a camera used withdrawal to detect damage to screen pipes. Here again, there is a strong interaction between in-situ observation and geological studies. Geologists studied core samples to determine the distribution of poorly consolidated zones with a low degree of compaction and cementing. Then, a logging method to pinpoint these zones was developed. It uses the responses of natural radioactivity, resistivities, acoustics and Neutron/Density porosity. The risk of non- consolidation is quantified according to the percentage of clay, the cementing factor (m of logs) and the compaction coefficient deduced acoustically. Software to process logging results on site now makes it possible to localize high-risk zones immediately. This method is currently being perfected jointly with partners in the petroleum industry (ARTEP) and in the short term, should make it possible to choose appropriate screen sizes for each level as soon as drilling is completed, so that sand ingress is limited.

Reservoir development can be planned in the longer term by means of numerical modelling. It is thus possible to determine accurately the sensitive points of operation at maximum stock and to outline measures for reinforcing the observation system (see fig. 13).

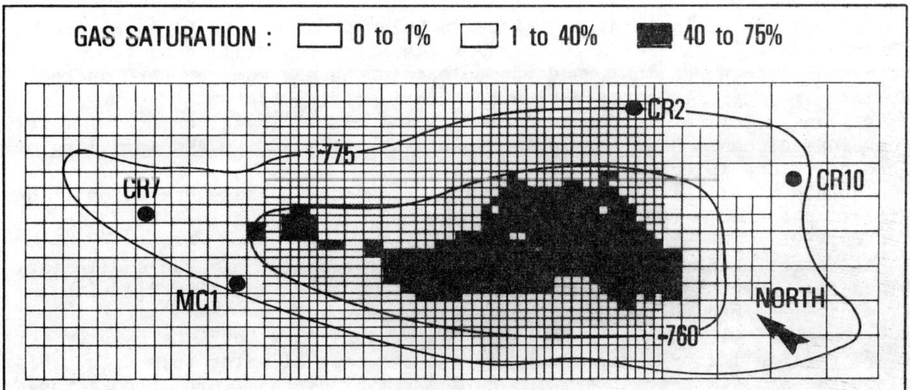

Fig. 13 : Previsional saturation at forecast maximum stock
(with isobaths on the roof of the Wealdian reservoir)

5. CONCLUSION

This reservoir is representative of problems which will be encountered in the future, on increasingly complex and heterogeneous sites. The development of such sites will only be possible if the multi-disciplinary and complementary skills of the geologist, reservoir and completion engineers are fully mobilized.

EL BORMA FIELD - SITEP - TUNISIA - AN INTEGRATED GEOLOGICAL, PETROPHYSICAL AND NUMERICAL APPROACH AIMED AT AN INFILL DRILLING PROJECT

C. ROSSINI, P. TERDICH, B. VOLPI, C. DESCALZI, F. GENOVESI and P.P. PIRAS

AGIP - Production Services - Reservoir Engineering Dpt.
Piazzale Vanoni 1, San Donato Milanese, ITALY

K. KADDOUR, R. GRIBAA, H. GAAYA, M. AZOUZ and C. ATTAYA
SITEP - Direction des Etudes et Developpement
92 - 94 Rue de Palestine - Tunis

Summary

The preservation of the reservoir internal heterogeneities highlighted in the geological model is fundamental for a reliable fluid flow simulation. For this purpose an innovative methodology based on a statistical algorithm of "cluster analysis" was developed to obtain a reservoir zonation in homogeneous subsets (lithofacies) with a univocal lithological and petrophysical meaning. This statistical approach is founded on a log and core data integration and is aimed at characterizing the log-lithofacies both from a qualitative and a quantitative point of view.
The "facies" concept was used as key parameter to define the main correlations between all dynamic variables (absolute permeability, relative permeability and capillary pressure) and to initialize the numerical model.

1. INTRODUCTION

This study is relevant to the Tunisian side of the El Borma field. Production was started in 1966; 137 wells have been drilled so far.
The main target of the study was to obtain the maximum recovery increase without modifying the production equilibria of the system.

The solutions proposed consist in water injection in a first development phase and then in the "pattern" optimization and infilling of producers and injectors.
To obtain a reliable tool for a correct planning of these operations, the first phase of the study was developed according to the following targets:

- to give a detailed description of the geometry of the sedimentary bodies composing the reservoir to locate future infill wells with the minimum hazard.
- to preserve the petrophysical heterogeneities of the system by respecting its extreme values (1). In the case of water injection, this would allow a correct simulation and timing of the fluid breakthrough for the best possible forecast of oil recovery.

The second phase of the study consisted in the definition of a model grid able to preserve and maintain the detailed description obtained from the static model.
The last generation simulation model utilized allowed the use of a "mixed" grid which can define subgrids in some areas granting the best "translation" of the geological model.

This obviously led to the set up of a very elaborate model as concerns communication schemes.
The construction and management of this model required long times and the cooperation between various professionals such as geophysicists, geologists, petrophysics, reservoir engineers and modelling engineers.
The result is however a very flexible and detailed tool able to verify

all the possible solutions both from the technical and economical point of view.

2. GEOLOGICAL SETTING

The El Borma field consists of Lower Triassic sediments of the Kirchaou Fm.
These deposits were laid on a Hercynian unconformity which eroded a Paleozoic sequence in this area.

The structure is a gentle anticline whose axis is Northwest-Southeast trending with a main culmination in the Northern side of the field.
Tectonic events deformed the reservoir only after its complete sedimentation.

Thus, chronostratigraphic panels allowed a detailed description of the geometry of the sedimentary bodies, removing the post sedimentary structuration.
The reservoir, resulting from the superposition of four main levels, namely A-B-C+D-E, is about 100 m. thick (Fig. 1).

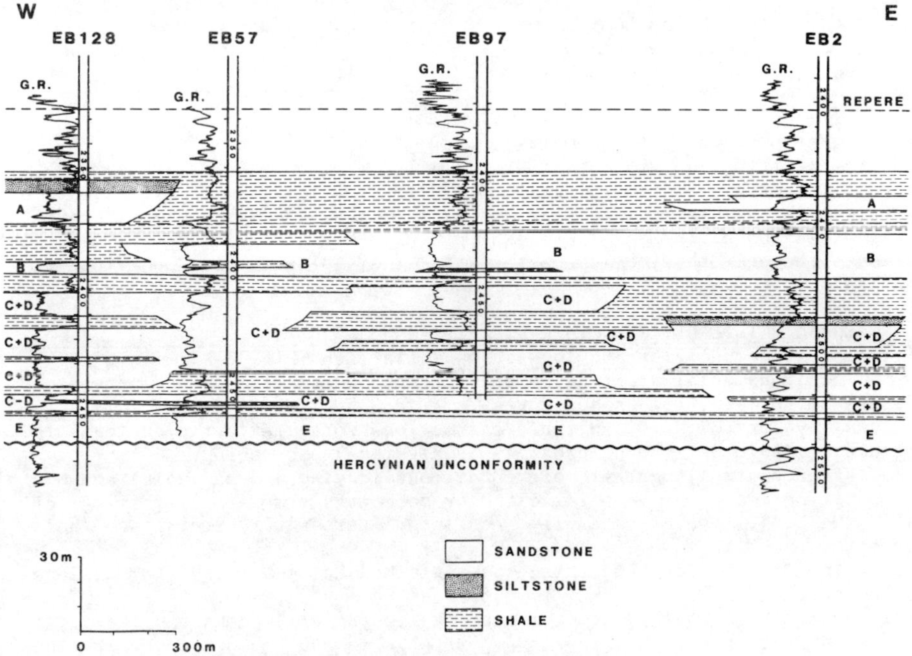

Fig. 1 Chronostratigraphic panel

Sandstones and siltstones deposited in fluvial environments form the reservoir rock.
Each level represents at least a complete cycle of fluvial aggradation.
Levels A and B were deposited in a meander-like environment.
The areal distribution of sandstones, siltstones and shales in both layers is strongly heterogeneous.
Sandstones deposits derive from the superposition and lateral migration of fluvial channels of which they represent the bed load.

In levels A and B the thickness of sandstones varies between 0 and more than 20 m. (Fig. 2).

In the upper part of the sequences the low energy deposits prevail, i.e. siltstones transported in suspension, crevasse deposits and alluvial plain shales which mark the end of the cycle.

Levels C+D and E were deposited in "braided" environments.

Their internal heterogeneity is especially related to the limited areal extension of the sandy bodies.

These two levels result from the juxtaposition of several lenses of sandstone, intercalated with non correlative shales (Fig. 1).

Alluvial plane shales continuous all over the field are present in the upper parts of the sequences.

As concerns the whole Triassic reservoir, the heterogeneity in the distribution of the different lithotypes and the discontinuity in the vertical and lateral extension of the sandy bodies affects the petrophysical characteristics and the production behaviour of the reservoir.

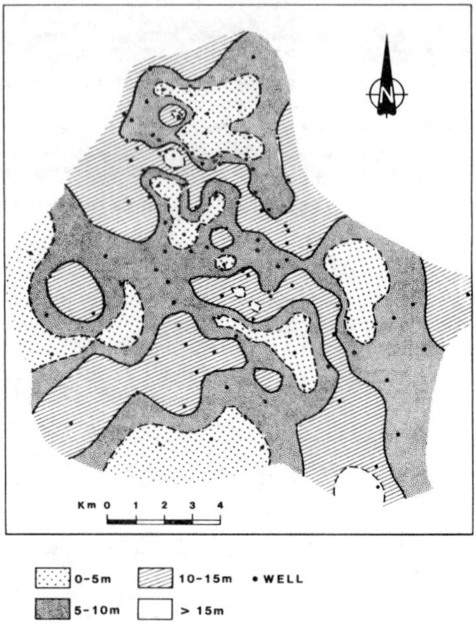

Fig. 2 - Level A - Sandstone thickness

3. RESERVOIR INTERNAL STRUCTURE DESCRIPTION

Methodology description and log-facies definition (Fig. 3)

The fundamental starting point was the construction of an integrated data base containing logs and cores.

A quality control and suitable corrections were performed on logs to obtain reliable data to be used in a statistical processing.

After controlling that hole conditions (caving and bad hole) and mud composition could not jeopardize one or more logs completely, the logs of all the wells were carefully corrected for environmental and hole effects.

Then an accurate depth matching was performed between the different logs of each well.

It was ascertained that sometimes a log (especially gamma ray) can give different readings when lithology, petrophysics and fluid content are the same.

For this reason, a suitable log normalization was performed in these cases.

Core data were depth matched with the logs of the same well and core porosities were corrected for the overburden pressure effect.

18 key wells were selected from the data base containing all the corrected core and log data.

These wells must fulfill the following requirements:

- to have the most complete set of cores
- to have the most reliable and complete set of logs

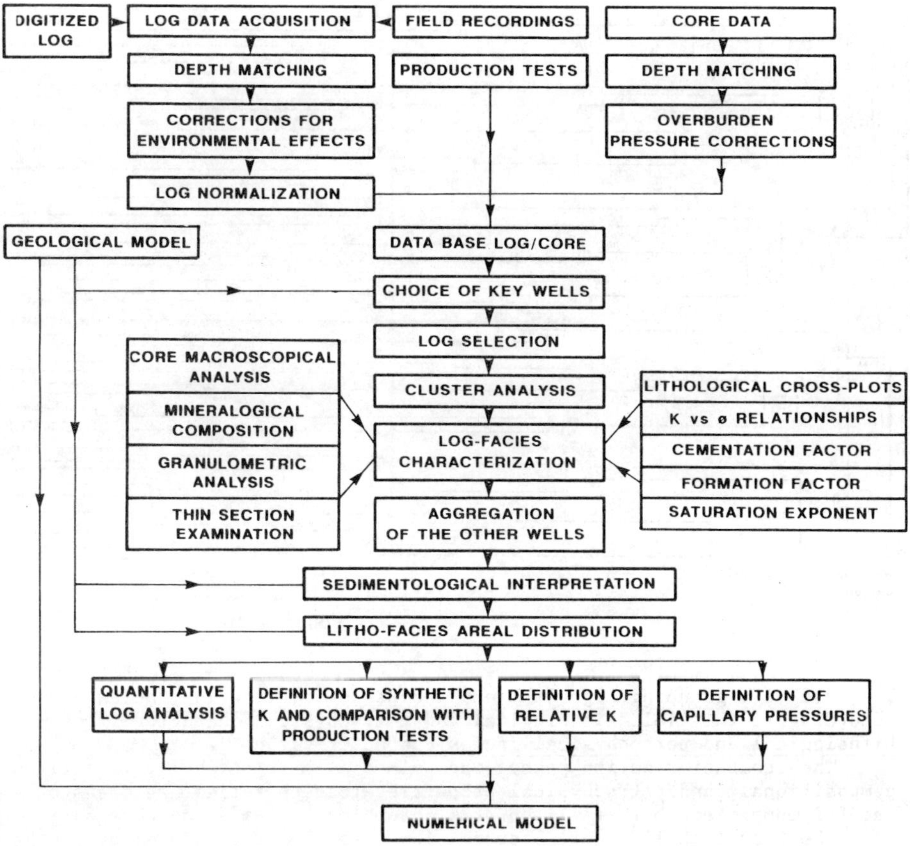

Fig. 3 - Methodology flow chart

- to cross the whole reservoir
- to be well distributed on the field extension.

These conditions are mandatory since the data of these wells, which are the basis of the statistical elaboration, must be fully representative of the reservoir.

In addition to the well selection, it is necessary to choose the best suite of log according to the objectives of the study.
Thus each chosen log response must reflect the formation lithological and petrophysical characteristics as much as possible.

Gamma ray and sonic log were selected.
These two logs for all the key wells were contemporarily submitted to a statistical processing based on a cluster analysis algorithm (2).
It minimizes standardized Euclidean distances in the n-dimension space of normalized log variables and groups the m considered samples (where m is the number of sets of logs readings on all the selected intervals of the chosen wells) organizing them in a hierarchical cluster succession until they are all composed in a unique set.

The dendrogram (tree) showing the distribution and evolution of recognized clusters is given in Fig. 4.

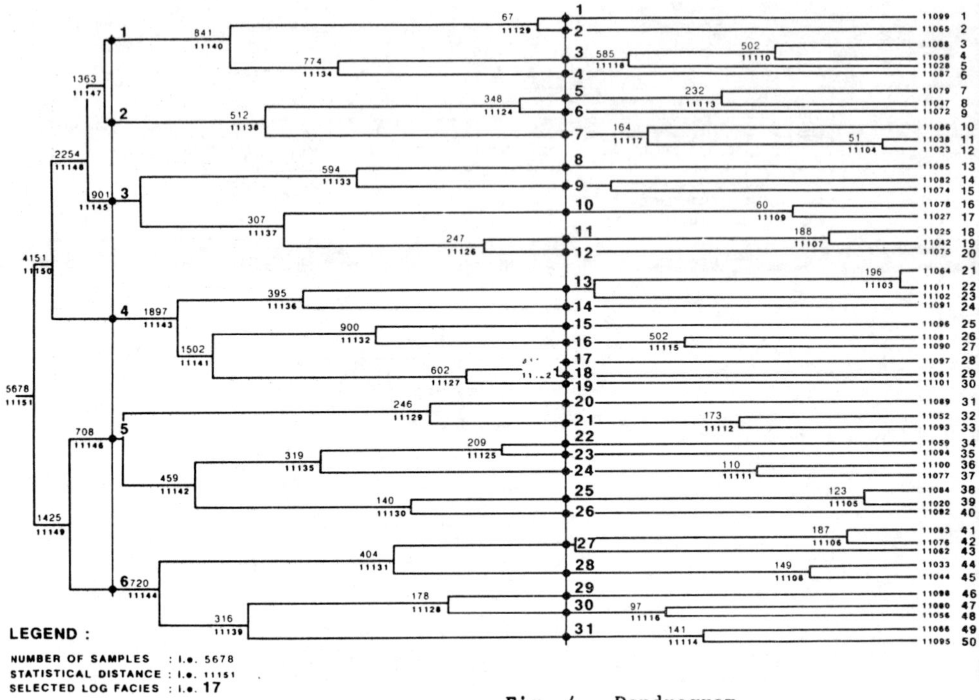

Fig. 4 - Dendrogram

The use of a grouping algorithm enabled us to vary the log-facies number according to the field lithological and petrophysical characteristics.

Lithological and petrophysical facies characterization

The characterization phase was aimed at attributing well defined compositional and petrophysical properties to each facies by means of a careful comparison between litho-stratigraphical columns and core data.

The availability of cores for the log-facies allowed the characterization of each class also by means of all the detailed analyses that can be performed on cores (i.e. mineralogical composition, examination of thin sections and analysis of grain sizes).

After this characterization process it was possible to identify the minimum number of facies necessary to describe all the lithotypes present in the reservoir.

Six facies were considered:

- class 1 : sandstone with high production potential
- class 2 : silt and thin sandstone with medium production potential
- class 3 : silt and shaly silt with poor production potential
- class 4 : sandstone with very high production potential
- class 5 : shale
- class 6 : shale

The comparison between log-facies and routine core analysis data allowed the attribution of a petrophysical meaning also to the log-facies.

The plots K vs PHI (core data) per log-facies are presented in Figs. 5-6-7.

They highlight the possibility of recognizing a permeability range characteristic of each of the four reservoir classes.

The existence of a correlation between litho-facies and permeability allowed a more detailed definition by selecting a higher number of classes.

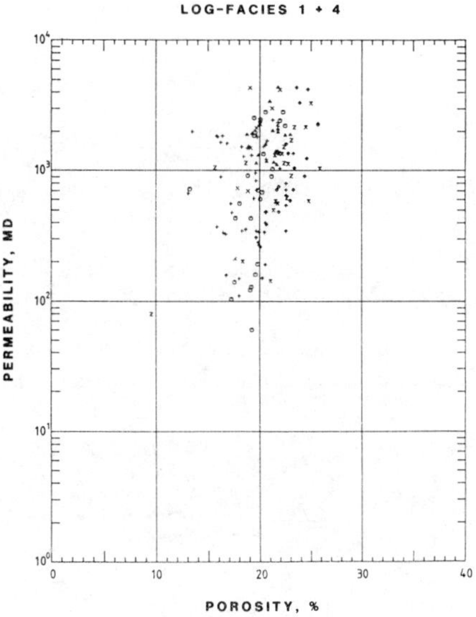

Fig. 5 - Plot K vs. Ø classes 1+4

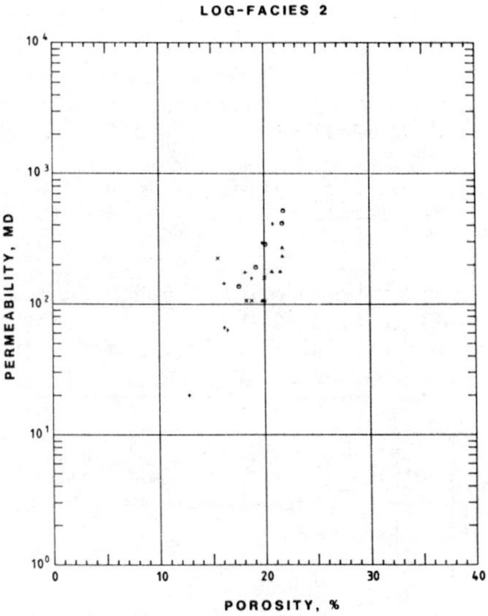

Fig. 6 - Plot K vs. Ø class 2

This possibility was of paramount importance taking into account that the above mentioned plots did not show a direct K vs PHI relationship.

<u>Aggregation phase</u>

In order to complete the data base, so far referred to the key wells, the data of the other wells of the field were aggregated.

The vertical distribution of the six lithofacies selected in the previous characterization phase could be obtained for these wells too.

The aggregation procedure utilized the same logs selected for cluster analysis (GR and DT).

The log readings at each single depth interval were compared with the whole population of each log-facies and attributed to the most similar one.

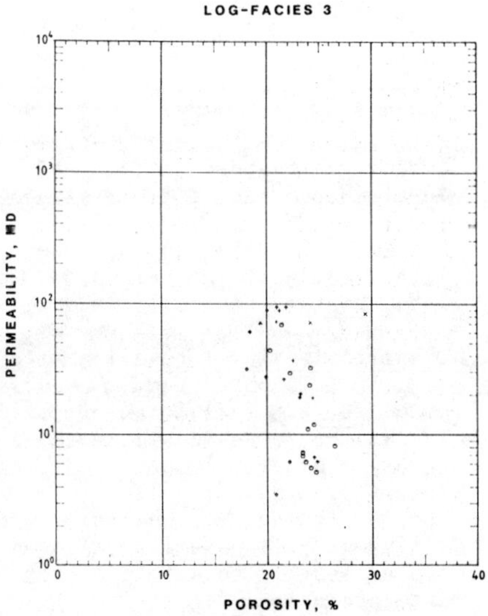

Fig. 7 - Plot K vs. Ø class 3

Then it was necessary to verify that the resulting facies were statistically consistent with the lithological and petrophysical meaning recognized in the key wells.

After the aggregation phase the field data base resulted complete, coherent and representative of the lithological and petrophysical description of the whole reservoir area.

Identification of sedimentary facies

In the El Borma field, due to the sedimentary environment, the single litho-facies are not correlative, but sequences of litho-facies representing sedimentary facies can be correlated.
The sequences of litho-facies were analysed in detail with a sedimentological approach.

The comparison between an ideal vertical sequence of a meandering system (layers A and B) and the result of the cluster analysis (i.e. well EB 77 layer B) is given in Fig. 8.

The comparison between an ideal vertical sequence of a braided system (layers C+D and E) and the lithostratigraphic column of the well EB 41 layer C+D is given in Fig. 9.
The lithofacies highlight each main component of a sedimentary sequence.

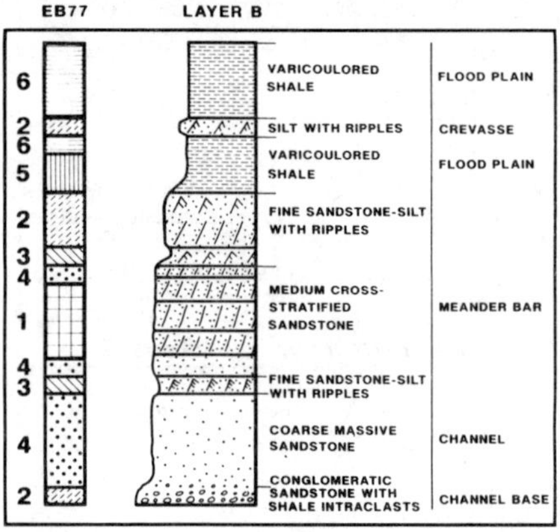

Fig. 8 - Comparison between a lithostratigraphic column and an ideal meandering vertical sequence

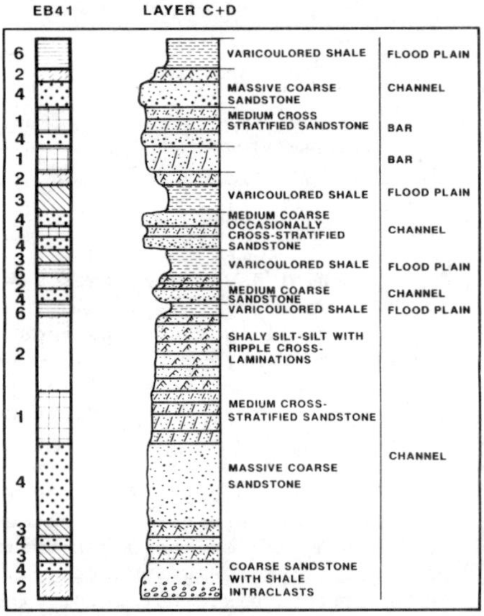

Fig. 9 - Comparison between a lithostratigraphic column and an ideal braided vertical sequence

Quantitative interpretation of petrophysical parameters

In this phase too, the facies concept represented the starting input point for the evaluation of petrophysical parameters.
Facies could be considered a new variable enabling us to build the input interpretative models and to verify the output of quantitative interpretation.
The mineralogical analysis, performed on core plugs corresponding to different litho-facies, allowed a complete compositional description of the reservoir rock. This represented a powerful tool for quantitative interpretation.
The cementation factor and the saturation exponent were calibrated for each class on the basis of special core analysis.
The quantitative log interpretation was performed by applying a linear programming algorithm based on the solution of a system of linear inequalities (3).
The solution of the system was obtained by minimizing the error function representing the sum of errors in each log response equation by means of

a linear programming technique.
The facies concept enabled us to establish a consistent interpretative model valid also for not cored wells.
The integration of the facies analysis into the process allowed a detailed control of the quantitative interpretation and, from a mathematical point of view, a decrease in the degrees of freedom of the system.

4. PETROPHYSICAL RESERVOIR CHARACTERIZATION FOR NUMERICAL PURPOSES

Synthetic permeability

The definition of porosity and fluid content profiles was not sufficient to evaluate the dynamic potential of reservoir rocks.
For this purpose it was fundamental to forecast absolute permeability variations too.

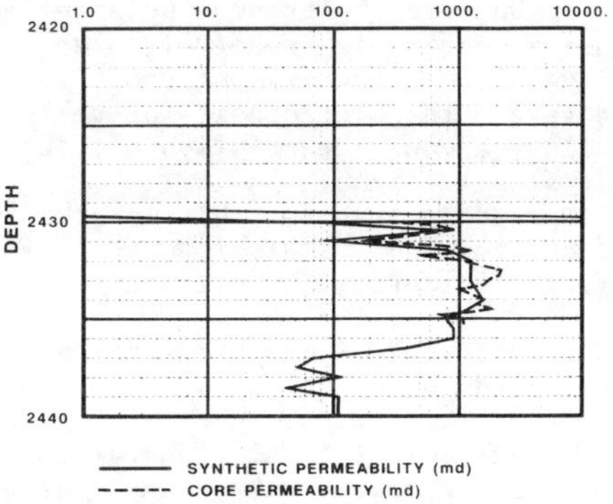

Fig. 10 - Synthetic permeability superimposed upon a log of core plug permeability

Porosity and permeability variations are usually connected by means of either correlations on core data or empirical formulae.
This approach can provide acceptable results in homogenous environments while it would lead to misinterpretations in a heterogeneous system like El Borma reservoir, where no correlation exists between porosity and permeability due to their texture differences.

By using the same concept of facies which univocally identifies not only porosity, but also lithology and texture, it was possible to reconstruct the permeability variations.

A high number of log-facies (31) (Fig. 5) were selected from the log-lithofacies data base, obtained in the same statistical analysis process, to achieve a more detailed description of the permeability profiles.

An average value of absolute permeability (from core analysis) was then attributed to each log-facies, by means of suitable operators, in order to obtain a "synthetic permeability" definition also for not cored wells.
As a check, the synthetic permeability logs were superimposed to a log of core plug permeability (Fig. 10).

Layering and gridding

Level A is dynamically separated from the remaining levels of the reservoir by means of a shaly horizon, continuous all over the field.
As a consequence, it was possible to construct two different simulation models with about 25,000 cells each.

The procedure utilized for the layering and gridding of level A model is presented like a case history.
The model set up and the choice of the averaging operators was made with the aim of preserving the reservoir heterogeneities.
Each well was zonated on the basis of its petrophysical characteristics to obtain a vertical sharing in homogenous subsets.
Three layers were defined for all the field wells.

The petrophysical discriminating parameter was essentially permeability (fig. 11).

A plot comparing average geometric permeability and build up permeability per layer is presented in fig. 12.

The satisfactory result of this comparison validated synthetic permeabilities also from a dynamic point of view, making them more reliable for the following phase of history match.

Due to the extreme heterogeneity of this reservoir, each layer is not areally uniform. However, the high density of wells allows a good description of the local situations.

In order to transfer this information to the simulation model a very fine grid was defined (Fig. 13).

The majority of the oil zone was detailed with cells of 100m. X 100m.

The use of a "last generation" simulation model enabled us to use a "mixed grid", with the possibility of having further subgrids in areas of peculiar interest.

The static distribution of the fluids (through capillary pressure curves) and their dynamic characteristics (relative permeability curves, end points etc.) were linked to the lithofacies concept by means of a detailed characterization of the dynamic behaviour of each lithotype.

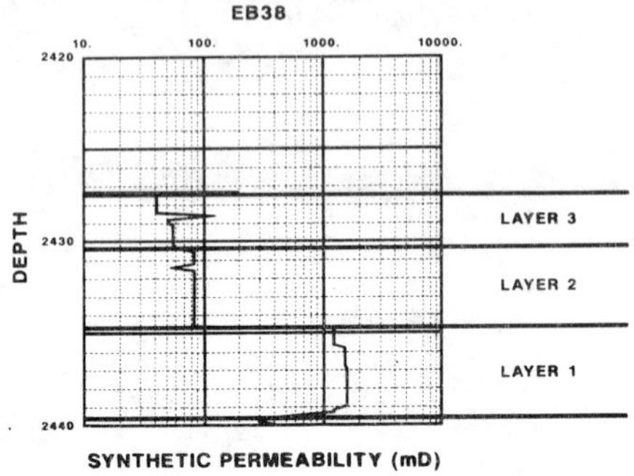

Fig. 11 - Synthetic permeability with model layers

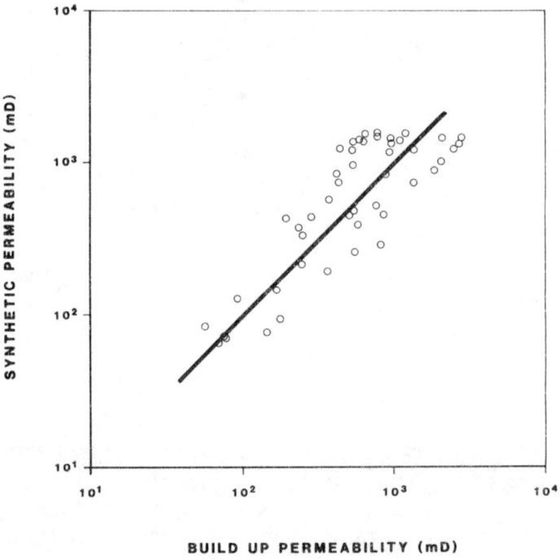

Fig. 12 - Average geometric permeability vs. build up permeability per layer

5. CONCLUSIONS

The success of this kind of study depends on a multidisciplinary approach since the dynamical model needs to be as "faithful" as possible to the geological description of the reservoir.

The final product, which is extremely complete and detailed, required remarkable efforts and was the result of an integrated, highly time consuming procedure.

However, this approach was the only one minimizing uncertainties and, thence, optimizing investments.

From the technical point of view, the model is stable and consolidated and, accordingly, easily updatable.

As a consequence, the results obtained both from the engineering and economical point of view rewarded the considerable efforts made during the static phase.

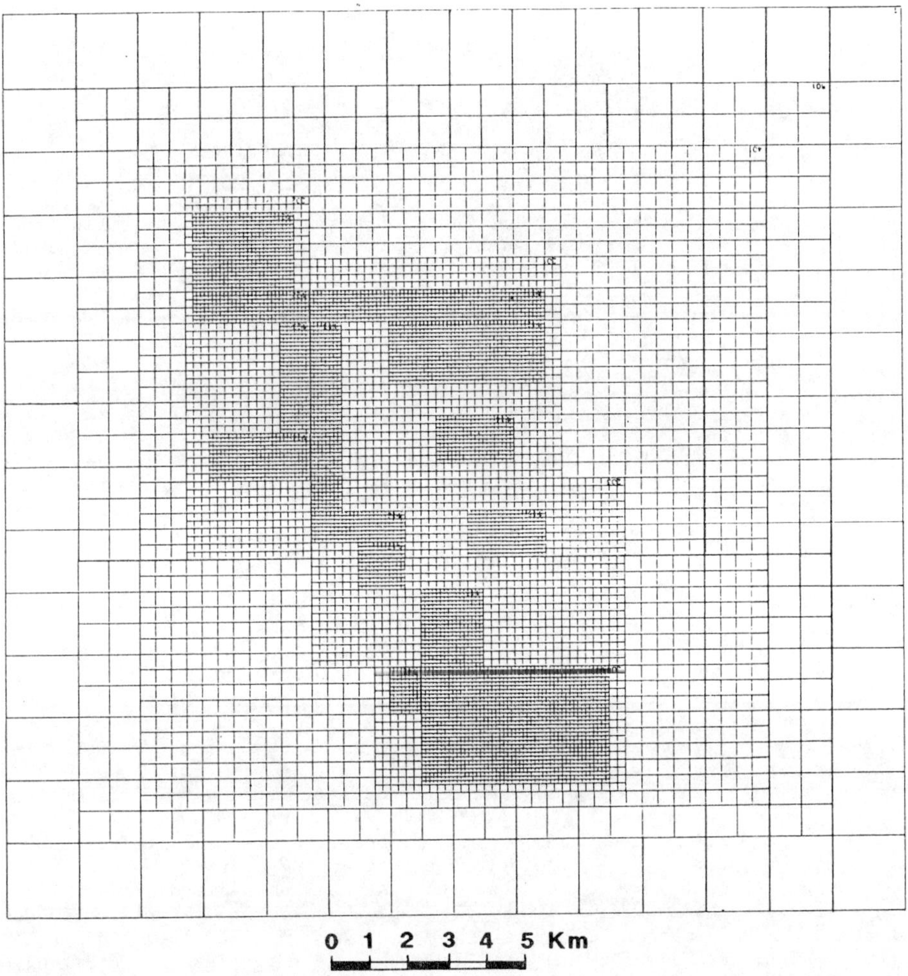

Fig. 13 - Level A - Model Grid

REFERENCES

(1) JOURNEL A.G. et al. (1988). Focusing on spatial connectivity of extreme-valued attributes: stochastic indicator models of reservoir heterogeneities. SPE 18324

(2) DAVIS, J.C. (1986). Statistical and data analysis in geology. II edition - John Wiley and Sons Inc, New York, Chichester, Toronto, Singapore.

(3) DESCALZI C. et al. (1988). Synergetic log and core data treatment through cluster analysis: a methodology to improve reservoir description. SPE 17637

GEOLOGICAL MODELLING OF TURBIDITE RESERVOIRS

J.D. SCHUPPERS, M.E. DONSELAAR and C.R. GEEL

Faculty of Mining and Petroleum Engineering
Delft University of Technology
P.O.Box 5028, 2600 GA Delft, The Netherlands

Summary

EOR requires detailed knowledge of permeability distribution in a reservoir. In reservoirs with a complex architecture, permeability distribution can be modelled with stochastic models which incorporate geological information such as width/thickness ratios of reservoir units and width/length ratios of shales layers. Reliable geological data can be derived from the reservoir itself or from analogue outcrops.
This paper presents the preliminary results of an outcrop study of a turbidite channel sediment body, the Lower Ainsa Channel. The deposits are of Eocene age and outcrop near Ainsa, central Spanish Pyrenees. The sediment body is at least 375m wide, has a maximum thickness of 30m and consists mainly of interbedded conglomerate, sandstone and mudstone. Five lithofacies are recognized. Their spatial arrangement is related to mass-transport processes. Data are presented on spatial distribution and geometry of the sandstone and mudstone layers. A shale length distribution curve is compiled from the outcrop data.

1. INTRODUCTION

Optimum recovery of hydrocarbon accumulations is, to a large extent, determined by the detailed knowledge of the spatial distribution of permeability within the reservoir. Permeability values can only directly be determined in wells in the reservoir. In the inter-well area the permeability distribution can be determined by drilling additional wells (at high additional cost), or it can be modelled by use of deterministic or stochastic methods. Deterministic correlation may be used when reservoir architecture is simple, i.e. when reservoir units are homogeneous and continuous between several wells, so-called 'layercake' reservoirs (1). When reservoir architecture is complex one has to invoke stochastic methods to generate possible models of the permeability distribution within the geological formation. Currently used models can be classified into object-based methods and sequence-based methods (2). Object-based methods generate discrete sand bodies or shale layers of random shape at random positions in space (3). Some geological knowledge is usually incorporated, such as width/height dimensions of sand bodies or width/length ratios of shale layers. Analogue outcrop studies provide this kind of information. Sequence-based methods generate distributions of permeability or lithofacies values that satisfy a certain variogram or conditioned probability distribution. These conditions can either be derived from the reservoir itself, or from analogue outcrop studies (4).

In this paper an example is presented of a geological outcrop study for modelling purposes. The outcrops comprise a turbidite channel complex. In reservoir terms the channel complex is a 'jig-saw puzzle' type of reservoir (1). The aim of this research is to establish a reliable quantitative data set which can be used in both object-based and sequence-based stochastic models. Various sedimentary aspects of importance to fluid flow models for turbidite reservoirs are dealt with:

(a) The definition of the different mass-flow layers that constitute the turbidite channel fill.

(b) The spatial distribution, geometry, and lithofacies distribution of the individual mass-flow layers.

(c) The spatial distribution and the geometry of the shale layers. In a mass-flow environment shales are prone to erosion resulting in amalgamation of sandstone beds. This enhances the effective vertical permeability of the reservoir.

In cases where recovery efficiency from turbidite reservoirs is low, enhanced production is likely to benefit from better, detailed knowledge of depositional architecture and reliable statistics on shale length. An example of low recovery rates of oil in place from turbidite reservoirs is given by figures from the Permian Spraberry-Dean trend of the Midland Basin, Texas. Recovery rates average 15 percent and may drop to 6 percent due to complex reservoir architecture and poor reservoir quality (5).

2. METHODOLOGY

In order to define and correlate the different lithofacies units sixteen vertical sedimentary log-sections were run across the outcrop width. Log scale is 1:50 and average log spacing 35m. The data gathered from these logs have been interpolated visually where possible, both on the outcrop and from photographic panels. In order to obtain fresh samples for laboratory permeability measurements and petrographical analysis each lithofacies has been cored using a 2" electrical drill. Permeabilities also have been measured on the outcrop with a portable permeability meter, constructed at Delft University of Technology.

3. GEOLOGICAL SETTING

One of the world's best exposures of turbidite channel fills is located in the southern Pyrenees, Spain, near the village of Ainsa. These turbidite deposits are part of the sediment fill of the Tertiary South-Central Pyrenean Basin, that trended parallel to the Axial Zone of the Pyrenean Uplift (Fig. 1). It is bounded to the east by the Segre Fault, whereas the western side was in connection with the Atlantic Ocean. The basin has a length of about 200 km, a width of 40-50 km, a maximum sediment fill of about 3500 m, and a total volume of some 25000 km^3.

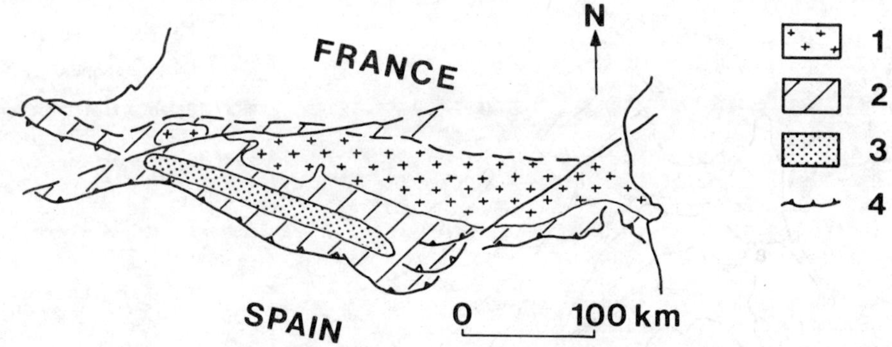

Fig. 1. Simplified geological map of the Southern Pyrenees. Modified after Souquet et al. (6). 1=Hercynian Basement; 2=Post-Hercynian sedimentary cover; 3=Tertiary South-Central Pyrenean Basin; 4=thrust.

The basin can be subdivided into three adjacent paleogeographic sectors (Fig. 2), where each sector is bounded by north-south trending structural ramps:

1) In the eastern sector, the sedimentary fill consists of fluviatile, deltaic and shallow marine deposits. This part of the basin is commonly referred to as the Tremp-Graus Basin (7).
2) The middle sector contains deep-marine mudstone and sandy turbidite channels (San Vicente Formation (8); Hecho Group (9)). It is sometimes referred to as the Ainsa Deep (7).
3) The western sector contains thick successions of sand-rich turbidite fan-lobes. To the west the lobes grade into mudstone and basin plain turbidites. The sediments in this sector are known as the Hecho Group (9).

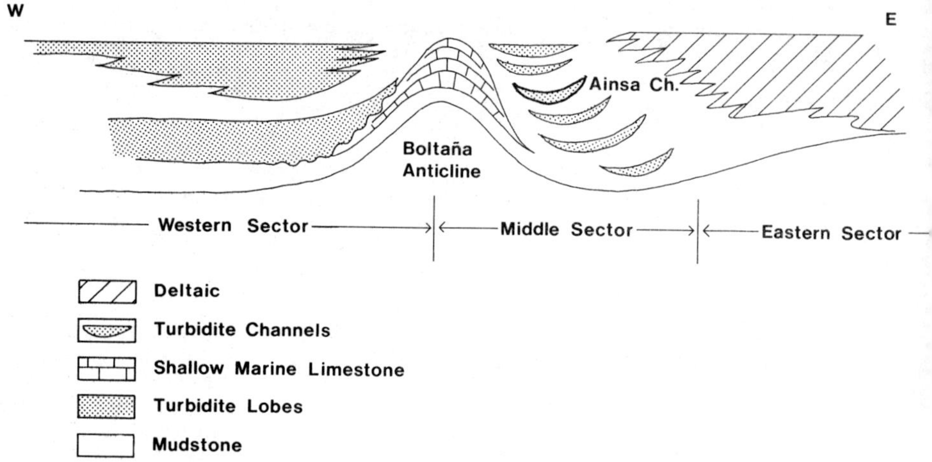

Fig. 2. Diagrammatic stratigraphic cross-section of the South-Central Pyrenean Basin (modified after Mutti (10)).

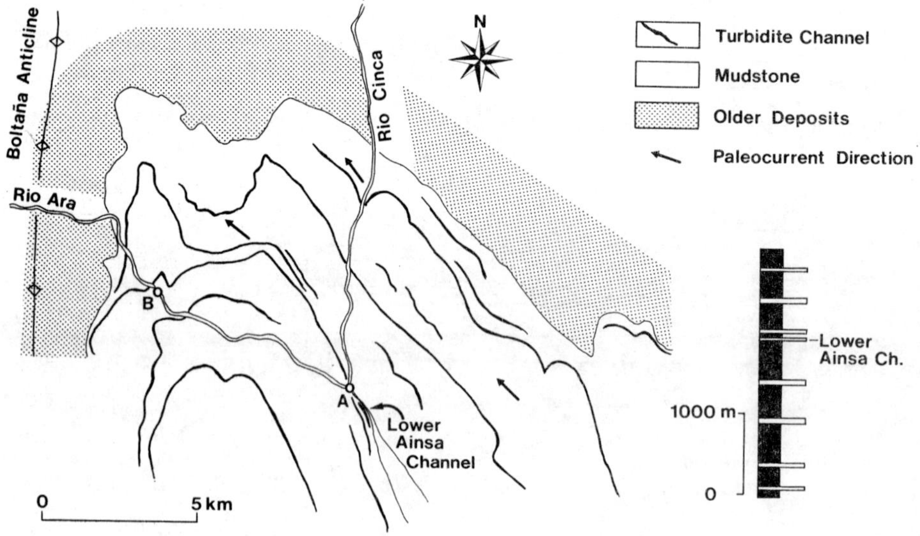

Fig. 3. Geologic sketch map of sandy turbidite channels in the middle sector of the South-Central Pyrenean Basin. A=Ainsa, B=Boltaña.

The present study focuses on one particular sandy turbidite channel in the middle sector, the Lower Ainsa Channel (Fig. 3). The channel is of middle Eocene age (10), and represents one of a series of vertically and laterally stacked turbidite channel complexes embedded in hemi-pelagic marine mudstone. The total thickness of the mudstone-sandstone succession in the middle sector is approximately 2500m, 10 percent of which are turbidite sandstone deposits.

The exposure of the Lower Ainsa Channel, south of Ainsa, is some 750 m wide and 30 m high. The left channel margin, the channel axis and a large part of the right channel margin are exposed.

4. LITHOFACIES UNITS

The sediments of the Lower Ainsa Channel are divided into five lithofacies units, based on lithology, grain size distribution, and physical and biological structures (Table I). These units are numbered in descending order of reservoir quality. Lateral transitions between lithofacies units are common both along the channel trend and across the channel. The most conspicuous transition (both lateral and downcurrent) is from conglomerate to pebbly sandstone (both unit 2) to sandstone (unit 1). Lithofacies units 1 and 3 occur channel-wide and are the most laterally persistent units (Fig. 4). Unit 1 has its thickest sandstone beds (bed-thickness 0.4-2.4m) in the channel axis. Units 2 and 4 occur mainly in the southern wing of the channel.

LITHOFACIES:	sst	sandy congl/ pebbly sst	bioturbated sh/silt/sst	pebbly mudst. muddy sst	mud
lithof. unit no.:	1	2 a/b	3	4 a/b	5
bed thickness in cm	30-130	5-200	2-95	50-100/200	n.a.[3])
max. lateral extent [1])	750	100/150	750	100/200	basin wide
base of bed	sharp	irregular	sharp	irregular	n.a.
grain size in microns	125-2000	125-20000	3-250 [2])	3-20000	3-60
grading	common	rare	rare	absent	absent
permeability	high	intermediate	low	zero	zero
amalgamation	common	common	rare	absent	absent
mudclasts	common	common	absent	rare	absent
armoured mudclasts	common	absent	absent	absent	absent
internal structure	absent or // or ripple	absent	//, ripple or wavy	absent	absent
slumping	rare	absent	rare	absent	common
bioturbation	common	absent	very common	common	rare

[1]) lateral extent (in meters) measured along outcrop
[2]) coarse sand (500 microns) in burrows
[3]) not applicable

Table 1. Characteristics of the turbidite channel lithofacies units.

Facies distribution and lateral transitions reflect the genetic relation of the lithofacies. The sandy conglomerates (lithofacies unit 2a) are residual deposits from highly charged, pebble and sand-laden density currents. Pebbly sandstones (unit 2b) are transitional to sandstone and reflect concomitant deposition of sand and pebbles. Weight separation may lead to downstream transition of pebbly sandstone to pure sandstone of lithofacies 1. The graded, medium to thick-bedded fine to very coarse-grained sandstones (lithofacies 1) were deposited by sand-laden, eroding and waning density currents. If erosive power of these currents was substantial, muddy substratum was eroded and mixed with the density current sand to form the muddy sandstone of lithofacies 4b. In marginal parts of the channel these density currents grade into smaller-volume low-density turbidity currents which produce the mudstone, siltstone and very fine-grained sandstone of lithofacies 3. During subsequent low-energy

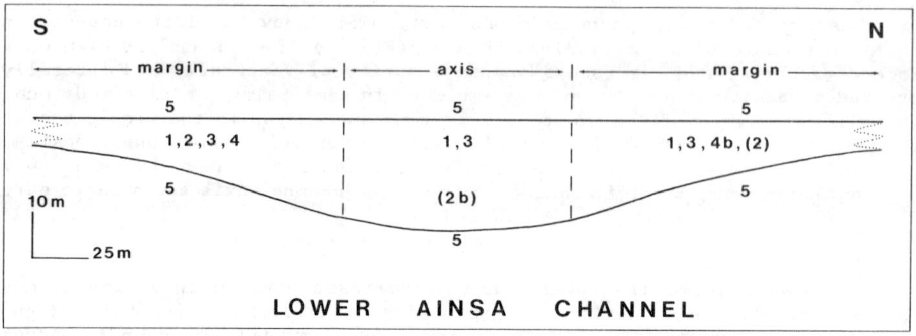

Fig. 4. Lithofacies distribution in the Lower Ainsa Channel. Numbers in brackets refer to the scarce occurrence of a lithofacies with respect to the other lithofacies.

periods these fine-grained deposits are bioturbated. During more prolonged periods of channel inactivity, bioturbated zones extend channel-wide. Pebbly mudstone (lithofacies unit 4a) reflect debris-flow deposits. During periods of normal marine sedimentation between successive channel-fills the dark-coloured mudstone (lithofacies 5) is formed by settling out of suspension.

Since lithofacies units 2 and 4 are concentrated in the southern wing of the channel, as are the observed occasional slumps, this suggests an asymmetric channel-fill. The coarser-grained, more "chaotic" left margin of the channel (looking downstream) might then correspond to the outer bend of a curved or possibly meandering channel.

5. SHALE LENGTHS

Data on lengths of 30 measured mudstone/siltstone levels are compiled in Fig. 5. The lengths were measured along the strike of the outcrop and therefore do not necessarily represent dimensions along or transverse to paleo-flow. From Fig. 5 it appears that the distribution of these levels ('shales') is log-normal and differs widely from the commonly used distribution curve for marine shales (11).

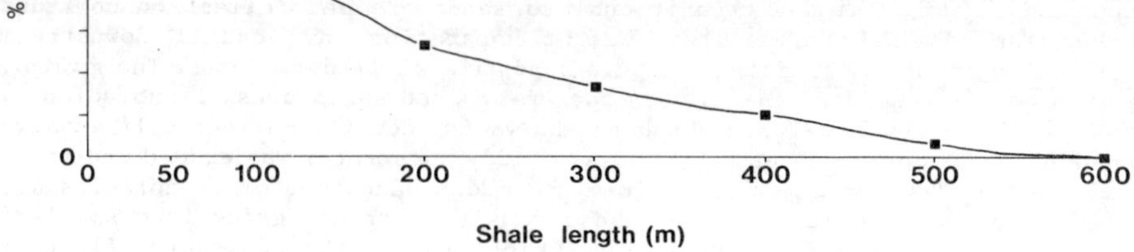

Fig. 5. Shale length distribution curve.

6. CONCLUSIONS

The deposits of the middle Eocene Lower Ainsa Channel, Spanish Pyrenees, are subdivided into five lithofacies units. Subdivision is based on lithology, grain size distribution, physical and biogenic structures. Lithofacies units are:

(1) graded, medium- to thick-bedded fine- to very-coarse grained sandstones,
(2) sandy conglomerates and pebbly sandstones,
(3) bioturbated mudstone, siltstone and very-fine grained sandstone,
(4) pebbly mudstone and muddy sandstone,
(5) mudstone.

Units 1 to 4 are interpreted as deposits from submarine density currents and debris flows. The mudstone of unit 5 settled out of suspension during periods without turbidite activity. Overall geometry of the sandstone beds in the turbidite channel approaches a sheet- to lenticular shape. The sand/shale ratio shows a conspicuous distribution pattern across the turbidite channel. In the channel axis sandstone layers are abundant and thick-bedded; sand/shale ratio is high. Towards the channel margins the bed thickness of the individual sandstone layers decreases and the amount and thickness of the shale layers increase, hence the sand/shale ratio of the channel margins is lower. Lengths of individual shale layers in the turbidite channel show a log-normal distribution curve which differs widely from commonly used distribution curves for marine shales.

This paper presents the first results of a larger study to construct a data base on turbidite facies. Future research focuses on the 3-D geometry and facies distribution of turbidite channels in a variety of basin settings. The turbidite data base will provide the reservoir geologist/engineer with a reliable outcrop-based data set that will enable him/her to carry out realistic reservoir simulations to predict fluid flow and production performance of turbidite reservoirs, and to select optimum well spacings and injection locations accordingly.

REFERENCES

(1) WEBER, K.J. and VAN GEUNS, L.C. (1989). Framework for constructing clastic reservoir simulation models. Paper SPE 19582, 64th Annual Technical Conference of the SPE, San Antonio, Texas, October 8-11, 1989.
(2) DUBRULE, O. (1989). A review of stochastic models for petroleum reservoirs. In: ARMSTRONG, M. (ed.) Geostatistics, vol.2. Kluwer Acad. Publishers, Dordrecht, Holland, 493-506.
(3) BEGG, S.H., CHANG, D.M. and HALDORSEN, H.H. (1985). A simple statistical method for calculating the effective vertical permeability of a reservoir containing discontinuous shales. Paper SPE 14271.
(4) DE FOUQUET, C., BEUCHER, H., GALLI, A. and RAVENNE, C. (1989). Conditional simulation of random sets - application to an argillaceous sandstone reservoir. In: ARMSTRONG, M. (ed.) Geostatistics, vol.2. Kluwer Acad. Publishers, Dordrecht, Holland, 517-530.
(5) TYLER, N., GALLOWAY, W.E., GARRETT, C.M. and EWING, T.E. (1984). Oil accumulation, production characteristics, and targets for additional recovery in major oil reservoirs of Texas. Geol. Circular 84-2. Bur. Econ. Geol., Univ. Texas, Austin, Texas, 31 pp.
(6) SOUQUET, P., PEYBERNES, B., BILOTTE, M. and DEBROAS, E. (1977). La chaine Alpine des Pyrenees. Geol. Alpine 53, 193-216.
(7) CUEVAS, M., DONSELAAR, M.E., and NIO, S.D. (1985). Eocene clastic tidal deposits in the Tremp-Graus Basin (provs. of Lerida and Huesca). In: MILA, M.D. and ROSELL, J. (eds.) Excursion Guidebook, IAS 6th Eur. Reg. Mtg., 217-266.

(8) VAN LUNSEN, H.A. (1970). Geology of the Ara-Cinca region, Spanish Pyrenees. Province of Huesca. Geol. Ultraiectina 16, 119 pp.
(9) MUTTI, E. (1977) Distinctive thin-bedded turbidite facies and related depositional environments in the Eocene Hecho Group (south-central Pyrenees, Spain). Sedimentology 24, 107-131.
(10) MUTTI, E. (1985) Hecho Turbidite System, Spain. In: BOUMA, A.H., NORMARK, W.R., and BARNES, N.E. (eds.) Submarine Fans and Related Turbidite Systems. Springer-Verlag, New York, 205-208.
(11) ZEITO, G.A. (1965). Interbedding of shale breaks and reservoir heterogeneities. Journ. of Petr. Techn., October 1965.

THE CLEETON RESERVOIR TEST : REDUCING UNCERTAINTY

IAIN W WRIGHT

BP Exploration
Southern North Sea Field Group
Dimlington Terminal,
Easington,
North Humberside. HU12 OSU
United Kingdom

SUMMARY

This paper illustrates how geological and geophysical uncertainties in reservoir description can be reduced by pressure transient testing using high resolution sensors and computer simulation.

The example used is a pressure transient test which was carried out on the Cleeton gas reservoir during summer 1989 in order to improve the description and technical understanding of the reservoir.

The reservoir was poorly defined by seismic interpretation and development drilling, but in order to maximise gas recovery, a good reservoir description was required. Before the test, the uncertainty in the volume of gas initially in place (GIIP) ranged from 366 Billion Standard Cubic Feet (BCF) to 170 BCF (10.4 to 4.8 BCM).

The test involved the creation, monitoring and analysis of pressure pulses moving from one end of the reservoir to the other. The test was successful, and has resulted in an improved reservoir description, which will allow optimum depletion of the gas reserves.

1. INTRODUCTION

The Cleeton gas reservoir is located in the Southern North Sea area of the UK Continental Shelf (UKCS) and has been producing since October 1988 as part of BP's "Villages" development. The gas bearing reservoir interval measures 8 km X 2 km, and averages 60m thick. The reservoir fluid is essentially dry gas, with a processed condensate/gas ratio of 4 Barrels Per Million Standard Cubic Feet BBL/MMSCF (0.018 CM/MMSCM).

The reservoir is very poorly defined by seismic interpretation due to the complex nature of the overburden, which suffers from large velocity contrasts. All the wells were drilled on the crest of the structure (to delay aquifer breakthrough), so consequently the reservoir information away from the crest is scant. There was great uncertainty regarding the volume of gas in the reservoir which had a range of 366 BCF to 170 BCF (10.4 - 4.8 BCM). The lack of data across most of the reservoir made good reservoir management impossible.

Figure 1 shows a plan of the reservoir based on seismic interpretation, after drilling all five development wells. The solid line shows the reservoir extent considered most likely after development drilling, and before the pressure test. This reservoir extent would equate to a volume of gas initially in place, of 280 BCF (7.9 BCM). The dotted line shows the minimum reservoir extent (170 BCF, 4.8 BCM GIIP) and the dashed line shows the maximum reservoir extent (366 BCF, 10.4 BCM GIIP).

FIGURE 1
MAP OF THE CLEETON RESERVOIR
AFTER DEVELOPMENT DRILLING

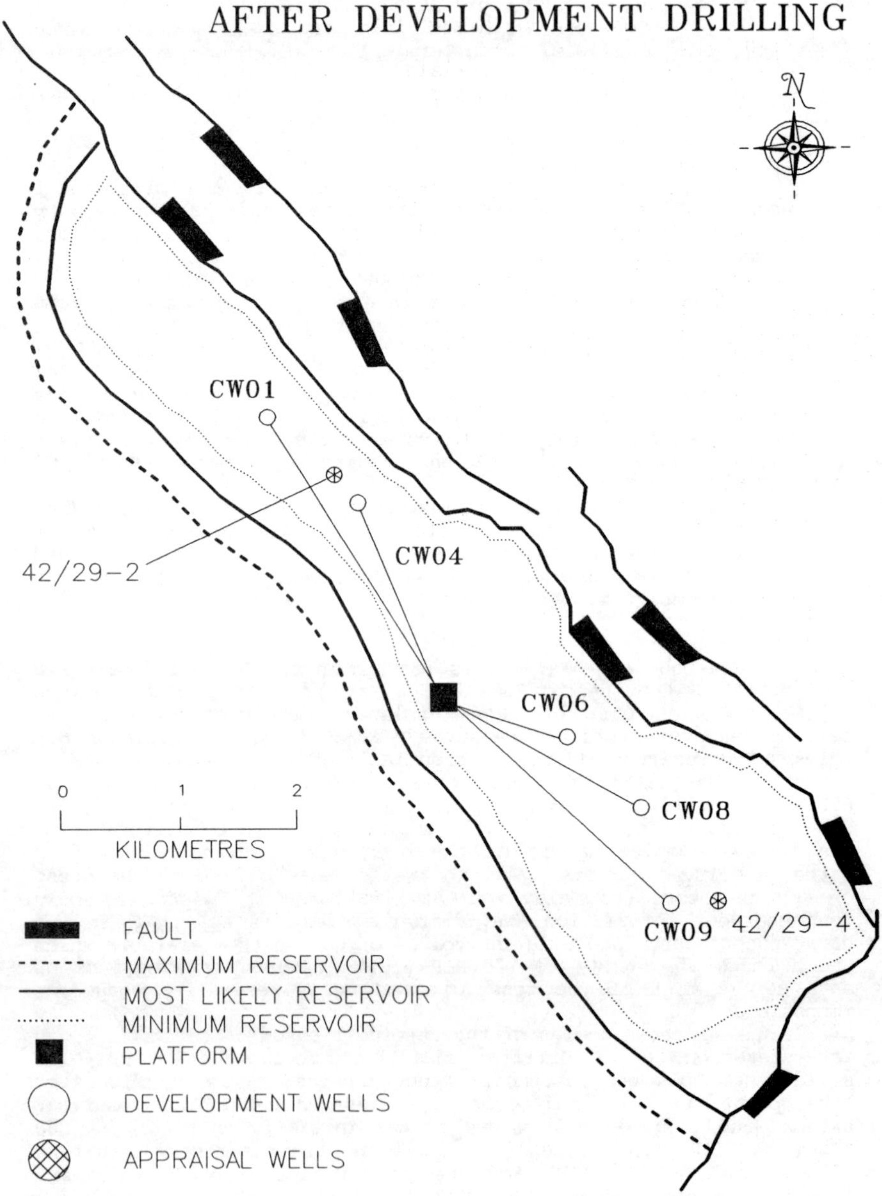

The reservoir rock comprises Rotleigendes sandstone whose deposition was in aeolian, fluvial and sabkha environments. The core permeability of the reservoir rock ranges from 20 to 500 mD, with the average being 200 mD.
This range was confirmed by pressure transient testing of individual wells, the results of which showed that well permeabilities range from 6000 to 64,000 mDft.

The gas reservoir is underlain by an aquifer the extent of which was largely unknown before the test, but whose influx characteristics were expected to govern the gas depletion.

2. TEST REQUIREMENT

The Cleeton reservoir forms part of BP's "Villages" project, which sells gas to two UK customers, British Gas and BP Chemicals. A feature of the sales contracts was the option for the operator (BP) to decide the optimum volume of gas to be sold from the project, for a period of seven years (the Plateau Period). This volume (the Plateau Nomination) had to be nominated before October 1989, and once made, could not be changed for seven years.

The gas produced from the Villages reservoirs is used largely for industrial and domestic space heating, so the demand for production is highly seasonal. Due to its relatively low volume and high permeability, the operator planned to use the Cleeton reservoir during times of peak demand only. Much of the risk associated with the plateau nomination was governed by the performance prediction of the Cleeton reservoir.

The volume of gas reserves recoverable from the reservoir will be governed by the extent and nature of the movement of the underlying aquifer. Preferential depletion of certain areas of the reservoir could maximise recovery, but only if aquifer performance could be accurately characterised.

The aquifer can invade the reservoir by (i) uniform vertical movement, (ii) vertical coning or (iii) preferential lateral influx along high permeability streaks. Vertical aquifer movement can be detected by time lapsed Thermal-Neutron Decay Time logging, but those logs cannot differentiate between uniform water movement and coning. Lateral water influx cannot be measured until the aquifer reaches the wells. In a high permeability gas reservoir, such as Cleeton, once the aquifer reaches a well, the well quickly dies, lacking sufficient energy to lift large amounts of water to surface.

Vertical aquifer movement is monitored annually in Cleeton, by Thermal-Neutron Decay Time logging, but as discussed above, that only gives part of the aquifer influx picture. Pressure depletion measurements and transient testing can indirectly detect and quantify aquifer influx.

The mapping of the reservoir has two areas of structural high (one in the north, and one in the south), separated by a saddle. Figure 2 is a cross section through the reservoir from North West to South East. The vertical exaggeration of Figure 2 is 1:25. Should aquifer influx kill off the two wells in the north of the reservoir, and the transmissibility of the saddle is low, then any remaining gas reserves in the north would no longer be accessible from the southern wells. Similarily, should the aquifer rise up quickly in the saddle (say by preferrential production from wells near the saddle), once the water level has reached the top of the reservoir, the two ends of the reservoir would be isolated from one another.

FIGURE 2
CROSS SECTION OF CLEETON RESERVOIR
ALONG DEVELOPMENT WELLS (MOST LIKELY CASE)

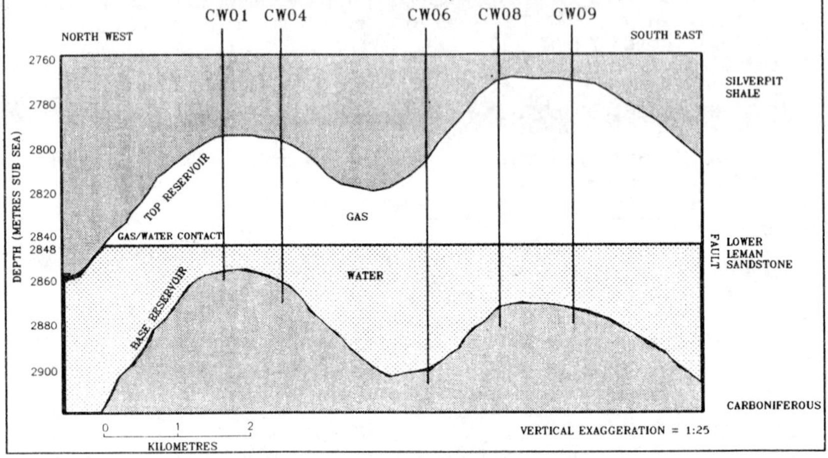

All these uncertainties combined to prevent the operator from defining the optimum reservoir depletion strategy. Without such a strategy, and a good technical understanding of the reservoir, much of the gas initially in place may not be recovered. The wrong depletion strategy could leave up to 50 BCF of gas reserves in the ground.

Therefore, in order to optimise the depletion of the reservoir, a better understanding of the reservoir and its behaviour was required.

3. TEST OBJECTIVES

The objectives of the test were:-

1. Reduce the uncertainty in the volume of gas in place.
2. Locate that gas relative to the production wells.
3. Identify transmissibility heterogeneities - in particular, those affecting pressure communication between the north and south.
4. Define aquifer characteristics and performance.

4. TEST PLAN

The test was designed to create pressure pulses in the reservoir, and monitor those pulses as they moved from one end of the reservoir to the other.

Figure 3 shows the planned production and shut-in regime for the test, the predicted effect on the pressure measured by gauges in two wells, and the information that could be gathered from interpretation of those measurements.

All field production was planned initially to be taken from the three wells in the south of the field (CW06 and CW08, shown in Figure 1) for as long as possible and at a uniform rate. Both wells were to be shut-in suddenly, and the whole field would be shut down for about two weeks. This would create a pressure pulse, moving from south to north in the reservoir, which would eventually be detected (a few days later) by the pressure gauges in the well in the north of the reservoir (CW04).

FIGURE 3
CLEETON RESERVOIR TEST PLAN

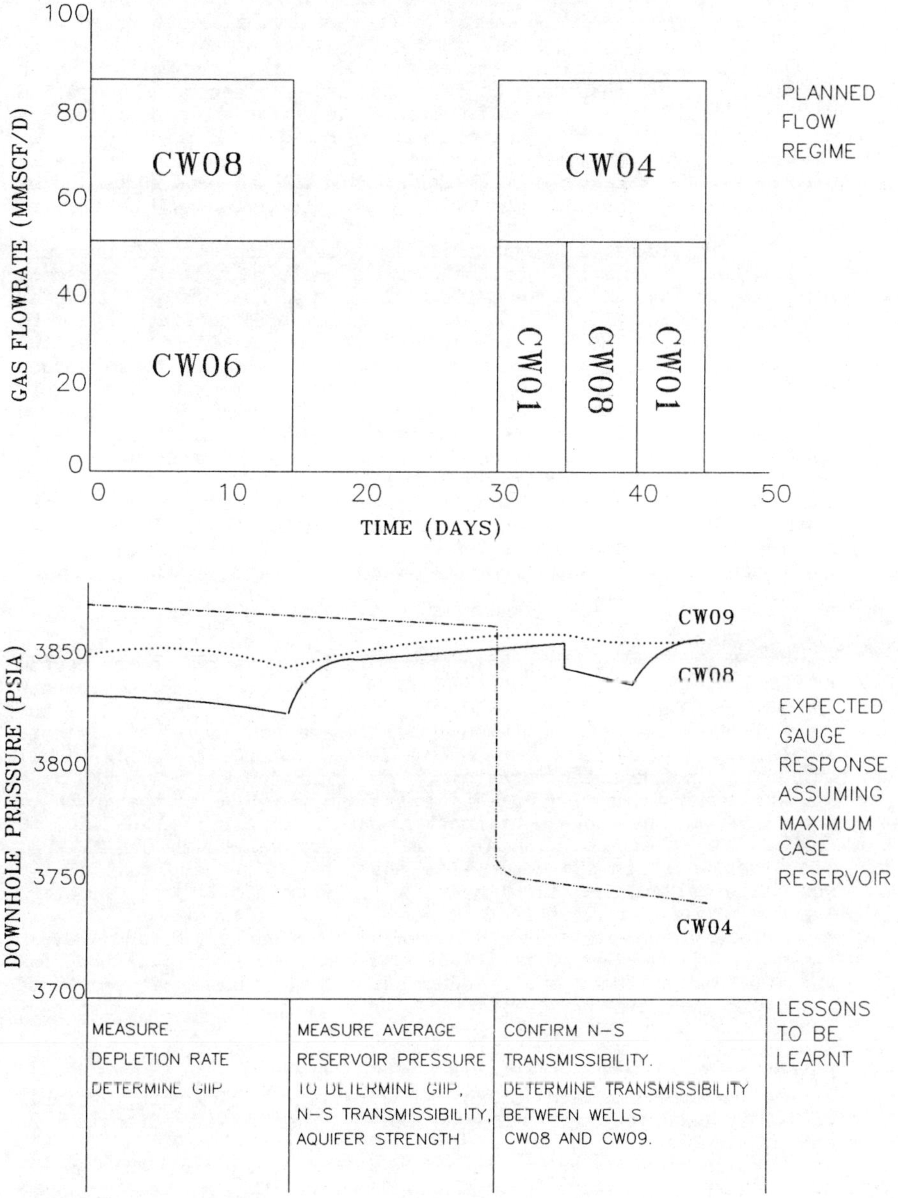

The field would remain shut-in for two weeks, and during this time the pressure build-up would be monitored to give an indication of aquifer influx. A much longer field shut-in (expected during July and August), would provide information to corroborate the test data, but the July/August shut-in period could not be guaranteed since it depended on gas demand from the customers.

After the two week shut-in, production would be restarted from wells in the north of the reservoir, thereby creating a pulse moving in the opposite direction to those from the first flow period (north to south). The measurements taken during the second flow period should corroborate those taken during the first flow period. A secondary objective of the second flow period would be to test for the presence of a suspected high permeability channel between wells CW08 and CW09. The presence of such a channel had been suggested by a previous test carried out during field start-up, but the results of that test were inconclusive.

Considerable work was done to select appropriate pressure gauges for this test, since the sensors would have to be extremely accurate for up to 50 days downhole and not drift while under pressure. Capacitance transducer gauges are considered to be extremely accurate (+/- 2 psi at 4000 psi), but analysis of previous work using this type of gauge had shown that some of these gauges can be prone to drift if left under pressure for a long time. This drift could be of the order of 20 psi for the type of test planned for Cleeton. Such drift would be impossible to separate from gauge measurements, and would render those measurements inaccurate. High resolution strain gauges were eventually chosen for the test, since their drift characteristics are better understood than the capacitance gauges. Subsequent to the Cleeton test, further work has led to an improved understanding of the drift characteristics of capacitance gauges, such that they would be used, should a similar test be carried out in future.

5. TEST EXECUTION

High resolution electronic memory gauges were run to the bottom of three wells (CW04, CW08 and CW09). Ideally, pressure gauges should have been run into all wells, but in order to minimise the cost of the test, gauges in three wells were considered sufficient. Wellhead pressures were measured daily throughout the test, in all wells, using a deadweight tester.

Starting on May 22nd 1989, wells CW06 and CW08 in the south of the reservoir were flowed at constant rates for a period of two weeks. The combined flowrate from the two wells was 85 MMSCF/D. All other wells in the field were shut in. After the initial two week flow period, the flowing wells (CW06 and CW08) were shut in. The whole field was left shut in for a period of two weeks.

Production recommenced on June 19th from wells CW01 and CW04 in the north of the reservoir. After five days, CW01 was switched off, and CW08 was switched on. Production was alternated between those two wells at five day intervals, for the remainder of the test (which was 15 days).

6. ANALYSIS

A computer simulator of the reservoir was built, using variable reservoir descriptions such as mapping, permeability distribution and aquifer extent.

Production history was available from field start-up (August 88), and the computer simulation was run, simulating the first four hundred days of field life, from start-up to the end of the test.

The model assumptions were varied until a good fit was achieved for all wells from start-up to the end of the test. Figure 4 shows the model prediction for the downhole pressure in well CW08, using the reservoir description considered most likely after development drilling and seismic interpretation. It can be seen that the data gathered during the test is considerably different from that predicted by this model of the reservoir. The model predicts much greater depletion than the gauges measured. Clearly, either the volume of gas in pressure communication with this well is considerably larger than modelled, or the volume of aquifer influx is greater than predicted.

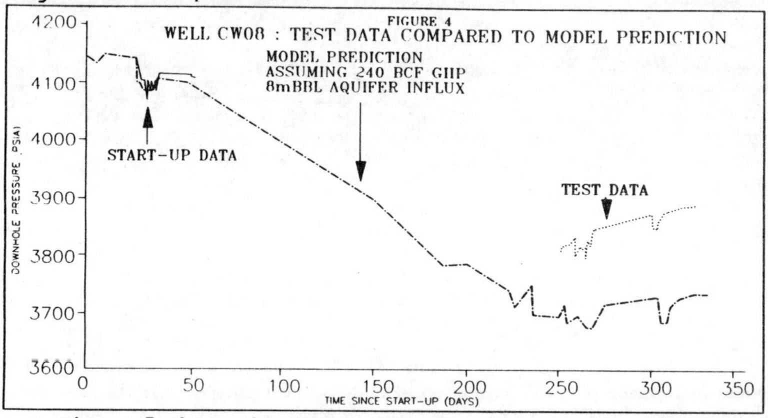

Figure 5 shows the computer prediction for well CW08 with the geological description of the reservoir which was considered (before the test) to be the largest possible. This model of the reservoir is a much better fit to the test data, than that shown in Figure 3. The volume of aquifer influx in this simulation is 8 million barrels over the entire reservoir. The portion inside the dotted box is expanded in Figure 6.

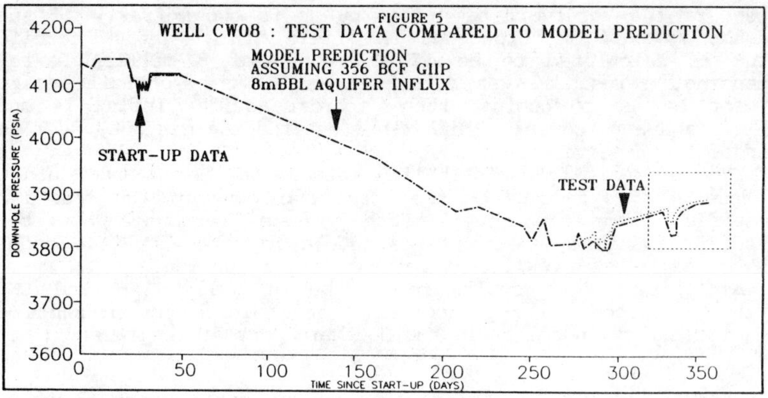

Figure 6 shows in detail the computer prediction for downhole pressure in well CW08 during the shut-in after the test, assuming no aquifer influx. It can be seen that aquifer influx is causing a pressure increase of 5 psi during the 20 days the field is shut-in. Comparison of this data with data gathered in other wells allows improved modelling of aquifer performance.

FIGURE 6

WELL CW08 : TEST DATA COMPARED TO MODEL PREDICTION

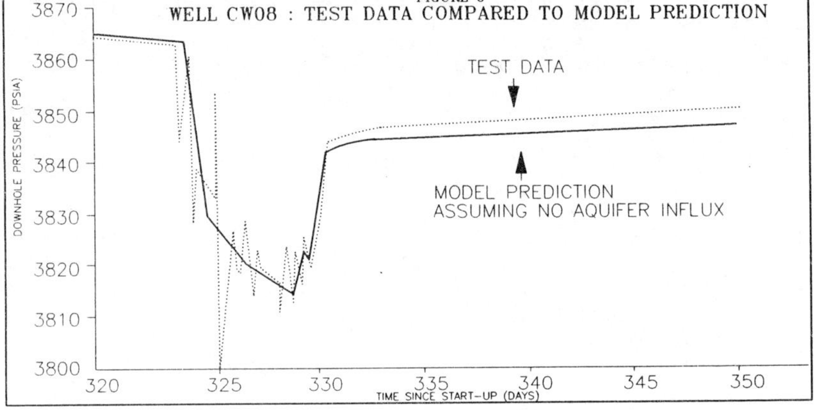

Analysis of the test data cannot determine the direction of aquifer movement (vertical or horizontal), but in conjunction with Thermal-Neutron Decay Time logging, it can give a good indication of the relative proportions of vertical and horizontal water movement.

The reservoir test detected 8 million barrels of aquifer influx. Thermal-Neutron Decay Time logging has shown that at the time the test was carried out, the water level in the near wellbore regions of the reservoir had risen by up to two metres. If such movement is uniform throughout the field, then vertical aquifer influx is calculated to be of the order of 8 million barrels (depending on residual gas saturation). If the vertical aquifer movement is due to coning, then vertical aquifer influx is much smaller, and any remaining influx has entered the gas zone from the sides of the reservoir.

The reservoir transmissibility between the North and South of the reservoir is reduced by the structural constriction caused by the saddle. However, analysis of the pressure responses shows that pressure communication across the saddle is good, and the saddle does not cause any significant permeability reduction.

Analysis of the cyclic production of wells CW01 and CW08 (during the second flow period) did not indicate the presence of transmissibility anomolies in the Southern part of the reservoir.

7. RESULTS

The volume of gas in pressure communication with the wells has been more accurately determined, and is larger than was considered before the test. The assessment of the most likely volume of gas initially in place has risen from 240 BCF to 356 BCF.

The location of the gas relative to the production wells has been established, and the structural mapping of the reservoir improved.

Transmissibility heterogeneities within the gas zone were not detected by the test, and therefore are considered to be small. North to South pressure communication within the reservoir is good.

Aquifer movement was detected, and aquifer performance characterised.

8. CONCLUSION

Analysis of the test data has given the operator an improved technical understanding of the reservoir. This has led to the optimisation of depletion strategy, which in turn will maximise gas recovery from the field.

ACKNOWLEDGEMENTS

H A Ahmad, BP EPS, Aberdeen
P M Stephenson, BP SNSFG, Dimlington
The Management of BP for allowing me to present this paper.

GEOLOGICAL MODELLING IN HC EXPLORATION: RECONSTRUCTION OF GEOMETRIES, TEMPERATURES, PORE-PRESSURES. PRINCIPLES AND APPLICATIONS

J.Burrus, F. Schneider, S. Wolf, W. Sassi, R. Zoetemeijer

Institut Francais du Pétrole, BP 311, 92506, Rueil-Malmaison, France

SUMMARY

This paper reviews the principles of geological modelling and stresses the importance of validation studies. Geological modelling aims to reconstruct the geometry of basins at geological time scale, taking into account the effects of sedimentation, erosion, compaction, large scale faulting, together with the history of temperatures, overpressures and fluid flow. Kinematic concepts permit today to address not only simple geometrical evolution (such as vertical burial of normally compacting basins), but also development of complex structures (thrusts). Thermal modelling poses practical problems; deducing the basement heat flow history by the inversion of inaccurate observations (temperatures, kerogen maturity, etc) is in general not possible. Modelling is rather used to test geological scenarii, and eliminate inconsistent hypotheses. Overpressures and fluid flow history is addressed by coupling conservation equations, flow equation (Darcy), and a stress-strain relation, generally approximated by an empirical relation between effective stress and porosity. These principles have been used to study overpressures in various basins containing clastics. Regional scale permeabilities are often free parameters in such studies, making it difficult to discuss the mechanical concepts. Additional work is needed to determine the contribution of chemical phenomena during compaction. Future developments include implementation of new algorithms (flow in 3D and through complex structures), and validation on well documented case histories.

1. INTRODUCTION.

Numerical basin models aim to describe the geological evolution of sedimentary basins through physical and chemical concepts. These concepts lead to partial derivative equations (most of them being transport and conservation equations) solved numerically with the help of computers. The typical space resolution of basin models is 1000 m horizontally, and 100 m vertically, whereas a typical time step is O.1 Ma. State-of-the-art basin models address three essential geological processes, the most important being the geometrical evolution of the basin. At the geological time scale considered, indeed, the processes of sedimentation, erosion, faulting and compaction change drastically the geometry of a basin, and need to be taken into account. This marks a significant difference with other models such as hydrogeological models, which consider transport of fluids through rocks with fixed geometry. The geometrical evolution of the basin goes along with a thermal evolution and the transport of fluids. Reconstructing the palaeotemperatures is achieved by considering the burial changes, the basement heat flow variations (which reflect the geodynamic situation) and the heat redistribution processes at intra-basinal scale (by conductive and possibly convective transport). Geometrical changes also cause the pore-pressures to be in disequilibrium, and fluid to flow across the basin. Pore-pressure disequilibrium may be due to compaction, or to topographic relief. During the early, generally marine, stage of basin development, insufficient drainage of compaction-related water flow is the dominant cause of overpressuring. Water velocities are then generally slow, below (or equal to) the sedimentation rate (frequently in the range 1-1000 m/Ma). Later, the basin often undergoes uplift, frequently associated with topographic relief and erosion. Regional circulations take then place from the most elevated zones (recharge areas), to the depressed zones (discharge areas). The range of water velocity can reach in aquifers much higher values (up to 1-10 m/a) than for the compaction-driven flow. Major heat redistribution and mass transport (mineral dissolution) may then take place.

The reconstruction of palaeo-geometries, palaeo-temperatures and palaeo-pore pressures forms the core of state-of-the-art basin models. Subsequent applications of these models address often the chemical consequences of the changes in pressure, temperature or geometry, such as : HC generation and its relations with HC migration, mineral diagenesis, etc. Based on the previous principles, several integrated basin models have been recently developped , in general in 2D (1 to 6). Many models do not address the mechanical aspect of compaction, which is physically and numerically more complex, and deal only with geometrical-thermal problems in 2D (7 to 11) or in 3D (12, 13).

This paper presents more in detail the principles used in modelling the evolution of geometries, temperatures, fluid flow (called here geological modelling). This paper does not address the principles used in geochemical modelling, such as the generation, expulsion and migration of HC, or the transport of mineral elements. The reason for this choice is that geochemical modelling is dependent on geological modelling, and poses particular difficulties most of which, being of chemical nature (validity of kinetic schemes, experimental determination of parameters, validity of thermodynamic equilibrium hypothesis in mineral geochemistry, etc), are beyond the scope of this paper. Examples will be discussed based on the THEMIS (5,6), CICERON (18,19), and THERKIN (34) numerical models developed at the IFP.

2. PRINCIPLES AND VALIDATION.

The principles used in geological modelling are provided by rock-mechanics, fluid mechanics and thermics. Applying these principles, which are mostly applied to solve engineering problems, to geological situations, poses the problem of changing the scale of time (from human scale to geological scale), and the scale of space (from the 0.1-10 m scale of engineering to the km scale in basin geology). Therefore, the validity of extrapolating the concepts and the parameters from the realm of engineering to that of geology needs to be established; this cannot be achieved except by studying well documented case histories.

2.1 Principles of geometrical reconstruction.

The decompaction of a pile of layers is easily achieved (Fig.1) if one assumes that compaction obeys given porosity-depth relations, following the backstripping technique (14). The existence of undercompaction processes (15), or of chemical loss or gain of mineral mass, such as during carbonate compaction, results then in errors in the reconstructed burial history which may be higher than 15 %.

More complex is the kinematic reconstruction when listric faults or thrusts control the evolution of the geometry. Techniques of kinematic reconstruction, initially introduced to balance cross-sections, may then be applied. The fault-bend folding theory (16) permits, among others, the development of computerized algorithms describing the evolution along a faulted section (Fig.2), (17 to 19). All these techniques assume the initial geometry, prior faulting, to be known, the sequence of faulting in case of multiple faults to be established, and the volume, or length, or thickness of layers to be preserved during deformation. In addition, the description of internal deformation of the blocks is purely kinematic (vertical, oblique or parallel to bedding shear), no mechanical stress-strain behaviour being introduced.

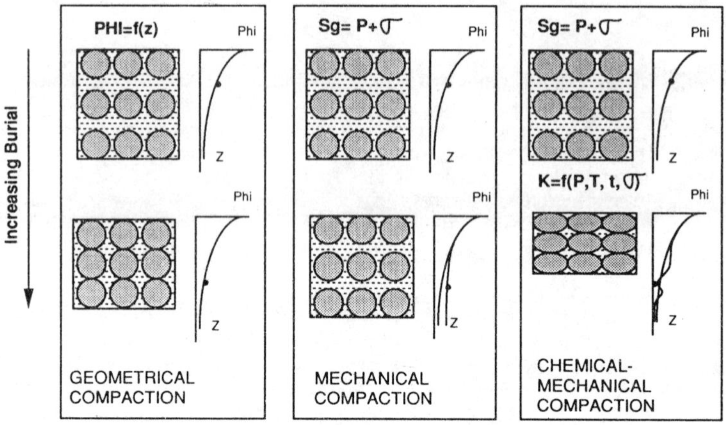

Fig. 1 : Conceptual models of compaction. Geometrical compaction considers porosity Phi as a function of depth z. Mechanical compaction introduces the effective stress (difference between geostatic stress Sg and pore-pressure P), and accounts for undercompaction processes. Chemical-mechanical compaction addresses the changes of grain mass due to chemical dissolution/precipitation reactions, which are complex functions of time, pressure, temperature, and stress. Grey pattern: grains; dashed pattern: interstitial fluids. The investigation of the significance of chemical processes during compaction is one target of future research in basin modelling.

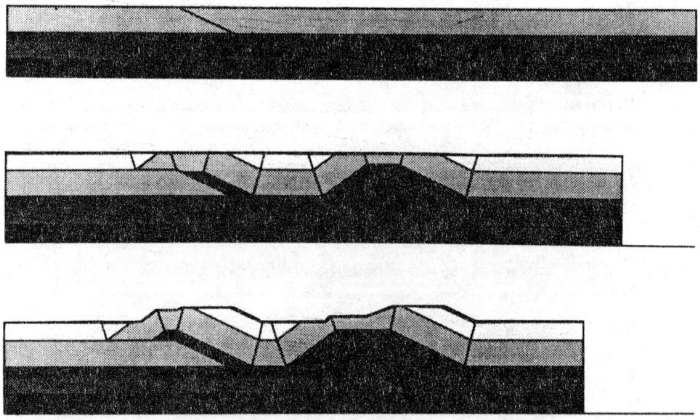

Fig.2: Forward computation of the development of fault-bent folds and piggy-back basins (using the CICERON program, after 19). The computation of related thermal and fluid flow history is a forthcoming objective of research.

2.2 Principles of thermal reconstruction.

Modelling past temperatures poses no theoretical difficulty; the heat equation is solved in conduction, and in convection if regional hydrodynamism is active. A transient solution is in general required, due to the inertia of thermal diffusion, and to the variability of boundary conditions with time. Relaxation time constants may be typically of the order of 0.1 to several Ma, depending on the depth at which the temperature of sediments is disturbed. Two important cause of disturbance are recognized. One is related to the sedimentation process, an effect that becomes significant for sedimentation rates above approximatively 100m/Ma (5,8,9,10). This blanketing effect was found to reduce the crustal heat flow effectively transferred into the sediments by 35% in the case of basins where 9 km of sediments were deposited in around 20 Ma (THEMIS studies, 5, 46). The second cause is regional hydrodynamism. The circulation of fluids in an dipping aquifer modifies the thermal gradient between the aquifer and the surface, and affects the subsurface temperatures below the aquifer by diffusion. Numerical modelling shows (20) that this thermal effect is significant when the number N, defined as :

$$N = 0{,}067 \,.\, \tan g(\text{dip}) \,.\, Q \quad \text{(dip: angle between aquifer and horizontal; } Q\text{ : water flow through the aquifer in } m^2/a)$$

is greater or equal to 1 (the 0,067 coefficient is linked to the mean thermal characteristics of water and sediments). As example, a flow at least as high as 1/ 0,067.tang(5), that is 170 m^2/a, is necessary to obtain thermal convective effects from circulation in an aquifer with a dip of 5°. If the aquifer is 50 m thick, this corresponds to a mean Darcy velocity of around 3 m/a (170/50). Such rapid velocities may be reached by regional hydrodynamical systems, but not by compaction-related water flow, which is much too slow to affect the thermal gradient, even in case of very rapid sedimentation rates (20). Higher time constants, of the order of 50 Ma, are associated with lithospherical thermal events, such as rifts.

The difficulty found in thermal modelling arises from the determination of the palaeo-boundary conditions (imposed temperature at the surface, imposed heat flux at the base), and of the thermal parameters, in particular the thermal conductivities. Boundary conditions may vary with space and time. The geodynamic context provides an idea about the variations of the mantle heat flow. Since studies of the early 80's, as (21), it is well established that rifted basins are associated with thermal events, whereas foreland basins have rather a constant mantle heat flow through time. Still, the relation between crustal thinning and magnitude of heating during rifting is controversial (see review in 22). In addition the lateral variability of the radiogenic heat flow (sourced in the upper crust), is always difficult to assess. It may vary at the scale of the terranes which often form the upper crust , i. e. at a scale of 10-100 km (23).

For these reasons, thermal studies rarely result in the determination of a unique thermal scenario; more than often, thermal modelling consists in testing several hypotheses on the heat flow. By comparison with observed temperatures and palaeo-thermal indicators (such as the maturity of the kerogen revealed by Rock-Eval analysis), the inconsistent scenarii can be discriminated (Fig. 3). Acceptable solutions are often not unique, but at least consistent with observations, and modelling enables sensitivity analysis.

Describing correctly the thermal conductivity of sedimentary rocks is not easy, because measurements on samples have frequently a precision of only 10%, and many parameters play a role (mineralogy, porosity, texture, temperature, etc). An error of 15% on the mean thermal conductivity commonly results in an error of about 10°C on the calculated temperature at a depth of around 3000m, and in an error of about 400m on the top of the oil window (parameters used in this estimation : heat flow 70mW/m^2, conductivity 1.7W/m/°C). Empirical relations between well logs and thermal conductivities have proved to have limited applicability (24 to 26). A more promising technique consists in deriving the composition of rocks (mineralogy and porosity) from well logs, and to use composition laws to obtain the bulk conductivities (27, 28). In many cases, variations of observed thermal gradients with depth may as well be due to lithological changes, and subsequent conductivity changes, or be indications of water convection (Fig. 4). There are numerous examples of contradictory interpretations as to the existence or not of convective thermal effects in sedimentary basins (see for example, (29,30) for the Paris Basin; (31,32) for the Alberta Basin).

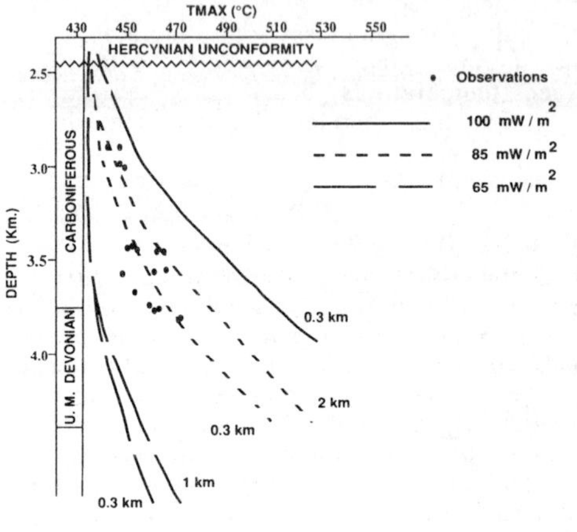

Fig.3: Example of calibration of the palaeo-heat flow (in mW/m^2) and thickness of sediments (in km) eroded at the Hercynian unconformity against observed maturity of kerogen (Rock-Eval Tmax) .The best fit is obtained with a mean $85 mW/m^2$ heat flow, and an eroded thickness of about 300m (based on the THEMIS model; well from the Algerian Triasic Basin). Courtesy Sonatrach. Note the inaccurate observed maturity trend, a frequent obstacle in calibration of the past heat flow.

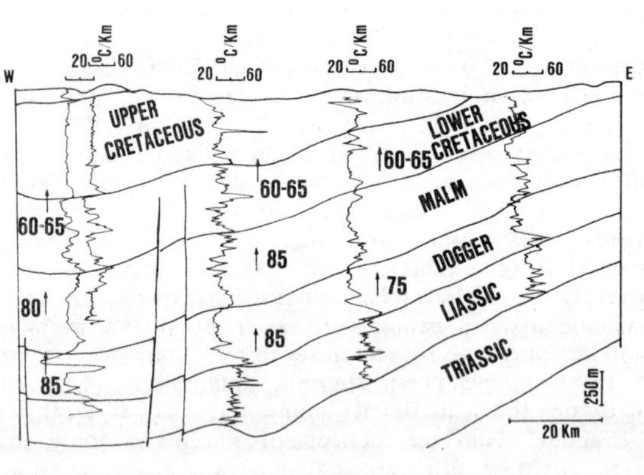

Fig. 4: Observed thermal gradients in five wells located in the Eastern Paris Basin, showing an increase from 20 °C/km to the surface to more than 50°C/km at depth. Detailed investigation of the thermal conductivities show that this increase is not entirely due to more shaly nature of the Liasic; losses of heat flow indicated by the vertical arrows (from 85 to 65 mW/m2) are linked with regional hydrodynamism (after 28, 30).

Coupling the kinematic evolution during thrust emplacement with the computation of temperatures has been attempted at least for simple structures (33,34). In addition to the previous difficulties, this approach presents a greater numerical difficulty (management of the mesh during thrust evolution). Such studies have shown that 2D effects are significant, which limitates the possibilities of previous 1D attempts (35,36). As hydrodynamic cooling, believed to be particularly active in foot-hills setting, is generally not yet accounted for, these models have still limited practical applicability.

2.3 Principles of over-pressuring and fluid flow reconstruction.

Compaction of poorly drained rocks results in overpressuring, which is a transient phenomenon, relaxed within time span between a year and several 100 Ma, depending on the permeability of rocks. Initial modelling of this process was attempted in soil mechanics, through the 'consolidation' theory (37). This theory relates the decrease of porosity phi with depth to the increase of frame pressure or effective stress (Fig.1). Compaction was proposed to be represented by three coupled, non-linear equations (38,39): conservation of water, Darcy equation, effective stress-strain relation. Such an approach permits to compute overpressure development as a response to loading (sedimentation) versus relaxation (fluid flow). Practical difficulties are encountered. How to define effective stress? As the difference Sg - P (Sg: total or geostatic stress, P: pore pressure) as proposed in soil mechanics, or by the difference

Sg - alpha . P (alpha: between phi and 1),

as in Biot's (40) theory of elasticity ? How to define the stress-strain behaviour? Sandstones are known in hydrogeology to follow an elastic behaviour, but not shales (41), at least at human time scale. Still, as elasticity is simple, it was considered by many, even for shales. More complex rheologies, such as the elasto-viscous rhelogies considered in soil mechanics, require many parameters to be calibrated, which is strictly impossible at geological time scale. To turn this problem, many models are based on empirical relations, easily calibrated from well logs and tests (such as measured overpressures, measured porosity, etc). Empirical relations between effective stress and porosity rank among these (2,6,38,39,42). This approach has been applied at IFP to model using the THEMIS code overpressures and undercompaction in various geological situations : a Tertiary passive margin with shales and salt, Gulf of Lions, France (5); extensional rifted basins with a significant Tertiary infill of clastics, Viking Graben, Norway (43 to 45); a Tertiary delta, Mahakam delta, Indonesia (46) (Fig. 5); an Atlantic margin, Venture field, Canada (47). A good agreement between calculated and measured overpressures was obtained. Interestingly also, these studies showed that observed overpressure gradients as high as 10 MPa/100m could not be supported by shales with vertical permeabilities above the order of the nanoDarcy (10^{21} m^2). It is only very recently that experimental permeability measurements confirmed that these values are encountered in nature (48), providing undirect checks of the validity of the approach. Although compaction appears as the dominant cause of overpressures, computations show that the generation of hydrocarbons may be an important additional mechanism (47, 49); but this is by no way a dominant and general mechanism.

In contrast with the previous approach, several authors emphasize the chemical effect of increasing stress during compaction, besides the mechanical effect of this increase. Grains are preferentially dissolved at grain contact (Fig. 1), and the dissolved mass may precipitate elsewhere, depending on the chemical conditions (P, T, chemistry of fluids) and on the occurence of regional fluid transport (50 to 52). This approach is theoretically not compatible with the existence of straight-forward stress-porosity relations, which still lead to correct overpressures computations in the examples listed above. Further work is today needed to investigate the reasons of this paradox. In particular, it would be interesting to document the kinetics of the chemical reactions involved. Most modelling studies have been devoted to basins where a significant Tertiary loading was the source of overpressuring. Could it be that older basins are more sensitive to chemical compaction?

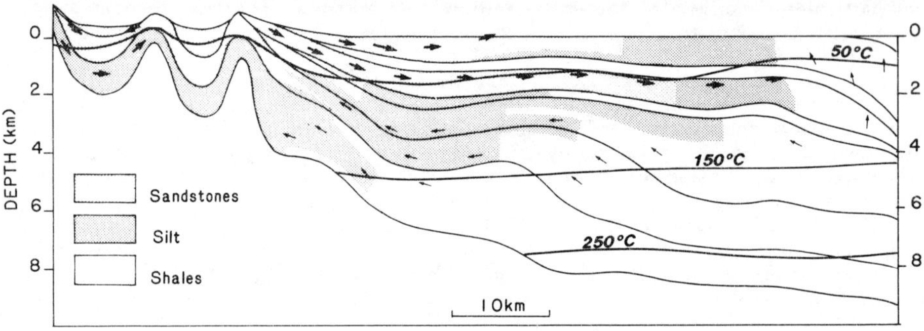

Fig. 5: Example of modelling in a Tertiary delta : integrated simulation of compaction evolution coupled with an uplift phase responsible of regional hydrodynamism; the geometrical changes are due to compaction, sedimentation, erosion,uplift. Temperatures are coupled with regional water flow (after 46, THEMIS model outputs, for Pliocene age). Heavy arrows: water flow due to topographic relief, concentrated in sanstones (order of magnitude: 1 m/a); light arrows: water velocity due to compaction-related overpressures.(magnitude: 1-10 mm/a. This type of computations permits to discuss the structural history (uplift) together with the fluid transport and thermal history.

Finally, the approach which combines the three equations mentionned above permit to account as well for regional water flow, together with compaction-driven flow. The magnitude of the palaeo-hydrodynamism is essentially a function of the palaeo-topographic profiles considered, and of the basin-scale permeability structure introduced (Fig. 5) (3,46,53). Determining these geological inputs is often very difficult. Here again, modelling can be viewed as a consistent way to test geological hypotheses, to discriminate between possible scenarii and inconsistent scenarii, rather to 'inverse' the observations.

3. CONCLUSION.

Geological modelling is an attempt to describe the evolution of sedimentary basin with the help of concepts derived from physics. Its objectives are to provide the geologist with reconstruction of geometry, temperatures and over-pressures/fluid flow histories. These geohistorical reconstructions are necessary to address the organic or mineral geochemical evolution of basins.

Geometrical reconstructions aim to account for sedimentation-compaction effects, and for large scale displacement due to faulting. Today algorithms permit without great numerical difficulty to account for vertical sedimentation-compaction (in 1D, 2D, or 3D), or to faulting in the absence of compaction (in 2D, not yet in 3D). Forward kinematic reconstruction of sedimentation coupled with compaction and large scale faulting will be, at least in 2D, the next step.

Reconstruction of palaeo-temperatures poses no theoretical difficulties. Practically, problems are encountered in the definition of the boundary conditions (imposed temperatures at the top, imposed heat flow at the base), which may vary with space and time. 'Inversion' of the scarce and inaccurate observations (temperatures, degree of maturity of organic matter) is in general not possible, and modelling appears more as a way to test the consistency of geological hypotheses, rather than a way to establish one unique scenario. Future developments include improving the knowledge of parameters such as thermal conductivities, and elaboration of new algorithms (3D, heat tranfer through complex active structures).

Overpressures and fluid flow modelling poses more conceptual problems. Most approaches are purely mechanical, and ignore the chemical evolution of minerals during compaction. Empirical relations account for the rigidity of the mineral frame, often through effective stress-porosity relations. This approach has been validated during several case studies, mostly dealing with Tertiary loading events. Additional work is required, both in the mechanical characterization in the laboratory of rocks like shales, in the sensitivity analysis to chemical factors, before more adequate concepts can be proposed. Palaeo-hydrological reconstructions pose essentially methodological problems: how to calibrate the basin scale permeability tensor, and how to constrain palaeo-topographies. Here again, modelling has to be viewed as a way to test geological hypotheses, and eliminate inconsistent scenarii. Future development include development of algorithms (3D, fluid flow through complex active structures: thrusts, listric faults).

Geohistorical modelling enables an improved understanding in various fields of petroleum geology, such as rock-fluid interactions and mineral diagenesis, generation, expulsion and migration of HC. It gives the possibility to explorationnists to integrate and interpret more data in basin evaluation. Still, validation of the concepts and parameters through the study of case histories appears as the only way to solve the difficulties posed by the change of time and space scales between engineering and basin geology.

REFERENCES.

(1) Welte D.H., Yukler M.A., 1981. Petroleum origin and accumulation in basin evolution, a quantitative model. Bull.AAPG, 65, p. 1387-1396.

(2) Ungerer P., Bessis F.,Chenet P.Y., Durand B., Nogaret E., Chiarelli A., Oudin J.L., Perrin J.F., 1984. Geological and geochemical models in oil exploration: principles and practical examples. AAPG Mem. 35, G.Demaison (ed.), p. 53-77.

(3) Bethke C.M., 1989. Modelling subsurface flow in sedimentary basins. Geologische Rundschau, 78, 1, p.129-154.

(4) Nakayama K., Lerche I., 1987. Basin analysis by model simulation: effects of geological parameters on 1D and 2D fluid flow systems with application to an oil field. Gulf Coast Ass. Geol. Soc. Publ., 37, p.175-184.

(5) Burrus J., Audebert F., 1990. Compaction and thermal processes in a young rifted basin containing evaporites, Gulf of Lions, Bull. AAPG, in press.

(6) Ungerer P., Burrus J., Doligez B., Chenet P.Y., Bessis F., 1990. Basin evaluation by integrated 2D modeling of heat transfer, fluid flow, HC generation and migration. Bull AAPG, 74, 3, p. 309-335

(7) Beaumont C., Keen C.E., Boutilier R, 1982. On the evolution of rifted continental margins; comparison of models and observation for the Nova Scotia margin. Geoph. Journ. Royal Astr. Soc., 70, p.667-715.

(8) De Bremaeker J.C., 1983. Temperature, subsidence and HC maturation in extensional basins: a finite element model. Bull. AAPG, 67, p.1410-1414.

(9) Lucazeau F., Ledouaran S., 1985. The blanketing effect of sediments in basins formed by extension. Earth. Plan. Sc. Lett., 74, p.92-102.

(10) Hutchison I, 1985. The effects of sedimentation and compaction on oceanic heat flow. Geoph. Journ. Royal Astr. Soc., 82, p. 439-459.

(11) Wygrala B.P.,1987. Integrated computer-aided basin modelling applied to analysis of hydrocarbon history in a Northern Italian oil field. Adv. Org. Chem., 13, 1-3, 187-197.
(12) Yalcin M.N., Welte D.H.,et al., 1987. 3D computer aided modelling of Cambay basin,India. A case history of hydrocarbon generation. In: Petroleum geochemistry in the Afro-Asian region, Kumar (ed), Balkema, Rotterdam, 417-450.
(13) Novelli L., Chiaramonte M., Mattavelli L., Pizzi G., Sartori L., Scotti P., 1987. Oil habitat in the NW Po Basin. In: Migration of HC in sedimentary basin, B. Doligez (ed.), Technip, p.27-57.
(14) Perrier R., Quiblier J. (1974). Thickness changes in sedimentary layers during compaction history. Bull. AAPG, 58, p. 507-520.
(15) Magara K., 1978. Compaction and Fluid migration; practical petroleum geology. In: Developments in petroleum science, Elsevier, 9, 319p.
(16) Suppe J.,1983. Geometry and kinematics of fault-bent folding. Am. Journ. Sc., 283, p. 684-721.
(17) Moretti I., Triboulet S., Endignoux L. Some remarks on the geometrical modeling of geological deformations. In: Petroleum tectonics in mobile belts, J. Letouzey (ed.), Technip, p.155-162.
(18) Endignoux L., Mugnier J.L., 1989. The use of forward kinematic model in balanced cross sections. Tectonics,
(19) Zoetemeijer R., Sassi W., submitted. The forward modelling of complex thrust evolution using the fault-bent fold method. In: Proceedings of Thrust Tectonics Conf., London, April 90.
(20) Vasseur G., Burrus J., 1990. Contraintes hydrodynamiques et thermiques sur la genèse des gisements stratiformes à Pb-Zn. In : Mobilité et Concentrations des métaux de base dans les couvertures sédimentaires, Documents BRGM, Orléans, 183, p.305-354.
(21) Beaumont C., Keen C.E., Boutilier R, 1982. A comparison of foreland and rift margin sedimentary basins. Philosophical Trans. Royal Soc. London, 305, p.295-317.
(22) Burrus J., 1989. A review of geodynamic models for extensional basins; the paradox of stretching in the Gulf of Lions. Bull. Soc. Geol. France, 8, 2, p.377-393.
(23) Jaupart C., 1983. Horizontal heat transfer due to radioactivity contrasts: causes and consequences of the linear heat flow relation. Geoph. Journ. Royal Astr. Soc., 75, p.1475-1501.
(24) Evans T.R., 1976. Thermal properties of North Sea rocks. The log analyst, 18, p.3-12.
(25) Houbolt J.J., Wells P.R., 1980. Estimation of heat flow in oil wells based on a relation between heat conductivity and sound velocity, Geol. Mijnb., 59, p.215-224.
(26) Vacquier V., Legendre E., Blondin E., 1988. An experiment on estimating the thermal conductivity of sedimentary rocks from oil well logging. Bull. AAPG, 72, p. 758-764.
(27) Brigaud F., Vasseur G., 1989. Mineralogy, porosity and fluid control on thermal conductivity of sedimentary rocks. Geophys. Journ., 98, p.525-542.
(28) Demongodin L., Pinoteau B., Vasseur G., Gable R., submitted. Thermal conductivity and well logs in the Paris Basin. Geology.
(29) Jessop A.M., 1989. Hydrological distorsion of heat flow in sedimentary basins. Tectonophysics, 164, p. 211-218.
(30) Pinoteau B., Gable R., Burrus J., submitted. Hydrogeological significance of detailed thermal profiles in the Paris Basin.
(31) Majorowicz J.A., Jones F.W., Lam H.L., Jessop A.M., 1985. Terrestrial heat flow and geothermal gradients in relation to hydrodynamics in the Alberta Basin. Journ. Geodyn., 4, p. 265-283.

(32) Bachu S., Burwash R.A., 1990. Geothermal regime in the Western Canada sedimentary basins. Basin Perspective Conference, Calgary, abstract.
(33) Shi Y., Wang C.Y., 1987. Two-dimensional modeling of P-T path of regional metamorphism in simple overthrust terrains, Geology, 15, p. 1048-1051.
(34) Endignoux L., Wolf S., 1989. Thermal and kinematic evolution of thrust basins: a 2D numerical model. In : Petroleum and tectonics in mobile belts, J. Letouzey (ed.), Technip, p.181-192.
(35) Angevine C.L., Turcotte D.L., 1983. Oil generation in overthrust belts. Bull. AAPG, 67, 2, p.235-241.
(36) Furlong K.P., Edman J.D., 1984. Graphic approach to determine the hydrocarbon maturation in overthrust terrains, Bull. AAPG, 68, 1818-1824.
(37) Terzaghi K., Peck R.B.,1948. Soil mechanics in engineering practice, John Wiley and Sons, NY, 566p.
(38) Rubey W.W., Hubbert M.K., 1959. Role of fluid pressure in mechanics of overthrust faulting. Bull. Geol. Soc. America, 70, p.167-206.
(39) Smith J.E., 1971. The dynamics of shale compaction and evolution of fluid pressure. Math. Geol., 3,3, p.239-269.
(40) Biot M.A., 1941. General theory of three dimensional consolidation. Journ. Appl. Phys., 12, p.155-164.
(41) De Marsily G., 1981. Quantitative Hydrogeology. Masson, Paris, 215 p.
(42) Palciauskas V.V., Domenico P.A., 1980. Microfracture development in compacting sediments: relation to hydrocarbon maturation kinetics. Bull. AAPG,52, p.57-67.
(43) Doligez B., Ungerer P., Chenet P.Y., Burrus J., Bessis F., Bessereau G., 1987. Numerical modelling of sedimentation, heat transfer, hydrocarbon formation and migration in the Viking Graben. in Petroleum Geology of NW Europe, Brooks and Glenie (ed.), Graham and Trotman, p.1039-1048.
(44) Ungerer P.,Bessereau G., Junca J., Rabiller P., 1987. Application d'un modèle de migration des hydrocarbures à l'évaluation des permis. 12th World Petroleum Congres, PD1,3,19-30.
(45) Burrus J., Doligez B., Ungerer P.,1990. Are numerical models useful to understand HC migration ? A discussion in the Viking graben. Marine and Petroleum Geology, in press.
(46) Burrus J., Brosse E., Choppin de Janvry G., Grosjean Y., Oudin J.L., in prep. Numerical modelling in a tertiary delta.
(47) Forbes P., Ungerer P., Mudford B., submitted. Overpressure modelling applied to the origin of overpressures and gas occurence in the Venture Field, Canada. Bull. AAPG.
(48) Morrow C.A., Shi L.Q., Byerlee J.D., 1984. Permeability of fault gouge under confining pressure and stress. Journ. Geoph. Res., 89, p.3193-3200.
(49) Doligez B., et al. in prep. The petroleum evolution of the Williston basin investigated by 2D numerical modelling.
(50) Merino E., Ortoleva P., Strickholm P., 1983. Generation of evenly-spaced pressure solution seams during diagenesis: a kinetic study. Contrib. Miner. Petrol, 82, 360-370.
(51) Gratier J.P., 1987. Pressure solution-deposition creep and associated tectonic differenciation in sedimentary rocks. In: Deformation of sediments and sedimentary rocks, Jones M. and Preston R. (ed.), Geol. Soc. Spec. Publ. 29, p. 25-38.
(52) Reuschlé T., Trotignon L., Gueguen Y., 1988. Pore shape evolution by solution transfer: thermodynamics and mechanics. Geoph. Journ., 95, p.535-547.
(53) Person M., Garven G., 1989. Hydrologic constraints on the thermal evolution of the Rhine Gràben. In: Hydrogeological regimes.and.their thermal effects, Beck A., Garven G., Stegena L. (eds), Geophysical Monogr. Series, 47, p.35-58.

FURTHER DEVELOPMENT OF HYDROCARBON POTENTIAL FORECAST SYSTEM IS A BASIS FOR EFFICIENT OIL AND GAS PROSPECTING AND EXPLORATORY WORKS

N.KULAKHMETOV, A.BREKHUNTSOV

Ministry of Geology of U.S.S.R.
Tyumengeologiya Concern,
Republic Street 55,
Tyumen 625000 U.S.S.R.

I.NESTEROV, A.RYLKOV, V.SHPILMAN

Ministry of Geology of U.S.S.R.
West Siberian Research Institute
of Petroleum Geology
Volodarsky Street 56,
Tyumen 625670 U.S.S.R.

Summary

The system of complex approach to the problem of development of quantitative methods of hydrocarbon potential forecast as the basis for sucessful prospecting and exploration of oil and gas is described in the given paper. The main blocks (directions) have been distinguished: data banks and multi-parameter modelling of the object structure; genetic, technogenic, simulation and local modelling. Characteristics are given to individual blocks, peculiarities of their interrelations, the results of modelling thetechnogenic and genetic processes, their relation to distribution of hydrocarbon reserves. The main concepts connected with effectiveness of prospecting -exploratory works heve been described.

Effective prospecting and exploration of hydrocarbon accumulations is possible , when a wide spectrum of theoretical and applied problems is solved. A special place among them belongs to development of the theory and methods of quantitative forecast of hydrocarbon potential of different objects (basins, districts, regions, individual sedimentary complexes, reservoirs, etc.). Theoretical and methodical approaches should provide not only the evaluation of hydrocarbon potential of geological sections but also make possible differentiation of these reserves by the phase state of hydrocarbons, their composition and properties, by the size of accumulations, etc. Only the complex approach to the problem of development of quantitative methods of hydrocarbon potential forecast (tasks of regional, zonal and local forecast) may be the base for effective prospecting and exploratory works in various geological conditions.

The given report describes in brief the methods for provision of forecasts with data base and for improvement of the models of hydrocarbon generation and accumulation. The main attention was given to development of the problems providing new data which help to correct and complete the previous forecasts.

The integrated system combining genetic studies and regularities of technogenic processes, seismic survey and drilling data, local and regional forecasts should lie at the base of scientific forecast and control of the prospecting-exploratory process. Five main directions may be distinguished in this system:

- data banks and systems of three-dimensional, multiple parameter modelling of object structure;
- genetic modelling of processes correlated with drilling data, which allows to make hydrocarbon potential forecast on the base of quantitative relationships;
- technogenic modelling allowing to estimate reserves on the data coverage;
- simulation modelling allowing to forecast the success ratio and to develop the programs of works;
- local modelling of the reserves base and of prospecting -exploratory works providing current control of prospecting and exploration.

Estimation of oil and gas reserves concentration is done on the base of geological-geochemical models, which use quantitative relationships describing the change of reserves concentration in oil-and-gas complexes versus the change of lithologic, tectonic, hydrogeochemical criteria of hydrocarbon potential, and on the base of these relationships the potential forecast is done for every small area.

Analysis of the change of effectiveness of works, the dynamics of transference of resources into the category of reserves depending on the level of study of oil-and-gas complexes (OGC) by drilling and on the discovered or supposed features of their geologic structure, primarily on the continuity of the reservoir, also allows to evaluate the unknown resources by the known reserves. The first and the second approaches to the forecast of reserves for the areas more or less explored by drilling should give (if both approaches are correct) similar results and, possibly, correct each other, which is essential within the limits of a special program using Monte-Carlo method. It should be stressed that in such a forecast of hydrocarbon potential geologic-geochemical and technologic aspects are taken into account when estimating the resereves.

For geologic-economical evaluation of the structure and the category of reserves it is necessary to consider in more detail distribution of reserves in accumulations of different types, the dynamics of effectiveness of works. With this goal in view the following directions are used:

1. Analysis of the time series of the actual effectiveness variation in different regions shows (which is confirmed by many studies) the regular character of these series. The initial and final stages of resource base development are characterized by the slow rate of effectiveness decline, and with transference of 40% to 60% of resources into the category of reserves this value reaches maximum which may be twice decreased, for example, during the 5-year period.

2. The study of the regularities of distribution of pools, fields, traps by size and type shows that these regularities are fairly strict and allow to predict reliably (as the experience of the last 15 years showed) the size and amount of new oil and gas objects. Correction of these data is done through simulation modelling. On the base of the established laws distribution of objects is given, then the present level of the data coverage of the region studied is simulated and the set of deposits which should be discovered is analysed and compared with the set of discovered objects , and on this base the initial set is corrected. Simulation modelling allows not only to determine precisely the distribution of undiscovered plays but to obtain one more estimation of the effectiveness of works. Combination of the methods allows to evaluate the effectiveness

of works for one, two, three years and more long periods, gives the set of objects for geologic-economical analysis. The present evaluation of the resource base envisaging the price of lease and reserves is impossible without such a reliable multi-parameter analysis.

With short-term change of the reserves base more important are various local characteristics but not the general trend of the reserves dynamics. In this case you may use such methods of study as "direct" modelling of plays, "direct" modelling of their prospecting and exploration. The first part of this problem is reconstruction of the contours of predicted plays on the reservoir maps, for this purpose various methods of the local forecast are used, and the result of all this is estimation of prospective hydrocarbon reserves with deliniation of promosing areas. The second part consists in location of wild-cats and exploratory wells in the plays with due regard for expected reliability of forecast and available data base. "Direct" modelling is one more (the last in the described scheme) method of correcting the reserves forecast and effectiveness, though in this case it is more appropriate to speak not about correcting but about distribution of regionally defined cheracteristics through smaller objects.

In this way five technological directions operate, mutually correcting and supplementing each other. This causes absolutely new attitude to the results of geologic exploration: all our conclusions, estimations should be of probabilistic nature, and just such a peculiarity as the estimation variance of reserves should be at the base of geologic exploration planning, which is reflected in the general scheme. The second significant feature of geologic exploration at a new stage is availability of powerful integrated data bases and programs of three-dimensional multi-parameter modelling of geologic objects. It should be stressed that data banks absolutely differ from archives. Bank is a live, continuously changing informative mechanism where information from different files is continuously corrected and refined. By this an important organisational task is achieved: a single bank unites not only information but specialists as well. Instead of being integrated into a single service they are combined by a common information bank and by common programs of 3-D modelling allowing to reconstruct the geologic-geophysical structure of a region.

Already today the developed 3-D models are no more purily geologic or purily geophysical. The classic concept of the stage-by-stage performance of works also disappears. Nobody thinks any more that first you should drill a well and then implement a complex of geophysical studies. In modern technologies the process od drilling and well-logging is simultaneous. The same tendency we see also in exploration geophysics. Beginning from prospecting of an area up to presentation of reserves for approval there is (should be) constant interrelation between drilling and seismic survey data. Drilling defines nature of seismic anomalies, seismic prospecting deliniates anomalies whose productivity was confirmed by drilling and deliniates new heterogeneities, and so on. A single seismo-geological model is formed and corrected both through reinterpretation of data and performance of new amounts of works. In the presence of integrated data banks and programs of 3-D modelling it is difficult to see what happens: whether geological models are corrected with regard for seismic data or seismometric models are refined with regard for drilling data. The simultaneous process conceals distinctions and stages similarly as it was with drilling stages and well logging. This tendency may be considered as the principal scientific concept of the modern stage of geologic exploration.

We shal describe in brief the main new scientific concepts for every of the main blocks of the system.

<u>Geologic-geochemical modelling.</u> Testing of different genetic situations on hundreds of complex models in different areas allowed to ascertain:

- Geologic-geochemocal parameters of an oil-bearing complex of rocks describe well the change of hydrocarbon concentrations in it. This is possible only in case when the oil-bearing complex is also oil-producing.

- The processes of hydrocarbon generation and migration are described most reliably by a set of curves characterising the relation of the composition of dispersed organic matter and hydrocarbons with different thermobaric conditions. The curves reflecting the changes in concentrations of carbon, hydrogen, heterocompounds with increase of catagenesis appeared to be complicated by different anomalous zones, at whose boundaries the molecular structure of kerogen is subjected to changes, and there is a significant shift in the proportions of generated hydrocarbons and non-hydrocarbon components. The dynamics of generation is unambiguously described by the curves of parameter variations for different types of dispersed organic matter. The similar approach allows to characterize gases dissolved in water and to show the leading role of their escape from water in very naroow and specific thermobaric interval when giant gas accumulations were formed in the Cenomanian sediments in the north of western Siberia.

- We have not observed the dependence of the reseres concentration change from the change of organic matter volumes (or any concrete class of organic matter), which indicates some incorrectness of many volumetric-genetic estimations. Probably, the main role belongs not to the mass but to the free surface of "particles" of dispersed organic matter. Studies showed that with up to 0.5-1.0% of organic carbon concentraion increase of concentration of dispersed organic matter coincides with increase of the total free surface, and with further increase of organic matter mass the free surface decreases. In a similar way behaves the concentration of paramagnetic centers. This results in formation of a new concept about hydrocarbon sources and changes our idea about the prospects of complexes with low concentration of organic matter.

- The role of the dynamics of sedimentary-tectonic processes in the change of hydrocarbon concentration is very great. It should be noted that with further study this process seems more regular. In particular, variations of rates of sedimentary basin formation appear to be regular. The great role belongs to regular networks of stresses resulting in "planetary" fracturing and many other phenomana in paleogeographic zoning, in deposit occurrences, in changes of reservoir properties.

The estimated quantitative dependences allow to make reliable forecast of hydrocarbon potential. But non-predictable variations of the possible parameter values relative to the mean value is inevitable: the number of effective quantitative dependences is always more than one; hence follows the inevitable probabilistic estimation of reserves with modelling of parameter changes by Monte-Carlo method. The result of this procedure is a curve describing the probability of virtual reserves being less than any predetermined value.

<u>Technogenic modelling.</u> The main task of this block of works is correlation of the actual results of geologic exploration with petroleum potential forecast. We know that depending on the structure of oil-and-gas-bearing complexes reserves in some of them are discovered more simply and easier than in others. Often the small intensity of transference of resources into the category of reserves is the base for their

reestimation. It was found that the main parameter which controls the dynamics of resource shift into reserves is the quality of reservoirs. The discovered quantitative dependences play a double role: on the one hand, they allow to estimate more precisely the potential of relatively well studied regions; on the other hand, they generate information about the effectiveness of works, which being corrected through the results of the next block intergrates the system into one unit.

S i m u l a t i o n   m o d e l l i n g. The regularities of distribution of geologic objects according to which the number of smaller objects (pools, traps) is greater than that of larger ones have been unambiguously confirmed in many regions. The data accumulated allowed to note in these regularities one peculiarity: discreteness of the geologic objects distribution. In the background of the total decrease of the frequency of occurrence of objects with their increase there are "minima" and "maxima". Their position on the axis of sizes is sufficiently stable. That is, there is good reason to speak about natural devision of objects into different classes by their sizes and to make by this our understanding of the resource base more precise. The new instrument - "geologic exploration filter" - allows to find relathionships in parameters of the discovered and undiscovered sets of geologic objects; to make more reliable both the evaluation of the resource structure itself and modelling of the prospecting-exploration process on its base.

The main conclusion of the numerous experiments on simulation modelling is the following: if wells were spudded ideally (zero rejects, penetration of the whole section, 100% effectiveness of well logging), then the percent of dry holes was 10-20% less than in reality, even if there were no seismic surveys or some other methods of forecast. Surely, there could not be an "ideal" well, but "non-ideality" of wells is such that it completely overlaps the positive effect of various methods of drilling ensurance.

Simulation modelling allows to develop short- and long-term plans and give the economic evaluation of reserves.

L o c a l   m o d e l l i n g. This direction provides short -term programs of prospecting-exploration works, being connected with other programs it corrects the effectiveness of works, develops the balance of reserves and resources, etc.

The initial blocks of this direction are similar to geologic-geochemical modelling, but already at the stage of object deliniation (traps) there appear principal differences consisting in the following:

- in specific approach to anomalies in various fields looked upon as objects subjected to drilling test after analysis;
- in specific character of objects with different contours and with non-zero probability of disappearance of the object itself. The main instrument for planning works in such objects become special maps of errors, maps of disperses, which determine the location of wild-cats and exploratory wells. In fact, these maps to some degree substutute the stage of simulation modelling typical of the previuos direction.

The modern state of geologic exploration has revealed a lot of principally new relations, regularities, interconnections. Hence there follows the appearance of new blocks, of new approaches. Practically everything that was described above has been experienced in practice. That is why the most important task is use of the above described complex of forecasting and performance of geologic exploration works in integration, as one unit, for increase of its effectiveness.

SEQUENCE ARCHITECTURE AND MODELLING IN THE WESTERN PROVINCE OF THE TROLL RESERVOIR, NORWAY - Part 1: SEQUENCES, SYSTEMS TRACT AND SEDIMENTARY FACIES DEVELOPMENT

S.D. NIO[1], K. GIBBONS[2], T. HELLEM[3], A. KJEMPERUD[4], K. VEBENSTAD[5] and C.S. YANG[1]

(1) INTERGEOS, Reaal 5Q, 2353 Leiderdorp, The Netherlands.
(2) STATOIL, P.O. Box 300, Stavanger, Norway.
(3) READ Production Geology Services A/S, P.O. Box 145, Sandvika, Norway. Present address: Grindstuveien 51, 1349, Rykkin, Norway.
(4) READ Production Geology Services A/S, P.O. Box 145, Sandvika, Norway. Present address: NOPEC, P.O. Box 88, 3478, Nærsnes, Norway.
(5) STATOIL, P.O. Box 300, Stavanger, Norway. Present address: BP Norway Ltd U.A., P.O. Box 197, Stavanger, Norway.

Summary

A sequence stratigraphic model is established and tested for the western provinces of Troll. Based on an integration of palynology, regional eustatic sea level changes, basin subsidence and sedimentary facies analysis from cores and wireline logs, an accurate reconstruction of the sequence architecture and a sequence stratigraphic modelling are possible. The results proved to be a powerful tool in mapping reservoir units and predicting geometric properties of permeability barriers. The datum line of the model was set at the lowstand at the beginning of the LZA-3 supercycle, while the top was set at the end of the LZA-4 supercycle. This period was characterized by a general eustatic sea level rise, which together with basin subsidence determined the development of accommodation. The third order eustatic sea level fluctuations and the fluctuating sediment supply were responsible for the development of sequences and the deposition of different sedimentary facies, including (1) deltaic sands (Highstand systems tract); (2) extensive inshore estuarine systems (Early Transgressive systems tract); (3) ebb-tidal deltas (Late Transgressive systems tract); and (4) condensed sequences and submarine hardgrounds with possible permeability barriers (maximum flooding period). The 'micaceous sands' and 'clean sands' were formed during different systems tracts.

1. INTRODUCTION

The Troll Field can be divided into three provinces according to hydrocarbon occurrence: (1) an eastern gas province with an average oil column of 3 m; (2) a western gas province with an average oil column of 12 m; and (3) a western oil province with an average oil column of 25 m. This study is based on well data in the western oil and gas provinces.

The main reservoir in the Troll Field, the Sognefjord Formation, was deposited during a general eustatic sea level rise in the LZA-3 and LZA-4 supercycles of Haq et. al.[1] from the Callovian to the Early Volgian. The Sognefjord Formation at Troll was deposited on the inner to middle shelf on the northwestern edge of the Horda Platform. Tidal action was present during the whole depositional period[2]. An important problem concerning the Sognefjord Formation is the spatial distribution and lateral extent of permeability barriers, such as calcite cemented horizons. The origin and geometry of calcite cemented horizons in the Troll reservoir have been discussed by many researchers[3,4]. More recently Gibbons et. al.[5] have established a model to distinguish between lateral extensive and discontinuous horizons[5]. This paper will concentrate on the sequence stratigraphic architecture and modelling in the Troll reservoir, which provide a framework for predicting the lateral extent of calcite cemented horizons.

2. SEA LEVEL FLUCTUATIONS AND SEDIMENTARY FACIES DEVELOPMENT IN SHORELINE-ATTACHED TO MIDDLE SHELF REGION

Sea level fluctuations have a major influence on the sedimentation pattern within the shallow marine environment (i.e. the inshore, shoreline-attached to outer shelf region). Depositional depths within these environments range from zero along the coast to a maximum of 200 m at the edge of the shelf. Generally, the shelf can be subdivided into the following zones[6]:

- Shoreline-attached zone (depth 0-15 m). Dominant processes are tidal currents, wave-induced currents and wave-storm processes.
- Inner shelf zone (depth 15-30 m). Dominant processes are tidal as well as wave-induced currents and wave-storm processes. Most of this zone is located within the fair weather wave base.
- Middle shelf zone (depth 30-100 m). Dominant processes are storm activities.
- Outer shelf edge zone (depth 100-200 m). Important processes are storm induced currents.

Deposition of the different systems tract depends on the rate of relative sea level changes which is controlled by both eustatic sea level changes and basin subsidence[7,8,9]. The magnitude of relative sea level fluctuations and sediment supply are important factors in the development of shallow marine facies on the shelf.

Nio and Yang[10] showed that different tidal sand bodies can be formed during the deposition of different systems tracts. The geometric variability of these tidal sand bodies is controlled by the sediment supply and tidal discharge of the depositional system (Fig.1). Changes in tidal discharge and sediment supply are related to relative falls and rises of sea level. A rise in relative sea level will enhance the effect of tidal wave propagation into the shallower parts of the basin, causing an increase of the tidal discharge. A fall in relative sea level, however, will strongly reduce tidal action. On the other hand, sand dispersal into the basin will decrease with a relative sea level rise, which will cause a flooding of the coastal plain. A large part of the sands will be trapped in the inshore depositional environments and only the fine-grained fraction will be transported basinward. However, during a relative sea level fall a large part of the shelf will be exposed and the eroded sediments will be brought into the deep parts of the basin.

The geometric variability of tidal sand bodies shown in Fig.1 therefore is very sensitive to relative sea level changes. The transition from fluvial dominated transverse sand bars (FB in Fig.1) into tidal pointbar complexes (TB in Fig.1) represents a change from low-sinuosity, high-gradient estuarine channels with high influxes of coarse sediments from fluvial systems into high-sinuosity, low gradient estuarine channels. This change is related to a relative sea level rise in the early stage of the Transgressive systems tract. During the later stage of the Transgressive systems tract an increase in erosion within the inshore estuaries will occur. This will cause a higher sediment influx into the basin and a prograding ebb-tidal delta will be formed. A constant high sediment supply will produce an offlap pattern of the different ebb-tidal delta lobes upon the shelf.

Nio and Yang[10] also showed the occurrence of various tidal sand bodies within a sequence stratigraphic framework (Fig.2). During the late stage of a Shelf Margin Wedge systems tract overlying a Type 2 sequence boundary, the gradual increase in the rate of relative sea level rise will cause an increase of the tidal discharge, resulting in increasing tidal reworking of the delta deposits belonging to the Shelf Margin Wedge or Late Highstand systems tract. The fluvial channels and adjacent floodplain system will be flooded and modified into an estuarine and tidal flat system (A in Fig.2). During the deposition of the Transgressive systems tract a rapid rise in the relative sea level causes a significant increase of tidal discharge. Different tidal sand bodies are formed depending on sediment supply and local palaeogeographic setting. Extensive high sinuosity, low gradient estuarine systems are formed (B in Fig.2). At a later stage of the Transgressive systems tract an offlapping ebb-tidal delta system will be formed (C in Fig.2). The Transgressive systems tract is concluded with the deposition of a condensed sequence or a

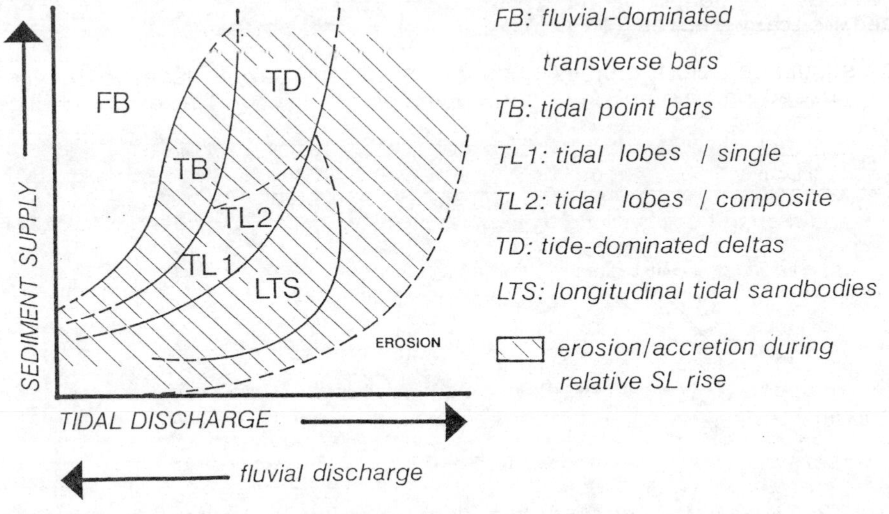

Fig. 1: The geometric variability of large scale tidal sand bodies as related to the sediment supply and the tidal discharge of the depositional system (after Nio & Yang, in press).

submarine hardground with a carbonate-cemented layer during the maximum flooding period (mfs in Fig.2).

These principles can be applied in the prediction of the different types of tidal sand bodies in a basin model. Although the rate of eustatic sea level change can be assumed constant across the shelf, the rate of basin subsidence is generally not, and usually increases from zero at the tectonic hinge line to a maximum subsidence rate at the basin centre[8,9]. The actual timing of the development of each systems tract and the formation of various tidal sand bodies at a specific location varies across the shelf with increasing subsidence rate, and thus with increasing rate of relative sea level change.

Figures 3 and 4 show two synthetic sections from the inner-shelf and shoreline-attached region and the middle shelf region. The inner-shelf and shoreline-attached region is the proximal location showing a low subsidence rate. Therefore this region is characterized by a long period of erosion during the Shelf Margin Wedge systems tract, or a long period of fluvial channel entrenchment during a Lowstand Fan systems tract stage (see Fig.3). The Transgressive systems tract begins at a much later period in this region and is characterized by the infilling of fluvial channels (A1 in Fig.3), followed by a tidal reworking and the formation of estuarine systems (C in Fig.3). At a later stage of the Transgressive systems tract ebb-tidal deltas (D in Fig.3) are formed, followed by the maximum flooding period (G1 in Fig.3). The subsequent Highstand systems tract is characterized by fine-grained sedimentation of distal shelf mud or storm layers (E in Fig.3), overlain by coarsening-upward sequences of prograding systems (B in Fig.3).

In the middle shelf region the synthetic section shows a higher subsidence rate which causes a significant change in the periods of deposition of the systems tract and different sedimentary facies (Fig.4).

3. SEQUENCE STRATIGRAPHY BOUNDARIES OF THE TROLL RESERVOIR

Based on palynology, the zonation of the Troll reservoir has

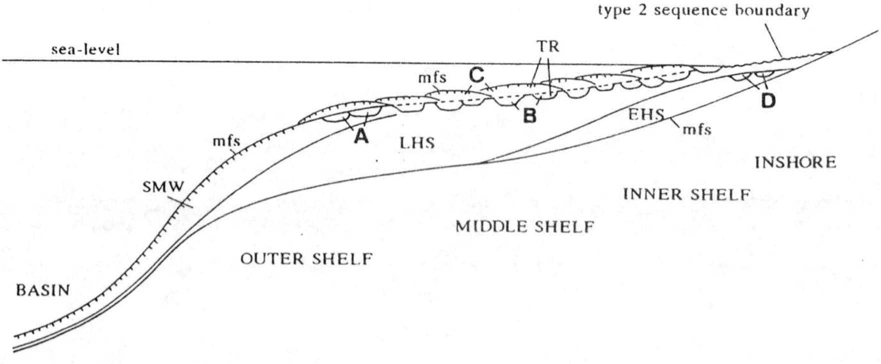

Fig. 2: A section across the shelf, showing a type 2 sequence boundary. See text for explanation (after Nio & Yang, in press).

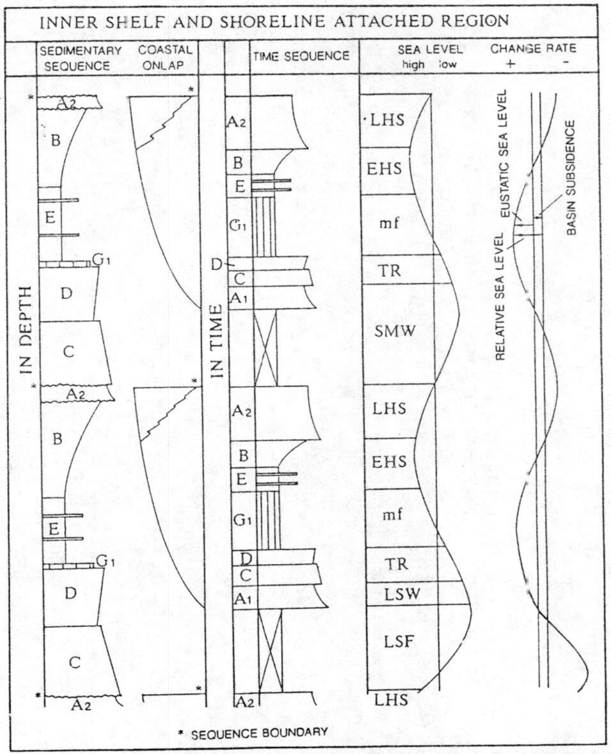

Fig. 3: Synthetic stratigraphic section from the inner-shelf and shoreline-attached region (after Nio & Yang, in press).

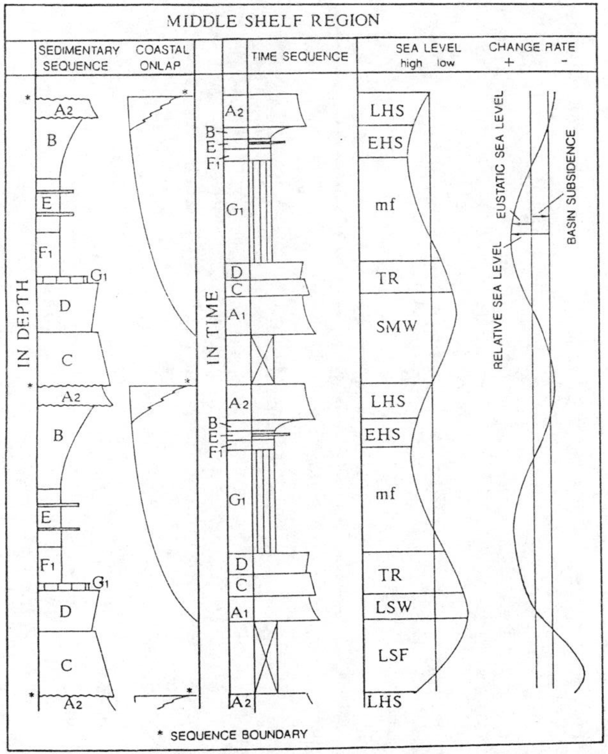

Fig. 4: Synthetic stratigraphic cross section from the middle shelf region (after Nio & Yang, in press).

A: Alluvial-fluvial facies; B: Delta facies; C: Estuarine facies; D: Ebb-tidal delta systems; E: Shelf mud with storm beds; F1: Shelf mud; G1: Condensed sequence.

been established and can be correlated to the third order cycles of the eustatic sea level curve of Haq et al.[1] This correlation is shown in Fig.5, and has been discussed by Gibbons et. al.[5]

Most of the zonation boundaries of the Troll reservoir can be correlated to the sequence boundaries or the maximum flooding surfaces. The zonation boundaries at SeD/SeC, Cr/Su and Su/Pu lie at sequence boundaries. All the rest are at or close to maximum flooding surfaces.

4. A SYNTHETIC SEQUENCE STRATIGRAPHIC CROSS SECTION THROUGH THE CENTRAL TROLL FIELD

The principles discussed earlier were applied in the construction of a synthetic sequence stratigraphic cross section through the Central Troll Field. The datum line used in the modelling was set at the lowstand at the beginning of the LZA-3 supercycle of the Haq et al. chart, while the top was set at the end of the LZA-4 supercycle. This period was characterized by a general eustatic sea level rise, which was more than 100 m (Fig.6). The third order eustatic sea level fluctuations during this periods, however, were less than 50 m, and in most cases only about 20 m, except for the dramatic sea level fall at the end of LZA-4.7 cycle. After digitizing the eustatic sea level curve of Haq et al.[1], the rates of eustatic sea level changes were calculated (Fig.6). The calculated rates of eustatic sea level fall are rather minor, usually less than 15 m/my. The major exception is the boundary between LZA-4.7 and LZB-1.1 cycles with a very high rate of eustatic sea level fall.

The subsidence rate at Troll is estimated based on the thickness of depositional sequences from well data. The effects of palaeo-water depth, eustatic sea level changes and hydro-isostasy have be corrected. The average minimum subsidence rate at Troll western provinces is 4 m/my. A higher subsidence rate is assumed for the shelf margin.

Figure 7 shows the synthetic sequence stratigraphic cross section through the Troll Field. This model is based on the data of eustatic sea level changes, basin subsidence, and all available well data from the western provinces of the Troll Field. The location of the Central Troll Field is indicated in the cross section. The cross section shows several interesting features.

(1) Most sequence boundaries in the cross section belong to type 2 sequence boundary as defined by Vail[7]. This was caused by small rates of eustatic sea level fall, which were less than the subsidence rates at the shelf margin. The rapid eustatic sea level fall at the end of LZA-4.7 cycle, however, produced a type 1 sequence boundary, which is characterized by a severe truncation of the LZA-4.7 and LZA-4.6 sequences and the development of the LZB-1.1 Lowstand systems tract. Such a remarkable truncation can be recognized in all wells.

(2) The Troll Field was located on the inner to middle shelf on the northwestern edge of the Horda Platform. Therefore the sequence at Troll Field consists mainly of Highstand and Transgressive systems tracts. No Shelf Margin Wedge systems tract can be observed at the Troll Field.

The position of the equilibrium point at the sequence boundaries (asterisk in Fig.7) shows a consistent basinward migration in time. This indicates a basinward progradation of the

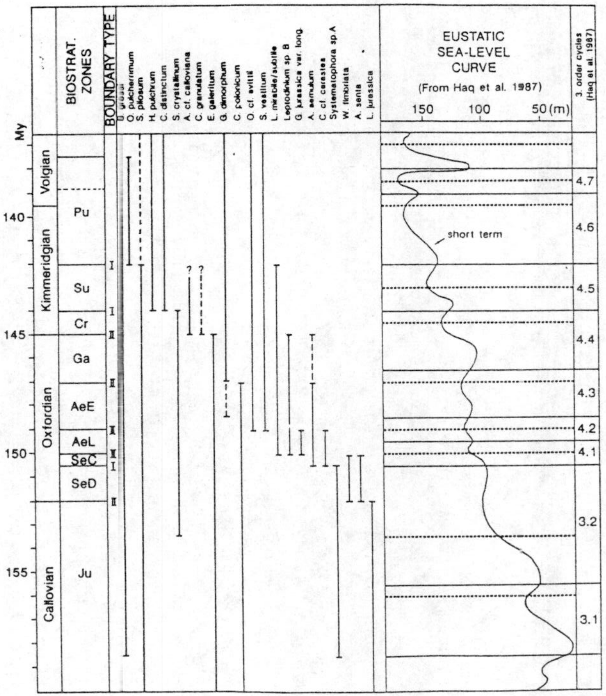

Fig. 5. Correlation between the bio-stratigraphic zones of the Troll reservoir and the third order cycles in the eustatic sea level curve of Haq et al.[1] (after Gibbons et. al., in prep.).

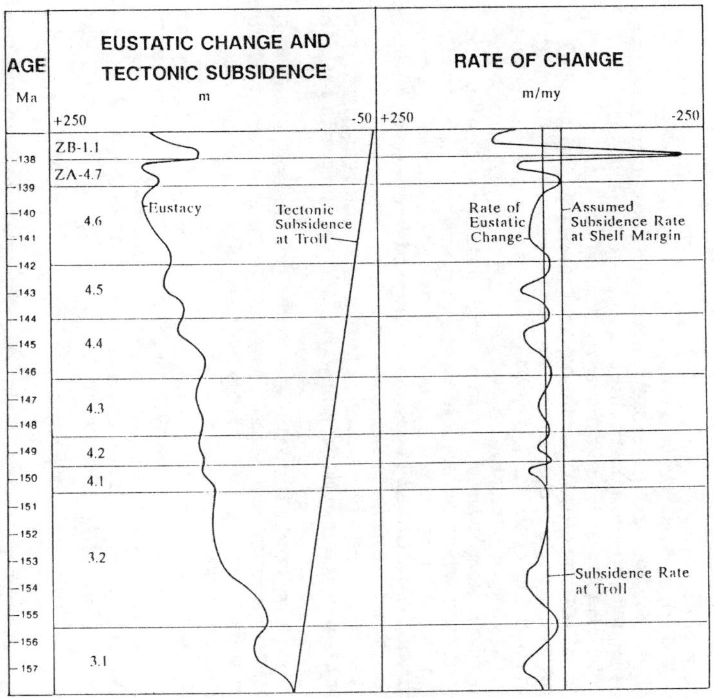

Fig. 6. The eustatic sea level changes and the rates of the change during the LZA-3 and LZA-4 super cycles.

systems tracts.

(3) The third order eustatic sea level fluctuations produced important changes in relative sea level. These changes together with the fluctuating sediment supply were responsible for the development of different sedimentary facies.

- A progradation of deltaic sands, including the distal to more proximal prodelta and/or the delta platform deposits, occurred during the Highstand and especially the Late Highstand systems tract.
- A severe to intermediate erosion and modification by tidal process occurred during the Early Transgressive systems tract during which extensive inshore estuarine systems were formed. An onlapping stacking of these estuarine channels characterized the spatial development.
- At the Late Transgressive systems tract extensive ebb-tidal deltas were formed, showing a distinct offlapping pattern.
- During the maximum flooding period, condensed sequences as well as submarine hardgrounds were formed. These possible permeability barriers may reach extensive sheet-like geometric proportions.

The predicted sedimentary facies can be recognized in the wells of the Troll Field. Beecham (pers. comm., 1985) described the repeated successions of clean, medium to coarse sands and micaceous, silty, very fine to fine sands at Troll. A single depositional cycle typically comprises a progradational and a

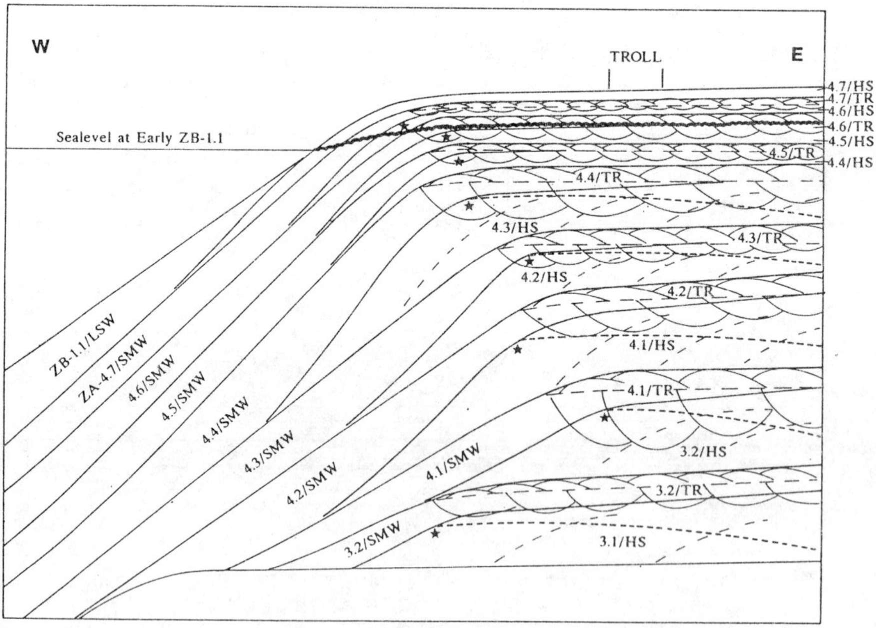

Fig. 7. Synthetic sequence stratigraphic cross section through the Central Troll Field.

transgressive component. From the base, low energy, frequently offshore, fine micaceous sediments coarsen upwards into nearshore, medium to coarse, clean sands which represents coastal progradation over a shallow shelf. This sequence is often overlain or partially replaced by a coarse, clean sand reworked from coastal sediments by high energy currents associated with an increase in transgressive activity. This description agrees with the sequence of the predicted sedimentary facies: the 'micaceous sands' corresponding to the Highstand systems tract, and the 'clean sands' to the Transgressive systems tract.

(4) The preservation of the Troll reservoir was controlled by the development of accommodation, which depended on the eustatic sea level rise and the basin subsidence. The net eustatic sea level rise during the LZA-3 and LZA-4 supercycles was more than 100 m, and the subsidence was about 80 m during the same period. Both factors contributed to the relative sea level rise and the accommodation development. This also accounts for the thicknesses of the parasequences and the depth of tidal erosion during different cycles.

The predicted sequences and sedimentary facies outlined by the synthetic sequence stratigraphic cross section could be recognized in all wells. With the help of the model, the presence of sequence boundaries, Transgressive systems tract, maximum flooding surfaces, and Highstand systems tract in each cycle has been defined in each well (Fig.8). This provided detailed

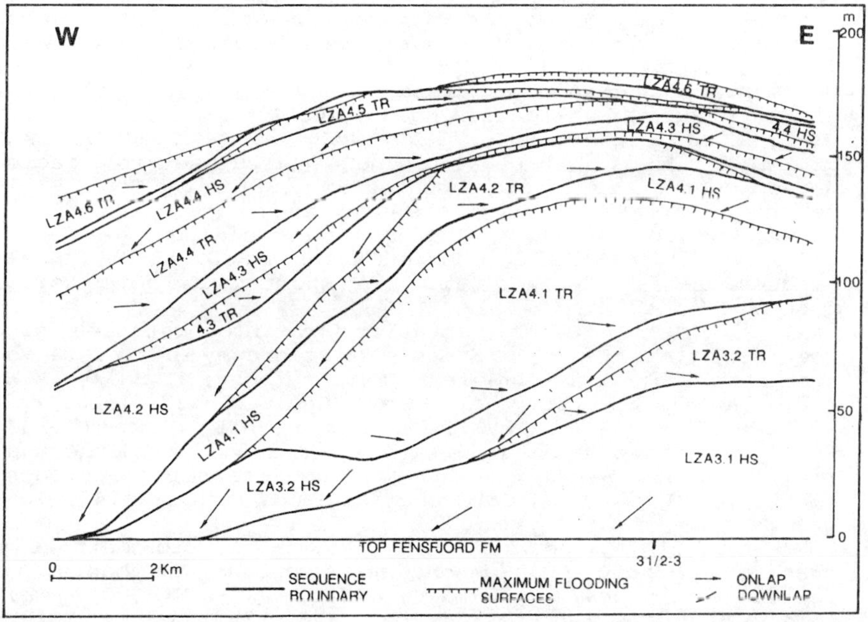

Fig. 8. Sequence boundaries, maximum flooding surfaces, and systems tracts in the Troll Field, with expected onlap and downlap pattern. TR: clean tidally-reworked sands; HS: micaceous, very fine-fine, bioturbated sands (after Gibbons, et. al., in prep.).

information about the geometry of the Troll reservoir sands and the distribution of laterally extensive carbonate cemented horizons which are associated with maximum flooding surfaces.

ACKNOWLEDGEMENTS

A large part of this paper is the result of a comprehensive sequence stratigraphic study of the Troll Field reservoir. The authors would like to thank the Licence Group of the Troll area, Statoil, Norsk Hydro, Saga Petroleum, Norske Shell, Conoco Norway, Elf Aquitaine Norway and Total Marine Norway for permission to publish this paper.

REFERENCES

(1) HAQ, B.U., HARDENBOL, J. and VAIL, P.R. (1987). The chronology of fluctuating sea level since the Triassic. Science, 235, 1156-1167.

(2) HELLEM, T., KJEMPERUD, A. and ØVREBØ, O.K. (1986). The Troll Field: a geological/geophysical model established by the PL085 Group. In: Habitat of Hydrocarbons on the Norwegian Continental Shelf, Norwegian Petroleum Soc., Spencer, A.M. (ed.), Graham and Trotman publishers, 217-238.

(3) IRWIN, H. and HURST, A. (1983). Applications of geochemistry to sandstone reservoir studies. In: Petroleum Geochemistry and Exploration of Europe, Brooks, J. (ed.), Blackwell Sci. Pub., 127-145.

(4) KANTOROWICZ, J.D., BRYANT, I.D. and DAWANS, J.M. (1987). Controls on the geometry and distribution of carbonate cements in Jurassic sandstones: Bridport Sands, southern England and Viking Group, Troll Field, Norway. In: Diagenesis of Sedimentary Sequences, Marshall, J.D. (ed.), Geological Society Special Publication No.36, 103-118.

(5) GIBBONS, K., HELLEM, T., KJEMPERUD, A., NIO, S.D., AND VEBENSTAD, K. (in prep.). Sequence architecture, facies development and carbonate cemented horizons in the Troll Field reservoir, offshore Norway. In: Advances in Reservoir Geology. Badley and Ashton (eds.), Conference Proceedings.

(6) TILLMAN, R.W., SWIFT, D.J.P. and WALKER, R.G. (1985). Shelf sands and sandstone reservoirs. SEPM short course notes No.13, 708 pp.

(7) VAIL, P.R. (1987). Seismic stratigraphic interpretation using sequence stratigraphy, 1. Seismic stratigraphy interpretation procedure. In: Atlas of Seismic Stratigraphy, 1. Bally, A.W. (ed.), Am. Assoc. Pet. Geol., Stud. Geol., No.27, 1-10.

(8) POSAMENTIER, H.W., JERVEY, M.T. and VAIL, P.R. (1988). Eustatic controls on clastic deposition I - Conceptual framework. In: Sea level changes: an integrated approach. WILGUS et al. (eds.), SEPM Special Publication No.42, 109-124.

(9) POSAMENTIER, H.W. and VAIL, P.R. (1988). Eustatic controls on clastic deposition II - Sequence and systems tract models. In: Sea level changes: an integrated approach. WILGUS et al. (eds.), SEPM Special Publication No.42, 125-154.

(10) NIO S.D. and YANG C.S. (in press). Sea level fluctuations and the geometric variability of tide-dominated sand bodies. Marine Geology.

PORE-PRESSURE-INDUCED FRACTURING OF PETROLEUM SOURCE ROCKS. IMPLICATIONS FOR PRIMARY MIGRATION

FLORIAN K. LEHNER

Koninklijke/Shell Exploratie en Produktie Laboratorium,
Postbus 60, 2280 AB Rijswijk Z.H.,
The Netherlands

Summary

A possible mechanism of expulsion of hydrocarbon fluids from mature source rocks relies on the generation of a network of microfractures which nucleate at kerogen/bitumen inclusions and are extended by excess pore pressures generated during the burial and transformation of kerogen. In this paper the stability and growth of isolated crack-like kerogen/bitumen inclusions are investigated, using concepts from fracture mechanics. Pore pressures exceeding lithostatic stresses are found to result mostly from the transformation of kerogen. It is shown how the critical depth for the onset of crack extension depends on burial rate, reaction kinetics, geothermal gradient and material properties. In competent source beds crack extension up to a multiple of the initial flaw size may occur suddenly, suggesting the possibility of a relatively abrupt increase in source rock permeability.

1. INTRODUCTION

Primary migration and the expulsion of hydrocarbons from mature source rocks has been the subject of many discussions, but of the more likely candidate mechanisms few have been examined quantitatively in any depth. This is true, in particular, for the often invoked mechanism of microfracturing, which is thought to be an effect of hydrocarbon generation in tight source rocks, providing for the necessary permeability in a compaction-driven primary migration of a hydrocarbon fluid phase[1–8] (see also the discussion in Chap. 2.4 of Tissot & Welte[9]). So far, however, there appears to exist no theoretical treatment of the process of pore-pressure-induced fracturing of source rock, that exhibits the combined effects of external loading and heating due to burial and internal pressure build-up due to the transformation of kerogen. A first step in this direction is outlined in this paper. Aimed primarily at qualitative insight, this investigation is based on strongly idealizing model assumptions and quantitative results reported in the following should therefore be viewed with some caution; it starts from a certain qualitative picture of the process of hydrocarbon release from kerogen and of the subsequent build-up of discontinuous, crack-like kerogen/bitumen strands, a picture reflecting observations made and ideas expressed by several investigators in the past.

It is generally agreed that hydrocarbons remain adsorbed to disseminated organic matter (kerogen) until the thermally activated breakdown reaction by which they are generated has advanced sufficiently to initiate their release from the shrinking source material. The liquid phase released in this *initial migration* step is usually referred to as bitumen; it constitutes a proto-petroleum or chemical precursor of oil. Microscopic observations made by Beiu (unpublished work cited by Momper[3]) have established that bitumen droplets in the 1–10 μm size range appear near peak oil generation at the surface of isolated kerogen material. According to Momper, the generation of bitumen creates microscopic "high pressure cells" within the organic matter. As the pressure reaches a certain critical level, the "thermochemically generated liquids are extruded or exuded from the interior of the organic matter. On reaching the surface of the parent material, the liquid gradually coalesces until a break-away (droplet) size is attained ..." At that stage, a mobile bitumen phase will surround the source material and occupy the space created by the shrinkage of the latter.

The mean specific volume of the kerogen/bitumen mixture phase contained in such a mature organic "inclusion" will be larger than that of the initial untransformed kerogen flake. This implies a certain build-up in the inclusion pressure, the immediate consequence of which may be an impregnation of the surrounding source rock ground mass with mobile bitumen. A comparative microscopic study of this process on laboratory- and naturally-matured samples has recently been reported by Lewan[10], whose observations suggest that the ground mass becomes impregnated with mobile bitumen as a result of the development of submicron parting separations parallel to the bedding fabric of the rock. The natural process appears to be quite reproducible in laboratory experiments. Core samples that were heated at 300°C for 72 hrs were found to contain fresh *en échelon* parting separations in the mm-size range, while no such "microcracks" were observed by Lewan on cores pyrolysed at 200°C and 250°C.

On the basis of these observations, the following model for the process of pore-pressure-induced microcrack growth in source rocks is considered here. In an early stage of the process, bitumen is extruded from transforming kerogen flakes and injected into the surrounding ground mass through submicron partings that are aligned preferentially parallel to bedding. This impregnation stage leads to the formation of extended strands or micro sills of bitumen by a process of coalescence of closely spaced neighbouring patches of bitumen. Eventually this process stops at barriers between neighbouring kerogen flakes in the form of large enough ligaments of cohesive rock to block further bitumen impregnation. A single strand of bitumen formed in this manner and connecting the sites of perhaps several actively generating kerogen flakes constitutes the basic entity studied in this report. It will be viewed as a crack-like kerogen/bitumen inclusion whose subsequent growth in response to heating, burial or uplift, and further bitumen generation will be investigated theoretically.

Our discussion will emphasise bedding-parallel "cracks" (although the analysis will remain applicable to arbitrarily oriented cracks), primarily because of the important contribution of bedding-parallel kerogen/bitumen laminae to the (lateral) permeability and yield of a source bed. (Bedding-parallel cracks in a mature source rock were described recently by Littke & Rullkötter[11], who emphasise the role of such cracks as migration avenues.) "Vertical" cracks—i.e., cracks oriented perpendicular to bedding—are not necessarily formed earlier in an overpressured rock than are bedding parallel cracks. This follows from the fact that preexisting crack-like kerogen/bitumen inclusions which act as initial flaws are preferentially aligned parallel to bedding, so that any vertical flaws may be much shorter and therefore - despite higher crack opening pressures - less likely to rupture according to standard fracture mechanics crack growth criteria. Also, because of their laminated fabric, the organic-rich portions of a source rock are likely to exhibit a strength anisotropy favouring bedding-plane-parallel crack extension. Moreover, as a result of an increased ductility, the state of stress in mature (especially shaly) source rocks may well be close to isotropic in many cases. Similar conclusions with regard to the difficulty of inducing vertical fractures by excess pore pressures were reached by Ozkaya[8], who has discussed the failure of cavities under external stress and internal pressure for otherwise more restrictive model assumptions than made in the following. Vertical cracks may indeed be propagated by distinctly different mechanisms , e.g., as "hydraulic intrusion fractures"[6], but possibly also as branch cracks that are driven by localized tensile stresses induced by slip along bedding-parallel cracks when these are subjected to combined normal and shear loads (see Ashby & Sammis[12] for a discussion of the latter mechanism, albeit not in the context of primary migration).

2. PORE PRESSURE RESPONSE TO STRESS, TEMPERATURE, AND KEROGEN TRANSFORMATION

In the following we analyse the behaviour of a single crack-like kerogen/bitumen inclusion from the point-of-view of linear elastic fracture mechanics. The analysis of this single-inclusion model may be viewed as a first step towards a more comprehensive theory of pore-pressure-induced microfracturing of source rocks that would have to account for crack interaction and crack coalescence phenomena.

In order to establish the principal effects of burial or uplift, heating or cooling, and hydrocarbon generation on the stability (i.e. tendency to grow or remain stationary) of a single kerogen/bitumen inclusion, we require first of all an expression for the inclusion pressure, i.e., for the pressure generated within the organic inclusion phase by these agencies. We consider, for this purpose, the volume of the assumed crack-like inclusion, which will be constrained in a twofold manner. First, since the inclusion is assumed to exchange no mass with its sourroundings, the (constant) mass m of the kerogen and its transformation products must occupy a volume in accordance with appropriate equations of state. In fact, only the bulk PVT properties of this inclusion mixture phase are needed in the following. These will of course depend upon the extent of the kerogen transformation and the simplest way of accounting for this dependence is by distinguishing at least three phases, i.e., kerogen, a bitumen fluid phase, and a coke residue. The actual spatial distribution of these phases within a closed crack-like inclusion will be irrelevant to all subsequent considerations. At any given instant of time the masses of all three phases satisfy the condition

$$m_k(t) + m_f(t) + m_c(t) = m = \text{const.}, \tag{1}$$

while the mass densities ρ_k, ρ_f and ρ_c are assumed to be governed by the equations of state

$$\rho_k = \rho_k(P,T), \quad \rho_f = \rho_f(P,T), \quad \rho_c = \rho_c(P,T), \tag{2}$$

where P denotes the (uniform) inclusion pressure and T the absolute temperature. Here a dependence of the density on the extent of the kerogen transformation has been disregarded for simplicity. The volume V of the inclusion is constrained by the relation

$$V = V_k + V_f + V_c = m_k/\rho_k + m_f/\rho_f + m_c/\rho_c, \tag{3}$$

where m_k, m_f and m_c are related through (1), while ρ_k, ρ_f and ρ_c are determined in terms of inclusion pressure and temperature by the equations of state (2).

We now wish to calculate the change in volume of the inclusion mass m due to changes in pressure, temperature and in the extent of the kerogen transformation. We linearize the equations of state (2) with respect to a reference state (P_0, T_0), but allow for arbitrary changes in the mass fractions $\mu_i = m_i/m$ $(i = k, f, c)$ of the individual phases, subject only to the constraint (1). Putting $q = P - P_0$, $\theta = T - T_0$, and $\mu_{i0} = m_{i0}/m$, one obtains the equations of state

$$\rho_i^{-1} = \rho_{i0}^{-1}(1 + \alpha_i\theta - \beta_i q); \quad (i = k, f, c), \tag{4}$$

where

$$\alpha_i = \rho_{i0}\frac{\partial \rho_i^{-1}}{\partial T}\bigg|_{P_0,T_0} \quad \text{and} \quad \beta_i = \rho_{i0}\frac{\partial \rho_i^{-1}}{\partial P}\bigg|_{P_0,T_0}; \quad (i = k, f, c) \qquad (5),(6)$$

are appropriate coefficients of thermal expansion and isothermal compressibility, respectively. The relative volume change of the three-phase inclusion material is thus given by

$$\frac{V-V_0}{V_0}=\frac{\sum_i(V_i-V_{i0})}{\sum_i V_{i0}}=\frac{\sum_i[\mu_i(\alpha_i\theta-\beta_i q)+\mu_i-\mu_{i0}]/\rho_{i0}}{\sum_i\mu_{i0}/\rho_{i0}}. \qquad (i=k,f,c) \quad (7)$$

For simplicity, but consistent with the equations of state (4), we shall make the further assumption that the kerogen transformation products m_f and m_c are produced in constant proportions throughout the entire reaction, as determined by the ratio

$$r_{cf}=m_c/(m_c+m_f)=\text{const.} \quad (8)$$

We therefore have the relationships

$$\mu_f=(1-r_{cf})\mu, \quad \mu_c=r_{cf}\mu, \quad \text{and} \quad \mu=1-\mu_k=\mu_f+\mu_c, \qquad (9),(10),(11)$$

where μ will be called the "transformation ratio". Making use of these definitions, one may write the relative volume change (7) as

$$\frac{V-V_0}{V_0}=\alpha\theta-\beta q+\gamma(\mu-\mu_0) \quad (12)$$

$$\alpha(\mu;P_0,T_0,\mu_0)=\frac{[\alpha_f(1-r_{cf})/\rho_{f0}+\alpha_c r_{cf}/\rho_{c0}]\mu+\alpha_k(1-\mu)/\rho_{k0}}{[(1-r_{cf})/\rho_{f0}+r_{cf}/\rho_{c0}]\mu_0+(1-\mu_0)/\rho_{k0}}; \quad (13)$$

$$\beta(\mu;P_0,T_0,\mu_0)=\frac{[\beta_f(1-r_{cf})/\rho_{f0}+\beta_c r_{cf}/\rho_{c0}]\mu+\beta_k(1-\mu)/\rho_{k0}}{[(1-r_{cf})/\rho_{f0}+r_{cf}/\rho_{c0}]\mu_0+(1-\mu_0)/\rho_{k0}}; \quad (14)$$

$$\gamma(P_0,T_0,\mu_0)=\frac{(1-r_{cf})/\rho_{f0}+r_{cf}/\rho_{c0}-1/\rho_{k0}}{[(1-r_{cf})/\rho_{f0}+r_{cf}/\rho_{c0}]\mu_0+(1-\mu_0)/\rho_{k0}}. \quad (15)$$

Here α and β are effective bulk coefficients of thermal expansion and compressibility for the inclusion phase.

A second constraint on the inclusion volume derives from the relationship between the cavity volume V, the internal pressure P, and the external or "remote" stress field σ_{ij}. Thus, consider the example, illustrated by Figure 1, of a crack-like cavity of length $2l$. Let this crack be subjected to an external stress field σ_{ij} with two principal directions in the plane of Figure 1, oriented perpendicular and parallel to the crack faces, respectively. Starting from a reference state (P_0,σ_{ij}^0,T_0) when the crack has the volume V_0, let $p=-n_i(\sigma_{ij}-\sigma_{ij}^0)n_j$ denote the magnitude of the excess compressive principal stress tending to close the crack,

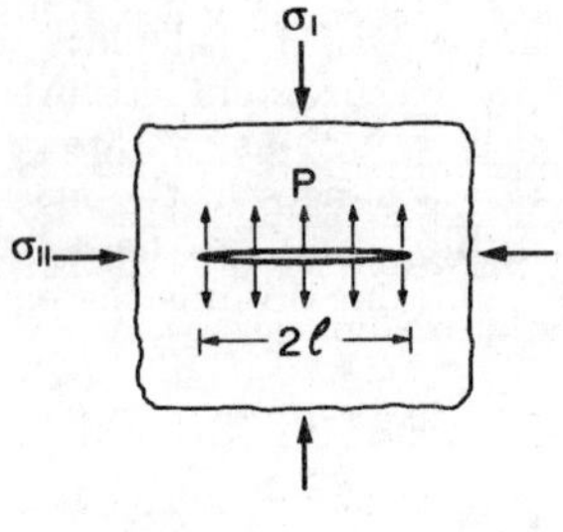

Fig. 1: Crack-like kerogen /bitumen inclusion.

the n_i being the components of the unit normal vector to the crack faces and σ_{ij} the components of the stress tensor in a Cartesian coordinate system (x_1, x_2), summation being implied for repeated indices. Also, let $q = P - P_0$ denote the counteracting excess inclusion pressure.

The response of the crack surroundings to the changes $-p$ and q in external stress and inclusion pressure from the reference state is then identical with the reponse to a net loading $q-p$ applied to the crack. Asuming that the rock surrounding the cavity is homogeneous and exhibits isotropic linear thermoelastic behaviour, the change in the cavity volume per unit length perpendicular to the plane of deformation in response to the changes q, $-p$, and θ in inclusion pressure, external stress and temperature will be determined by the relation

$$V - V_0 = \frac{2\pi(1-\nu^2)l^2}{E}(q-p) + V_0\alpha_r\theta, \tag{16}$$

where E is Young's modulus, ν is Poisson's ratio, and α_r a coefficient of thermal expansion of the rock. Note, that V and V_0 denote volumes per unit depth perpendicular to the plane of Figure 1, equation (16) pertaining to conditions of plane strain. The actual depth of the crack therefore remains undefined in the present analysis. Of course, there is no reason other than simplicity for making this assumption of plane strain and a 3-D analysis of a penny-shaped crack could have been carried out along identical lines.

Relation (16) may be written, in view of possible generalizations, as

$$\frac{V - V_0}{V_0} = \beta_r q + \beta_r n_i n_j(\sigma_{ij} - \sigma_{ij}^0) + \alpha_r\theta \tag{17}$$

$$\beta_r = \frac{1}{V_0}\frac{\partial V}{\partial P}\bigg|_{P_0,\sigma_{ij}^0,T_0} = \frac{2\pi(1-\nu^2)l^2}{EV_0} \tag{18}$$

Since (12) and (17) must be compatible, the relative volume changes may be eliminated from these equations to give

$$q = \frac{\alpha - \alpha_r}{\beta + \beta_r}\theta - \frac{\beta_r}{\beta + \beta_r}n_i n_j(\sigma_{ij} - \sigma_{ij}^0) + \frac{\gamma}{\beta + \beta_r}(\mu - \mu_0) \tag{19}$$

This relation governs the "undrained" pore pressure response to external loading, heating, and kerogen transformation. When $\theta = \mu - \mu_0 = 0$, it is reduced to a type of relation, well known from Biot's theory of poroelasticity (see Rice & Cleary[13] ; there the undrained response of a porous medium is often taken to be isotropic in which case the coefficient of the second term in (19) is the same as Skempton's[14] pore-pressure coefficient $B = \beta_r/(\beta + \beta_r)$. Indeed, if one visualizes a rock containing randomly oriented non-interacting cracks with an uncorrelated, random distribution of lengths, then the dyadic product $n_i n_j$ in that term would be replaced in an appropriately averaged form of (19) by the isotropic unit tensor δ_{ij}, the stress term reducing to $B(\sigma_{ii} - \sigma_{ii}^0)$ in that case.

3. STABLE VERSUS UNSTABLE CRACK EXTENSION

We now consider the growth of a single crack under mode-I loading conditions. For this purpose we rewrite (19) in the form

$$q - p = \frac{\alpha - \alpha_r}{\beta + \beta_r}\theta - \frac{\beta}{\beta + \beta_r}p + \frac{\gamma}{\beta + \beta_r}(\mu - \mu_0), \tag{20}$$

where the excess opening pressure on the inclusion (as compared with the opening pressure $P_0 + n_i\sigma^0_{ij}n_j$ in the reference state) is now shown explicitly to result from thermal and phase transformation expansion effects which are counteracted by the effect of an increase p in the external load.

A build-up in inclusion opening pressure may eventually lead to crack propagation. The crack propagation criterion which we adopt here from linear elastic fracture mechanics (see, for example, Atkinson & Meredith[15], Chap.1) requires that the *stress intensity factor* K_I, generated by the inclusion opening pressure at the crack tips be equal to the critical value K_{Ic}, i.e.

$$K_I = (P + n_i\sigma_{ij}n_j)\sqrt{\pi l} = K_{Ic}. \tag{21}$$

In order to exploit this condition in the simplest manner, let us now chose the reference state $(P_0, \sigma^0_{ij}, T_0)$ such that

$$P_0 = -n_i\sigma^0_{ij}n_j. \tag{22}$$

This zero effective stress condition marks the point at which the kerogen becomes fully "load-bearing". The criterion (21) for the onset of pore pressure induced microfracturing thus becomes

$$K_I = (q - p)\sqrt{\pi l} = K_{Ic}. \tag{23}$$

The value of the crack opening pressure $q - p$ determined by this condition must be attained by the right-hand-side of equation (20) before crack growth can occur. Thus, the equality

$$(\alpha - \alpha_r)\theta - \beta p + \gamma(\mu - \mu_0) = (\beta + \beta_r)K_{Ic}/\sqrt{\pi l} \tag{24}$$

must be satisfied continuously during crack extension. The function on the right-hand-side is seen to possess a critical point (minimum) at $l = l^*$, which is determined by the condition

$$\beta_r(l^*) = \frac{2\pi(1-\nu^2)l^{*2}}{EV_0} = \frac{1}{3}\beta. \tag{25}$$

We use l^* together with (18) and (25) to normalize the function $(\beta + \beta_r)K_{Ic}/\sqrt{\pi l}$ and rewrite equation (24) in dimensionless form as

$$\theta' - p' + \mu' = (l/l^*)^{-1/2} + \frac{1}{3}(l/l^*)^{3/2}, \tag{26}$$

$$\theta' = \frac{\alpha - \alpha_r}{\beta}\frac{\sqrt{\pi l^*}}{K_{Ic}}\theta; \quad p' = \frac{\sqrt{\pi l^*}}{K_{Ic}}p; \quad \mu' = \frac{\gamma}{\beta}\frac{\sqrt{\pi l^*}}{K_{Ic}}(\mu - \mu_0). \tag{27}$$

Figure 2 shows a plot of the (normalized) "load" $\theta' - p' + \mu'$ as a function of l/l^*, which attains the minimum value of 4/3 at $l/l^* = 1$. The existence of this minimum implies that a crack of initial length $2l_0$ will support an increasing load up to the value $(l_0/l^*)^{-1/2} + (1/3)(l_0/l^*)^{3/2}$. As soon as this value of the load is reached, crack extension will proceed stably (demanding continuously increasing loads), if $l_0 > l^*$, but unstably, if $l_0 < l^*$.

As may be seen from Figure 2, the latter case arises, because the equilibrium crack length is a decreasing function of the load whenever $l_0 < l^*$. Such a crack, when kept under a constant load $\theta' - p' + \mu'$, will therefore grow spontaneously from l_0/l^* at point A to a stable length ratio l_0'/l^* at point A'. In the former case, when $l_0 > l^*$, crack extension follows the stable branch of relation (26) from point B upwards. The physical reason for this behaviour lies in two opposing effects of crack extension upon the value of the stress intensity factor. The first is the $l^{1/2}$ proportionality of K_I at fixed crack opening pressure $q - p$, as seen from (23); acting alone, this dependence would invariably lead to unstable crack propagation. However, a second opposing effect stems from the l^2 proportionality of β_r which tends to reduce the crack opening pressure and thereby the stress intensity factor. According to (20), this effect will amount to a lowering of the inclusion pressure q, if θ, $-p$, and $\mu - \mu_0$ are kept constant, thereby allowing for the necessary expansion of the kerogen and hydrocarbon inclusion phases during crack growth. As long as β_r is smaller than $(1/3)\beta$, the first effect outweighs the second, but beyond this value of β_r the pressure in the inclusion phase drops steeply enough with increasing crack length to stabilize crack extension.

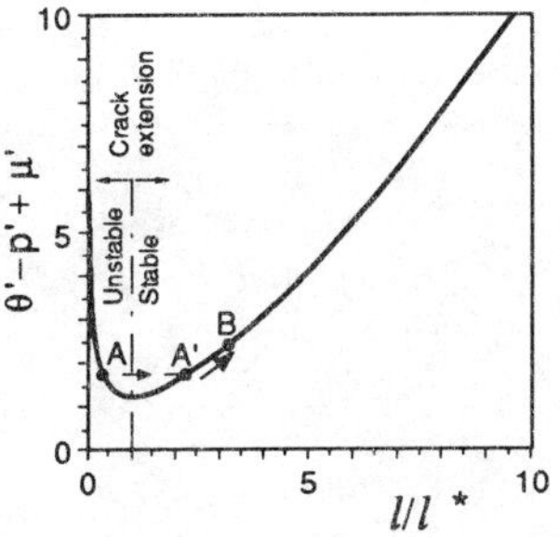

Fig. 2: Normalized load vs. reduced crack length (Eq. 26).

To assess the importance of spontaneous, unstable crack extension in practice, we substitute the expression $V_0 = 2l_0^2/\xi_0$ for the initial volume of a crack into (25), to obtain

$$l_0/l^* = \sqrt{3\pi(1-\nu^2)\xi_0/\beta E} \tag{28}$$

Here $\xi_0 = l_0/h_0$ denotes an initial aspect ratio of the inclusion, $2h_0$ being its initial mean cross-sectional width. The product βE will vary considerably with the amount and composition of the hydrocarbons generated, with the organic content of the source rock and with the nature (lithology) of its inorganic mineral matrix (i.e. shale vs. carbonate). Thus, Young's modulus E might range between 2×10^9 Pa (for a rich shale) and 5×10^{10} Pa (for lean carbonates), while β, according to definition (14), will vary between its extreme values at $\mu = 1$ and $\mu = \mu_0$. If $\mu_0 = 0$, then for the higher molecular weight (liquid) hydrocarbon components the value of β at $\mu = 1$ will be of the order of 5×10^{-9} Pa, while for methane at 40 MPa pressure and 100°C the same quantity is approximately 25 times larger. It follows that the product βE may vary over three orders of magnitude. For an aspect ratio $\xi_0 = 10$, which may be viewed a setting a lower limit for "crack-like" inclusions, the ratio l_0/l^*may therefore attain values somewhere between 1/4 and 10. From this it can be concluded that unstable crack extension may indeed be expected to occur.

4. CRACK GROWTH IN TIME. EXAMPLES

Consider now the case of a kerogen inclusion subjected to a burial history which is characterized by a constant (possibly vanishing) rate of temperature increase $\dot{T}$. The left member of equation (28) will then change with time, not only because μ' will increase, but also because θ' and p' will in general increase with continuing burial. For simplicity, we shall take $p = \eta\bar{\rho}g\dot{T}t/T_z$, where η is a stress ratio expressing the crack closing pressure as a fraction of the overburden, $\bar{\rho}g$ denotes a thickness-averaged unit weight of the overburden, T_z the geothermal gradient, and t the time elapsed during continuous burial counted from the reference state (P_0, T_0, μ_0) to which (24) applies. Also, $\theta = \dot{T}t$ and it remains to determine μ'. The quantity $\theta' - p' + \mu'$ may then be calculated from the definitions (29) and can be used together with (30) to predict the change of the fractional crack length l/l_0 with time, beginning with the instant at which the load $\theta' - p' + \mu'$ has reached the critical value $(l_0/l^*)^{-1/2} + (1/3)(l_0/l^*)^{3/2}$.

To obtain μ we shall assume here that the kerogen transformation may be characterized quantitatively by a single first order bulk reaction, its rate being given by

$$\frac{d\mu}{dt} = k(1-\mu). \tag{29}$$

Here k satisfies the Arrhenius relation

$$k = Ae^{-\frac{Q}{RT}} \tag{30}$$

in which frequency factor A and effective activation energy Q are assumed to be constants. $R = 8.31$ J mol^{-1}K^{-1} is the universal gas constant. Putting $T = T_0 + \dot{T}t$, ($\dot{T}$ = const.) and integrating (29) by parts, we obtain

$$ln\frac{1-\mu}{1-\mu_0} = -\frac{AT_0}{\dot{T}}x_0\Big\{\frac{e^{-x}}{x} - \frac{e^{-x_0}}{x_0} - Ei(x) + Ei(x_0)\Big\}.$$

with $x = Q/RT$ and $x_0 = Q/RT_0$. The exponential integrals appearing on the right-hand-side have the series representation $Ei(x) \equiv \int_x^\infty e^{-\xi}d\xi/\xi = (e^{-x}/x)(1 - 1/x + 2!/x^2 - 3!/x^3 + ...)$. For $T_0 = 373$ K and $Q = 2.15 \times 10^5$ J mol^{-1} one finds $x_0 = 69.36$ and for x-values of this magnitude only the first two terms in the series need to be taken into account. Hence, in terms of the original variables,

$$\frac{\mu-\mu_0}{1-\mu_0} = 1 - exp\Big\{-\frac{RT_0}{Q}e^{-\frac{Q}{RT_0}}\frac{AT_0}{\dot{T}}\Big[\Big(1+\frac{\dot{T}t}{T_0}\Big)^2 e^{\frac{Q}{RT_0}\frac{\dot{T}t/T_0}{1+\dot{T}t/T_0}} - 1\Big]\Big\}. \tag{31a}$$

In the limit $\dot{T} \to 0$, when $T = T_0$ is constant, this reduces to

$$\frac{\mu-\mu_0}{1-\mu_0} = 1 - exp\{-Ae^{-\frac{Q}{RT_0}}t\}, \tag{31b}$$

a result, which also follows directly from (29) and (30).

Given now μ as a function of time, the left-hand member of equation (26) can be calculated, using definitions (27) and the above expressions for p and θ. Thus, after some rearrangement, one can write

$$\theta' - p' + \mu' = \Big[\frac{\gamma}{\beta}(\mu-\mu_0) - \Big(\frac{\eta\bar{\rho}g}{T_z} - \frac{\alpha-\alpha_r}{\beta}\Big)\dot{T}t\Big]\frac{\sqrt{\pi l_0}}{K_{Ic}}\Big(\frac{l_0}{l^*}\Big)^{-1/2}, \tag{32}$$

where l_0/l^* and $\mu - \mu_0$ are given by (28) and (31) respectively, and α, β and γ are allowed to depend upon μ according to the definitions (13) – (15). It is seen that the normalized load $\theta' - p' + \mu'$ will either increase monotonically with time or attain a maximum depending on the sign of the coefficient of t. As long as $\dot{T} > 0$, i.e., during burial, the sign of this coefficient reflects the relative importance of the crack closure force by gravitational loading over the crack opening force by thermal expansion. In the majority of situations, the former tends to outweigh the latter, so that the load $\theta' - p' + \mu'$ will indeed reach a positive maximum at a certain time. Crack propagation in this case results entirely from the gain in specific volume of the kerogen/bitumen mixture. Of course, for crack extension to become possible at all, the load $\theta' - p' + \mu'$ must at some time attain the level required by the constraint relation (26), as plotted in Figure 2. The evolution of the crack length with time may then be obtained by eliminating $\theta' - p' + \mu'$ from (26) and (32) with the aid of relation (28). To illustrate this, we shall now consider some examples for the following parameter values:

$A = 10^{13}\text{sec}^{-1}$; $Q = 2.15 \times 10^5$ J mol^{-1}; $T_0 = 373$ K; $E = 10^{10}$Pa; $\nu = 0.25$; $K_{Ic} = 10^6$ Pa m$^{1/2}$; $\xi_0 = 10$; $l_0 = 0.005$ m; $\alpha_f = 10^{-3}$ °C^{-1}; $\alpha_k = 5 \times 10^{-4}$ °C^{-1}; $\alpha_c = \alpha_r = 10^{-4}$ °C^{-1}; $\beta_f = 3 \times 10^{-9}$ Pa^{-1}; $\beta_k = 5 \times 10^{-10}$ Pa^{-1}; $\beta_c = 2.5 \times 10^{-10}$ Pa^{-1}; $\rho_{k0}/\rho_{f0} = 1.43$; $\rho_{k0}/\rho_{c0} = 0.77$; $r_{cf} = 0, \frac{1}{3}$; $\bar{\rho} g = 2.6 \times 10^4$ Pa m^{-1}; $\eta = 1$; $T_z = 3.10^{-2}$ °C m^{-1}.

Figure 3 shows the calculated dimensionless crack length as a function of burial depth $z - z_0 = \dot{T}t/T_z$, as measured from the reference z_0 (at which $\theta' - p' + \mu' = 0$), for two different heating rates, for $\mu_0 = 0$, and for two values of the mass fraction r_{cf} of coke to coke plus fluid, the latter being taken constant for simplicity. The depth below z_0 at which l/l_0 starts to grow from its initial value 1 corresponds to the point at which a graph of the function $\theta'(l/l^*) - p'(l/l^*) + \mu'(l/l^*)$ meets the ascending branch of the critical curve plotted in Figure 2 (e.g. at A'). Crack extension is seen to proceed stably in this example up to a maximum length, corresponding to the completion of the kerogen transformation. Beyond that point, the crack opening force will decline under the influence of the growing overburden. This is reflected by the decline in stable crack length shown in Figure 3, which however should be viewed as a theoretical artifact resulting from a reversibility property built into the basic crack model. The features of practical interest are the time at which crack growth commences, the period of growth up to a maximum crack length, and the magnitude of the latter. In the example of Figure 3 the heating (or burial) rate is seen to affect essentially only the timing of crack growth, while the magnitude attained by l/l_0 depends strongly on the mass fraction r_{cf}.

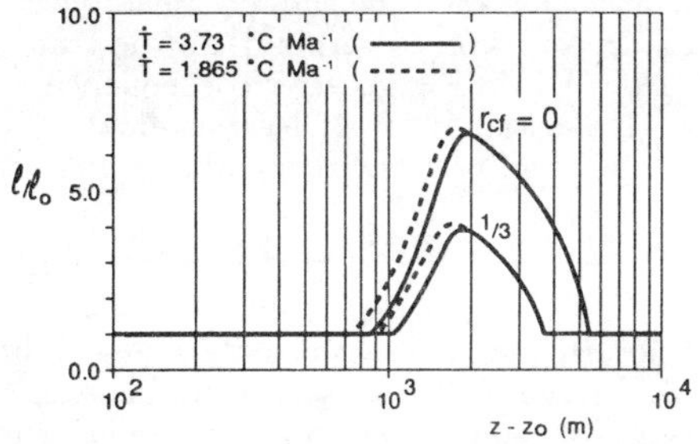

Fig. 3: Evolution of crack length with depth of burial beyond initial depth z_0 (corresponding to zero effective opening pressure at $T_0 = 373$ K); for two different heatig (burial) rates and coke fractions r_{cf}.

A different behaviour is found for the examples shown in Figure 4. Here the initial state of zero crack opening (or closing) pressure was assumed to have been attained, after an initial episode of burial and kerogen transformation, at $T_0 = 403$ K and $\mu_0 = 0.3$; no further subsidence was assumed to occur thereafter, so that Figure 4 traces the evolution of crack length with time measured from that initial state at a constant temperature T_0. When all other parameters are given the same values as in the previous example, the crack is predicted to extend stably to about eleven times its initial length (case A). The process starts earlier, but proceeds at a much slower rate than in the example of Figure 3. The monotonic increase in crack length reflects the same behaviour of the crack opening force in the absence of further burial. However, this smooth behaviour is altered drastically in case B for values for E and β_f that correspond to a more competent source rock generating a lighter mix of hydrocarbons. As is evident from Figure 4, the onset of crack extension is delayed in case B with respect to that of case A by the increase in E and β_f; but once the critical load is reached, the crack will at first extend spontaneously, tripling its length, and continue with stable growth thereafter to attain a maximum length equal to about seven times its initial length (for $r_{cf} = 0$). The total duration of this growth period is the same as in the previous case, being governed solely by the rate of kerogen transformation. The sudden, unstable crack extension corresponds to the transition AA' in Figure 2. A consequence of this instability that deserves further study could be a relatively abrupt increase in source rock permeability.

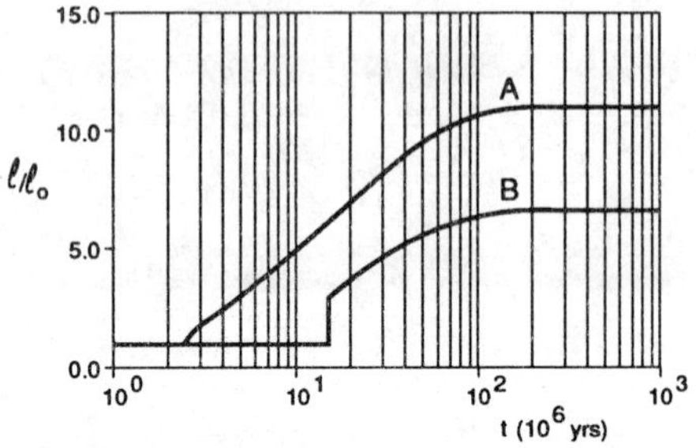

Fig. 4: Continuous vs. discontinuous crack growth at $T_0 = 403$ K from initial state of zero effective opening pressure, $\mu_0 = 0.3$, and $r_{cf} = 0$; effect of Young's modulus and fluid compressibility:
A: $E = 10^{10}$ Pa, $\beta_f = 3 \times 10^{-9}$ Pa^{-1};
B: $E = 5 \times 10^{10}$ Pa, $\beta_f = 8 \times 10^{-9}$ Pa^{-1}.

5. CONCLUDING DISCUSSION

This paper has dealt with only one aspect of a theory of pore-pressure-induced fracturing of source rocks, i.e., the conditions governing the growth of a single crack-like kerogen/bitumen inclusion; bedding-plane parallel growth of such cracks has been emphasized as an essential prerequisite to the drainage of the organic-rich laminae of a source rock. In order to gain insight into the basic mechanism through some explicit results, several simplifying model assumptions on the material behaviour (linear equations of state, linear elastic fracture mechanics) and the kinetics of the kerogen transformation (single first-order bulk reaction) had to be made. It could be established in this manner that maturing kerogen flakes in the mm size range, viewed as crack-like kerogen/bitumen inclusions, may be expected to grow up to a multiple of their size either in a stable, gradual fashion or—especially in competent gas-prone source rocks—as an unstable, sudden event.

In most cases of continuing burial the kerogen transformation appears to be the principal source of excess fluid pressures; the rate of gradual crack extension will then be controlled by the progression of the breakdown reaction. During the initiation phase of this process, fluid pressures are predicted to rise significantly above lithostatic values by a microfracturing model that assumes linear elastic fracture mechanics to be applicable. Thus, for the examples of Figures 3 & 4 the pressure in excess of the lithostatic value in the reference state can be determined directly from relation (23) in terms of the current crack length and the facture toughness K_{Ic}; for curve A of Figure 4, for example, this yields a fluid pressure of 8×10^6 Pa above lithostatic at the onset of crack growth, which would drop to about one third of this value as a result of the subsequent crack extension. Although these hypothetical fluid pressures pertain to *undrained* conditions prior to the establishment of source rock permeability and therefore cannot be verified directly by in situ measurements, the present analysis suggests that they could nevertheless account for the generation of the source rock permeability necessary to accomplish the expulsion of hydrocarbons along with the dissipation of excess pressures.

We have not discussed the fracture behaviour of mature source rocks during episodes of relatively rapid tectonic uplift, when the removal of overburden will lead to a corresponding increase in the effective crack opening pressure while the kerogen transformation may be virtually arrested. Nevertheless, the interesting possibility of the generation of unloading fractures in a source rock during such an episode could be fully explored with the aid of equation (20) and condition (21) or (23), assuming the temperature and erosion histories to be given.

Further theoretical studies of source rock fracture processes are needed, but will require better quantitative information on the PVT properties of the mixture phase consisting of kerogen and its breakdown products. Finally, there is a clear need for some direct laboratory studies of pore-pressure-induced fracturing of petroleum source rocks, as there is for more field data on in-situ stresses and fluid pressures in source rock sequences.

REFERENCES

(1) Snarsky, A.N. (1970). The nature of primary oil migration. Neft Gaz, Vol. 13/8, 11–15 (Transl. by Assoc. Technical Serv. Inc.; with references to earlier work by the author).

(2) Tissot, B. and Pélet, R. (1971). Nouvelles donneés sur les mechanismes de génese et de migration du pétrole: Simulation mathematique et application a la prospection. Proc. 8th World Petr. Congr., Vol. 2, 35–46.

(3) Momper, J.A. (1978). Oil migration limitations suggested by geological and geochemical considerations. In "Physical and Chemical Controls on Petroleum Migration": AAPG Continuing Education Course Note Series, No. 8, B1–B60. Am. Ass. Petrol. Geol., Tulsa.

(4) Meissner, F.F. (1978). Petroleum geology of the Bakken Formation Williston Basin, North Dakota and Montana. 24th Ann. Conf. of the Montana Geological Society, Billings, Montana.

(5) Du Rouchet, J. (1981), Stress fields, a key to oil migration. AAPG Bull., Vol. 65/1, 74–85.

(6) Mandl, G. and Harkness, R.M. (1987). Hydrocarbon migration by hydraulic fracturing. In: Jones, M.E. & Preston, R.M.F. (eds.), "Deformation of sediments and sedimentary rocks". Geol. Soc. Ld. Special Publ. No. 29, 39–53.

(7) Ozkaya, I. (1988). A simple analysis of primary oil migration through oil-propagated fractures. Marine and Petroleum Geology, Vol. 5, 170–174.

(8) Ozkaya, I. (1988). A simple analysis of oil-induced fracturing in sedimentary rocks. Marine and Petroleum Geology, Vol. 5, 293–297.

(9) Tissot, B.P. and D.H. Welte (1984), "Petroleum Formation and Occurrence" 2nd Ed., Springer-Verlag, Berlin etc.

(10) Lewan, M.D. (1987). Petrographic study of primary petroleum migration in the Woodford shale and related rock units. In: B. Doligez (ed.), "Migration of Hydrocarbons in Sedimentary Basins", 113–130. Editions Technip, Paris.

(11) Littke, R. and Rullkötter, J. (1987). Mikroskopische und makroskopische Unterschiede zwischen Profilen unreifen und reifen Posidonienschiefers aus der Hilsmulde. Facies, Vol. 17, 171–180.

(12) Ashby, M.F. and Sammis, C.G. (1990). The damage mechanics of brittle solids in compression. Pure and Applied Geophysics, Vol. 133/3, 489–521.

(13) Skempton, A.W. (1954). The pore-pressure coefficients A and B, Géotechnique, Vol. 4, 143–147.

(14) Rice, J.R. and Cleary, M.P. (1976). Some basic stress diffusion solutions for fluid-saturated elastic porous media with compressible constituents. Rev. Geophys. Space Physics Vol., 14/2, 227–241.

(15) Atkinson, B.K. and Meredith, P.G., (eds.) (1987), "Fracture Mechanics of Rock", Academic Press, London.

A MULTIDISCIPLINARY APPROACH TO PRIMARY MIGRATION PATHWAYS OF PETROLEUM - THE LOWER TOARCIAN OF NW-GERMANY AS A MODEL SOURCE ROCK-

Ulrich Mann

Institute of Petroleum and Organic Geochemistry at the Research Centre (KFA) Jülich, P.O. Box 1913, D-5170 Jülich, Federal Republic of Germany

Summary

Revised exploration concepts have to be applied in well explored basins with major emphasis on excellent data integration.

This paper integrates conceptual modelling considerations for primary migration pathways with a case study from the SE Lower Saxony Basin, field data from logging and laboratory data from petrographical, petrophysical and geochemical analyses in order to demonstrate that from Lower Toarcian Shales the pore network influences the efficiency of petroleum expulsion.

It was found that petroleum expulsion occurred at highest efficiencies via diagenetically formed migration avenues. Therefore, conceptual models for petroleum expulsion should integrate chemical diagenesis in order that the formation of such predominating expulsion pathways can be traced during basin evolution. This will allow a better prediction of the petroleum expelled from a source rock.

1. INTRODUCTION

A generally revised exploration concept has to be applied to recognize and mobilize additional petroleum reserves in already well explored basins (1). There, the classical approach as used in less explored areas has to be changed. A more appropriate and integrated concept has to be applied and deliberately adjusted to the exploration stage. In addition, the concept has to be extremely adaptable to the rapidly changing economic situation in production and sales (Fig. 1). Therefore information from low-cost methods such as landsat imagery, geochemistry, mathematical modelling or probabilistic prospect assessment has to be maximized, whereas all efforts for expensive 3D-seismic processing and especially for drilling should be minimized. However, it then becomes even more essential that all data are fully integrated with each other.

This paper aims to demonstrate the integration of part of the relevant information from geology, petrography, geochemistry petrophysics as well as from concepts for numerical modelling in order to locate and quantify pathways through the pore network of a source rock during the primary migration process of petroleum. This will then allow a better prediction of drainage area, timing, and volume of the expelled petroleum.

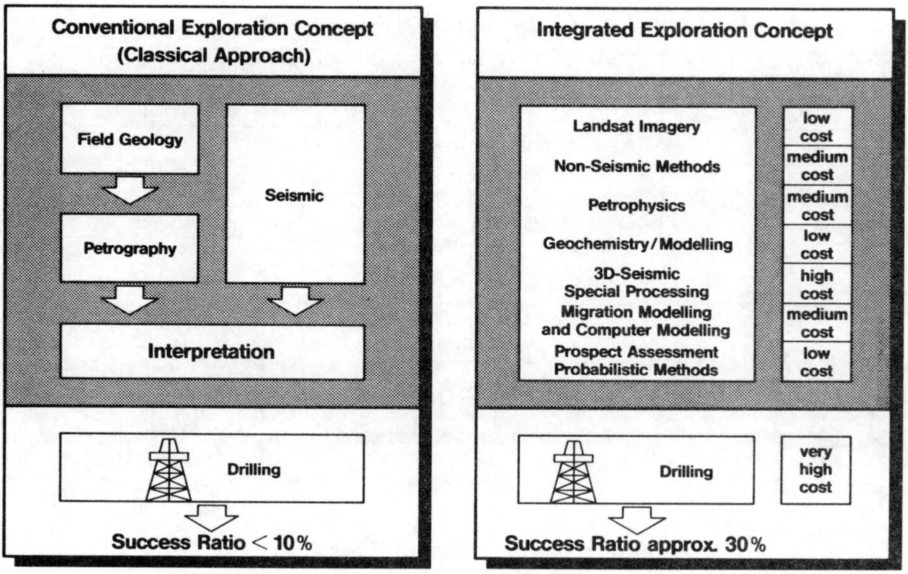

Fig. 1: Conventional vs integrated exploration concept

2. PETROLEUM EXPULSION FROM SOURCE ROCKS

Influencing Parameters and Possible Pathways

Petroleum expulsion from a source rock is influenced by two processes, source rock deposition and basin evolution (2). Each process holds a scenario of quite variable parameters (Fig. 2). Litho- and organofacies as well as texture and bedding of the sedimentary rock are the primary controlling parameters determined by and inherited from the specific depositional environment. Petroleum generation, petroleum composition, chemical diagenesis, the process of fracture formation as well as the large-scale development of three-dimensional compartments with different fluid flow conditions are the secondary controlling parameters during basin evolution. They are governed by the pressure-temperature regime of the subsiding basin and its specific tectonic history.

Several pathways can be considered which enable petroleum expulsion from a source rock (Fig. 3). This includes original attributes of the source rock inherited from the depositional environment during sedimentation, and secondary ones, created during diagenesis and catagenesis. To the original ones belong the kerogen network and a pore system only altered by compaction ("primary pores"). To the second group belong fractures in association with tectonic stress or due to hydrocarbon generation, "secondary pores" altered or newly created by dissolution and/or re-precipitation reactions, and stylolites which form due to re-distributions of both organic and inorganic material in deeply buried carbonate source rocks.

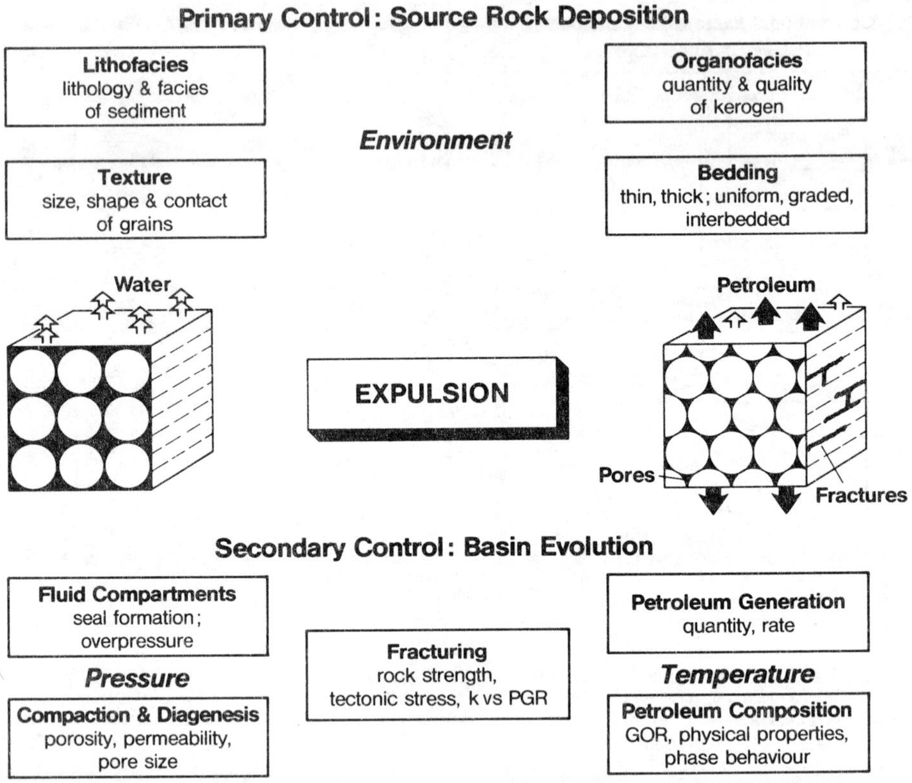

Fig. 2: Primary and secondary controlling parameters for petroleum expulsion

Conceptual Models

To allow all types of pathways (see Fig. 3) to contribute to petroleum expulsion, a source rock has to be regarded as a multifold porous medium. Indeed, it can be compared to an artificial lithology which combines a reservoir rock (porous sandstone or limestone) and a coal seam, and which could be fractured. Hence, permeability of a source rock should be described as the total permeability k_T through the kerogen, pore and fracture networks:

$$k_T = k_{Kerogen} + k_{Pores} + k_{Fractures}$$

Permeability through a fracture network with laminar flow is proportional to fracture width (3). However, no single mathematical equation can exist to calculate the permeability through a source rock because two overlapping transport processes participate: laminar flow which dominates in the macropore size-range, and diffusion which dominates in the micropore size-range. Accordingly, individual pore volumes

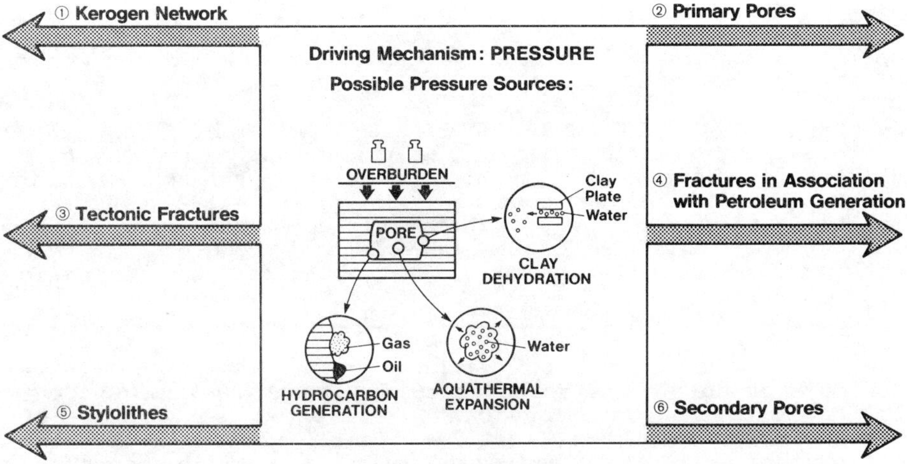

Fig. 3: Possible pathways for petroleum expulsion from source rocks.

contributing to the two different transport mechanisms have to be differentiated which can be achieved with an empirical cut-off value in the mesopore size-range. Similar to the micropores, diffusion should be the predominant petroleum transport mechanism via the kerogen network.

The efficiencies of kerogen, pores and fractures to act as petroleum expulsion pathways can be rated as follows:

$$E_{Kerogen} << E_{Pores} << E_{fractures}$$

This results directly from the type of processes involved and from the wide range of permeabilities of source rocks over more than 6 orders of magnitude from about 10^{-18} to 10^{-24} m^2 (4). As a consequence first numerical expulsion models consider the pore and fracture network (5,6), and neglect the kerogen path as being inefficient. However, in certain sedimentary basins with geologic parameters like an extremely tight source rock, low basal heat flow, with slow heat conduction, slow subsidence and tectonic quiescence, a transport process like activated diffusion of dissolved bitumen molecules through the kerogen network (7) may become quantitatively important. The same concerns the diffusion part of petroleum through the meso- and micropore network.

Fluid-flow is generally quantified by an extension of the well-known Dary law (8) to two-phase fluid-flow which relates relative permeabilities of petroleum and water to petroleum saturation of the source rock, a concept commonly applied in reservoir engineering. However, this law can strictly be applied only for the macropore network. Furthermore various electrokinetic forces become operative during the flow of water through the clayey sediments, because of the presence of fixed and mobile double layers, and invalidate Darcy's equation. Osmosis, reverse osmosis, and filtration through clay membranes further complicates the picture (9). Therefore, it is of primary importance to learn about petroleum expulsion effects in respect to the pore networks of the source rocks.

3. CASE HISTORY: THE LOWER TOARCIAN OF NW-GERMANY

To evaluate the effects of a single pathway like the pore network on petroleum expulsion requires that many other parameters which also control this process are constant. Usually this general scientific approach cannot be consistently followed in the geosciences, however, in this case history we were able to optimize our scenario considerably.

Ideal Setting

Effects of permeability differences within a single source rock unit on petroleum expulsion should become evident if the following conditions can be fulfilled: early phase of petroleum expulsion, no fractures, fairly uniform litho- and organofracies. At an early phase of petroleum expulsion (=early stage of petroleum generation = early mature stage of organic matter) differences in petroleum expulsion should be most pronounced, because even small differences of capillary resistance, as defined by the pore throats of the rock, may allow expulsion within one specific rock interval, but not from another. Fractures should not be present, because otherwise they would be the predominating pathway and not the pores. Therefore, the case study has to be from an area, with no or low tectonic stress, and with a time-temperature evolution suitable for a low petroleum generation rate which could not have caused fracturing. Lithofacies should be more or less the same throughout the source rock to exclude possible catalytic mineral matrix effects. Organofacies should be uniform, too, because otherwise differences in the petroleum generation potential could be mistaken for differences in the expulsion volume.

Among several boreholes specifically drilled for scientific research purposes (10) borehole Dielmissen in the Hils syncline at the south rim of the Lower Saxony Basin in NW-GERMANY is unique and provides these ideal conditions. The maturity stage is 0.68 % vitrinite reflectance (11). Vertical tectonic fractures are present only very sparsely (12). Horizontal fractures in association with hydrocarbon generation occur much less frequently than within the adjacent boreholes with higher heat-flow regimes (13,14). Lithofacies ranges from calcareous clay-shale (upper interval) to marlstone (lower interval) with occasional limestone beds (15). Variation with depth of the organofacies is also low (14).

Locating Expulsion Pathways through the Pore Network by Logging Techniques

Quite recently (16,17) reported that it is possible to locate petroleum expulsion pathways as "high" porosity zones in the Lower Toarcian, and by low resistivity values due to reduced hydrocarbon saturation. Therefore, rock samples from borehole Dielmissen were analysed for pore volumes in the laboratory. Then pore volume vs. depth profiles were compared to various log traces. It was found that high-porosity intervals fall together with intervals of a reduced uranium content (Fig. 4, borehole A). This can be explained by two ways. First, uranium reduction can take place directly after sediment deposition during bioturbation and/or during bacterial reworking (18). Second, uranium leaching can occur in the course of diagenesis and catagenesis during kerogen maturation due to a weakening of the uranium-organic matter association (19). Nevertheless, in both cases high porosity intervals are favoured for an uranium reduction due to an increased water circulation and therefore a better chance for an ion-exchange. Analysis of a further borehole profile which includes the Lower Toarcian, provided identical results (Fig. 4, borehole B).

Analytical Results

- Potential vs Residual Product

Organic geochemical analyses reveal discrepancies within the profile of borehole Dielmissen between petroleum generation potential and the generated products. The quantity and quality of organic matter does not agree with the transformation ratio (pyrolysis) or the yield of C_{15+}-soluble, organic matter (solvent extraction). Especially the distribution of total organic carbon as well as of the hydrogen index within the borehole profile does not reflect the two-fold division of the transformation ratio and of the C_{15+}-soluble, organic matter content (Fig. 5). Furthermore, this bipartition is in no way identical with the upper clay-shale and lower marlstone interval (cf. boundary of lithofacies, Fig. 5). The conclusion at this point is that the lower source rock interval between 60 and 70 meters depth has expelled much more hydrocarbons that the upper interval between 40 and 60 meters.

According to (20), no catalytic effect for petroleum expulsion has to be assumed within the lower source rock interval because the type of minerals in the matrix is the same throughout the total thickness of the source rock, only their relative amounts vary (15). Petrophysical analyses of the micro-, meso-, and macropores show, however, that petroleum depletion is greater in those source rock samples which possess a considerable number of pore throats with a radius of 5 nm or greater.

Petrographical criteria classify these source rock samples as three diagenetically different facies of marlstones: an unmodified marlstone facies, a partly carbonate-cemented marlstone facies representing today a limestone, and a marlstone facies with carbonate dissolution, representing today a clay-shale (21). All diagenetic carbonate re-distribution effects were determined qualitatively by changes of texture and particle morphology via SEM-studies (Fig. 6), and quantitatively by changes of carbonate content and pore volumes.

- Residual Product vs Lithological and Petrophysical Criteria

Representative samples from the clay-shale, the unmodified marlstone, the cemented marlstone, and the

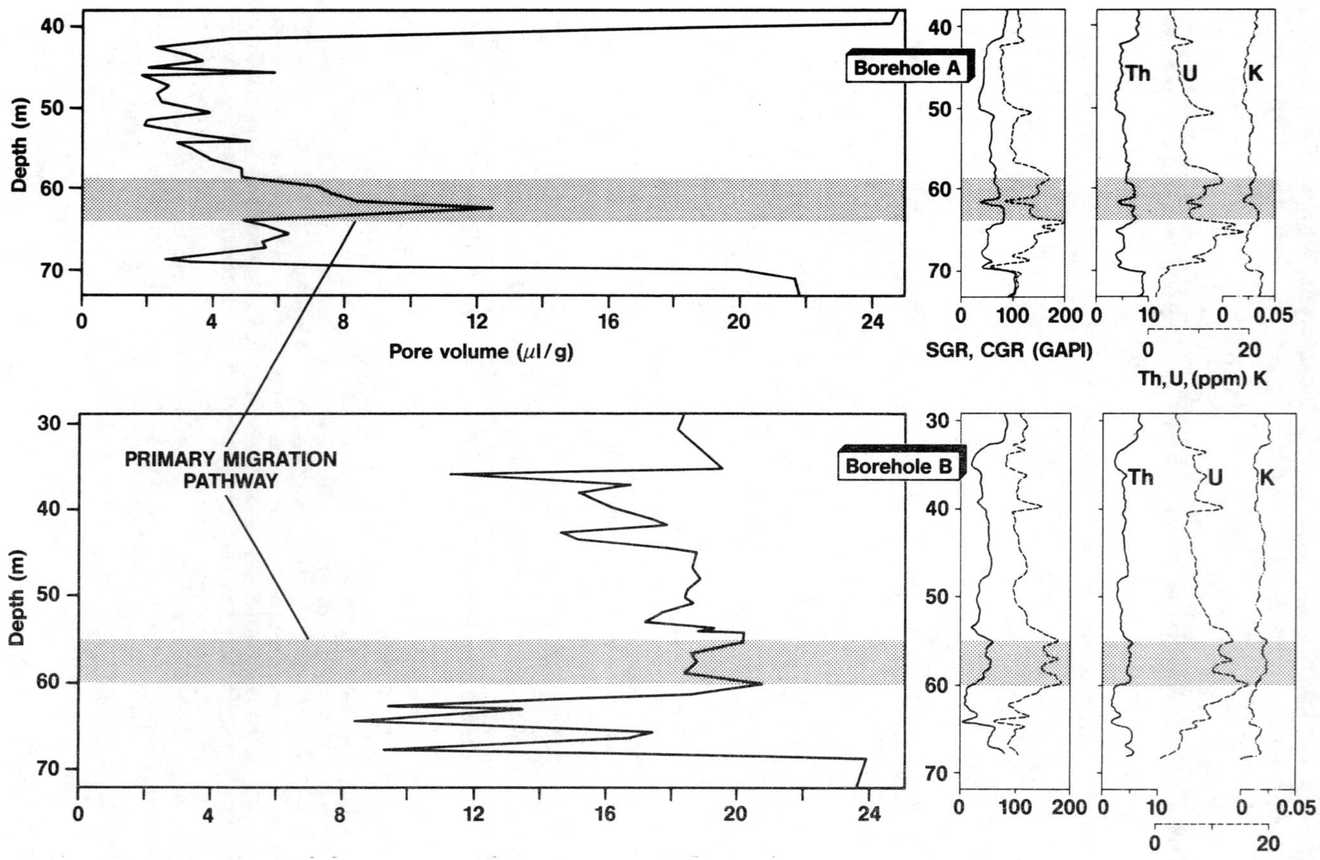

Fig. 4: Locating primary migration pathways from well-log data.

marlstone with secondary porosity were investigated in detail. Three selected source rock properties which give an indication of the direct influence of chemical diagenesis on petroleum expulsion are numerically presented in Figure 7: concentrations of the C_{15-27}-n-alkanes, specific surface area, and cummulative pore volume vs pore throat equivalents. Concentrations of the C_{15-27}-n-alkanes from the clay-shale facies were used as reference and compared to the concentrations of the diagenetically different marlstones (Fig. 7, upper line). For the same samples expulsion efficiencies were calculated relative to the clay-shale, but based on data normalized to organic carbon (Tab. 1).

	CLAY SHALE FACIES	MARLSTONE FACIES Unmodified	Cemented	Secondary porosity
Dominant Pore Radius(nm)	<2	6	10	20
pristane	0	19.3	(47.4)*	24.6
phytane	0	23.3	(50.9)*	24.2
C_{27}-n-alkane	0	36.5	50.4	64.5
C_{15+}-SOM	0	53.8	64.5	75.9
C_{15}-n-alkane	0	54.5	71.6	81.4

* probably influenced by early diagenetic transformation of organic to carbonate carbon

Tab. 1: Expulsion efficiencies (%) of the marlstone facies relative to the clay-shale facies.

These numbers clearly indicate the preference for expelling molecules of shorter chain length at the early expulsion stage which has also been claimed by other authors (for example 22,23). A comparison of the isoprenoids like pristane and phytane with the C_{15-27}-n-alkanes reveal that they are expelled with a lower efficiency. Nevertheless, the expelled quantity of all types of hydrocarbons follows the magnitude of available pathways through the pore network. The availability of such pathways is expressed by the pore volumes and by the size of the pore throats present in the rocks of the different facies (Fig. 7, lower line). A second observation concerns the process of primary migration itself. By determining separate pore volumes for the micro-, meso-, and macropores, and this twice before and after extraction, very similar residual C_{15+}-concentrations were found for all lithofacies in the micropore size range. However, concentrations in the macropores were inversely correlated to amount and size of the macropores present in the individual lithofacies. This supports the idea that primary migration progresses from the smaller to the larger pores, but expulsion takes place primarily form the large pores. At the same time, this observation implies that petroleum expulsion through the pore network can generally be best detected by analyzing the content of macropores. Furthermore, this provided the base for a conceptual model to describe

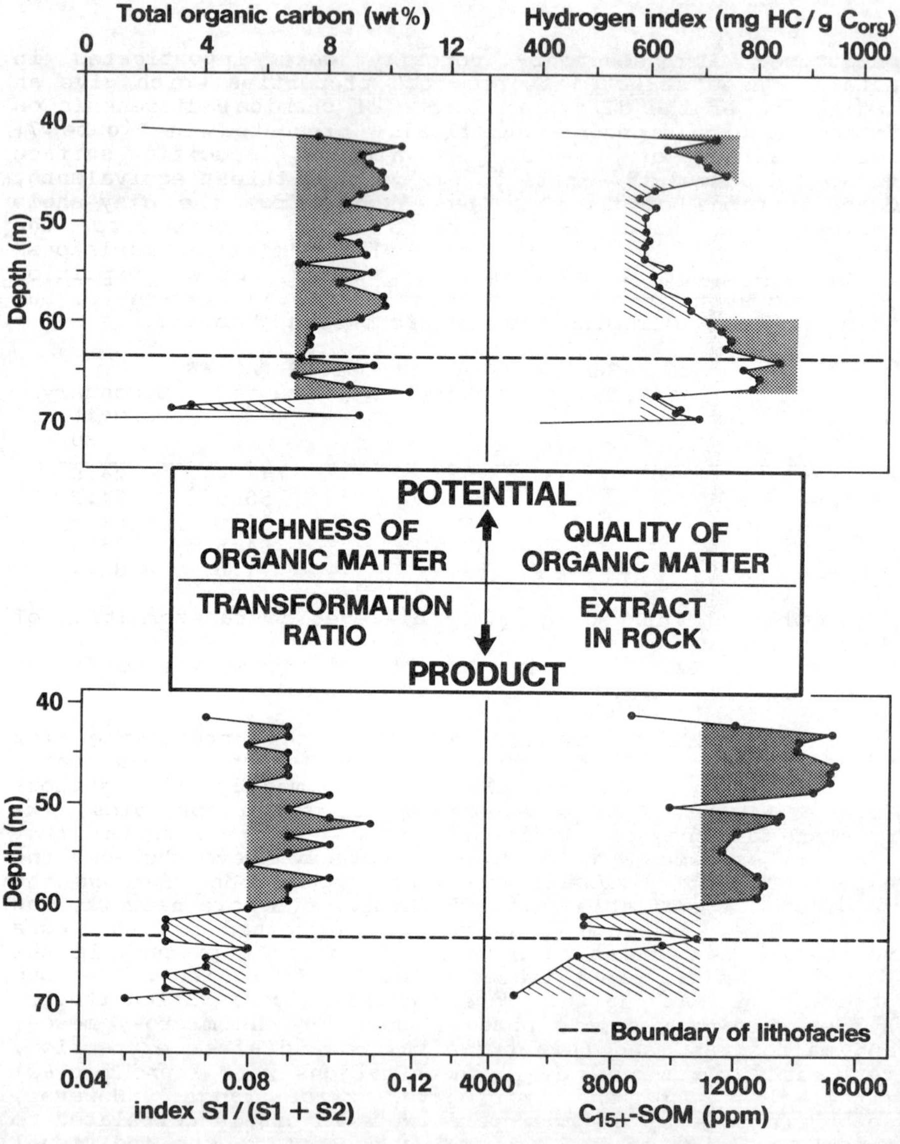

Fig. 5: Petroleum potential vs residual product of Lower Toarcian shales.

quantitatively petroleum expulsion through the pore network (and potential fractures) during the primary migration process (24).

Another interesting observation concerns the wetting conditions within the source rock for the different

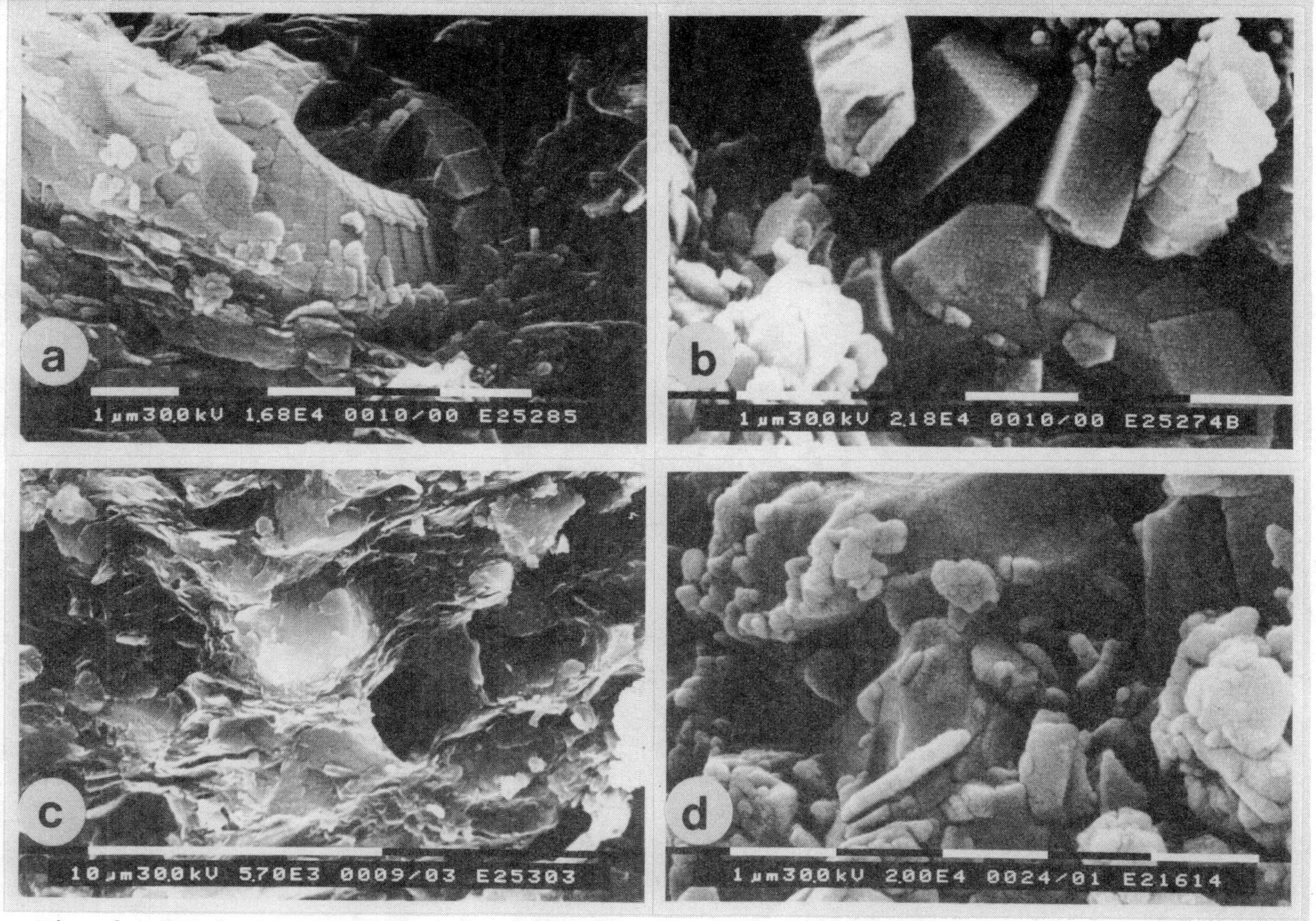

Fig. 6: SEM-photomicrographs of carbonate particle morphology, depiciting original (a), re-crystallized (b), partly dissolved (c), and cemented (d) stage.

lithofacies. The increase of specific surface area after exhaustive solvent extraction by a factor of about two for all lithotyps of the marlstone facies, and by a factor of about five for the clay-shale facies (see Fig. 7, middle line, for individual values) indicates that the specific surface area of the clay-shale is much more covered by C_{15+}-soluble, organic matter, i.e. oil-wet than the marlstone facies. Similar to reservoir conditions, where water-wet rocks allow for better oil-recovery, a better drainage should occur in the marlstone than in the clay-shale. This could already be confirmed by the residual C_{15+}-concentrations as revealed by the detailed organic-geochemical analyses (see above).

4. CONCLUSIONS AND PERSPECTIVE

The efficiency of early petroleum expulsion from the Lower Toarcian source rock is tied to the permeability of the different lithofacies as expressed by different pore volumes and pore radii. Petroleum expulsion occurred at higher efficiency in the marlstone than in the clay-shale. It occurred at highest efficiency via diagenetically formed migration avenues. These migration avenues can be located as uranium-minimum zones on the natural gamma-ray log, and by this way considered for the drainage direction.

Due to the predominance of those migration avenues during petroleum expulsion, more appropriate conceptual models should integrate chemical diagenesis. Thereby, also a possible enlargement of the pore network can be considered, and not only the general reduction due to compaction. In order to trace the formation of such pathways during basin evolution, fluid flow and chemical reactions have to be coupled. This will then allow a better prediction of timing and volume of the expelled petroleum.

Acknowledgement

I thank H.S. Poelchau for reviewing the first draft of the manuscript and all other colleagues form KFA/ICH-5 for organic-geochemical analyses. Typing of the manuscript by M. Sostmann is greatfully acknowledged. I am indebted to BEB, Erdgas und Erdöl GmbH, Hannover and to Wintershall AG, Kassel for advice and for providing the funds for logging, and to D.H. Welte for continuous interest and support.

References

(1) BLOHM, M. (1990). Integrated hydrocarbon exploration concepts in the sedimentary basins of West Germany. In: Heling, D., Rothe, P., Förstner, U. and Stoffers, P. (eds.) Sediments and environmental geochemistry. Springer, 134-151.

(2) WELTE, D.H. (1987). Migration of hydrocarbons, facts and theory. In: Doligez, B. (eds.) Migration of hydrocarbons in sedimentary basins. Technip, Paris, 393-413.

(3) NGHIEM, L.X., FORSYTH Jr., P.A. and BEHIE, A. (1984). A fully implicit fracture model. J. Petrol. Technol., 1191-1189.

(4) BRACE, W.F. (1980). Permeability of crystalline and argillaceous rocks. Int. J. Mech. Min. Sci. & Geomech. Abstr. 17: 241-251.

(5) UNGERER, P., BURRUS, J., DOLIGEZ, B., CHÉNET, P.Y. and BESSIS, F. (1990). Basin evolution by integrated two-dimensional modeling of heat transfer, fluid flow, hydrocarbon generation, and migration. Am. Assoc. Petrol. Geol. Bull. 74: 309-335.

(6) DÜPPENBECKER, S. (1990). Genese und Expulsion von Ko-hlenwasserstoffen in zwei Regionen des Niedersächsischen Beckens unter besonderer Berücksichtigung der Aufheiz-raten. - Ph. D. Thesis RWTH Aachen (in preparation).

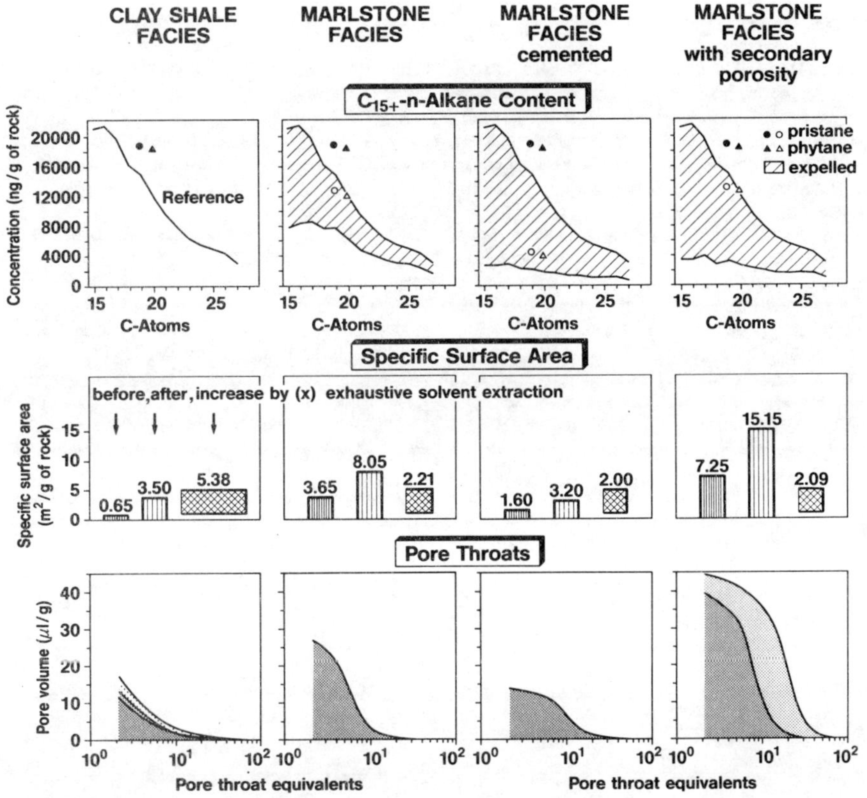

Fig. 7: C_{15+}-n-alkane content, specific surface area and pore size for the different lithofacies of the Lower Toarcian source rock

(7) STAINFORTH, J.G. (1989). Primary migration of hydrocarbons by diffusion through organic matter networks. 14th International Meeting on Organic Geochemistry. Paris, September 1989, abstract no. 54.

(8) DARCY; H.P.G. (1856). Les fontaines publiques de la ville de Dijon. In: Henry Dary - engineer and benefactor of mankind. J. Petrol. Techn., 8 (October 1956), 12-14.

(9) FERTL, W.H. (1976). Abnormal formation pressures. Developments in Petroleum Science, 2. Elsevier Scientific Publishing Company.

(10) MARZI, R. and MANN, U. (1985). Entnahme von Proben für geochemische Untersuchungen aus der Flachbohrung Dielmissen-1001, Hilsmulde 1985. KFA/ICH-5 Interner Bericht Nr. 501785.

(11) LITTKE, R. and RULLKÖTTER, J. (1987). Mikroskopische und makroskopische Unterschiede zwischen Profilen unreifen und reifen Posidonienschiefers aus der Hilsmulde. Facies 17, 171-180.

(12) MANN, U., DÜPPENBECKER, S., LANGEN, A., ROPERTZ, B. and WELTE, D.H. (1989). Evolution of the network of a clastic petroleum source rock during catagenesis (Lower Toarcian Posidonia Shale, Hils Syncline, Northwestern Germany). Am. Ass. Petrol. Bull. 73, 385-386.

(13) JOCHUM, J. (1988). Untersuchung zur Bildung und Karbonatmineralisation von Klüften im Posidonienschiefer (Lias Epsilon) der Hilsmulde. Diplomarbeit, RWTH Achen, 76pp.

(14) LITTKE, R., BAKER, D.R. and LEYTHAEUSER, D. (1988). Microscopic and sedimentologic evidence for the generation and migration of hydrocarbons in Toarcian source rocks of different maturities. Org. Geochem. 13, 549-559.

(15) MANN, U. (1987). Veränderung von Porosität und Porengröße eines Erdölmuttergestein in Annäherung an einen Intrusivkörper. FACIES 17, 181-188.

(16) MANN, U., DÜPPENBECKER, S., LANGEN, A., ROPERTZ, B. and WELTE, D.H. (1989). Petroleum pathways during primary migration: evidence and implications (Lower Toarcian, Hils Syncline, NW-Germany). 14th International Meeting on Organic Geochemistry. Paris, September 1989, abstract no. 128.

(17) MANN, U., DÜPPENBECKER, S., LANGEN, A. ROPERTZ, B. and WELTE, D.H. (1990). Pore network evolution of the Lower Toarcian Posidonia Shale during petroleum generation and expulsion-a multidisciplinary approach. Zentralblatt für Geologie und Paläontologie, Teil I, Schweizerbart (in press).

(18) BAKER, D.R. (1990). Personal communication.

(19) DENHAM, M.E. and TIEH, T.T. (1989). Geochemistry of thorium and uranium during late-stage diagenesis - potential new method fore Exploration. Am. Assoc. Petrol. Geol. Bull. 73, 394.

(20) TANNENBAUM, E., BRADLEY, J.H. and KAPLAN, J.R. (1986): Role of minerals in thermal alteration of organic matter - II: A material balance. Am. Assoc. Petrol. Geol. Bull. 70, 1156-1165.

(21) MANN, U. (1990). Sedimentological and petrophysical aspects of primary petroleum migration pathways. In: Heling, D., Rothe, P., Förstner, U. and Stoffers, P. (eds.). Sediments and environmental geochemistry. Springer, 152-178.

(22) MACKENZIE, A.S., LEYTHAEUSER, D. SCHAEFER, R.G. and BJOROY, M. (1983). Expulsion of petroleum hydrocarbons from shale source rocks. Nature, London 301, 506-509.

(23) LEYTHAEUSER, D., MACKENZIE, A.S., SCHAEFER, R.G. and BJOROY, M. (1984). A novel approach for recognition and quantification of hydrocarbon migration effects in shale-sandstone sequences. Am. Assoc. Petrol. Geol. Bull. 68, 196-219.

(24) DÜPPENBECKER, S., DOHMEN, L. and WELTE, D.H. (1990). Nummerical modelling of petroleum expulsion in two areas of the Lower Saxony Basin, Northern Germany. Proc. Meeting on Petroleum Migration 1989, Geol. Soc. London (submitted).

ULTRASONIC VELOCITY :
A BREAKTHROUGH FOR OIL AND GAS CHARACTERIZATION.

S.YE, B.LAGOURETTE, J.ALLIEZ, H.SAINT-GUIRONS, P.XANS
(L.P.M.I. Université de Pau et des Pays de l'Adour, FRANCE)
and F.MONTEL
(ELF AQUITAINE, CSTJF, PAU, FRANCE)

Summary

Compositional modelling of reservoir fluids has been widely studied during the last fifteen years and a tremendous number of equations of state have been proposed. But little has been done to define the best data set required for characterizing such fluids.

In this paper, we propose to integrate ultrasonic velocity in the usual database in order to increase the predictability of the model.

Ultrasonic velocity measurements have been performed on pure hydrocarbon compounds, binary mixtures and reservoir fluids at different temperatures and pressures. The measurements are much more accurate than for any other thermodynamical property. The experimental results are compared to the values calculated using different thermodynamic models and different mixing rules.

Results so far show that the ultrasonic velocity in synthetic mixtures is a severe test for the equations of state and the associated mixing rules. Furthermore, phase transitions could be determined easily by this method even in the case of liquid/liquid equilibria.

This new approach of reservoir fluid characterization would increase the accuracy of all the thermodynamical properties computed with the model.

1 INTRODUCTION

Sound velocity "U" is related in a simple way to the various coefficients of compressibility, isentropic, isenthalpic and isothermal :

$$\beta_S = \frac{1}{\rho \cdot U^2} \qquad \beta_H = \frac{1}{\rho \cdot U^2} + \frac{\alpha}{C_p} \qquad \beta_T = \frac{1}{\rho \cdot U^2} + \frac{\alpha^2 \cdot T}{C_p}$$

where α : thermal expansivity
ρ : density
C_P : specific heat at constant pressure

These coefficients of compressibility are fundamental for the characterization of the behaviour of reservoir fluids at the various production stages : within the reservoir where phenomena are practically isothermal, at the level of chokes where expansion is isenthalpic and for compressors and turboexpanders where the process is mainly isentropic.

Hence the importance of measuring the acoustic velocities in temperature and pressure conditions encountered by the fluid.

More generally speaking, a coherent thermodynamic model corresponds to the expression of the Helmoltz free energy and the various thermodynamic functions applied in the oil industry are derived from such expression. The identification of the fluid characteristic parameters introduced in modelling is therefore of importance. Matching or calibration is currently referred to ; this critical operation is based on volumetric behavior measurements, or "PVT" measurements. The sound velocity can also be deduced from the model, it thus appears as an useful data to complete the fluid characterization.

Acoustic velocity sensitiveness to the fluid characteristics is also regarded as useful for direct use in the detection of phase transitions in reservoir fluids. As opposed to the visual detection methods, that technique is not subjected to the severe pressure limitations resulting from the use of a transparent material. It can also be more easily automated, which is a considerable asset to the global reduction of cost for the PVT measurements. In this paper we also show that this method can be extended to the liquid/liquid phase transitions, for which it is much more reliable than the visual method, especially in the case of heavy fluids.

Many works performed so far tend to substitute the measurement of acoustic velocity as a function of pressure for those of other thermodynamic properties more critical or less accurate. For example, from the theory developed by GORDON and DAVIES [1], MURINGER ET al. [2] derived various thermodynamic properties for toluene and n-heptane from the measurement of ultrasonic velocity and so deed SUN et al. [3] for cyclohexane and benzene. MILLS et al.[4, 5] combine experimental data of ultrasonic velocity and density to define an equation of state from which they derive several thermodynamic variables; WANG et al. [6] mention that the detection of liquid/vapour phase transitions in oils could be possible from ultrasonic velocity measurements.

The importance of these works, still marginal, have not thus far been recognized by the oil specialists and pushed them into completing PVT data with sound velocity measurements.

2 SUMMARY OF THE PROPOSED APPROACH

We propose several converging approaches, each of them alone justifying the plotting of sound velocity data to improve thermodynamic modelling procedures.

- Sound velocity is related to the other thermodynamic properties of fluids ; its measurement is reliable and accurate, and is more easily accessible in pressure and temperature than the other properties. It can therefore serve as an useful database to discriminate the thermodynamic models.
- Sound velocity measurements in reservoir fluids can also be useful to complement the experimental data (PVT) necessary for the characterization of the parameters of the selected thermodynamic model. They can here be substituted for the analytical data, which are unfit for the description of heavy fractions.
- Sound velocity is sensitive to phase transitions liable to occur in reservoir fluids. This is, of course, the case for liquid/vapour phase transitions but also for liquid/liquid phase transitions. The phase ranges thus obtained are valuable data for matching the models, while lacking in the conventional PVT measurements for obvious reasons of cost.

3 EXPERIMENTAL RESULTS

3.1 Brief description of the measurement method

The experimental system and measurement method have been previously described in detail [7]. The measurement cell is equipped with two transducers, one serving as emitter and the other as receiver, isolated from the sample by two cylindric bars to keep the temperature of ceramics close to the ambient temperature. The frequency applied is 5 MHZ, the ultrasonic pulse is subjected to successive reflections on the interfaces and the signal picked up consists in a series of echoes. The PAPADAKIS [5] method called "echo overlap method" is applied for the accurate measurement of travelling time through the fluid.

From checks carried out at different pressure and temperature values in water, n-decane and n-hexadecane compared to the results obtained by WILSON [9], BADALYAN [10] and BOELHOUWER [11], the accuracy of our determination of U was about +/- 0.5 m/s.

3.2 Measurements in synthetic mixtures

3.2.1 Methane/n-hexadecane mixtures

Isothermal measurements of U have been made in mixture of various concentrations. The curves U(P) obtained (as in figure 1) are described as being of decreasing pressure to the bubble point, which causes a very significant decrease in the amplitude of echoes within the fluid. Figure 2 displays the results relative to the bubble points with 3 distinct concentrations.

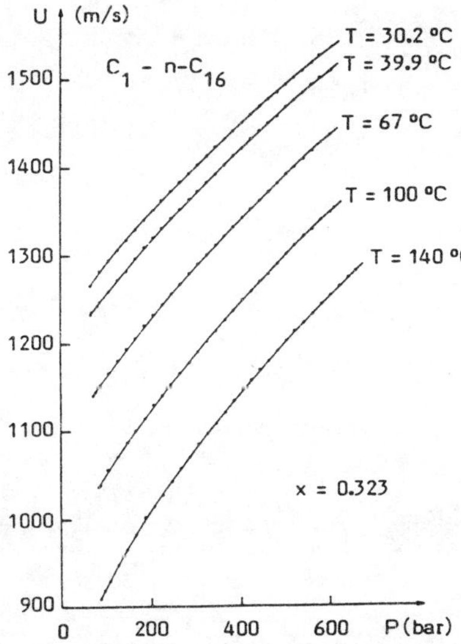

Fig.1 : Variation of sound velocity in C1-nC16 mixture as a function of pressure for different temperatures : Methane fraction = 0.323

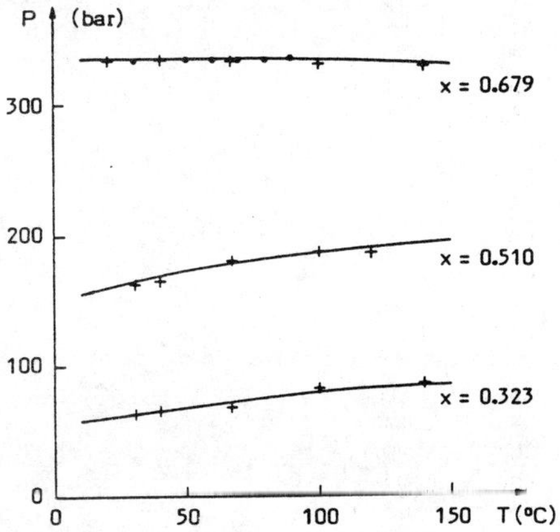

Fig.2 : Bubble point pressure in different C1-nC16 mixtures as a function of temperature : (•) visual, (+) ultrasonic determination.

nC6-nC16 mixtures of different concentrations have also been studied by the ultrasonic technique within the same temperature and pressure ranges. The values of U obtained will be used for restitution testing as in the case of C1-nC16 mixtures.

3.2.2 Carbon dioxide/n-hexadecane mixture

Velocity U has been measured for four CO2-nC16 mixtures as a function of P and T (CO2 mole % : 0.25,0.50, 0.80, 0.90). The curves U(P) for the two mixtures poorest in CO2 are similar to those displayed in figure 1. For the two mixtures richest in CO2, the isothermal lines corresponding to T = 40°C and 60°C are also regular. On the contrary, distinct alterations can been observed on the curves with T = 20°C and T=32°C (figure 3).

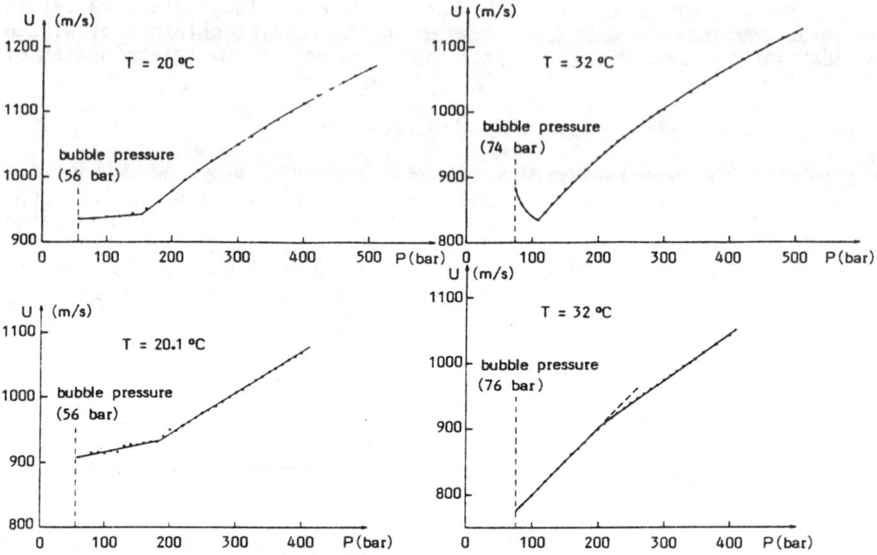

Fig.3 Variation of sound velocity in CO2-nC16 mixture as a function of pressure for different temperatures : CO2 fraction = 0.80 and 0.90.

Co-ordinates (x,P) of these angular points can be related to the liquid/liquid phase transition characteristic of the mixtures rich in CO2 and are consistent with LARSEN et al. [12] phase diagram.
The agreement as concerns the bubble point pressures is also excellent.
These conclusions have been confirmed by direct observations made in a full visibility PVT cell.
Bubble point pressure variations as a function of concentration x for different temperatures (figure 4) reveal evidence stages for the strong values of x at T = 20°C and 32°C, which is a result consistent with the existence of a two-phase system at the liquid state.

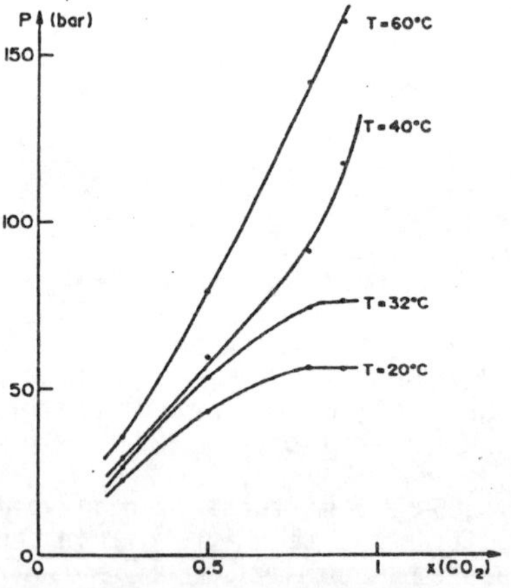

Fig.4 : Bubble point pressure variation in CO2 n-C16 mixtures as a function of CO2 fraction for different temperatures.

3.3 Reservoir fluids

Three reservoir fluids have been studied. Their main characteristics are given in table I hereafter :

	FG1	FG2	FG3
Reservoir pressure P_f (bars)	260	264	245
Reservoir temperature T_f (°C)	139	154	97
Methane content (molar%)	42.5	45.5	38.2
C11+ content (molar%)	19.6	15.8	27.1
C11+ molecular weight (g/mole)	294	276	305
Density at T_f,P_f (kg/m^3)	621	577	710
Bubble point at T_f (bars)	255	262	187
Total GOR (Stdm3/m3)	163	215	103
Stock oil density (kg/m3)	833	829	850

Table I : Main PVT characteristics of reservoir fluids

For two of the fluids, the curves for ultrasonic velocity as a function of pressure within the range of temperature investigated are regular, as was experimental with some binary mixtures (figures 5 and 6). The curve is stopped when the bubble point is reached, at the lowest pressure value.

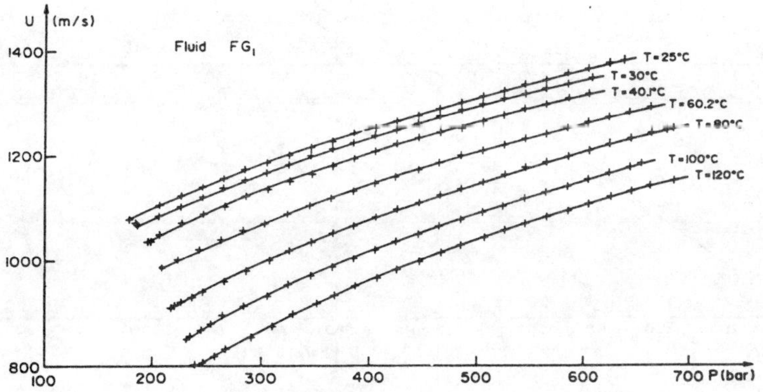

Fig.5 : Variation of sound velocity in FG1 reservoir fluid as a function of pressure for different temperatures.

On the contrary, the behaviour of the third fluid differs from the other two by significant changes of slope at given temperatures in the curves U(P) (figure 7). This phenomenon can be compared to that observed in the case of CO2-nC16 samples with high CO2 content.

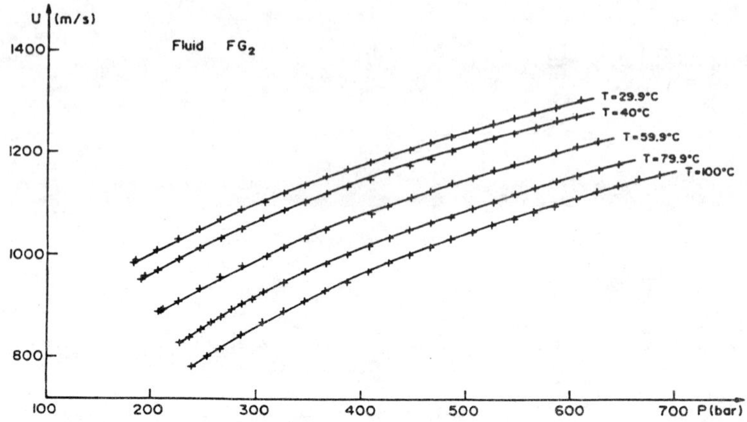

Fig.6 : Variation of sound velocity in FG2 reservoir fluid as a function of pressure for different temperatures.

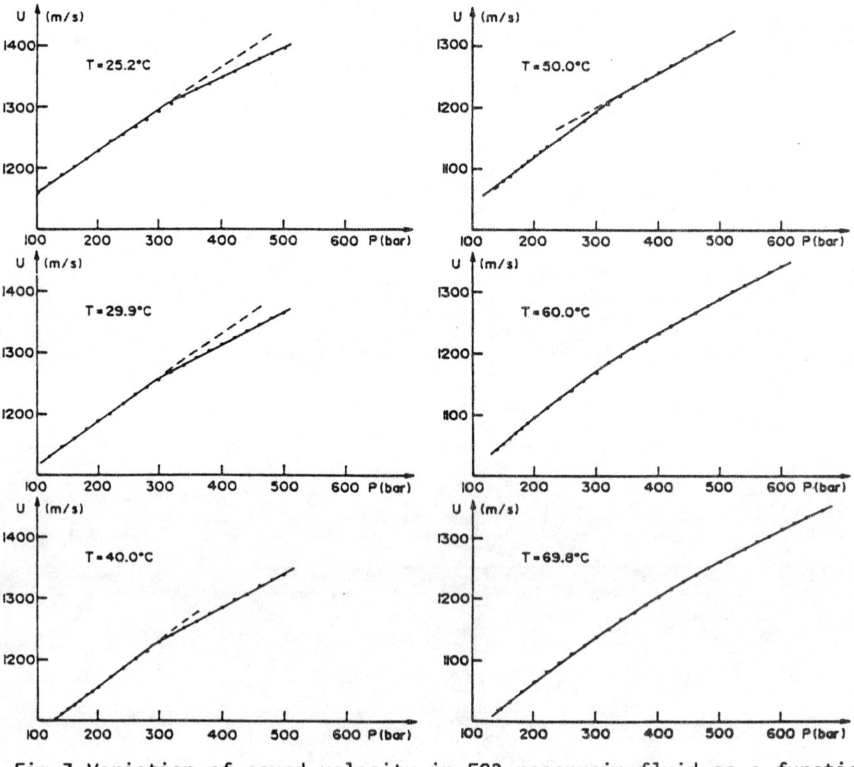

Fig.7 Variation of sound velocity in FG3 reservoir fluid as a function of pressure for different temperatures.

4 MODELLING

4.1 Calculation procedure

Sound velocity can be expressed from the system free energy A by the following formula :

$$U = \frac{V}{M^{1/2}}\left(\left(\frac{\partial^2 A}{\partial V^2}\right)_T - \frac{(\partial^2 A/\partial V\partial T)^2}{(\partial^2 A/\partial T^2)_V}\right)^{1/2}$$

But, in the practice, a free energy analytical expression is not always available to perform direct calculation.

In the numerical tests, for which results are further reported, the calculation of velocities lies on the formula :

$$U^2 = \frac{1}{\rho\cdot(\beta_T - \alpha^2\cdot V\cdot T/C_p)}$$

and the procedure applied, at given P and T values, is the following :

- we first solve the equation of state, which gives the values of V et ρ and the partial derivatives $\left(\frac{\partial P}{\partial T}\right)_V$ and $\left(\frac{\partial P}{\partial V}\right)_T$.
- then coefficients α et β_T are calculated,
- we proceed in the same way for the departure function for the heat capacity at constant pressure from :

$$C_p - C_p^* = \int_\infty^V T\left(\frac{\partial^2 P}{\partial T^2}\right)_V dV - T\frac{(\partial P/\partial T)_V^2}{(\partial P/\partial V)_T} - R$$

Cp being deduced after calculation of the ideal part by means of ALY and LEE [13] or PASSUT and DANNER [17] representation.

Four equations of state have been retained for the restitution tests :

- PENG-ROBINSON (PR) EOS in its form of origin [15],
- PENG ROBINSON (PR-RP) EOS as modified by RAUZY-PENELOUX method [16,17],
- The SIMONET BEHAR RAUZY (SBR) EOS [18],
- and finally LEE KESLER (LK) EOS [19], application of the law of corresponding states, based on the selection of two reference compounds (methane and octane for hydrocarbon compounds).

All these equations are currently used for thermodynamic calculations in reservoir fluids.

4.2 Application to pure compounds

The equations selected served to calculate the values of U in P, T conditions identical to those of experimental data reported in the literature (2077 points), for a certain number of hydrocarbon pure compounds. The most reliable correlation (mean deviation of 3.8 % between calculated and experimental values) is LEE-KESLER's, the other equations leading to much greater deviations ($\geq$ 10 % as an average). LK's description of the behaviour of U within very extended pressure ranges (0.2 - 24 000 bars) is absolutely remarkable compared to the other models.

4.3 Mixtures

The calculation of parameters from the equation of state for mixtures implies the application of mixing rules to the parameters of pure compounds. Such rules are empirical but of prime importance on the quality of results. Among the equations applied, the LEE KESLER's model is based on corresponding states. There is a distinct calculation step to get the fluid pseudo-critical coordinates by application of the mixing rules to the critical coordinates of the compounds. It is therefore reasonable to seek for the mixing rule best adapted to the restitution of optimum pseudo-critical properties with respect to the thermodynamic properties to be restituted.

Such approach has been extended to the other equations of state with regard for homogeneity, although these equations are not used in practice with other mixing rules than those set by the authors on the parameters involved.

4.3.1 Mixing rules studied

Six mixing rules have been studied :

1 : PEDERSEN [20] 3 : TEJA [22] 5 : LEE-KESLER [19]
2 : SPENCER [21] 4 : HANKINSON [23] 6 : PLOCKER [24]

Comparison between the various mixing rules was made over a data set of U relative to alkane binary mixtures : our own measurements of binary mixtures C1-nC16, nC6-nC16 along with TAGAKI and TERANISHI's [25] experimental data on mixtures of various nC6-nC10, n-C8-nC12, nC10-nC14 concentrations. Results are collated in table II below :

			Mixing rules					
EOS	Mixtures	N	1	2	3	4	5	6
PR	nC6-nC10	15	17.5	18.1	17.3	17.1	17.4	17.2
	nC8-nC12	15	22.3	22.7	22.0	21.7	22.1	22.3
	nC10-nC14	15	28.6	29.3	28.1	27.8	28.4	28.0
	C1-nC16	438	5.9	35.0	13.6	13.9	8.3	8.8
	nC6-nC16	346	14.7	19.1	12.8	12.2	13.8	12.5
PR-RP	nC6-nC10	15	16.0	16.5	15.8	15.6	15.8	15.7
	nC8-nC12	15	18.4	18.8	18.1	17.8	18.2	18.3
	nC10-nC14	15	19.8	20.1	19.6	19.4	19.7	19.8
	C1-nC16	438	5.2	32.9	14.2	14.7	7.7	6.5
	nC6-nC16	346	8.6	10.8	8.1	7.8	8.1	8.7
SBR	nC6-nC10	15	21.8	22.1	21.6	21.5	21.6	21.7
	nC8-nC12	15	21.5	21.7	21.3	21.0	21.4	21.7
	nC10-nC14	15	20.1	20.2	20.0	19.8	20.0	20.1
	C1-nC16	438	12.0	37.5	6.4	6.4	6.2	10.1
	nC6-nC16	346	12.6	14.4	11.7	11.4	12.0	12.2
LK	nC6-nC10	15	5.6	4.5	6.0	6.1	5.7	6.5
	nC8-nC12	15	9.4	9.0	9.7	9.8	9.5	9.8
	nC10-nC14	15	12.7	12.5	12.8	12.9	12.7	12.9
	C1-nC16	438	5.7	18.8	13.8	14.1	8.4	4.4
	nC6-nC16	346	7.5	5.5	9.4	9.6	8.1	10.4

Table II : Ultrasonic velocity deviations for various equations of state as a function of the mixing rule (N : number of experimental points).

This table calls for the following comments : for the TAKAGI-TERANISHI's experimental mixtures, whose two constituents differ only by four CH2 groupings, it is found that whatever the equation of state used, the influence of the mixing rule is negligible.

As in the case of pure compounds, LK equation is that leading to the most accurate restitution. For C1-nC16 mixtures, whose two compounds display very far apart critical properties, we note that the mean deviations vary very significantly following the compositional rule considered and for a same equation of state.

Associations of equation of state/mixing rule more compatible with velocity description emerge from the table (leading to mean deviations of about 6 %).

4.4 Reservoir fluids

The identification and quantitative analysis of reservoir fluid constituents cannot be possibly performed in routine beyond n-undecane, at least in an exhaustive way. Even with such limitation 150 compounds are generally present in the analysis.

It is therefore necessary to distinguish the heavy fraction including all C11+ compounds and the other compounds, identified individually, for which we know the mole fraction Xi, mole weight Mi and the characteristic triplet Tci, Pci, ω_i.

The C11+ heavy fraction is assimilated to a standard molecule of an average molecular weight equal to that of the fraction (determined by gel permeation chromatography - GPC -) and constituted of aliphatic (CH3 and CH2) ; cyclanic (CH2) and aromatic (CH) series in C6-C10 intermediate fraction proportions.

The characteristic triplet of the fraction is then estimated by means of group contribution methods : FEDOR's method for the evaluation of the critical temperature, JOBACK's for the critical pressure, NATH's for the acentric factor. In addition, the BENSON's method is referred to for the calculation of the ideal heat capacity.

These various methods are depicted in the REID et al. book [26]. Calculation of pseudo-critical properties and acentric factor of reservoir fluids is then performed by applying the various mixing rules previously cited. Huge deviations between the values of coordinates Tcm, Pcm obtained with each mixing rule are evidenced (figure 8). Systematically, the SPENCER-DANNER's rule leads to the highest Tcm, the PLOCKER et al. rule to the lowest Tcm and that of LEE-KESLER to the strongest Pcm.

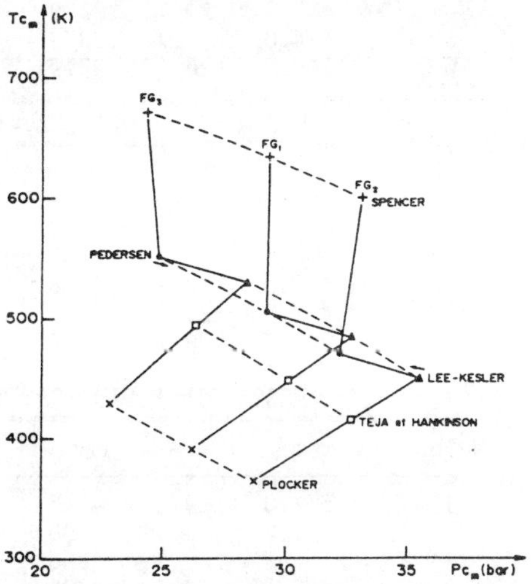

Fig.8 : Comparison of pseudo-critical coordinates obtained with the different mixing rules for the 3 reservoir fluids.

Results of the comparisons performed between the calculated values of U and the experimental values relative to the fluids are summarized in table III. These restitution tests have the advantage of underlining unquestionably the most satisfactory couplings : equation of state/mixing rule. These optimum couplings, identical for the three fluids are : the LEE-KESLER's rule with cubic equation of state, PLOCKER's rule with equation SBR, PEDERSEN with correlation LK.

A few general conclusions emerge from table III.

Correlation LK, which appeared to be the most adequate to the case of pure compounds and a few alkane binary mixtures, looses here its interest, although its restitution capacities when coupled with the PEDERSEN's rule, remain very acceptable (7.2 % out of the 503 experimental points as an average). This reduction in efficiency comes from a lesser adequation of the two reference compounds (methane and octane) compared to the overall fluid components.

The most accurate restitution of the overall data is supplied by equation PR - RP when the evaluation of the characteristic triplet lies on the LEE-KESLER's compositional rule. The application of the equation of state to these complex mixtures results in a remarkable improvement of the behaviour description of U, compared to the poor results obtained for the pure compounds tested.
The BIAS value calculated for the 503 experimental points is 0.7 %, furthermore satisfactory result which shows a relatively fair and alternated distribution of the experimental points from either part of the calculated curve.

RESERVOIR FLUID FG1, 193 experimental points

Rules	Tc	Pc	ω	PR	PR-RP	SBR	LK
1	504.8	29.2	.3379	10.5	10.5	44.9	**6.4**
2	634.2	29.3	.3379	57.7	58.1	65.8	21.4
3-4	448.5	30.1	.3379	9.9	9.8	28.3	19.9
5	484.0	32.7	.3379	**2.6**	**2.6**	35.2	13.1
6	390.5	26.2	.3379	24.7	24.7	**12.0**	30.8

RESERVOIR FLUID FG2, 136 experimental points

Rules	Tc	Pc	ω	PR	PR-RP	SBR	LK
1	469.6	32.2	.2808	6.6	9.2	49.6	**4.4**
2	600.2	33.1	.2808	56.8	60.9	84.0	29.3
3-4	415.8	32.7	.2808	13.5	11.6	23.8	18.8
5	449.8	35.5	.2808	**3.9**	**2.2**	36.6	11.3
6	364.9	28.7	.2808	26.8	25.3	**7.1**	29.6

RESERVOIR FLUID FG3, 174 experimental points

Rules	Tc	Pc	ω	PR	PR-RP	SBR	LK
1	551.0	24.8	.4396	14.4	9.5	26.5	**10.2**
2	672.2	24.3	.4396	56.9	50.1	39.6	11.0
3-4	494.2	26.3	.4396	6.0	9.8	18.3	21.7
5	529.9	28.4	.4396	**4.3**	**2.9**	19.7	15.8
6	428.8	22.8	.4396	22.4	25.7	**11.6**	32.7

Table III : Mean deviations (in %) on velocity U in the three reservoir fluids.

5 CONCLUSION

The results we presented show that the ultrasonic velocity measurements in pure compounds and in synthetic mixtures are a severe test for the equations of state and associated mixing rules. We could see that cubic equations, poorly adapted to the description of pure compounds and binary mixtures, were very reliable to restitute the behaviour of complex mixtures such as reservoir fluids.

Ultrasonic velocity is related to the coefficients of compressibility, which are involved in the various stages of oil production, it will therefore serve as an excellent means of selection for thermodynamic models.

In the reservoir fluids such measurements will complement the conventional PVT data used for the identification of the model parameters, more particularly heavy fraction characteristics. We also confirmed that ultrasonic velocity measurement makes it possible to detect liquid/vapour phase transitions under conditions which make the visual methods aleatory.

The characterization of oil fluids by ultrasonic velocity measurements should greatly develop in the coming years. A lot of work has still to be achieved : in the experimental field to complete the databases necessary for testing the models and, in the modelling field, for evidencing defects and qualities in the thermodynamic models to be greatly improved thanks to this new means of investigation.

AKNOWLEDGEMENT

The authors thanks the management of ELF AQUITAINE, for permission to publish this paper.

REFERENCES

(1) DAVIS L.A. GORDON R.B., J.Chem. Phys. (1967), 46, p.2650
(2) MURINGER M.J.P., TRAPPENIER N.J., BISWAS S.N., Phys.chem. Liq. (1985), 14, pp.273-296.
(3) SUN T.F., KORTBEEK P.J., TRAPPENIER N.J., BISWAS S.N., Phys.Chem.Liq. (1987), 16, pp.163-178.
(4) MILLS R.L., LIEBENBERG D.H., BRONSON J.C., SCHMIDT L.C., J. Chem. Phys. (1977) Vol 66, N° 7, pp. 3076-3083
(5) MILLS R.L. LIEBENBERG D.H., LE SAR R., PRUZAN PH. Mat. Res. Soc. SYMP. PROC. (1984) Vol 22 published by Elsevier Science Publishing Co.Inc..
(6) WANG Z., NUR A.M., STANFORD M., BATZLE M.L., Society of Petroleum Engineers, (1988) Paper N°18163, pp. 571-585.
(7) YE S.,ALLIEZ J., LAGOURETTE B., SAINT-GUIRONS H., ARMAN J. And XANS P. Revue Phys. Appl. (1990) 25.
(8) PAPADAKIS E.P., J. Appl. Phys. (1964) 35 p 1474.
(9) WILSON W.D., J.Acoust. Soc. Am. (1959) 31 pp 1067-1072
(10) BADALYAN A.L. OTPUSHCHENNIKOV N.F., SHOYTOV U.S., Izv Akad. Nauk. Arm. SSSR Fiz. (1970), Vol.5, p.448.
(11) BOELHOUWER J.W.M., Physica (1967), 340, p. 484.
(12) LARSEN L., SILVA M.K., TAYLOR M.A.. Soc. Pet. Eng. (1986) Paper N°15399
(13) ALY F.A., LEE L.L. Fluid Phase Equilibria (1981) 6 , pp.169-179
(14) PASSUT C.A. DANNER R.P. I & EC, Process Des. Dev. (1972) 11, p.543
(15) PENG D.Y., ROBINSON D.B. Ind. Eng. Chem. Fundam. Vol.15 n°1 pp.59-64 (1976)
(16) PENELOUX A., RAUZY E., FREZE R. Fluid Phase Equilibria, (1982) 8, pp.7-23.
(17) RAUZY E. Les méthodes simple de calcul des équilibres liquides vapeur sous pression, Thèse de doctorat d'état Université Aix Marseille II (1982).
(18) BEHAR E. SIMONET R. RAUZY E. Fluid Phase Equilibria (1985) 21, 237-255
(19) LEE B.I., KESLER M.G. AIChE Journal (1975) Vol.21, N°3 pp.510-527
(20) PEDERSEN K.S., FREDENSLUND A., CHRISTENSEN P.L., THOMASSEN P. Chem Eng. Sci. (1984) 39 (6) pp.1011 1016.
(21) SPENCER C.F., DANNER R.P. J. Chem. Eng. Data (1972) 17 (2) pp 236-241
(22) TEJA A.S. AIChE Journal (1980) (26) 3 pp 337-345
(23) HANKINSON R.W. THOMSON G.M. AIChE Journal (1979) (25) 4 pp 653-663
(24) PLOCKER U. KNAPP H. PRAUSNITZ J. Ind. Eng. Chem. Process Dev. (1978) Vol.17 N°3 pp 324-332.
(25) TAKAGI T., TERANISHI H. Fluid Phase Equilibria (1985) 20 pp.315 320.
(26) REID R.C. PRAUSNITZ J.M. POLING B.E. The properties of gases and liquids 4th Edition, Mc. GRAW HILL Book Company 1987.

GRAVITY-ASSISTED INERT GAS INJECTION: MICROMODEL EXPERIMENTS AND MODEL BASED ON FRACTAL ROUGHNESS

ROLAND LENORMAND
Institut français du Pétrole,
BP 311,
92506 Rueil Malmaison,
France

Summary
Displacement of oil by inert gas is an efficient recovery process. Experiments have been performed in micromodels in order to study the basic mechanisms and to develop a model. They show three kinds of flow taking place at different time and length scales: drainage in the bulk of the pores, drainage of trapped blobs and drainage of the fluid filling the roughness of the walls. A fractal model is then developed to account for the variety of scales. The main prediction concerns the production rate as a function of time in the form t^{D-4}, where D is the fractal dimension of the internal surface area of the medium.

1. INTRODUCTION

Pore level mechanisms during displacement by gravity drainage have been elucidated,[1] at least in the case of water-wet media: the oil phase spreads between water and gas, this oil film maintaining hydraulic continuity. Consequently, any production of oil (or water) is due to the fluid flowing along the roughness of the solid walls or spreading over the other fluid.

The purpose of our study is to model this flow at low saturation and to obtain the relationship between oil production and the microscopic stucture of the rock sample and fluid properties. This approach should lead to a model of oil production, without the need for any flow experiments.

The pore geometry and the roughness of the grains can be obtained by using image analysis of thin sections. However we need a model for fluid transport in such complex geometries.

This paper presents our first results concerning the drainage of only one liquid (either oil or water) when an immiscible gas is injected. The first part of this paper presents experiments in transparents micromodels. They show that several mechanisms are involved at various time and length scales. This variety of scales suggests the use of a *fractal* model which is described in the second part. Finally, we propose a procedure for measuring, on a real rock, the various parameters of the model.

2. EXPERIMENTS IN MICROMODELS

Experiments in two-dimensional artificial porous media (micromodels) provide information on the physical mechanisms that take place at the pore level and, secondly, they test the validity of the model by comparing the results on a large scale.

Micromodels

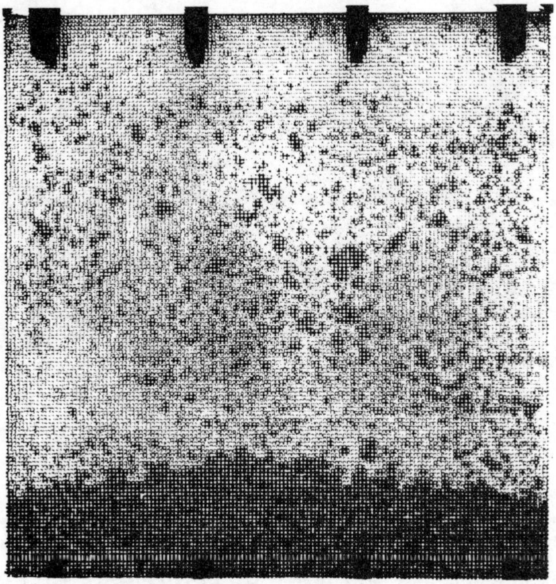

Fig. 1. Photograph of the micromodel during gravity drainage (oil in black).

Fig. 2. Close-up of the micromodel shown in fig.1: trapped clusters left behind the front.

We have developed a molding technique[2] using a transparent resin and a photographically etched mould. The cross-section of each duct is rectangular with a constant depth (1mm) and a width which varies at random from pore to pore. For this study we used a large network (135 x 150 mm) containing about 42 000 ducts with seven classes of width (from 0.1 to 0.6 mm), distributed with a log-normal law. The distance between two sites of the square network is 1 mm and the permeability is approximately 10^{-9}m^2 (1000 Darcy).

The wetting fluid is a paraffinic oil (Soltrol) dyed in red (density $\rho = 800$ kg m^{-3}, viscosity $\mu = 4\ 10^{-3}$Pa s).

Experiments

The micromodel is saturated under vacuum. Then, inlet and outlet valves are open and oil production is recorded. Fig. 1 is a photograph of the micromodel during an experiment (oil in black). It shows the end effect at the bottom due to capillary forces and the trapped blobs left behind the front. Fig. 2 represents a close-up of theses blobs.

Oil production (Fig. 3a in log scale and 3b in linear scale) is similar to the one obtained with a rock sample:[3] a linear part followed by a long tail. The linear part corresponds to the drainage of the bulk of the pores. The long tail production is due to the drainage of all the trapped blobs. This drainage is possible by the presence of roughness on the walls of the micromodel. In addition, the change in color in the upper part of the model shows the disappearance of the film of red oil which surrounds the grains.

From these observations we can distinguish three mechanisms for the production of the wetting fluid:

1. drainage of the *bulk* of the pores.

2. drainage of the*disconnected blobs.*

3. drainage of the liquid filling the *roughness.*

A physical model should account for these three different mechanisms which take place at different time and length scales. However, we will follow another way, by using a fractal model, an approach which can globally solve problems involving many length scales.

3. FRACTAL MODEL OF ROUGHNESS

Several recent studies have used a fractal approach for solving problems of flow in porous media. We will review briefly how this fractal approach is used to describe the geometry of pores, the partial filling by a wetting fluid (static),the capillary pressure and the conductivity. A description of the theory of fractal geometry and applications to porous media can be found in reference 4.

The fractal structure of porous materials have been investigated with different techniques. Avnir and coworkers[5] have probed surface areas of various materials by adsorption isotherms obtained with molecules of different sizes. Katz and Thompson[6] have studied geometrical properties of images obtained with electron microscopy and

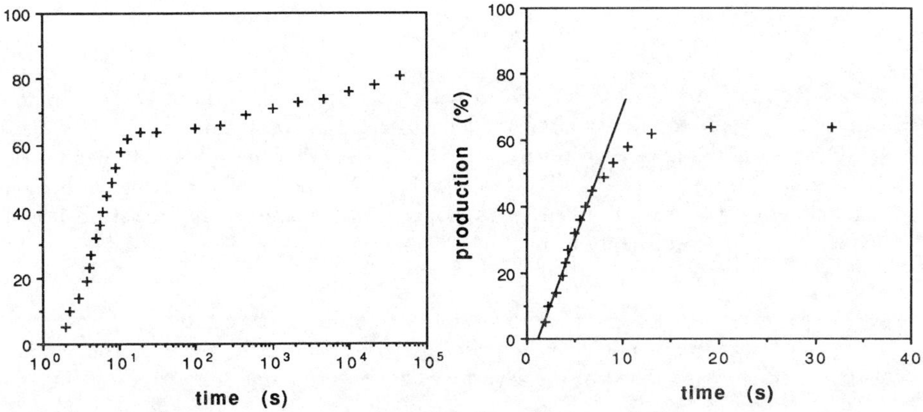

Fig. 3. Oil production during gravity drainage in a micromodel: a) long time production (log scale), b) short time production (linear scale).

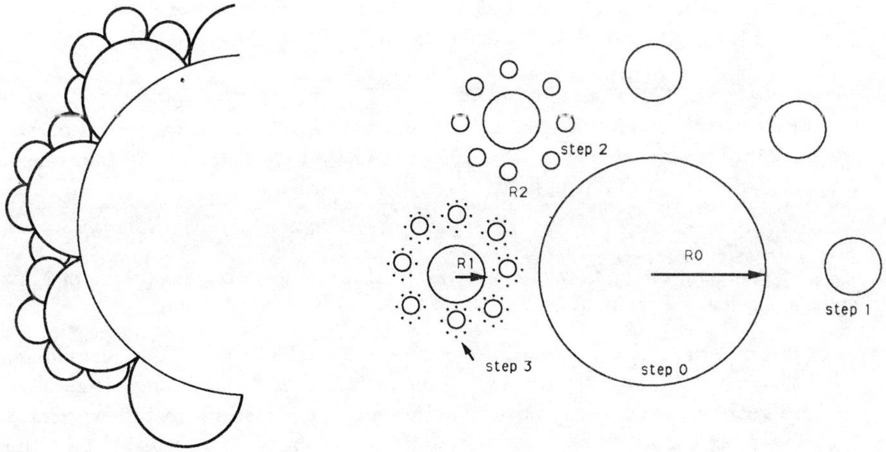

Fig. 4. Fractal model of roughness (a) and equivalent tube model used for flow calculations (b).

Hansen and Skjeltorp[7] by using optical microscopy. The main results of these studies is that the internal surface area A depends on the linear scale of observation ℓ:

$$A \propto \ell^{2-D_S} \tag{1}$$

where D_S is the fractal dimension. In the case of a rough surface, ($D_S > 2$), the surface area tends to infinity when the length scale tends to zero.

The partial filling of a fractal surface by a wetting fluid have been studied by De Gennes,[8] by using simple iterative models of pits and flocs. The wetting saturation S is related to the fractal dimension D_S of the surface and to the curvature $1/r$ of the interface between the fluids by:

$$S \propto r^{3-D_S} \tag{2}$$

From this result, we can deduce that the capillary pressure P, which is proportional to the curvature, is related to the saturation by:

$$P \propto S^{1/D_S-3} \tag{3}$$

This result have been validated by Davis,[9] with a series of experimental results published by Melrose.[10] In these experiments performed in a Berea sandstone, the water saturation ranges from 0.101 to 0.033 and the fractal dimension is found to be $D_S = 2.55$.

Now, modeling a transport property, such as relative permeability, is more complicated. De Gennes distinguishes between disconnected pockets where the conductivity is controlled by thin films (Van der Walls forces) and connected pockets. In this later case, the permeability is given by:

$$K \propto S^{\beta} \tag{4}$$

with $\beta = (5 - D_S)/(3 - D_S)$.

Following this fractal approach, we propose now a geometrical model based on capillary tubes, which allows us to calculate explicitely the static (S, P) and transport properties and leads to a simple fractal model for gravity drainage.

<u>Fractal model</u>

Our modeling of the flow of a wetting fluid inside a porous media under gravity forces consists in 3 main steps:

1. the internal surface is assumed to be a fractal surface. As a consequence, the wetting fluid is always *continuous in bulk* if the contact angle is zero[11] and there is no residual saturation. This assumption is justified by the experimental results of Dullien *et al*[12] and Melrose.[10] In addition we can neglect the disjoining pressure if the flow takes place in roughness as it has been shown by O'Meara *et al.*[13]

2. *geometrical model:* the isotropic fractal surface is then modelled by a bundle of identical parallel tubes with a *fractal cross-section* (fig. 4a). The direction of these tubes corresponds to the main direction of flow during gravity drainage.

The cross-section of a tube is constructed by a simple iterative process, by dividing the half perimeter in η parts and replacing each part by half a circle. This model is quite similar to the iterative pits model developed by de Gennes and it is used for the calculation of geometrical and static properties. However the calculation of the flow inside a groove is not straighforward[14] and we prefer go a step further in modeling each groove by a capillary tube.

3. *flow model*: for the calculation of flow properties, we use a model made of a bundle of capillary tubes deduced from the recursive construction of the fractal tube (fig. 4b). In fact, we assume that the flow in a groove is identical to the flow in a tube of the same radius.

At each step k of the construction corresponds N_k grooves with radius R_k (fig. 4a) and total cross-section area A_k. The main tube ($k = 0$) corresponds to the bulk of the pores and the following grooves to the roughness of the walls.

We obtain easily the following values at step k:

$$R_k = R_0(\pi/\eta)^k \tag{5}$$

$$N_k = (\eta)^k \tag{6}$$

$$A_k = \frac{1}{2}\pi R_0^2(\pi^2/\eta)^k \tag{7}$$

The condition of finite surface area implies $\pi^2/\eta < 1$.

Static properties

Now, we can calculate the properties of this model of porous medium. At step k, the perimeter L_k in a cross section is equal to the total perimeter of all the smallest grooves (fig. 4a):

$$L_k = N_k \pi R_k \tag{8}$$

Eliminating k between L_k and R_k leads to the standard equation for a fractal curve:

$$L/L_0 = (R/R_0)^{(1-D_L)} \tag{9}$$

where the fractal dimension D_L associated with the perimeter is given by:

$$D_L = \frac{\lg\eta}{\lg\eta/\pi} \tag{10}$$

The internal surface scales as the perimeter (the dimension in the longitudinal dimension does not depend on the microscopic scale R). By using the definition of the surface fractal dimension D_S (equ. 1), we verify the standard relation $D_S = D_L + 1$.

Now, we considere the filling of the geomety depicted in fig. 4a by a wetting fluid at the capillary pressure P_k. Due to capillary effects, all the roughness of sizes smaller or equal to R_k are filled. The cut-off value is given by Laplace's law: $P_k = 2\gamma/R_k$, γ being the surface tension. Consequently, the saturation S_k of the wetting fluid is equal to the fraction of invaded grooves:

$$S_k = (A_k + A_{k+1} + ...)/(A_0 + A_1 + ...) \tag{11}$$

which leads to:

$$S = (R/R_0)^{2-D_L} \tag{12}$$

and:

$$P = \frac{2\gamma}{R_0} S^{1/(D_L-2)} \tag{13}$$

Relative permeability

Any two-phase displacement, including gravity drainage, can be modeled by generalised Darcy's law (Buckley Leverett equations[3]) and solved numerically. The main problem is to obtain values for the relative permeability at low saturation (film flow). This is the main application of our fractal model.

Let us first calculate the monophasic permeability of the fractal tube by using Poiseuille's law in each tube of length H:

$$\Delta P/H = \frac{128\mu Q_k}{\pi R_k^4} \tag{14}$$

All the tubes are in parallel, so the total flow Q is the sum of all the flow rates Q_k. We obtain the monophasic permeability K from Darcy's law:

$$\Delta P/H = \frac{\mu Q}{\Sigma K} \tag{15}$$

The cross-section area Σ is equal to the pore area divided by the porosity ϕ.

$$K = \frac{R_0^2\phi}{128} \times \frac{(1-\pi^2/\eta)}{(1-\pi^4/\eta^3)} \tag{16}$$

Limiting now the sum to the fraction of filled tubes corresponding to the capillary pressure P_k leads to the permeability K_k:

$$K_k = K\eta^k(\pi/\eta)^{4k} \tag{17}$$

At any step k, the relative permeability $K_r = K_k/K$ is a function of the radius of curvature R:

$$K_r = (R/R_0)^{4-D_L} \tag{18}$$

and, by using equ. 12:

$$K_r = S^{(4-D_L)/(2-D_L)} \tag{19}$$

The results obtained with our fractal tube model are similar to those obtained by scaling considerations (equ. 1 to 4). The main advantage is the possibility for the calculation of the prefactors. In addition, it can be used for modeling much complex flow, such as gravity drainage.

Gravity drainage

Oil production by gravity drainage can be calculated numerically by using capillary pressure curves and oil relative permeability calculated in the previous paragraph. However we can use the tube model to obtain directly an analytical law when capillary effects are negligible.

We assume that drainage of all the tubes are independant and, as a first approximation we neglect capillary effects in a tube. Consequently, the production rate of each tube is constant and there is no capillary trapping at the bottom of the tube.

The flow rate Q_k in a tube is given by writing the balance between viscous and gravity forces:

$$Q_k = \frac{\rho g \pi R_k^4}{128\mu} \tag{20}$$

and the corresponding time to drain this tube of length H is

$$t_k = \frac{128\mu H}{\rho g R_k^2} \tag{21}$$

At time t_k, all the tubes of radius larger or equal to R_k are empty, and all other tube n have produced a fraction t_k/t_n of their volume. Oil production V_k, ratio of the produced volume by the total volume of all the tubes, is given by:

$$V_k = 1 - (\pi^2/\eta)^{k+1} \times \frac{(1 - \pi^2/\eta)}{(1 - \pi^4/\eta^3)} \tag{22}$$

The production of the main tube R_0 is proportional to time and ends at $t_0 = (\rho g R_0^2)/(128\mu H)$ (drainage of the bulk of the pores). For $t > t_0$ we obtain the scaling law for oil production:

$$1 - V \propto (\frac{t}{t_0})^{D_L - 2} \tag{23}$$

and, for the production rate (dV/dt):

$$Q \propto (\frac{t}{t_0})^{D_L - 3} \tag{24}$$

4. DISCUSSION

We have calculated the production rate during gravity drainage in a micromodel. The plot of the production rate versus time in log-log scale gives an exponent close to -1.5 (fig. 5). This leads to a fractal dimension $D_L = 1.5$, in agreement with fractal theory ($1 < D_L < 2$). However, this experiment must be seen as an illustration of the method rather than a validation of our model. A series of experiments with consolidated and unconsolidated porous media will be used to validate and improve the model.

In order to obtain the relative permeability and then to calculate oil production, the following measurements will provide data for the model:

1. measurement of the fractal dimension of perimeter of the grains by using SEM or optical microscopy and image analysis.

2. mercury porosimetry which can provide the relationship between partial filling and capillary pressure. In addition, using a pore size distribution (instead of a single value R_0) can improve the model. We prefer Hg porosimetry to capillary curves because of the very long time needed for displacement of a liquid in the roughness.

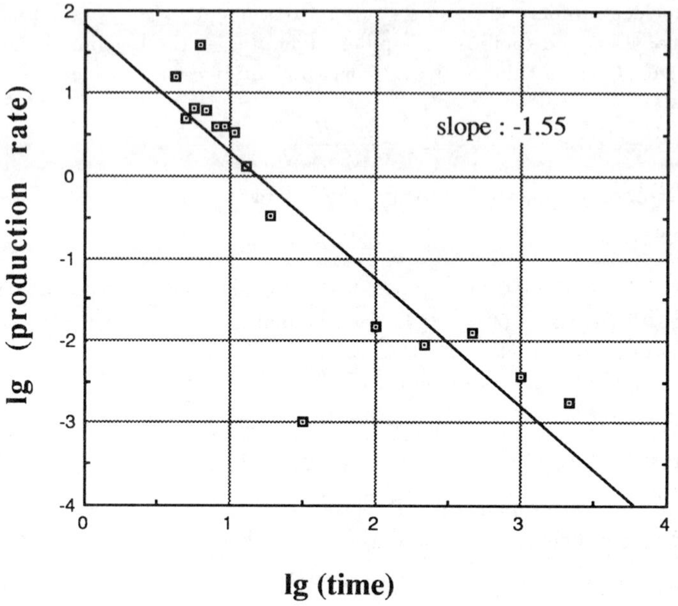

Fig. 5. Oil production rate in a micromodel (log-log scale).

The model will be compared to direct measurements of the relative permeability. For instance, by studing transient regimes in capillary desorption with very thin samples (porous plate method) or by using centrifugation techniques.

ACKNOWLEDGEMENTS

This work has benefited from conversations with D. Pavone. I also thank S. Undreiner and P. Bruzzi who assisted with the experimental work.

REFERENCES

1) A. Kantzas, I. Chatzis and F.A.L. Dullien, SPE 17506, 1988.

2) R. Lenormand and C. Zarcone, SPE 13264, 1984.

3) D. Pavone, 5th European Symposium on Improved Oil Recovery, Budapest, 1989.

4) J. Feder, Fractals, Plenum Press, New york, 1988.

5) D. Avnir, The fractal approach to heterogeneous chemistry, John Wiley and Son, 1989.

6) A.J. Katz and A.H. Thompson, Phys. Rev. Lett. 54, 1985.

7) J.P. Hansen and A.T. Skjeltorp, Phys. Rev. B, 38, 1988.

8) P.G. de Gennes, in Physics of Disordered Materials, ed. by D. Adler, H. Fritzsche and S. Ovshinsky, Plenum Pub, Corp., 1985.

9) H.T. Davis, Europhys. Lett., 8 (7), 1989.

10) J.C. Melrose, SPE Res. Eng., feb. 1990.

11) A.J. Katz and S.A. Trugman, J. of Colloid and Interface Science, 123, 1988.

12) F.A.L. Dullien, C. Zarcone, I.F. Macdonald, A. Collins and R.D.E. Bochard, J. of Colloid and Interface Science, 127, 1989.

13) D.J. O'Meara, G.J. Hirasaki and J.A. Rohan, SPE 18296, 1988.

14) E. Raphael, J. Phys. France, 50, 1989.

PREDICTION OF HYDROCARBON PHASE STATE AT GREAT DEPTHS FOR OIL & GAS PROVINCES AND AREAS OF THE USSR

N. NEMCHENKO and A. ROVENSKAYA

All Union Petroleum Exploration Research Institute (VNIGNI),
36 Shosse Entuziastov, 105118 Moscow, USSR
Institute of Geology and Exploitation of Fossil Fuel (IGIRGI),
50 Fersman St., 117312 Moscow, USSR

Summary

Peculiarities of the phase state of hydrocarbon systems have been considered on the example of petroleum provinces and regions of the USSR. The discovery deposits prediction of different phase state has been done. On the complex of geologic–geochemical indicators of different prediction of hydrocarbons. New type of hydrocarbons systems – high temperature undersaturated gas – condensate systems with high contents of condensate and extraction ratio of liquid hydrocarbons equal or approximating is established and being predicted. Prediction of distribution Zones these systems has been done.

1. INTRODUCTION

The future of oil and gas industry greatly depends upon successful exploration of deeply buried formations. In old oil and gas producing regions, where oil and gas pools have been already discovered at depths to 4 km, the deeper horizons will help to maintain the production at levels achieved thus far; while in new regions the deep exploration should aid to properly evaluate oil and gas potential resources.

Integrated approach to studying the regularities in distribution HC of various phase state (oil, gas, gas–condensate) at great depths with a due account for various natural factors has permitted to elaborate criteria for separately predicting oil and gas occurrence and compile respective prognostic maps for the USSR territory. For the first time gas systems unknown before in explorational practice are predicted for the great depths, such as gas systems with abnormally high condensate percentages (1000 g or more per m^3). Their formation is believed to be conditioned by high in situ temperatures (more than 120°C) and abnormally high pressures.

Some territories have good prospects oil in the deep. These include bastine filled with young weakly metamorphosed deposits of great thickness (young platforms, Alpean foredeeps and intermountain depressions).

The present investigation can contribute to a more directed exploration for desired type of hydrocarbon crudes in various oil and gas regions of the world.

Prospects of the oil and gas industry development in the USSR are closely connected with exploration of deeply buried formations. In old oil and gas producing regions where all the HC pools are practically developed up to the depth of 4 km the exploration in the deep must help to maintain the current level of production, while in new still understudied areas it will help to correctly evaluate potential resources. The experience shows that the great depths are rich in oil and gas, for the HC pools have been known at depths below 4.5 km in majority of oil and gas basins all over the world (USA, Mexico, Venezuela, Libia, Lomar, Bu–Atirel fields, etc.). Discoveries of the kind were made in the Soviet Union in Precaspian,

North Caucasus, Ukrain. However, the problem of oil and gas distribution at great depth is still not completely solved.

A well is named "deep", "superdeep" or "Ultradeep" (deeper than 4.5; 6.0; 7.5 km) depending upon a technological possibility of drilling. Meanwhile a number of problems related to formation and preservation of pools at great depths remain unsolved:

1. Generation of HC (hydrocarbons), pressure–temperature (P–T) conditions of their existence, interaction with host rocks and formational waters, physical–chemical processes at media–phase interfaces.
2. Possibility of the existence of reservoirs and seals and their limiting in oil and gas basins of various types.
3. Traps at great depths, their structural features and size.
4. Methods of exploration and development.

Determining the oil and gas prospects at great depths is an important and complex problem. It is no multi–espectual, that a solution of only some of the aspects will undoubtedly enhance a more founded direction of explorational activity.

At present there is no unanimous opinion on the character of phase state of hydrocarbons at great depths, which is very important for correctly directing oil and /or gas exploration. The majority's opinion on a gaseous state of HC at great depths is based on the belief "rigid" P–T conditions in the deep responsible for the abundance of mainly gaseous HC (predominantly methane) both in the process of generation and in distribution of liquid HC. However, the factual information accumulated thus far on the HC occurrence in the deep shows that in various oil and gas provinces there are discoveries of gas, gas–condensate and oil pools as well.

To improve the oil and gas exploration efficiency it is necessary to find a proper direction for the activity on the basis of a reliable prediction of oil and gas potential.

It is well known that for many years the oil and gas prediction was based on a general evaluation of HC resources in terms of amount of conditional liquid fuel. Differentiated prediction of oil and gas–rich areas was made by analogues or empirically with a due account for the results of exploration already effected. The lack of reliable criteria for separately predicting oil and gas can lead in some cases to serious miscalculations when planning crude bases for growing oil or gas industries.

As shown by the analysis of distribution of oil and gas fields, there is a certain regularity and zonation in their displacement relative to each other, both vertical and areal, and even a complete separation in major sedimentary basins. Overwhelming part of natural gas and gas–condensate pools within the zones of a predominant gas accumulation. The most of the oil resources is concentrated in pure oil, oil–gas–condensate and oil–gas pools within zones of predominantly oil accumulation.

Based upon generalisation of experience accumulated in the USSR and other countries in oil and gas exploration as well as upon theoretical investigations in the field of HC generation and pool formation one has already made certain steps towards solving the complex problem of differential prediction of oil and gas resources. To solve the problem of pointing out zones and sedimentary sequences predominantly rich of gas or petroliferous (both in the producing regions and those under exploration) it is necessary first of all to determine the conditions controlling the predominant generation and accumulation of oil and gas even if the general balance of hydrocarbons is observed. The most important conditions controlling the generation, accumulation and, finally, conservation are as follows (from the point of view or organic theory of oil and gas origin):

- composition of organic matter in the oil and gas source rocks (humic or sapropelic) determining the prevailing generation of liquid or gaseous HC;
- initial organic matter (OM) catagenesis controlling the specific of HC generated at various stages of the OM transformation;
- different ability of oil and gas to migrate vertically and internally thus explaining specific features of oil and gas fields and their distribution, both areal and vertical.

The problem of predicting zones of predominantly oil and gas accumulation is closely related with studying the vertical and spacial zonation.

HC generation is a result of biological and thermocatalytic processes of transformation of OM known to consist of two major genetic types – sapropelic and humic. The sapropelic matter accumulates in seas and lakes and mainly consists remains of plankton, benthos and, to a lesser extent, highest algae. In case of enrichment of the rock within sapropelic OM (about 40%) the combustible shales are formed. The humic matter accumulates mainly at continental conditions and is composed of remains of land plants. Complex facial conditions of sedimentations are often responsible for simultaneous accumulation of various types of OM, which results in mixed types of OM: humic–sapropelic and sapropelic–humic. Experimental observations show that transformation of the humic matter is accompanied by generation of large quantities of methane and smaller amounts of its gaseous homologue. Transformation of sapropelic matter leads to generation of both gaseous and liquid HC.

The initial OM composition controls the generation of liquid or gaseous HC and their relative percentage. Gas generating formations are those containing predominantly humic OM and predominant generation of oil by the sapropelic matter is known to take place at certain stages of catagenesis. The old generation is limited by a rather narrow gap and does take place within the zone of mild catagenesis during the OM transformation stages known as long–flame, gaseous and lipid stages of coalification. Predominant or exclusive generation of gas goes on within some intervals of two zones: the upper one up to the depth of 2 – 2.5 km, corresponding to initial stages of the OM transformation (peat, brown coal, beginning of long–flame) and the lower one at great depth (deeper than 5 – 6 km) corresponding to the high degree of catagenesis of OM (from the end of lipid to the beginning of caking coal stage and higher), where gas generation is explained by a deep thermocatolitic transformation of the OM.

This scheme is to a certain extent idealised. At real geological conditions the intervals of prevailing oil or gas generation can vary. Study of regularities in oil and gas distribution as dependent of various natural factors in geological basins of different types helps to reveal and scientifically substantiate the criteria for defining zones and sequences of predominant gas or oil accumulation, which can be used in predicting the oil and gas distribution.

Characteristics and geothermal environments of deeply buried formations

Predcarpatskaya NGO. Deeply buried zones within the Pre–Carpathian oil and gas area are related to the Pre–Carpathian foredeep occurring between the platform and Carpathian geosyncline known as a northern branch of the Alpian folded area. There are two structural–tectonic zones in the foredeep: The Outer Zone adjacent to the platform and the Inner one (Borislavsko–Pokutskaya) bordering the Outer Carpathians in the South–East. The deeply buried rocks of the Outer and Inner zones are represented by Mesozoic and Caenozoic terrigenous deposits including the entire section from Upper Jurassic to Neogene. In the North–West part of the Outer zone there are Palaeozoic terrigenous–carbonate formations. Total thickness of the Mesozoic and Palaeozoic deposits is 14 – 15 km. Thickness of the

Neogene and Pro–Neogene deposits in the Outer zone of the foredeep is 1 – 6 km (counting from the – 5000 m plane), while that of the Cretaceous and Palaeogene deposits in the Inner zone of the foredeep is about 2 – 3 km. A somewhat weakened temperature field is characteristic of great depths in the Pre–Carpathian foredeep. Formation temperature measured in the Pocut well Lugi–1 while testing the intervals of 6018 – 6045 m is +131°C. In the Borislav–Pocut zone relatively low temperatures are observed: at depths of 4500 – 5000 m they are from 105°C to 110°C.

Commercial flows of oil from Palaeogene, Eocene–Oligocene deposits at the Borislow–Pocut zone fields: Pasechnia, Novoskhodnia, Ivannic and others from depths up to 5500m. Some oil and gas shows and kicks are found even deeper. Water production with considerable amount of gas and oil film is obtained from 5490 m and 6260 m (Well Lugi–1). In superdeep well Shevchenko–1 drilled in a part of the Skib cover adjacent to the Borislav–Pocut zone a good flow of gas within condensate was observed from 6280 m, and the well reached 7520 m a gas kick took place.

Zonation in distribution of HC phase state on the –5000 m horizontal section

Analysis of distribution of pool types over the –5000 m horizontal section for oil and gas basins of the USSR helped to reveal a certain zonation in the HC phase state and outline zones of predominantly oil, gas and gas–condensate pools.

Zone of predominantly oil pools is developed in the Pre–Caspian provinces: oil pools are found in the East and South–East marginal parts of the Pre–Caspian depression in carbonate and clastic carbonate deposite of Carboniferous and Lower Permian.

Zone of predominantly oil–gas–condensate pools is revealed in the North–Caucasus–Manguishlak (excluding Manguishlak area), South–Caspian, Timan–Pechora and Pre–Caspian provinces and Pre–Carpathian and Fergana Areas.

In the North–Caucasus there are 23 oil–gas–condensate pools among the total 25 pools discovered at depths greater than 5 km. Majority of them (21) are found in carbonate deposits of the Upper Cretaceous within the Terek–Caspian foredeep and two pools are in Jurassic terrigenous reservoir in the platform of West–Pre–Caucasus.

In the South–Caspian province 22 such pools were discovered in clastic reservoir of the Paleogene Productive Sequence both in Azerbaijan and West–Turkmenia.

In Pre–Carpathian area the oil–gas–condensate pools are discovered in terrigenous Palaeogene deposits, in Fergana they are found in Palaeogene rocks and in Timan–Pechora province such pools are established in Upper–Devonian carbonates as well as Silurian–Lower–Devonian carbonates in the North–East of the province. In Pre–Caspian province such pools are discovered in Lower Carboniferrous carbonates in the North–East.

Zone of predominant development of gas–condensate and gas pools includes the Dnieper–Pripiat and Volga–Ural provinces. In the first there are 22 such pools in Carboniferrous terrigenous deposits and Devonian clastic–carbonates. In the second the majority of pools belongs to this type (in the Lower Povolzhie in Devonian clastic–carbonates 5 pools out of 7 are gas and gas–condensate ones).

In the process of analysis of pool distribution on the –5000 m horizontal section one can observe that the oil pools are found in provinces related to the belt of Alpian, geosynclinal folding, Alpian foredeeps and epiplatform orogenic areas (intermountain depressions), to syneclises within young platforms with great thickness of sedimentary cover and in the Pre–Caspian depression, while the gas and gas–condensate pools are concentrated in provinces and areas related to ancient platforms.

Zonation in the distribution of catagenesis and paleotemperatures on the –5000 m section.

Oil and gas producing regions and their structural elements as studied on the –5000 m horizontal section are subdivided (with respect to the stage of catagenesis) into three groups as follows: with elevated catagenesis characterized with grades MK_3/Zh/ and MK_4/K/; high catagenesis with grades MK_5/OS/ and MK_1/T/, and higher one – with grades AK_2/PA/ and M_3/A_1/.

Analysis of the catagenic stages and paleotemperatures has shown that the deep horizons in the major provinces and areas are essentially different from each other with respect to the catagenic (paleotemperatural) zonation. Elevated catagenesis (grades MK_3/Zh/ and MK_4/K/) is characteristic of areas related to the Alpian geosynclinal belt, foredeeps, areas of epiplatform orogenesis and young platforms. These regions have relatively low paleothermal gradients (0.22 – 0.26°C/100 m).

High (M_3/OS/ and MK_1/T/) and higher (AK_2/PA/) catagenesis are typical for ancient platforms. These regions are distinguished by elevated paleothermal gradients (0.32 – 0.45 and 0.62 – 0.64/100 m).

Comparison of zonations shown for the HC phase state in pools and for the distribution of catagenesis stages on the –5000 m horizontal section demonstrates their coincidence: the zone of prevailing occurrence of oil pools occupies the areas where the catagenesis stage for the OM in hydrocarbon–rich rocks corresponds to grade MK_4/K/. The zone of prevailing gas and gas–condensate pools corresponds to stage MK_5/OS/AK/T/, and the zone of exclusively gas accumulations coincides with grade AK_2/PA/.

the coincidence of the HC phase zones with catagenesis stage zones gives a possibility to use the catagenesis as the principal criterium for determining the phase state at great depths, i.e. to use the prognostic map of grades on the –5000 m section (with a due account for the formation types) as a prognostic map of HC phase state for great depths.

For exploration for oil pools one should choose the areas with relatively low paleothermal gradients with catagenesis grades MK_3, MK_4 where marine sand–shale and carbonate formations are developed (Upper Jurassic carbonates of the epi–Paleozonic plates in the USSR South, Upper Jurassic shales of the West Siberia plate, Carboniferous and Lower Permian carbonates of the Pre–Caspian depression).

Good prospects for gas pools should exist in areas characterised by elevated paleothermal gradients where marine facies (Upper Jurassic carbonates of the Amu–Daria province and Manguishlak area), subcoal and continental facies (Lower Carboniferous of DDD, Lower and Middle Jurassic of the North Caucasus and Manguishlak epi–Palaeozoic plates, etc.) are developed.

Analysis of the HC phase state at great depths has shown that the zonation in question is different for young and ancient platforms, which is a function of differences in their catagenic (paleothermal) histories. The analysis also helped to distinguish (among the wide variety of HC systems existing temperatures and pressures in situ) a new type of HC pools that was not known before. This is a gas pool with extremely high content of liquid hydrocarbons. Such HC systems are undersaturated, their temperature is above the point of beginning of condensation. At such conditions up to 1000 g or more of liquid HC can dissolve in the gas in situ. In other words it is a gas–condensate system with extremely high percentage of condensate that stays in a gaseous state.

The pools of this type are found in Pre–Caspian depression, West Siberia, Dnieper–donets Depression, their prediction is also possible on the basis of the analysis of P–T conditions and HC composition for various provinces and areas. Such pools are related to zones of abnormally high formation pressure, they are predicted at temperatures 120°C and higher and pressures exceeding 40MPa. Such zones are supposed to exist in Pre–Carpathian, Fergana, North–Caucasus–Manguishlak, South–Caspian, Amu–Daria, Dnieper–Donets (Northern part), Pre–Caspian (marginal parts) and West Siberia provinces.

The high–temperature gas–condensate pools have a specific feature: they are undersaturated, the percentage of liquid HC being high, which will enable to use such a production procedure that the condensate loss in situ will be at minimum with the recovery factor close to 1.

Discovery of high–temperature gas–condensate pools with high content of liquid HC will permit to considerably increase production of liquid HC from such pools owing to a low loss of condensate during extraction.

The prediction of the HC phase state helps to evaluate potential possibility of discovering pools of liquid and gaseous HC in individual zones within oil and gas provinces and areas, and conduct exploration for oil and gas at great depths more confidentially.

FINES MIGRATION IN ARGILLACEOUS SANDSTONES

D C Buller & T R Harper

BP Research Centre
Chertsey Road
Sunbury-on-Thames
Middlesex
United Kingdom

SUMMARY

Stress state and fluid seepage forces can have a seriously detrimental influence upon the permeability of argillaceous sandstone reservoirs in the near-well region. Because of stress concentrations around perforations, horizontal permeability in the critical area around perforation tunnels is particularly susceptible to impairment.

Stress difference induced significant loss of permeability in about 40% of 7 reservoir sandstones tested (total of 12 individual core tests). In some of these cases, the damage was attributable to fines mobilisation. Weaker reservoir rocks were most susceptible to such damage. Rock wettability, nature of the mobile phase and the magnitude of associated seepage forces were also found to be influential.

1. BACKGROUND

Dispersion and migration of formation fines have long been recognised as a potential source of permeability impairment and productivity decline in sandstone reservoirs[1,2]. Migrating fine particles can concentrate at pore constrictions, causing reductions in rock permeability. The extent to which this formation damage mechanism occurs in argillaceous sandstone reservoirs is not clear from the published literature. The principal purpose of this research was to attempt to assess this for selected reservoirs in the North Sea region.

Two basic mechanisms of fines migration are known: chemically-induced and mechanically-induced. Chemical effects include chemical incompatibility between an invading wellbore fluid and the formation, causing dispersion of pore-lining minerals. Our concern here is mechanically-induced particle movement within the rock associated with hydrodynamic forces and the influence of in-situ stresses, and principally formation damage governed by the latter. No attempt is made to quantify the influence of chemical control of fines migration.

Analysis of mechanically-induced fines migration has focused almost exclusively upon the influence of fluid seepage forces[3-9]. These authors reported the presence of critical or threshold pressure gradients or flow velocities for plugs taken from reservoir and outcrop sandstones, below which fines migration does not cause permeability damage. Critical fluid velocities of between 0.004 - 0.25 cm/s have been measured. Apart from Muecke, most investigators usually used brines as the mobile fluid phase and have ignored possible wettability effects.

It has been known for over 30 years that the permeability of sandstone cores can be reduced significantly by an applied stress[10]. The suggested mechanisms have invariably centred upon elastic deformations of the rock matrix causing reduction of pore throat sizes. Gobran et al. and Jones have reviewed the prior literature[11,12].

Apart from this well-known effect, however, a few researchers[13-15] have postulated that stress changes can initiate fines migration leading to **irreversible** permeability impairment. Dey proposed that clay particle rearrangement and migration within high aspect ratio (slot) pores as a result of hydrostatic confining stress was responsible for impairing rock permeability. Harper and Buller[14] proposed that fines may be liberated in a reservoir as a result of effective stress differences around a well or perforation. Aggour et al.[15] proposed that loosely cemented formation fines within Berea sandstone can be mobilised by the influence of hydrostatic confining stress leading to permanent impairment of rock permeability.

Recently, Holt[16] reported significant losses of absolute permeability (>60%) in a relatively weak outcrop sandstone subjected to non-hydrostatic stress conditions close to rock failure. He explained this damage by invoking grain crushing and/or a change of tortuosity caused by microcracking. He suggested that the phenomenon is a significant mechanism of formation damage.

Because of stress concentrations, the potential for microscopic failure of stressed rock is greatly increased around openings. The stress difference, $\sigma_V - \sigma_H$, is increased around a perforation. This represents an increase of deviatoric or shear stress. Deviatoric stress causes macroscopic rock failure e.g. wellbore breakout or sand production. However, microscopic local failure (microcracking) of stressed rock occurs prior to macroscopic failure[17]. This may result in either local enhancement of permeability or, alternatively, fines release leading to formation damage. Reservoir depletion exacerbates the problem by increasing the in-situ effective rock stresses. Failure around a perforation tunnel can be thought of as analogous to the phenomenon of wellbore breakout.

In reality, stress distribution and rock strength distribution around a perforation tunnel will also be affected by jet perforation. This promotes microscopic failure and generates fines, influencing permeability and the potential for permeability change during flow to the well. However, current knowledge of the condition of reservoir rock adjacent to a perforation is at best extremely limited. This effect is ignored for the purposes of this paper.

It is helpful to estimate the likely stresses around a perforation tunnel to set in context this experimental work. Rocks fail in response to effective stress (total stress minus pore pressure), so it is appropriate to estimate stress differences at typical reservoir depths. Assuming the minimum horizontal total stress equals 0.6 of the vertical total stress, then the maximum stress difference calculated for a linear elastic material at the shoulder of a perforation tunnel could be about 170 MPa at a depth of 3000 m. Stress differences of up to about 55 MPa were used during the experimental programme. Thus, stress conditions around perforations at typical reservoir depths, calculated using the conventional oversimplified model of stress state, are theoretically more conducive to rock breakdown than those simulated in this work. Indeed, in reality, failure local to perforation tunnels analogous to wellbore breakout is strongly implied. Such failure and associated permeability changes will vary with orientation around the perforation tunnel according to the effects of the interplay of stress concentration and seepage forces (see Hoek and Brown[18]).

2. TEST MATERIAL AND PROCEDURES

2.1 Reservoir Cores

Preserved core material was selected from 7 reservoir sandstones, principally from the North Sea or Dorset Coast. The plugs contained a wide variety of detrital and authigenic minerals (quartz, feldspars, mica, chert, ankerite, kaolinite, illite, smectite, chlorite, calcite, dolomite, siderite, haematite, pyrite and anhydrite). Core pieces were analysed by CATScan prior to plugging to avoid significant heterogeneity.

Plugs of typically 3.8 cm diameter by up to 7.6 cm in length were cut in the plane of bedding from whole cores. Considerable care was taken to ensure that the plug end-faces were trimmed square. The cores were plugged and stored under simulated formation water (SFW) or a sodium chloride brine. Prior to testing, the plugs were evacuated for 24 hours under brine to remove trapped gases. Pore size determinations were conducted on the plugs or plug trims. Most of the material was water-wet; core from one reservoir was intermediate to oil-wet.

2.2 Test Fluids

Two types of core flow tests were carried out: oil as mobile phase and water as mobile phase. Premium grade kerosene (additive-free) and various water-based fluids (1-5% NaCl, 3% wt. KCl and SFW) were used. KCl and SFW brines were used in those tests in which the specimen was predicted (on the basis of the mineralogy) to be particularly water-sensitive. Fluids were filtered to 0.45μm prior to use, and were passed through an in-line, 0.22 μm membrane filter immediately upstream of the core plug.

2.3 Core Flow Tests

Twelve cores were tested during this study, applying independent confining and axial loads to each sample. A maximum value of about 55 MPa confining and axial pressures were applied. Spherical seat end-platens were used to ensure uniform application of axial load. An ability to achieve uniformity of axial loading of plugs was confirmed prior to the test programme using strain-gauged stainless steel dummy core plugs.

All experiments were conducted at a constant temperature of 303°K and atmospheric pore pressure. The influence of stress on permeability was assessed over a series of constant flow rates, commencing the test at a low flow rate and increasing it in stages. The effective stress difference applied to a plug was increased in stages at each constant flow rate. Fluid differential pressures typically varied from a minimum of about 0.17 MPa to a maximum of 3.45 MPa. Injection was continued in one flow direction at a particular stress until steady-state conditions were attained; often this required prolonged flow periods. The rock stress or fluid flow condition was then altered. Data are presented as permeability change of the plug (ratio of measured permeability K to initial permeability K_i) with volume of injected fluid. DP represents the measured differential pressure across the core.

Particulate material flushed from the plugs was collected on an in-line, 0.22μm membrane filter. They were analysed by quantitative scanning electron microscopy to characterize the nature and size of the particles.

2.4 Rock Strength Tests

Strengths of the core samples were quantified by the Vickers hardness test[19] and ISRM point load test[20]. Measurements were made on the specimens after flow testing, the samples being saturated in the test fluid.

3. INTERPRETATION OF RESULTS

3.1 Basis For Interpretation

Identification of fines movement in core is a somewhat subjective process, invariably reliant upon the investigator's expertise based upon prior work. Primarily, three criteria (outlined below) were used. Only if at least 2 of these were clearly satisfied was fines migration concluded to have occurred. (This basis for interpretation will tend to set a lower bound estimate for the frequency of fines migration.)

1. Significant (>10%) changes of core permeability associated with either effective stress or flow rate changes.

2. Reversal of fluid flow direction through a sample[6,21,22]. If fines are present, the sample permeability is predicted to increase signficantly immediately upon reversal of flow and to decline gradually to the original permeability. In the author's experience, such behaviour is observed infrequently in reservoir core, only one clear instance being recorded in this study. It may be that the migration of fines across pore bodies is so rapid that the pressure perturbation cannot be measured.

3. A correlation of the nature and frequency of expelled particulate material with experimental conditions. However, migrating fines may not be flushed from the rock.

3.2 Nature of Fines Expelled From Cores

Much of the prior research implies that moveable formation fines are comprised solely of authigenic clay minerals (see review by Egbogah[8]). In particular, the presence of kaolinite in a reservoir is often taken as an indication of fines migration potential (chemically or mechanically induced). This is because of its inferred "loose" attachment to pore walls under microscopic examination[8]. Several researchers[3,8,23] have also claimed that fibrous illite in reservoir rocks is susceptible to mechanical fragmentation by fluid seepage forces. These suggestions are highly speculative because no direct measurements have been made of the bonding of authigenic minerals to a rock substrate, or of clay particle strength. In

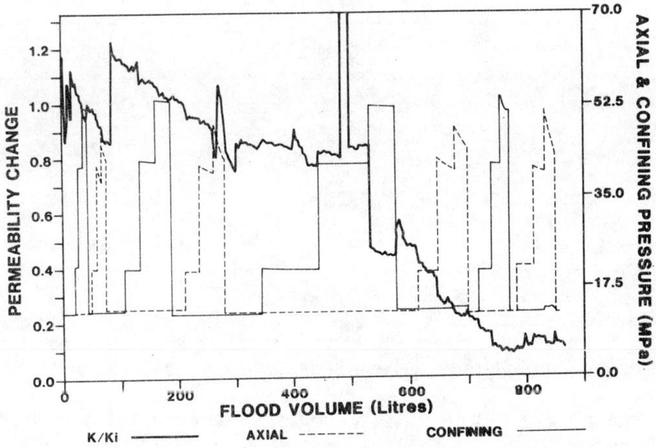

Fig. 1 Influence of effective stress on permeability of Sherwood sandstone 2166.69m.

addition, most prior work has used outcrop sandstone, not reservoir rock.

It is quite clear from our research that non-clay minerals are also mobilised by fluid flow through argillaceous sandstones. Significant amounts of quartz, pyrite, calcite, as well as illite and smectite were expelled from plugs during this series of tests. In addition, only minor quantities of kaolinite, chlorite and apatite were present in core effluent. This confirms previous sparse findings of the potential for non-clay sandstone minerals to be dispersed and transported by fluid flow[4,6,22,24].

Interestingly, in our work kaolinite was not flushed from a plug from the Sherwood reservoir (2166.69 m), although the specimen contained 2% of this authigenic clay and the material was damaged by brine injection (Figure 1). Previous researchers[9,22] have reported a similar finding - **kaolinite in reservoir rock is not necessarily mobile under high fluid seepage forces**. It must be stressed, however, that the lack of a particular mineral in core effluent need not imply that it has not been mobilised. Damaging particles may not travel far within the rock. In an extreme case, mobilised pore-lining minerals may not escape from a single pore.

The sizes of the fines eluted during each experiment were invariably less than the maximum effective pore diameter of the host rock. The majority of particles were much smaller (roughly 5-fold) than the upper pore size bound (Table 1). This is hardly surprising. No correlation existed between the mineralogy of the samples (as analysed by standard petrographic techniques) and the magnitude of permeability damage suffered by the rocks. It was impossible to predict the nature of the expelled minerals by these techniques.

Table 1: Summary of tests conducted with flowing wetting phase

SANDSTONE FORMATION	DEPTH (m)	INFLUENCE OF EFFECTIVE STRESS	PORE DIAMETER RANGE (μm)	FINES PRESENT	POINT LOAD STRENGTH (MPa)	VICKERS HARDNESS (MPa)
Bunter	1069.11	N	0.1-10	N	0.51(6)	282
Rotliegendes	1751.27	M	4-15	M	0.5(6)	451
Devonian	1777.80	T	0.2-40*	N	0.17(5)*	-
Bridport	959.55	Y	2-15	Y	0.04(1)	22
Sherwood	2166.69	T	0.1-40	Y	0.14(2)	48
Lower Leman	3095.15	N	0.01-8	N	0.63(5)	450
	3291.21	M	1-50	M	0.02(2)	-

Key: T Triggered progressive decline; Y Yes; N No; M Minor

Notes: Flowing phases were either a chemically compatible brine or kerosene.
Number of Point Load Strength measurements given in brackets.
* at 1777.76m

3.3 Kerosene as Mobile Phase

Influence of Stress Difference

Increasing stress difference caused negligible permeability damage with oil flow in 2 out of the 5 reservoir rocks tested. In addition, increasing stress difference caused minor irreversible permeability increases in core from another reservoir.

However, effective stress changes caused significant irreversible losses of permeability to oil in plugs from 2 of the 5 reservoirs assessed (Devonian sandstone and Bridport sandstone). Both of these specimens had low strength (<0.17 MPa by Point Load). (Interestingly, only minor losses of permeability occurred during injection of brine into identical Devonian rock under similar stress conditions. This reservoir is intermediate to preferentially oil-wet and fines migration would only be expected to occur within this rock when oil is the mobile phase[4].)

Increasing stress difference caused a major, largely irreversible loss of permeability to oil (86%) in the weak Bridport sandstone (Figure 2). Flow reversals indicated the absence of mobile fines prior to stressing the core and the presence of moveable material in pores after loading and associated permeability loss (Figures 3 and 4). In addition,

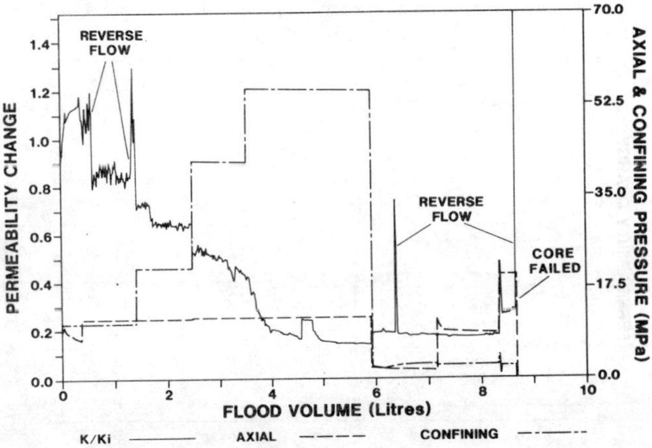

Fig. 2 Influence of effective stress on permeability of Bridport sandstone 948.89m.

quartz and aluminosilicate minerals (up to 3.8 μm diameter) were filtered from the core effluent during and after loading. The size of these particles was roughly similar to the lower bound of the effective pore diameter.

Influence of Seepage Forces

Oil injection pressure was found to be an important factor controlling permeability in only 1 out of the 5 reservoir sands assessed. Only the permeability of a specimen from the Bunter sandstone (1068.99 mm) was **reduced substantially** (between 29-36%) at zero effective stress by increased fluid seepage forces. There was no evidence for significant particulate material flushed from this sample during the test.

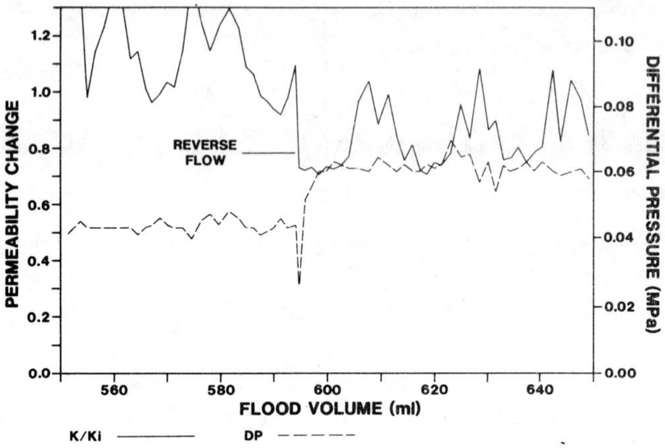

Fig. 3 Influence of reverse flow on permeability of Bridport sandstone 948.89m before stress.

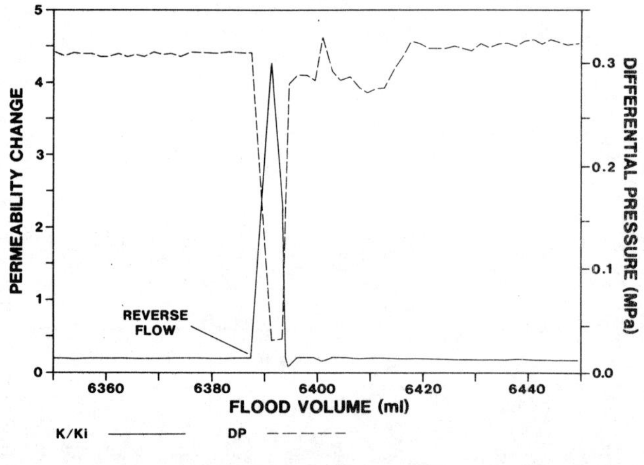

Fig. 4 Influence of reverse flow on permeability of Bridport sandstone 948.89m after stress.

3.4 Brine as Mobile Phase

Influence of Stress Difference

Only 2 of the 7 reservoir sandstones tested were influenced significantly by effective confining stress when brine was the mobile phase. Both of these were from the weak (0.14 MPa) Bridport and Sherwood reservoirs. A 48% irreversible loss of permeability was measured in Bridport sandstone when subjected to increased stress difference (Figure 5). Reverse flow stages suggested that the applied stress had initiated fines movement within this rock. Interestingly, the permeability of the

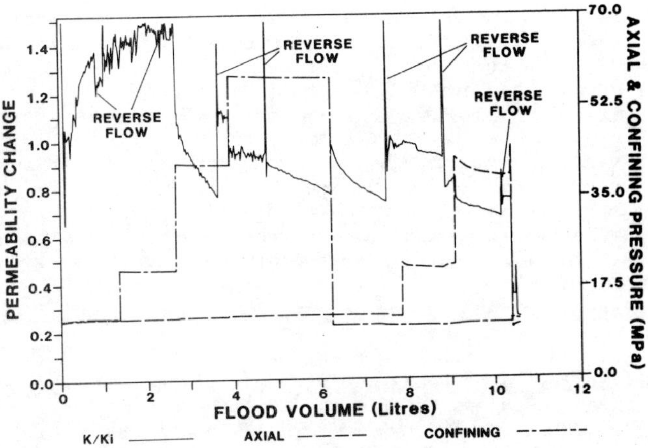

Fig. 5 Influence of effective stress on permeability of Bridport sandstone 959.55m

Sherwood sandstone was also impaired irreversibly (43%) by increased stress difference, but only at a relatively high fluid injection pressure of 1.38 MPa acting on the 7.6cm long plug (Figure 1). Stress difference had no obvious influence on this latter material at the lower fluid pressure gradients. This suggests that initiation of permeability damage in this material could be dependent upon a coupled effect of high fluid differential pressure and effective stress, or stress cycling. Again, the permeability damage measured in this latter test was probably caused by fines migration, although it was not possible to link the amount of eluted fines to stress changes.

However, the permeabilities of the reservoir sandstones to brine were less affected by applied stress difference when the axial stress exceeded the confining stress.

Influence of Seepage Forces and Flow Volume

In all but one (6 out of 7) of the tests conducted using brine as a mobile phase, the permeabilities of the cores were reduced significantly by fluid throughput, irrespective of the magnitude of the injection pressure or applied effective stress. The experiment in which brine injection failed to induce permeability damage was conducted on an intermediate to oil-wet Devonian sandstone (1777.76 m).

Typically, the permeability to brine declined gradually with the volume of fluid injected. It can be argued that gradual blocking of the injection face of the samples by corrosion products (rust) from reaction of the brine with the equipment caused this phenomenon. Indeed, some rig corrosion material (iron, nickel and chromium oxides) was detected on most membrane filters collected from downstream of the samples. Even so, core face blocking is not considered to be the principal cause for the following reasons:

1. In every test the brine injected into the plug was passed through an in-line, 0.22 μm membrane filter immediately before being injected into the core. Visual examination of the pre-filters, filters and

inlet faces of the specimens indicated that this precaution resulted in the removal of particulate contaminants. In only 1 of the 12 experiments was the injection face of the specimen found to be slightly discoloured by invading particulate material.

2. The plug of Devonian sandstone suffered no loss of permeability during injection of about 28 litres, even though rig corrosion products were detected on the downstream filters. This suggests that the corrosion products that passed through the membrane filters were small enough to pass straight through the rock. (However, it is clearly dangerous to generalise and sub-micron sized fines may become trapped within tight rocks which contain small pore throats.)

Several authors[6,7,8,9] have reported the existence of critical fluid injection pressures for initiating fines migration. In this study only 1 reservoir sandstone of the 4 water-wet specimens tested for this phenomenon exhibited this effect with brine as the mobile phase. The rock was weak (0.14 MPa by Point Load) but was not affected to any great extent by stress or brine throughput at fluid injection pressures below 1.38 MPa. However, at a differential pressure of 1.38 MPa (0.18 MPa/cm), high effective stress difference caused a serious loss (43%) of permeability (Figure 1). Continued fluid injection at pressures of up to 3.44 MPa (0.45 MPa/cm) resulted in additional significant permeability damage.

4. CONCLUSIONS

1. Stress difference during flow of oil or brine induced significant permeability damage in about 40% of the reservoir sandstones tested. The four weakest rocks (Point Load Strength <0.2 MPa) all showed some response to increasing stress difference. The least responsive of these four rocks was that with the highest maximum pore diameter. Clearly, application of shear stress may induce significant formation damage in weak reservoir rocks. Increased drawdown will exacerbate the problem.

2. The tests confirmed that rock wetted condition is an important factor controlling the susceptibility of sandstones to permeability impairment by fines movement. Flow of a chemically compatible brine revealed impaired permeability in over 80% of the sandstones, while injection of oil was associated with a loss of permeability in only 33%. This is a reflection of the bias towards water-wet sandstones in the test samples. Significant gradual and irreversible losses of permeability occurred, irrespective of the fluid pressure gradient or applied stress difference.

3. Rock strength was found to be a guide to the susceptibility of a sandstone to mechanically-induced permeability damage by fines movement (in response to applied stress difference or fluid seepage forces).

4. Significant formation fines were expelled from most cores during fluid flow. These materials were not restricted to clay minerals. The predominant mobile mineral was found to be quartz particles.

5. Reversible stress-dependent permeability changes were insignificant.

6. Although the literature suggests that threshold fluid pressure gradients exist in outcrop and reservoir sandstones, in all but one of these tests using reservoir material, critical injection pressures above which damage (fines movement) was initiated were not evident.

7. Prediction of the potential for fines migration from inspection of thin sections is typically, if not inevitably, misleading.

REFERENCES

1. NOWAK T J and KRUEGER R F (1951). The effect of mud filtrates and mud particles upon the permeabilities of cores. API Drilling and Production Practice.
2. BERTNESS T A (1953). Observations of water damage to oil productivity. API Drilling and Production Practice.
3. KRUEGER R F, VOGEL L C and FISHER P W (1967). Effect of pressure drawdown on cleanup of clay or silt blocked sandstone. JPT. (March) 397-403.
4. MUECKE T W (1978). Formation fines and factors controlling their movement in porous media. Paper SPE 7007.
5. BERGOSH G L and ENNISS D O (1981). Mechanisms of formation damage in matrix permeability of goethermal wells. Paper SPE 10135.
6. GRUESBECK C and COLLINS R E (1982). Entrainment and deposition of fines particles in porous media. Paper SPE 8430.
7. GABRIEL G A and INAMDAR G R (1983). An experimental investigation of fines migration in porous media. Paper SPE 12168.
8. EGBOGAH E O (1984). An effective mechanism for fines movement control in petroleum reservoirs. Paper 84-35-16 presented at 35th Anniv. Techn. Meet. of Petr. Soc. of CIM, June 10-13.
9. LEONE J A and SCOTT E M (1987). Characterisation and control of formation damage during waterflooding of a high-clay-content reservoir. Paper SPE 16234.
10. FATT I and DAVIES D H (1952). Reduction in permeability with overburden pressure. JPT. (Dec) 4, 16.
11. GOBRAN B D, BRIGHAM W E and RAMEY H J (1981). Absolute permeability as a function of confining pressure, pore pressure and temperature. Paper SPE 10156.
12. JONES S C (1988). Two-point determinations of permeability and PV vs net confining stress. SPE Formation Evaluation (March), 235-241.
13. DEY T N (1986). Permeability and electrical conductivity changes due to hydrostatic stress cycling of Berea and Muddy-J sandstone. J. Geophysical Research, 19, B1, 763-766.
14. HARPER T R and BULLER D C (1986). Formation damage and remedial stimulation. Clay Minerals, 21, 735-751.
15. AGGOUR M A, MALIK S A and HARARI Z Y (1989). Effect of cyclic formation-pressure changes on permeability. SPE Reservoir Eng. (February), 91-96.
16. HOLT R M (1989). Permeability reduction induced by a nonhydrostatic stress field. Paper SPE 19595.
17. SCHOLZ, C H (1968). Microfacturing and the inelastic deformation of rock in compression. J. Geophys. Res., 73, 4, 1417.
18. HOEK E and BROWN E T (1986). Underground Excavations in Rock, 1.M.M.
19. O'NEILL H (1964). Hardness measurement of metals and alloys, 2nd ed., Chapman and Hall.
20. ISRM Point Load Test (1985). International J. Rock Mechanics, Minerals Science and Geomechanics. Abstr. 22, 2, 53-60.
21. HEWITT C H (1963). Analytical techniques for recognising water sensitive rocks. JPT. 15, 813.
22. GRAY D H and REX R W (1965). Formation damage in sandstones caused by clay dispersion and migration. Proc. 14th Natl. Conf. on Clays and Clay Minerals. Pergamon Press, NY, 355-366.
23. MUNGAN N (1965). Permeability reduction through changes in pH and salinity. Paper SPE 1283.
24. REED M G (1977). Formation permeability damage by mica alteration and carbonate dissolution. JPT. (Sept) 29, 1056-1060.

A MULTIDISCIPLINARY APPROACH IN RESERVOIR DESCRIPTION AND PRODUCTION PLANNING

A. COSTA E SILVA, J. CARVALHO

Partex, CPS, Av. 5 de Outobro 160, 1000 Lisboa, Portugal

Summary

Nowadays Reservoir Description has become a critical step in the whole Reservoir Engineering subject. It has been understood that numerical simulation results of fluids flow are dependent on the accuracy of Reservoir Description Models. Furthermore, given the greater uncertainty about the subsurface conditions, the prediction of reservoir architecture between the wells justifies the use of a stochastic approach to deal with the uncertainties and quantify them.

This paper summarizes the results obtained by Partex in the development of stochastic modelling techniques applied to oil reservoirs based on a multidisciplinary approach due to parallel developments in petroleum geology, stochastic techniques and computer science. This research project has been conducted in cooperation with the Technical University of Lisbon (IST/CVRM) and is supported by the EEC Hydrocarbons Commission.

Case studies are presented showing that a combination of multivariate statistical techniques, geostatistics and appropriate geological reasoning can provide a reliable basis for the improvement of reservoir description and modelling with positive consequences for the field development strategy and oil recovery optimization.

The heterogeneity problem is approached by subdividing the reservoir into Reservoir Quality Zones (RQZ) by Multivariate Statistical techniques. The simulation of reservoir and/or zones boundaries is approached by a geostatistical method based on indicator Kriging. Oil–in–place volume is calculated and distribution of reserves per zones is achieved.

1. INTRODUCTION

Reservoir Description is a conceptual model that describes the spatial distribution of fluid and rock properties and the internal architecture of oil reservoirs.

A major difficulty in building reservoir description models is, almost always, the limited data available given that the reservoirs are sampled by core and log data which represent an infinitesimal fraction of the total reservoir volume.

Interpolation and extrapolation are required in the description phase, always under geological guidance and interpretation, in order to discretize the reservoir into subunits such as layers or grid blocks and assigning values of relevant reservoir properties to these blocks. This task is critically given the scale difference between the sampling supports and simulator grid blocks (1).

The development and application of stochastic modelling techniques to reservoir description problems (Fig. 1) combined with the integration of geology and reservoir engineering, can lead to more accurate models in order to improve the prediction of reservoir architecture specially in the interwell scale.

According to this issue oil companies and research centers have concentrated efforts in the last years aiming at the improvement of Reservoir Description and Quantitative

Geological Models. Within this framework it is felt that a Multidisciplinary approach, involving a close cooperation between different disciplines and techniques, is a major guideline for improving the knowledge of the internal architecture of oil reservoirs. The economic consequences for the oil industry can be very positive at different levels specially in areas like simulation of fluids flow, reservoir management, production planning and secondary or enhanced oil recovery programmes.

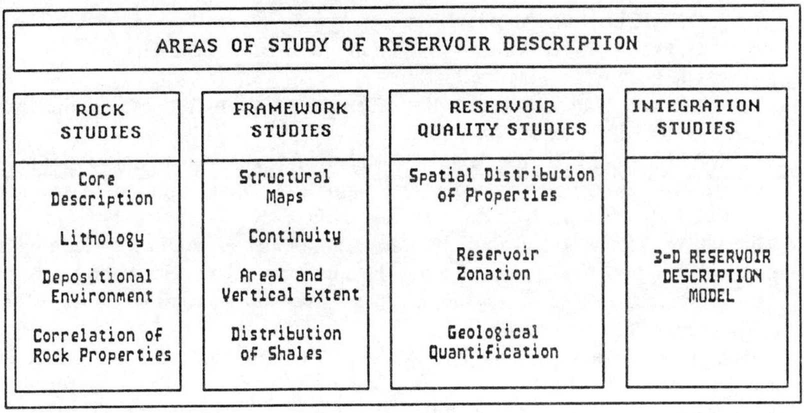

Fig. 1

2. MULTIDISCIPLINARY APPROACH

The multidisciplinary approach was developed on the basis of synergism between the different disciplines involved in the acquisition and interpretation of reservoir data – geology, seismic, petrophysics, reservoir engineering – combined with the use of probabilistic models and computer science (Fig. 2). This is a major technological issue and allows to bridge the gap between descriptive approaches (geology) and numerical models (simulator). The use of probabilistic models can cope with the uncertainties about the subsurface conditions and quantify them leading to better predictions of reservoir internal architecture and properties variations.

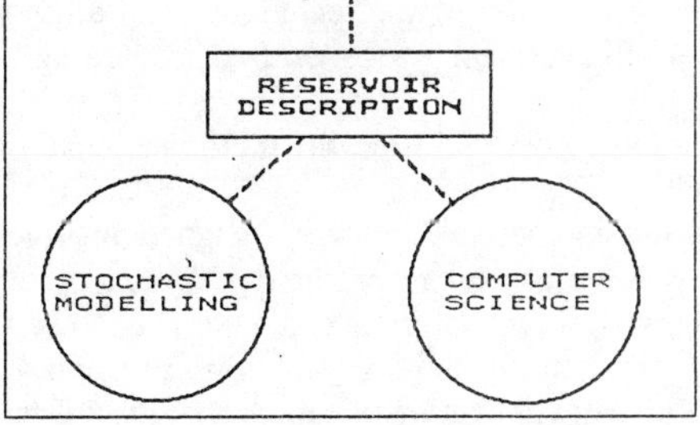

Fig. 2

The multidisciplinary approach is based as well on a synergism between different models and techniques which include Multivariate Data Analysis, Geostatistics and Discrete Models emerging from the application of Indicator Kriging and Morphological Simulation (2).

The different techniques have been articulated according to the main steps of the proposed methodology summarized as follows:

- ZONATION
- STRUCTURAL ANALYSIS
- GEOSTATISTICAL ESTIMATION AND SIMULATION
- DRILLING OPTIMIZATION

The application of geostatistics in the reservoir modelling process requires always a previous structural analysis, based on variography (3).

The VARIOGRAM is the basic geostatistical tool, which conveys the spatial correlation into the estimation procedures (Kriging) by modelling the variance of the Regionalised variable increments at different spatial lags.

The statistical inference on a reliable variogram model, as required by Kriging, entails the assumption of stationarity of increments at the appropriate scale. Hence, in heterogeneous fields, where stationarity does not hold, it is necessary to define zones of similar characteristics, prior to structural analysis.

This is the rationale behind the ZONATION TECHNIQUES used in this project, both in vertical and horizontal directions of the reservoir. The zonation was performed using multivariate statistical techniques like Principal Components Analysis and Correspondence Analysis, which rely on cross–correlation between variables.

Once a zone is defined on samples, either by a specific algorithm or by raw indicator data the problem of boundaries estimation or simulation arises. In fact, analogously to continuous regionalised variables, the spatial arrangement of indicator values can be modelled via a structure function – the transitive variogram or geometric covariance –, which allows the interpolation of sample information into the spatial domain. If the estimation variance is minimized, morphological kriging provides the optimal boundaries from zone to zone; when it is required a set of images honouring the geometrical variability of the zones for the initialization of production planning models, the morphological simulation is the appropriate technique for mapping heterogeneities.

Whenever an estimation is performed using geostatistical methods, it is always possible to assess the sensitivity of the error involved in the estimation with respect to the number and configuration of any information pattern, since the kriging variance depends only on the variogram model and on the location of existing and predicted wells. So, OPTIMIZATION OF THE SUPPLEMENTARY RECOGNITION pattern can be performed, by maximizing the decrease on the estimation variance, entailed by each new configuration.

3. METHODOLOGY

Once defined the articulation of the techniques and areas of study, a representative data set of a Middle East oil reservoir was organized, according to the requirements of the processing modules.

Hence, the methodology derived from the multidisciplinary approach was extensively tested as described in Fig. 3.

A step–by–step iterative procedure was designed in order to screen the available processing techniques to solve critical reservoir description problems in different areas of interest: zonation, geometrical modelling, estimation of reservoir properties, reserves calculation, simulation of heterogeneities and drilling optimization. Once the steps accomplished, the conditions of application of each technique are determined, their adjustment to the specific characteristics of oil variables was performed and, eventually, new approaches are developed.

Examples of case studies based on this methodology and calling for integration of different techniques and areas are presented below.

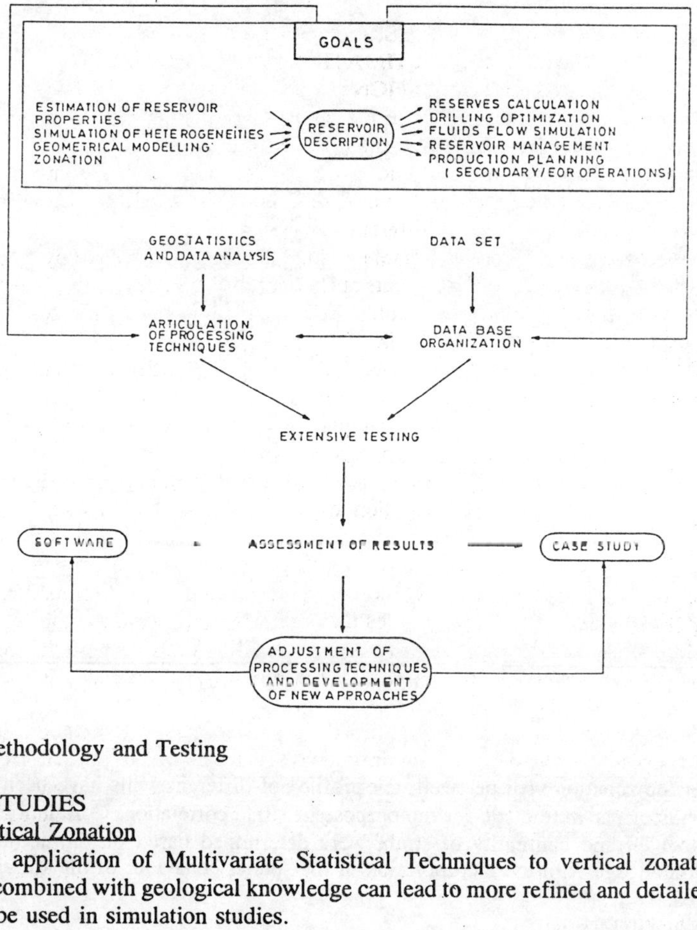

Fig. 3 Methodology and Testing

4. CASE STUDIES

4.1 Vertical Zonation

The application of Multivariate Statistical Techniques to vertical zonation of oil reservoirs combined with geological knowledge can lead to more refined and detailed layering models to be used in simulation studies.

In this case Principal Component Analysis (PCA) was applied to multivariate information on reservoir properties (porosity, elevation, lithologies, fluids saturation, grain density, etc.) in order to define meaningful petrophysical layers for input in simulation studies.

The relations and interdependencies among variables and their influence in the characterisation of the different strata can be viewed by the study of their projections onto the factorial planes and geometrical arrangement.

The effects of lithologies in the Zonation and the appraisal of heterogeneities within the facies can be evaluated from the example shown in Fig. 4.

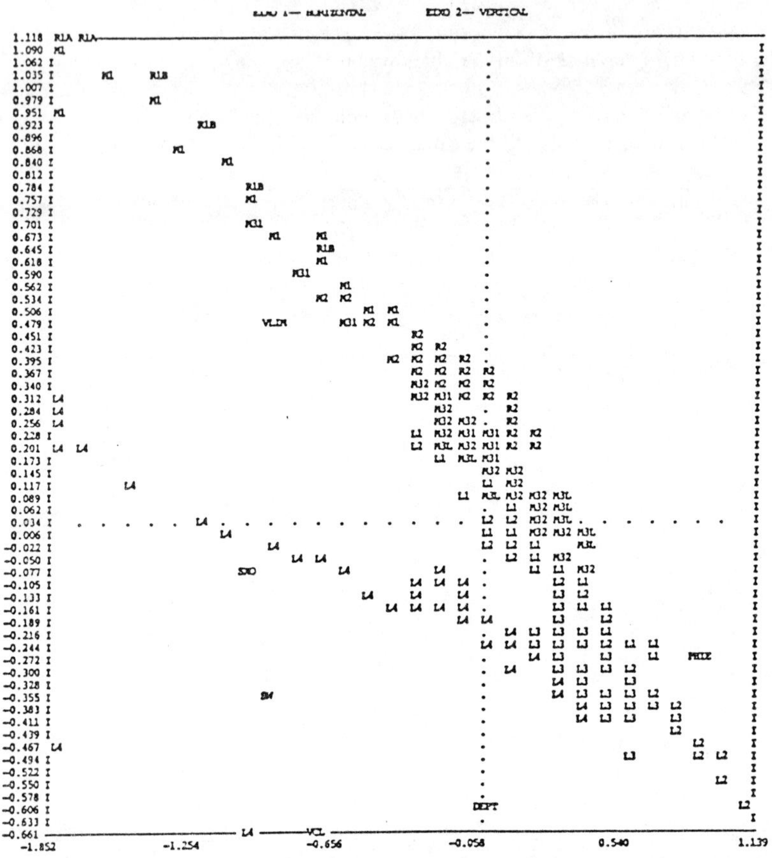

Fig. 4

After the zonation well per well, the profiles of different wells have been compared and local adjustments were made for the purpose of strata correlation. Correlations of zones from well to well and continuity of strata were determined based on the combination of geological knowledge with the incorporation of the spatial behaviour of the variables in the interwell area.

4.2 <u>Horizontal Tesselation</u>

From the layering model which result of the vertical zonation, a study was conducted for the most important layers in order to define Reservoir Quality Zones (RQZ) regarding the fluids saturations and spatial distribution of the net oil column. This study provides the knowledge of the internal architecture of the reservoir, which is a key point for the planning of production and recovery operations.

The heterogeneity problem was approached by a tesselation of the reservoir into RQZ which are set up by using multivariate statistical techniques reflecting the current understanding of correlations involving the available geological and petrophysical information.

After a global test involving all the information available, a case study was run based on selected variables which control oil distribution: elevation and water saturation. Using different class limits in the selected variables, a series of two–way contingency tables was built and the Correspondence Analysis programme (CA) was run for each contingency table, taking the other attributes (facies, porosity, transmissivities) as supplementary variables. The "final" class limits, which define the groups of wells retained as RQZ, were chosen using the criterium of maximum similarity of supplementary variables within the groups, combined with the geological knowledge of the reservoir.

The spatial two–dimensional arrangement of groups is displayed in Fig. 5. Their characterization in terms of supplementary variables is useful for production planning purposes. Furthermore correlation of CA zonation results with facies distribution are analysed, in order to understand the role of facies in reservoir continuity.

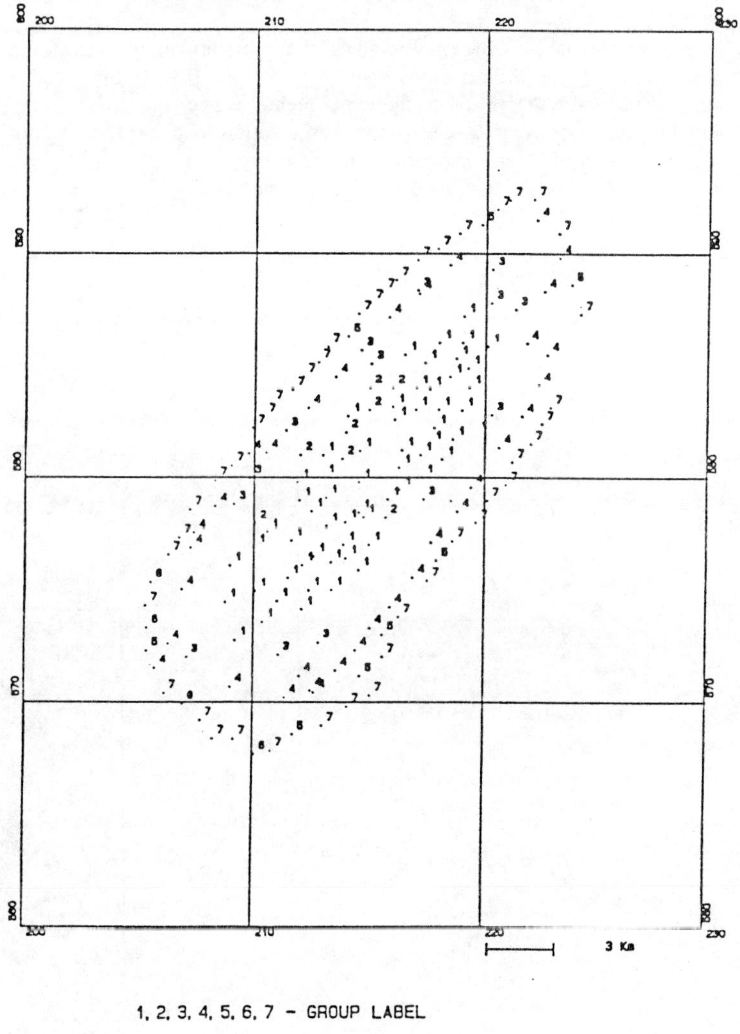

Fig. 5

4.3 Geometrical Modelling

The boundaries of each zone and the reservoir geometry can be modeled using a technique developed on the grounds of transitive Kriging of Indicator Data (4).

The basic idea underlying the method is that the limits of the closure area can be obtained by Kriging an indicator variable, which can only take two values: 1 if a sample belongs to the zone or 0 otherwise. Using the indicator data as an input, the transitive variogram is calculated and the Kriging system is solved, producing a map of Kriged indicator values, contained in the interval (0,1) which can be interpreted as the probability of belonging to the entity. This probability map is converted into a binary map, giving, for the dense grid of points, the sub-set which is most likely to be considered as the selected zone. In order to illustrate the methodology, an example is presented of the boundaries estimation of the richest oil zone.

The indicator variable, which is the basis of the methodology, was defined and the transitive variogram of the indicator variable is shown in Fig. 6 for the NS and EW directions. A power model was fitted to the experimental curve.

A regular grid was superimposed to the well location map and the Kriging system is solved producing a map of kriged indicator values.

The estimated boundaries of the Zone 1 are shown in Fig. 7.

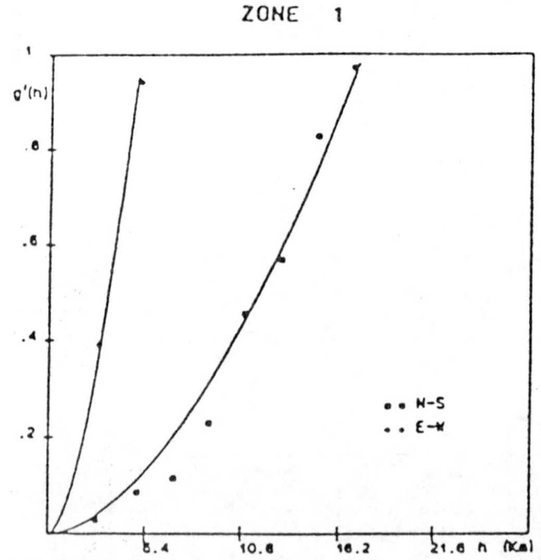

Fig. 6

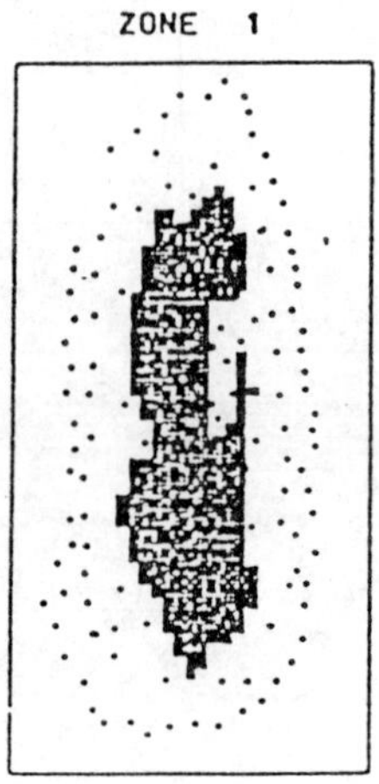

Fig. 7

4.4 Estimation of Net Oil Column and Reserves Calculation

Following the estimation of each zone boundaries next step is the estimation of average net oil column inside the boundary which accounts for the distribution of the oil–in–place per zone. This enhancement is an important factor for the planning of future recovery operations.

Given a certain zone, limited in 2–D by the boundaries estimated in the previous step, the oil–in–place contained in the zone is given by the product of the area by the mean kriged value of the variable net oil column calculated inside the same boundaries.

Variograms of net–oil–column were calculated and interpreted for each one of the zones. Example for zone 1 is given in Fig. 8. In this case the fitted model is spherical, parameters of the model and results of tests for this case are presented in table 1.

VARIOGRAM PARAMETERS	
C0 $(ft)^2$	.55
C1 $(ft)^2$	.24
C2 $(ft)^2$	.24
a1 (Km)	4.5
a2 (Km)	1.5
anisotropy ratio	1:3
VALIDATION CRITERIA	
ME	–.031
MRSE	–.999

Table 1

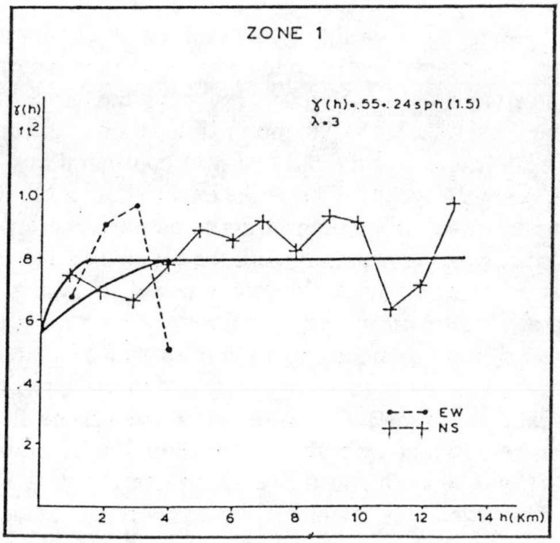

Fig. 8

For estimation purposes and according CA zonation results the reservoir is split into 6 embedded zones. Zone 1 corresponds to the highest oil content and quality. If zone 2 is considered, group 2 is added to zone 1 and the overall oil quality decreases. Finally zone 6 contains the global amount of the recoverable oil in the reservoir.

After the modelling and interpretation of the variagrams Universal Kriging was used as a geostatistical method of estimation of the average value of net oil column in the zones (5).

For oil–in–place calculations geostatistical methodology was adapted to the specific characteristics of oil variables and the 3–D geometric problem was split into two parts: Estimation of the net oil column variable and goemetrical estimation of the zone boundaries.

In this regard estimates and associated errors of net oil column variable were composed with estimators and associated errors of surface (boundaries) for each zone and results are presented in Table 2.

ZONES	Area (Sq.Km)	Average Net Oil Column (ft)	Volume Error (%)	Oil Volume (STB)X10^9	Percentage of Total Oil–in Place (%)
1	80.784	8.405	6.96	0.81359	52.60
2	93.258	7.922	6.73	0.88525	57.20
3	128.304	8.030	4.60	1.23450	79.90
4	182.655	6.844	4.19	1.49790	96.95
5	189.486	6.492	4.66	1.47400	95.40
6	204.336	6.310	4.43	1.54500	100.00

Table 2

4.5 Drilling Optimization

Being an estimation performed by geostatistical methods it is possible to assess the sensitivity of the error involved according to the number and configuration of drilled wells.

In this regard, a case study was conducted involving the determination of the optimal drilling sequence in a certain stage of the field's life using the Kriging variance as a precision measure which enables us to calculate to optimal location of supplementary data points comparing the Kriging Variances of the old and new configurations.

Taking into account the wells drilled in the exploration phase (Fig. 9) and a variogram model fitted to the variable net oil column, a fictitious point sweeps over the field and the new estimation variance is calculated along with the relative variance reduction. The results show that the optimal zones of the field to locate new wells, leading to a larger reduction of the kriging variance, are located in the western flank.

Furthermore a drilling sequence was optimized and compared with the real drilling history of the field corresponding to the 17 wells drilled in the appraisal phase (Fig. 10). The results show that the first wells (A–B–C) are located in the western flank; the optimal drilling sequence is significantly different from the real drilling history and if it was followed, the objective of a satisfactory knowledge of the field at the early stages could be attained, with less wells and at lower costs. This feature is very important because the critical decisions concerning the development and management of the field are taken at early stages and the drilling of each new well should be optimized.

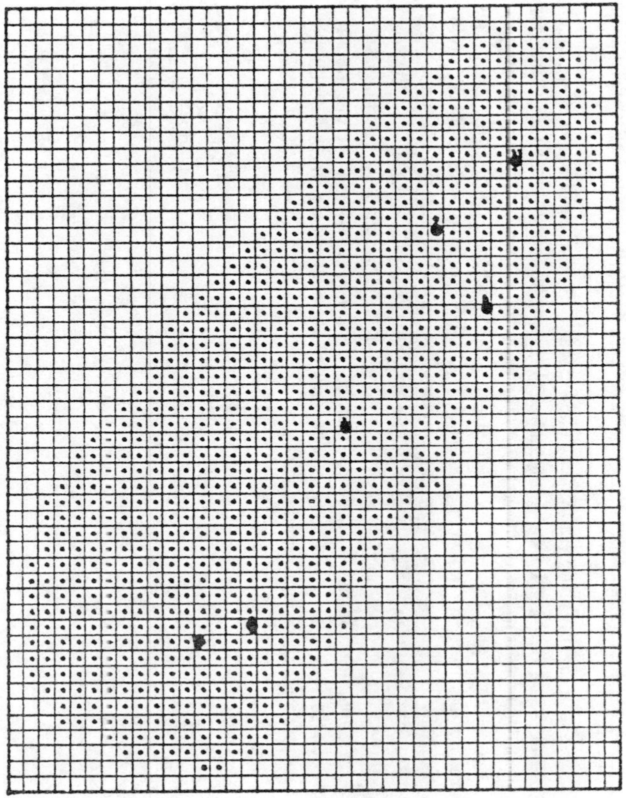

Fig. 9 Exploration Phase (Wells location)

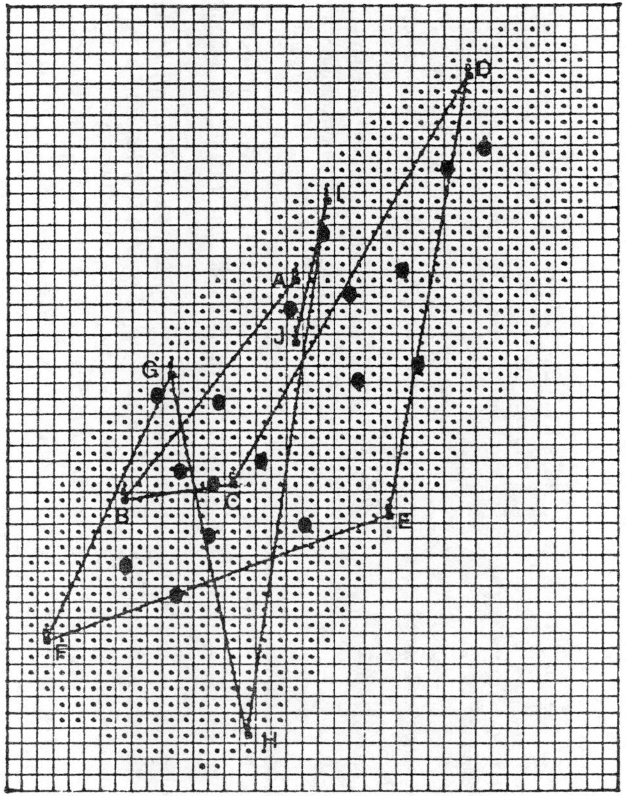

Fig. 10 Optimal Drilling Sequence

5. CONCLUSIONS

A combination of geostatistics, multivariate data analysis and geological knowledge is an important technological issue in order to bridge the gap between descriptive approaches and numerical models, providing more reliable estimates of reservoir properties and improving the input for simulators.

Multivariate Statistical Analysis offers a practical method of dividing a reservoir into physically – meaningful zones and ascertain which zones are likely to be continuous between adjacent wells.

The horizontal tesselation of reservoir layers is a useful basis for the optimization of oil recovery programmes; the definition of Reservoir Quality Zones and their correlation with facies distribution can highlight problems like reservoir continuity, recovery efficiency and fluids displacement paths.

From the case study results, the conclusion was reached that zonation outputs are resistent to the decreasing on the number of variables, provided that a strong correlation exists, linking the overall set of variables.

Given a set of samples, which were previously assigned to a certain type by an indicator function, the problem of spatial extension arises when boundaries from zone to zone are estimated. This problem is of paramount importance in the modelling process, since reserves and recovery depend strongly on the quantification of geometrical characteristics of the reservoir. Hence, a new methodology – morphological Kriging – was developed in order to take into account the spatial structure of the available geometrical information, through an indicator variogram. The advantage of this technique is that uncertainty levels are associated with the estimated boundaries. Hence, the geometrical estimation errors can be included in the decision supporting models.

The geostatistical estimation procedures for generation of reservoir properties profiles are preferred to the classical ones because they take advantage of the spatial structure of the regionalized variable.

The problem of oil–in–place volume calculation was solved according to the argument that geometric and morphological issues play usually a determinant role in the estimation procedures. In fact, the estimation of the net oil column on the nodes of the appropriate grid depends on the closure area definition, which is a geometric problem, calling for boundaries optimal design. Furthermore, the concept of Reservoir Quality Zones was introduced, giving rise to the classification of reserves according to a quality index, which accounts for fluids saturation, spatial distribution of net oil column and petrophysical characteristics of the zone.

The methodology developed for drilling optimization appears to display some potentialities for application in the oil industry and can be a useful tool for the optimization of drilling sequences and it is based on the criterium of the estimation variance reduction according to different drilling configurations.

REFERENCES

(1) HALDERSON, H.H., 1986 – "Simulator Parameter Assignment and the Problem of Scale in Reservoir Engineering", in Reservoir Characterization, p.p. 293 – 340, L. Lake and H. Carrol (ed.), Academic Press, Inc.

(2) HALDERSON, H.H., DAMSLETH E., 1990 – "Stochastic Modelling", JPT, April 1990.

(3) DA COSTA E SILVA, A., 1985 – "A New Approach to the Characterization of Reservoir Heterogeneity Based on the Geomathematical Model and Kriging Techniques". SPE paper 14275, Las Vegas.

(4) SOARES, A., 1988 – "Conditional Simulation of Indicator Data", Quantitative Analysis of Mineral and Energy Resources. Ed. Chung et al, Reidel, Dordrecht, pp 375–384.

(5) PEREIRA, H., DA COSTA E SILVA, A., RIBEIRO, L., GUERREIRO, L., 1989 – "Estimation of Reserves at different Phases in the History of an Oil Field", in Geostatistics, Vol. 2, 543–555, M. Armstrong (ed), Kluwer Academic Press.

FACIES ESTIMATION USING WIRELINE LOGS (ELECTROFACIES): A CASE HISTORY FROM SOUTH-EAST SICILY

P. BALOSSINO & A. VALDISTURLO
AGIP SpA, P.O. Box 12069, I-20120 Milano (Italy)

M. LOMBARDINI
Western Atlas International, Via G.S. Bondi 2, I-48100 Ravenna (Italy)

1. INTRODUCTION

The recognition and characterization of stratigraphic units in a well holds a key role in the reconstruction of the geologic framework of a given area. Cuttings and core data may not be representative of the many aspects of a stratigraphic sequence. Well log data is thus preferred for detailed stratigraphic analysis.

2. METHODOLOGY

A newly developed statistical method has been used to predict facies automatically through a probabilistic approach. It requires the definition of classes as clusters of points in n-dimensional space (n: number of logs employed) using well log data derived from stratigraphic intervals chosen from key wells. With this method clusters, independent from the natural distribution of points but reflecting a precise strategraphic and/or environmental significance, can be established.

Log responses within the sample intervals are used to construct a data base (DB) in the shape of a multidimensional histogram. In each dimension, the range of the values and the number of steps can be adequately chosen to establish the discriminant capability of each parameter used. A transitional matrix is used to assess the discriminant capability of the DB and to emphasize, for each class, the purity and the degree of contamination by other classes. The DB can then provide a continuous layer by layer estimation of the occurrence of the defined classes.

3. DISCUSSION AND CONCLUSIONS

Late Triassic to late Oligocene carbonates were analyzed in eight wells of the Iblean Plateau. Two DB were created: a "lithological" DB, where classes are defined by lithology and porosity, and a "facies" DB where classes are defined according to their stratigraphic and /or sedimentologic significance.

Raethian to lower Lias

The identification of the boundary between the Noto and Streppenosa Formations has often resulted problematic, since they are both composed of shales and interbedded carbonates. Shales from the Noto Formation are characterized by a high content in organic matter, which can be detected on wireline log. The carbonates of both formations have similar characteristics. The boundary is thus better defined examining the shaly lithologies. Carbonate facies with high porosities were also recognized within the Noto Formation.

Lower to middle Lias

Within the Siracusa, Modica and Rabbito Formations log responses were used to differentiate platform carbonates from pelagic carbonates, with the result of a better definition of the lithofacies evolution. Reworked bodies characteristic of transitional facies were also recognized.

Upper Lias to Cretaceous

Within the Buccheri Formation the following elements are present: a marly-shaly basal layer absent in the more condensed sequences, nodular limestones well distinguishable from the prevalent micritic limestones, volcanic rocks discernible in basalts and tuffs. The Chiaramonte Formation is mainly composed of argillaceous carbonates, with typical and constant vertical distribution. The Hybla Formation is characterized by an increase of the marly-argillaceous components from bottom to top.

Upper Cretaceous to late Oligocene

Two zones can be distinguished within the Amerillo Formation: a lower zone consisting of low porosity carbonates and an upper one consisting of calcareous-marly interbeds. A better definition of the boundary between the Amerillo and Ragusa Formations has been gained through the definition based on logs of the carbonates and marls of the two formations.

RESERVOIR ROCK CHARACTERIZATION USING COMPUTED TOMOGRAPHY

B.Goričnik[1], Z.Krilov[1], M.Marotti[2]

1) INA-Naftaplin, Zagreb

2) University Hospital "Dr.M.Stojanović", Zagreb

X-ray computed tomography (CT) technique has been increasingly used in petroleum industry for rock petrophysical and mineral property evaluation and studies of fluid saturations in cores under static or dynamic conditions in order to determine the effect of core heterogeneity on the saturation distribution, the extent of formation damage and saturation changes during various (im) miscible displacement tests.

This work deals mainly with the CT use as a tool for qualitative and quantitative rock mineral composition and core anisotropy determination. X-ray attenuation in a number of minerals, found in sandstone formations, was measured in terms of Hounsfiels units (HU). Analogous measurements were performed on samples of crude oils, mud systems and brines, all having different composition.

Mineral densities were correlated with their respective HU values and compared with available published data. Full diameter core samples from different formations were examined by CT-technique for mineralogy identification and determination of rock texture anisotropy. The experimental data were verified by comparison with associated XRD and SEM analysis data. Core CT-scanning data were also correlated with corresponding well log data (Figure 1).

X-ray scanning technique was also applied to evaluate formation damage in terms of permeability impairment caused by mud invasion into sandstone rock samples from a gas well.

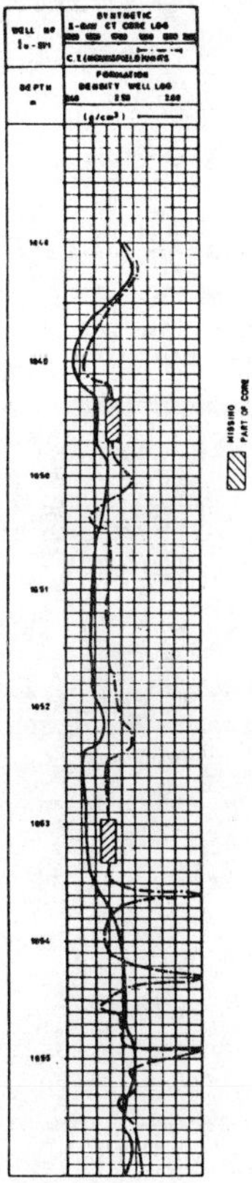

Fig.1 Correlation between rock density, generated from well neutron log and CT core data

References:

1) Wellington,S.L.,Vinegar,H.J.,J.Pet.Techn.,Aug.1987

2) Withjack,E.M., SPE Form.Eval.J.,Dec.1988

SCLEROGLUCAN RHEOLOGICAL PROPERTIES AND COREFLOOD PERFORMANCES AT HIGH TEMPERATURE

C.Noïk and J.Lecourtier (Institut Francais du Petrole), R.Rivenq and A.Donche (Elf Aquitaine)

Scleroglucan is an viscosifying nonionic polysaccharide attractive for several applications such as drilling fluids or improved oil recovery. Due to its molecular structure in triple helix strands and its high molecular weight, scleroglucan has an intrinsic rigidity that gives it its high viscosity properties, even at high temperatures up to 130°C. The rheological properties of scleroglucan in coreflood experiments have been analyzed as regards its solution behavior at various temperatures from 30°C up to 90°C.

Experimental :

A new equipment has been developed to measure solution viscosity at different shear rates and at high temperatures. The experimental principle is based on pressure drop measurements across a capillary viscometer during polymer flooding at a constant flow rate and controlled temperature.

Variations in flow rate during experiments and dimensions of capillaries permit to enlarge the range of shear rate investigation. To avoid any corrosion problem and metallic contamination of the solution, the calibrated capillaries are nickel tubes.

Capillaries differ by their length and inner diameter, which allows to determine rheological curves of concentrated polymer solution between $0.01s^{-1}$ to $1000s^{-1}$ shear rate and, consequently, to work in the Newtonian as well as the shear-thinning regimes.

In case of dilute polymer solutions for which viscosity and pressure drop are low, the pression transducers are selected to measure very low pressure drops, down to 1 millibar.

Viscometer calibration and feasability have been tested with brine solutions and give 1% absolute precision for measurements at 30°C and at 130°C in comparison to previous values in the literature. For low flow rates, polymer viscosity measurements are reproducible within a precision of less than 2%.

Scleroglucan was manufactered by Sanofi Bio Industries (France). Polymer solutions have EOR quality which means high viscosifying properties as well as good filterability without any problem of aggregation in time. The main solution properties of this improved scleroglucan have been presented in different papers (Ref.).

Results :

Relative viscosities of concentrated scleroglucan solutions in 20g/l NaCl and in seawater brine measured at 90°C, 120°C and 130°C show an increase with polymer concentration according to a powerlaw in log-log coordinates. Intrinsic viscosity variation with temperature indicates that scleroglucan retains its structure up to 120°C. The slight decrease of viscosity and the constant value of the Huggins coefficient are due to the increase of the flexibility of the molecular chain with temperature. At around 130°C, the scleroglucan solution loses its viscosity, which indicates a conformational change.

In relation to solution properties, scleroglucan behavior has been studied in two model porous media. Flow experiments have been performed at different temperatures, through an unconsolitated medium constitued with a mixture of pure sand and sodium clay, and through a Berea sandstone core with a permeability of 100 millidarcy. Alternating slugs of polymer and water were injected successively.

In both cases, flooding behavior shows that mobility reduction is close to relative viscosity as measured in the capillary viscometer, even at high temperature.

The adsorption level evaluated both from the delay in breakthrough of the effluent and mass balance decreases with temperature.

References : R.Rivenq, A.Donche and C.Noïk, SPE 19635 (Oct.1989)

J.Lecourtier, C.Noïk, G.Chauveteau and P.Barbey, 4th Eur.Symp.on EOR,Hamburg (Oct1987)

CHARACTERIZATION OF SOURCE ROCK AND RESERVOIR QUALITIES USING INFRARED ANALYSIS

Sh.N.Ganz[1], W.Kalkreuth[2], F.Öner[1], M.J.Pearson[3] & H.Wehner[4]

[1] TU Berlin, SFB 69, Ackerstr. 71-76, D-1000 Berlin 65, FRG
[2] ISPG Calgary, 3303-33rd str.NW, Calg. Alberta T2L2A7, Canada
[3] Marishal College, The University, Aberdeen AB91AS, Scotland
[4] BGR Hannover, Stilleweg 2, 3000 Hannover 51, FRG

During the last years a geochemical routine for the determination of petroleum source rock characteristics using IR-spectroscopy has been developed. The spectra of isolated kerogen reveal information about kerogen type, thermal maturation and hydrocarbon potential. Thermal maturation of associated bitumen may be predicted by IR-aromaticity and quantitative mineralogical composition is determined in the original homogenized sample.

So far all data and parameter are calculated by hand using several intensity ratios and extinction coefficients. In this paper first results are presented about a computerized version of IR-kerogen, and -mineral analysis. Instead of intensity ratios, factor analysis is applied after calibration with large sample sets of known composition. The sets include about 200 kerogens of different type and thermal maturation and about 100 varieties of different minerals. A method for mineralogical analysis generally consists of about 150 synthetic mixtures including clay minerals, quartz and carbonates. The error is generally below 5 % (relative error).

Factor analysis on IR spectra of organic-rich gas proone sediments (type III kerogen) allows prediction of kerogen properties such as Rock Eval S_1+S_2, T_{max}, OI, HI, ash-content, % volatile matter, A/C-Factor, W_{min}, without need of prior kerogen isolation and bitumen extraction. Further development of the technique will probably avoid time consuming kerogen isolation also for prediction of petroleum source rock properties (type I and II kerogen).

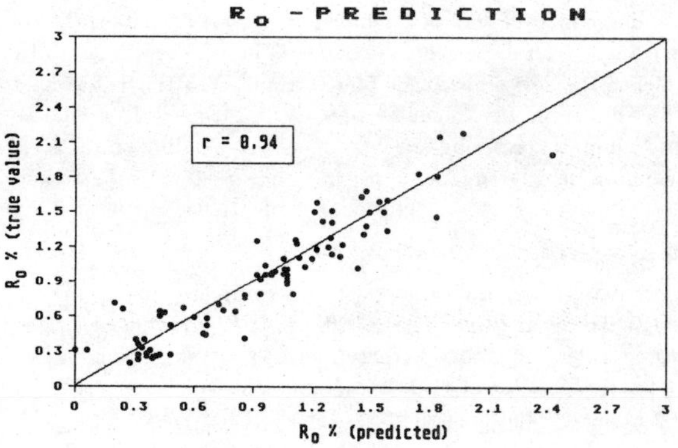

Fig.1: Prediction of vitrinite reflectance from IR-analysis

A NUMERICAL STUDY ON THE EFFECT OF CLAY DISTRIBUTION ON SHALY SAND CONDUCTIVITY

X.D. Jing, J.S. Archer and T.S. Daltaban
Imperial College of Science, Technology and Medicine, London

1. INTRODUCTION

One reason for the uncertainties associated with the use of existing shaly sand conductivity models is that there is not enough detailed study on the effect of clay distribution on the conductive behaviour of shaly sands. Therefore, this paper presents some latest results of our systematic study on the effect of clay on the electrical properties of shaly rocks. We define a term 'effective clay concentration', $Qv_e = \tau Qv$, which may be experimentally determined by the 'Multi-Salinity Method[(1),(2)]' or the 'Membrane-Potential Method[(3)]'. Qv is the clay concentration derived from the conventional conductometric titration measurements and τ is a constant relating the above two clay concentrations, which is a function of clay distribution mode.

2. DESCRIPTION OF THE PETROPHYSICAL MODEL

A 3-D pore space network model has been developed to simulate the effect of clay distribution on shaly sand conductivities. Three basic types of clay distribution (dispersed, laminated and structural) have been considered in the model. For each type of clay distribution, values of clay concentration (Q_v) are generated for all the pore tubes using random, normal and log-normal mesh generators. Both Waxman-Smits[(1)] and Dual-Water[(2)] models are used to calculate the equivalent solution conductivities for all the pores. Rock conductivities at various pore fluid salinities are calculated by solving sets of linear equations. The numerical results are then compared with the experimental results of 14 synthetic shaly rocks with various clay distribution modes.

3. RESULTS AND DISCUSSION

As shown quantitatively in Fig.1, the distribution type of clays significantly influences the conductivities of shaly sands. If values of Qv derived from the conventional conductometric titration measurements are used in the interpretation of the shaly sand (A3), which ignores the influence of clay distribution, total clay effect can be underestimated by 33% for laminated clay distribution with the layers parallel to the flow direction, while for dispersed clay, total effect of clay minerals can be overestimated by up to 400% depending on the degree of dispersion (*DOD*) which is defined here as the fraction of the number of pores containing clay over the total number of pores. Therefore, the constant τ as mentioned earlier changes from 0.2 to 1.0 for dispersed clays as *DOD* increases from 0.2 to 1.0, while for laminated clays with layers parallel to the flow direction, τ ranges from 1.0 to 1.5. For laminated clays with layers perpendicular to the flow direction, the value of τ is always less than 1.0. Similar to dispersed clays, the value of constant τ for structural clays ranges from 0.0 to 1.0 depending on the local distribution. Finally, for a given clay distribution mode (laminated, dispersed or structural), the actual distribution of clays (random, normal or log-normal) also affects shaly sand conductivity, but with a second-order magnitude.

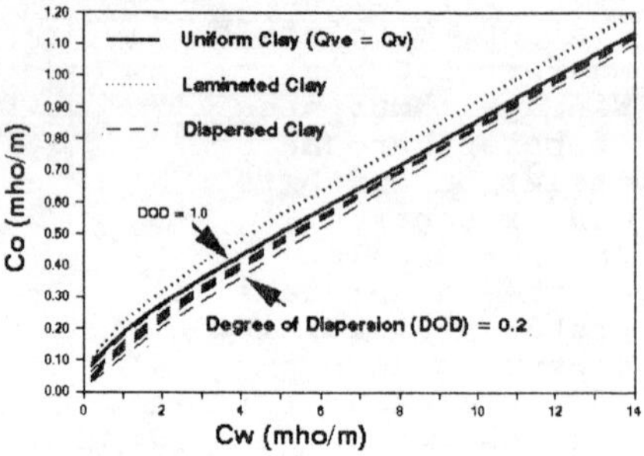

Fig. 1: The effect of laminated and dispersed clay distribution

REFERENCES

(1) Waxman, M.H. and Smits, L.J.M. Electrical Conductivities in Oil-Bearing Shaly Sands. *SPEJ* (June 1968), p.107-22; *Trans., AIME,* Vol. 243.
(2) Clavier, C., Coates, G.R. and Dumanoir, J. Theoretical and Experimental Bases for the Dual-Water Model for Interpretation of Shaly Sands. *SPEJ* (April 1984), p.153-165.
(3) de Waal, J.A. Influence of Clay Distribution on Shaly Sand Conductivity. *SPE Formation Evaluation* (Sept. 1989), p.377-383.

Gamma-ray Tomography Inspection Technique for Flooding Experiments in Porous Medium.

JANN-RUNE URSIN
ROGALAND UNIVERSITY CENTRE
P.O. BOX 2557 ULLANDHAUG, 4004 STAVANGER, NORWAY

Introduction

In order to tell the difference between possible flooding regimes in laboratory experiments, information about the fluid saturation inside the porous medium is often required. This situation frequently occur when one fluid is displacing a less mobile one. For this reason we have developed a tomographic inspection system based on gamma attenuation where each tomographic image is represented as a discrete function giving the water saturation on a cross-section surface of the porous medium[1].

Choice of radioactive isotope

In gamma-tomography no "beamhardening effects" will distort the measured data. This advantage to common X-ray tomography, enables inspection of more massive objects. Choosing between different isotopes, the optimum choice is decided by considering the average density and the diameter of the porous medium. We have chosen a Cesium—137 (3Ci) source (660 keV) as our radioactive isotope. Expanding such a system to an array of detectors, scanning time will reduce considerably, hence allowing more dynamic experiments.

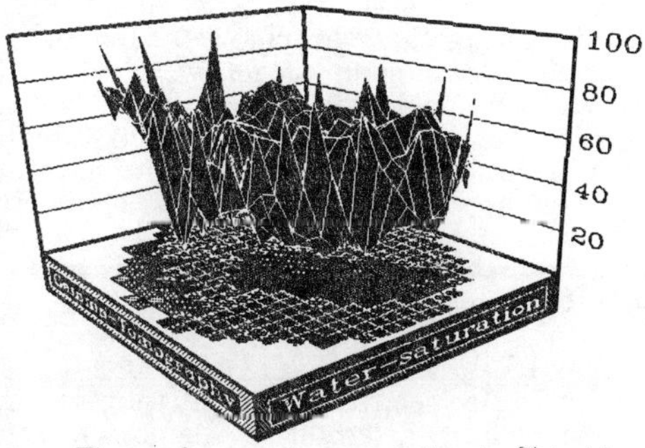

Figure shows water saturation in % inside a 140mm thick porous medium.

Tomographic reconstruction technique

A personal computer is analyzing the measured data. The reconstructing method adopted is based on *backscattering of convoluted data*[2] using a mathematical filter function designed and optimized particularly to our experimental condition. This optimization reduces the uncertainties in the calculated water saturation and consequently allows shorter scanning times.

Experiments

Gamma-attenuation at Cesium energies is purely density dependent. In two-phase flow, where the density difference between the two phases could be small, increased difficulties arises in differentiating the two phases. A remedy in these situations would be adding f.ex. a salt to the solution in one of the phases, making it "thicker".

[1] J.R. Ursin. *Industriell Tomografi*, Høgskolesenteret i Rogaland, 89/1988

[2] W.G. Gilboy. *Tomographic gamma-ray scanner for industrial applications*, NIM 193(1982) 209-214

CHARACTERIZATION OF OIL COMPONENTS SORBED ONTO CLAY MINERALS AND QUARTZ GRAINS IN OIL SAND

A. Fendel and K. Schwochau

Institute of Petroleum and Organic Geochemistry
at the Research Centre (KFA) Juelich, D-5170 Juelich,
Fed. Rep. of Germany

INTRODUCTION

The wettability state of reservoir rocks often determines the feasibility of conventional oil production and enhanced oil recovery (1). Thus the understanding of processes responsible for oil or water wettability appears to be of high significance. In particular, the insight into the composition and properties of rock-oil interfacial layers has to be improved. The objective of our investigations is to gather information on the quantity, variety, and functional groups of oil components sorbed onto clay minerals and quartz in reservoir rocks.

EXPERIMENTAL

Unconsolidated oil sands (Cold Lake, Athabasca, Utah) were exhaustively extracted with dichloromethane to remove the free oil. The clay minerals and quartz grains were separated by sedimentation and wet sieving. After extraction up to 4 wt-% of organic carbon still remained on the clays, but only up to 0.2 wt-% on the quartz grains; the highest organic carbon content showed kaolinite/illite samples. The clays and quartz grains were successively extracted with solvent mixtures of increasing polarity for releasing the sorbed oil components. These extracts were separated by liquid chromatography into aromatic and saturated hydrocarbons, high polar compounds, acids, bases, and alcohols. The fractions were characterized by means of GC, GC/MS and IR-spectroscopy.

RESULTS

Solid state IR-spectroscopy of clay fractions revealed CH_2-, CH_3-stretching vibrations around 2900 cm^{-1}, carbonyl frequencies between 1750 and 1680 cm^{-1}, C=C-vibrations of aromatic compounds at 1620 cm^{-1}, a weak shoulder of the asymmetric COO^--stretch at 1580 cm^{-1} and a stronger absorption of the symmetric COO^--stretch at 1380 cm^{-1}. The occurence of these assigned groups was confirmed by ^{13}C CP/MAS NMR-spectroscopy exhibiting in addition an aliphatic C-O-signal at 73 ppm. The extractable fraction of sorbed oil proved to be significantly depleted in saturated and aromatic hydrocarbons and highly enriched in hetero-compounds. In particular monocarboxylic acids, alcohols, and esters (C_{12}-C_{30}) were found to be concentrated by factors of 5 to 50. The sorbed oil percentages of saturated hydrocarbons turned out to be distinctly higher than that of the aromatics. Whereas in the free oil no n-alkanes could be detected because of heavy biodegradation and water-washing effects, the gas chromatograms of the sorbed oil extracts surprisingly demonstrated the existence of more or less complete n-alkane (C_{15}-C_{39}) patterns suggesting that the sorbed oil was protected against microbial attack (2). In general the composition of sorbed and free oil appeared to be not only quantitatively but also qualitatively different.

REFERENCES

(1) Anderson, W.G., Wettability Literature Survey - Part 1: Rock-Oil-Brine Interactions, and the Effects of Core Handling on Wettability. J. Petrol. Technol. 1125-1144 (1986)

(2) Fendel, A. and Schwochau, K., Protection of Crude Oil N-Alkanes against Biodegradation by Sorption onto Mineral Surfaces (to be submitted to Nature)

THE EFFECTS OF BRINE MIGRATION AND SALT CEMENTATION ON HYDROCARBON RESERVOIRS

H.DRONKERT

Faculty of Mining and Petroleum Engineering
Delft University of Technology
P.O.Box 5028, 2600 GA Delft, The Netherlands
&
International Geoservices B.V. Reaal 5, 2353 TK Leiderdorp, The Netherlands

As part of a larger study on the prospectivity of the Buntsandstein in the Netherlands Offshore (North Sea), the effects of halite cementation within potentially porous reservoir rocks has been investigated. The Buntsandstein contains good reservoir sandstones (10-1500 mDarcy) and is the second important gas target in The Netherlands. However, production problems arose which are related to patchy, or complete cementation of the reservoir.

Models have been developed for the provenance, the migration and the precipitation of the salt. Seven drive mechanisms for brine migration have been analyzed, and four scenarios for halite precipitation have been worked out. These models have been applied to various potential gas field for different companies. A good area control (seismic profiles, structural maps) and well control (wire line logs, core descriptions and thin section analyses) are prerequisites for reliable calculations and mapping of hydrocarbon reservoir boundaries.

Depending on the burial history (diagenesis), potential reservoirs can be completely cemented or have patchy cements causing compartmented reservoirs. In some cases remobilization of the cements can create high secondary porosities that provide good gas targets.

Keywords: reservoir characterization, cementation, salt brine, migration

MEASUREMENT OF GAS TRACER RETENTION UNDER SIMULATED RESERVOIR CONDITIONS

Ø. DUGSTAD and T. BJØRNSTAD

Institutt for energiteknikk, N–2007 Kjeller, Norway

1. INTRODUCTION

Tracer tests are widely used to improve reservoir characterization. This work presents results from a comparative study of a selection of radioactive and new non–radioactive tracers. Special laboratory equipment is developed to carry out dynamic flooding experiments under reservoir conditions. Automatic sampling systems are constructed to enable continuously recording of tracer responses.

2. SLIM–TUBE EQUIPMENT

The porous media applied in these studies consist of packed columns of length 6 meter or 12 meter where the packing material was Ottawa sand under residual oil saturation. Experiments have been carried out at pressures ranging from 100 bar to 250 bar. The temperature has been between 50°C and 120°C. The columns are mounted in a heat cabinet and the pressure has been maintained by a back pressure regulator. Two separate analytical techniques are applied to measure the radioactive and the chemical tracers in the same effluent. A sample of the effluent gas is injected into scintillation vials for counting of radioactivity and another part of the effluent is automatically injected into a gas chromatographic column for separation and detection on an electron capture detector.

3. TRACER RESPONSES

The tracers studied are tritiated methane, ^{14}C labelled ethane, ^{85}Kr, perfluoromethylcyclopentane (PMCP) and perfluorometylcyclohexane (PMCH). The perfluorocarbons have so far only been used as tracer in reservoir studies in one single pilot project.

A good tracer shall follow the traced phase closely. Under methane injection tritiated methane will be an ideal tracer. However, all the tracers even methane, will have a partitioning into the oil phase and the tracer will therefore be retained with respect to the average gas flow velocity in the reservoir. Our results generally show that the retention is increasing in the order $CH_3T < {}^{85}Kr < {}^{14}CH_3CH_3 < PMCP > PMCH$. At low pressure (100 bar) the perfluorocarbons are considerably retained, but at more realistic reservoir pressure the retention decreases and becomes more like that of ethane. In some cases, i.e. at low temperature and high pressure, the PMCP will even appear before ethane. The behaviour of ^{85}Kr is almost like that of methane. For identification of channelling and preferred flow directions all the presently tested tracers show behaviour which make them feasible. The retention is a function of pressure and temperature, but it is also a function of oil saturation. With accurate data on partitioning of tracer into the oilphase it may, however, also be possible to estimate the residual oil saturation in the swept zone.

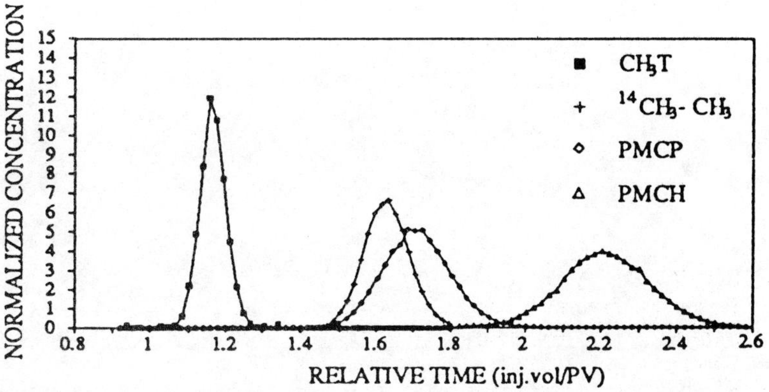

Figure 1: Tracer retention as a function of injected pore volume with the following parameters: Temp. 50°C, pressure 150 bar, flow rate 0.24 ml/min, Sor 30%

4. CONCLUSION

The two chemical tracers PMCP and PMCH have shown properties which make them possible as gas tracers. The developed slimtube equipment has been used to measure the retention of the tracers under different conditions. This information may be used to estimate residual oil saturation in the swept zones.

REFERENCES

1) CALHOUN, T.G., and TITTLE, R.M., "Use of Radioactive Isotopes in Gas Injection". SPE-2277, 1968

2) DIETZ, R.N., "Perfluorocarbon tracer technology", BNL-38847, DE87

OPERATIONS

Session Chairman: H.J. DE HAAN, Univ. Delft (NL)

ENVIRONMENTAL PROTECTION IN AGIP HYDROCARBONS E & P

CEFFA L. DI LUISE G.
DOSSENA G.

Geodynamics & Enviroment,
AGIP SpA,
S.Donato M.

Summary

These notes review some of the most important experiences and studies that AGIP has carried out in the environmental protection field and geodynamics with respect to E & P activities in the last two decades. These are part of an environmental policy that the Company follows in order to operate towards minimizing impacts on the enviroment, thus making a considerable contribution to environmental science progress in Italy and foreign countries in which it operates, and contributing to an improved land protection.

1. INTRODUCTION: AGIP's ENVIRONMENTAL POLICIES

In the last 20 years AGIP has confronted the problematic nature of environmental Protection, in the context of oil E & P, by operating on certain base points that contribute to defining the "environmental policy" of the Company.

The first step has been to establish a "Geodynamics and Environmental" Department, at the Company's headquarters, that is presently operated by technical staff employed by the Vice-president. They are composed of a group of specialists from various fields (geology, environmental engineeering, environmental modelling, biology, chemistry, natural sciences etc.) falling into the most general context of applied Ecology and Geodynamics.

The principles of environmental policy that the Company intends to apply at all levels and in all branches are transferred in the various operative realities as a result of activities carried out by the Company. They orientate around the following fundamental points:

- Study and direct solution of operative problems connected to exploration and production projects of particular complexity as far as environmental effects are concerned.
- Definition of operative procedures for industrial activities in which constraints and environmental protection norms represent critical points.
- Assistance and collaboration in environmental protection with Italy's and AGIP's operational branches abroad.

- Auditing activities carried out on the plants and installations of a potential environmental impact in Italy and abroad.
- Training and "on job" formation of italian and foreign technicians; teaching with environmental, theoretical material and applied matters and applied to AGIP's training course of Cortemaggiore (Drilling and Production assistants and workers) and to the "Petroleum Engineering" course.
- Sensitization to environmental protection techniques of the operative units of AGIP's exploration and Production involved in the various aspects and activity moments of a possible environmental impact.
- Promotion of Scientific research activity to the point of recognising critical aspects of the environmental protection activity.
- Preparation, updating and implementation of contingency plans onshore and offshore for national and foreign zones.

The most significant experiences, of considerable environmental interest, brought to a head in these years by AGIP in certain E & P acitivity fields, are controlled in these notes and thus, like some prevention and intervention procedures, they are part of a comprehensive action of analysis and research into minimizing impact risks.

2. REAL ENVIROMENTAL PROBLEMS IN THE E & P

A map of Italy locating AGIP's E & P's activities of stresses their homogenous distribution standing out from the North to the extremes of south Sicily, passing through environmental and land settings that, at times, are extremely diverse and include flat areas of the Pò valley, the adriatic, pre-appenine hills, the southern appenine mountain system and the historic oil sites of south-east Sicily: parallely, offshore activities across the Adriatic, the Ionian sea up to the Sicily Channel: recently the Sardian research area in the Oristano Gulf has been added.

The natural surroundings that are made up of various scenarios, are each marked by a natural, environmental sensibility in which, one by one, naturalistic, landscape, anthropical and sometimes artistic aspects are accordingly noted.

In a few countries in the world the expression, "land protection" assumes, therefore, as in Italy, connotation and meanings equally full.

The environmental scenarios, in which the most important areas of AGIP's activity abroad fall, is not inferior, moreover, as far as complexity is concerned: for example, the Niger Delta area , the middle-east offshore, the indonesian seas and the China sea.

In this context, for an important Company like AGIP, everything can be interpreted as a whole as lawfull commitments and obligations that influence the planning and operative aspect of its activities.

Therefore, in the meantime, the need has been born to place side by side the traditional exploration and production mining techniques with a series of procedural and operative measures orientated towards the prevention and minimization of environmental risk.

Geo-physical surveys

This problem comes from the use of particular techniques at sea and on land.

At sea one can say that the impact on the ecosystem is actually minimized by methods used for more than 20 years (air gun, vapor shock, etc.) that have taken over the use of the explosive.

On land, particularly in the Po plain, the eventual environmental impact is linked to the risk that the little wells, established to arrange the explosive, can create a vertical proximity and , therefore, the mixture, of water coming from the different quality ground-water bodies, with the risk of allowing pollution episodes; this prospective is even more negative if there, it refers to geo-physical operations carried out in the zones of typical lombard-piemont rice fields.

The problem has been tackled and resolved using the closure scheme of the little wells by cement-clayey plug system and integrating this technique with particular attention to the natural structure of the subsoil and of the relative hydraulics.

Drilling site wastes

The management of these materials has been a major environmental problem for a long time for the oil industry, not so much for their chemical characteristics and treatment, but also for the difficulty in managing them in respect to the limits set by the laws in force, also taking into account the variation in volume and relevant typology.

The solution to the problem has been solved concentrating on two aspects: reduction in volume and optimization of the treatments.

The first objective has been achieved, after numerous experiences on pilot and industrial scale, intervening on the primary drilling wastes circuit (vibrating screen, desilter, desander, centrifuge, etc.). The results obtained have limited the drilling wastes to 1 cubic metre for every metre drilled during a drilling within 3500 metres deep, (see wells of the Malossa Field).

The second objective achieved has been that of rationalizing and improving the treatment systems of every type of site waste, in particular, the cuttings and drilling wastes exhausted.

The following have been achieved:

- confining and separating the various types of refluents achieving, inside the site, a series of basins of different capacity and constructive formalities;
- periodically treating and disposing of the refluents produced during the drilling;
- employing purified chemicals with the scope of reducing the ecotossicological wastes capacity.

The exposed criteria have lead to arrange, today, on site, five types of effluents with predefined characteristics:

- exhausted drilling fluids;
- cuttings;
- slightly polluted waters;
- special fluids (oils and acids);
- civil wastes.

The first are treated periodically in mobile units by means of chemical process and succesive aqueos phase draining, using centrifuges or filter presses.

The aqueous phase becomes successively neutralised and treated with activated carbons before being unloaded in superficial water bodies.

Likewise, the possibilty of a diversification of the destination of solid waste, using these wastes as additives in the brick industry has been investigated: the relative feasability study has made it clear that this method is usable for part of the quantatives produced on site.

With reference to the reduction in volumes obtained, a corresponding reduction in treatment costs has not been verified, essentially due to a notable increase in disposal expenses, as a result of the limited number of plants available on national land. (Fig. 1 - 2)

Subsidence

The subsidence constitutes a possible reflex of production activity of hydrocarbons: this is a natural phenomenon know in Geology and explained as a result of a complex of physical situations (tetonic movements, grains sediments compaction, etc.) of land dynamics.

However, a second subsidence imputable to "man-induced" factors more or less easily identifiable (heavy groundwater extraction, land reclamation, building load, hydrocarbons exploitation) can be superposed to this "geological" defined subsidence.

In both cases the physical mechanism is reduced, essentially, to grains squeezing of the geological formation making up the geological formation subsident with consequent deformation, slipping, re-arrangement and fluids expulsion (particularly important in the case of clayey rocks) and successive reduction of the original volume that is transposed to differential soil lowering.

One element that has emerged from the studies carried out on the phenomenon over 20 years is the

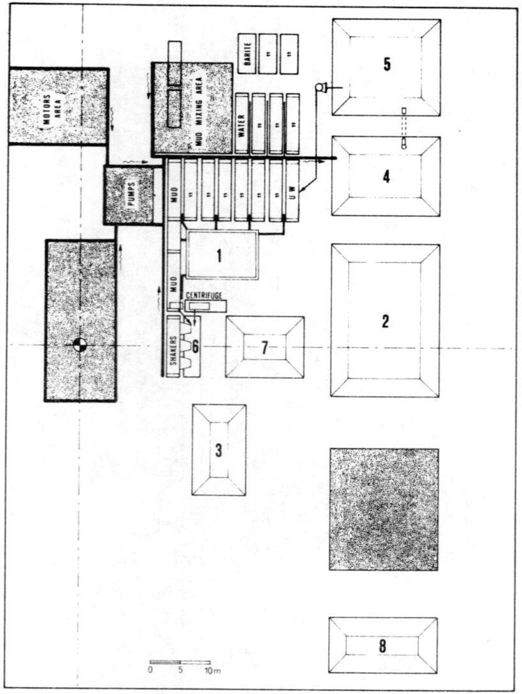

Fig. 1 - Location layout for new method of waste management

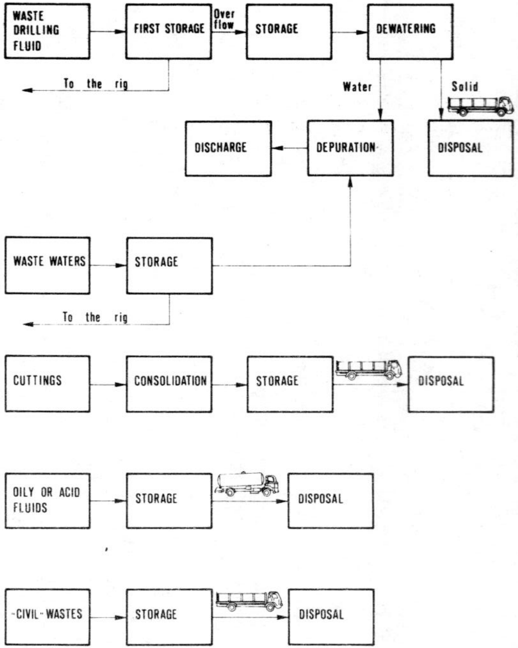

Fig.2 - Wastes treatment and disposal flow-chart

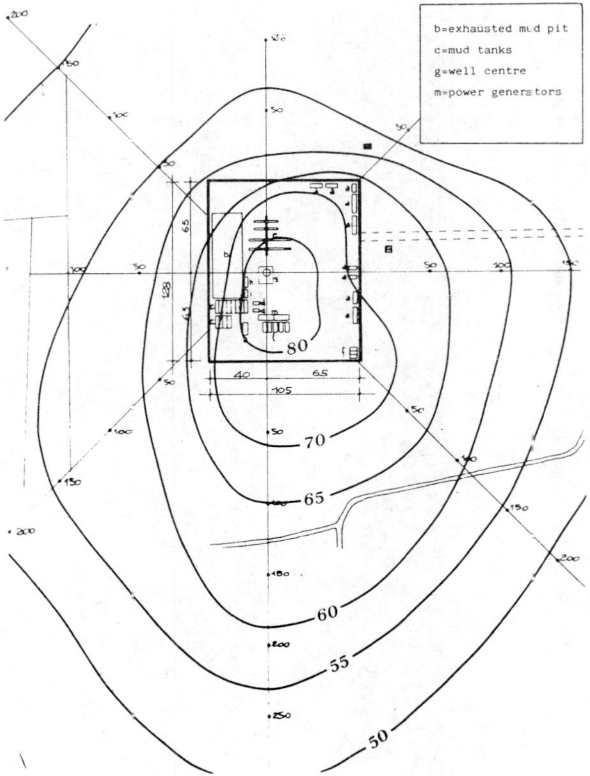

Fig. 3 - Seregna 3 location - Saipem emsco C3 drilling rig
Sound power level curves - Drilling phase

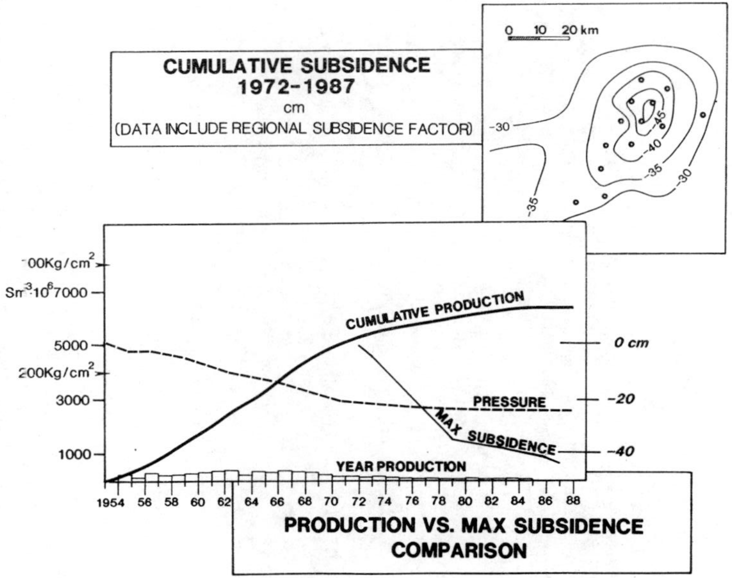

Fig. 4 - Correggio gas - field

exsistence of a regional trend of soil lowering in the Po Plain, the seat of a large part of AGIP's E & P activity, as a result of natural causes.

The knowledge acquired in this period of time by AGIP has permitted certain elements to be identified on which it has been possible to carry out an effective prevention and minimization of the negative effects that the phenomenon can have on the land, as can be seen in the following paragraph.

It has been noted that the the highest value of the lowering of the soil is normally verified in correspondence of the oil production field top , gradually lowering up to weakening, towards the regional subsidence values.

Another element that systematically appears is the correlation between the oil field depressurization - function of the production rate imposed on the reservoir - and the subsidence.

Referring to the Correggio field case (lowering at the top by 45cm in 15 years) and Ravenna-Terra (50cm of subsidence in 12 years) clearly show both observations.

The importance of the phenomenon, in these cases and other studied italian cases is, however, far from that noted in certain famous world cases (Galveston, Maracaibo, Groningen, Ekofisk, etc.) in which cases the order of the greatness of the lowering is in metres, and has been verified in less years. (Fig. 4 - 5)

3. ANALYSIS, PREVENTION AND INTERVENTION METHODS

<u>Enviromental Impact Assessment</u>

The E.I.A. procedure, as in Italy and many european countries, has become an obbligation for law for approving projects of particular environmental importance.

The possible and more frequent environmental impacts, connected to the activities of the oil and gas gathering centres and site operations, are represented on the one hand by disturbance made to the population situated in the areas near to the plants as well as consistent noise, vibrations, dust contamination and gaseous emissions and on the other hand, from the influence of the "habitat" from the point of view of naturalistic values that can be damaged as a result of road construction and deforestation.

The E.I.A. executed in 1989 for the project of the development of the oil field of Villafortuna-Trecate was an opportunity for implementing knowledge and intervention techniques in this field, given the complexity of the problems affronted.

The field was discovered in 1983-84 on the border between Lombardy and Piemont in a triasic age research theme, at a depth of 5500 and 6500 metres. The project forsees the establishment of 26 drilling sites of which 17 are in the constraint areas of the regional park: the sites can receive up to 3 wells each and are optimized in accordance with oil exploitation. The conveyance of oil towards 5 satellite sites is foreseen, while the satellites will be connected with the gathering Centre at Trecate by a closed circuit.

The E.I.A. study carries and emphasizes various influential factors that result from the final sensitive analysis.

With reference to single influential factors, those concerning site operations have been considered together with those resulting from electric current motor drives (immission into the atmosphere, noise, vibrations) and landscape interference (as a result of project works in the study area). These factors have been represented by iso-path maps (for example for the iso-acoustics curves and night iso-lightening of the site) or other maps and tables (road map, number of journeys and relative load of the motor vehicles, in relation to the traffic inconvenience, etc.). Precise details have been acquired on the location of arboreal and shrubby vegetable fields and on the different forms of soil use, following a network of irrigation ditches and small, artificial channels as a result of the field and air surveys. The same information has been useful for dealing with the argument of the environmental restoration of the parts of land no longer used by work projects.

The fundamental result of the work is the identification of the environmental guide-lines of the whole of the project, becoming a discussion point for the Park Administration. (Fig. 6)

Control and prevention of subsidence

The subsidence monitoring is usually carried out by means of high precision geometric levelling on the bench-marks stationed along the lines that cover the land that the phenomenon concerns: the levelling frequence of the height of the bench-marks can go from 2 to 10 years.

The monitoring of the subsidence of the sea bed cannot be carried out using the same method on the edge of the offshore production platform for difficulty represented by the distance between the structure and the coast. The problem has been overcome resorting to the measurement of the heights via satellite, using the GPS system (Global Positioning System) that exploits the path of the special satellite motor vehicles to monitor the heights of certain antenna- points at frequent intervals.

AGIP's levelling network that, in 1968, consisted of 176 km and essentially covered the middle-east part of the Po Plain to the south of the Po Delta, has arrived at the point, today, of growing more than 4200 km; this growth has taken place at the same time as the discovery of the new hydrocarbon fields.

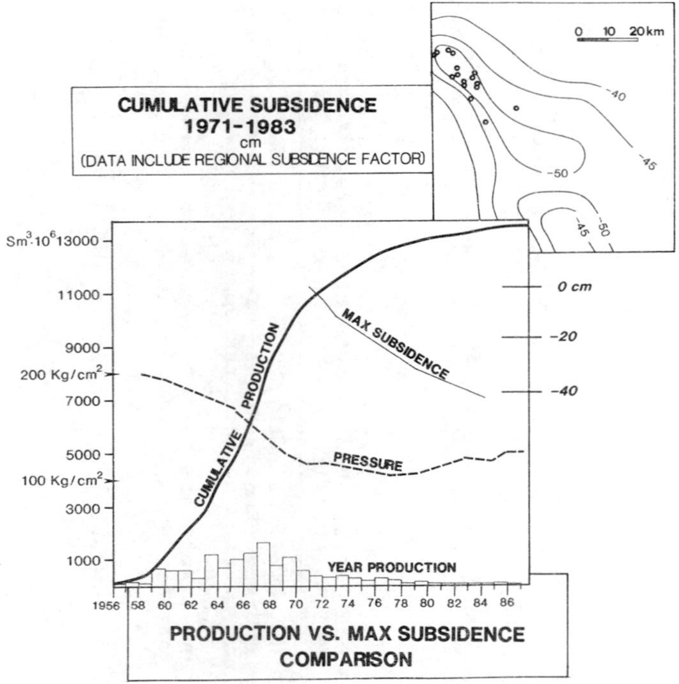

Fig. 5 - Ravenna - Onshore gas field

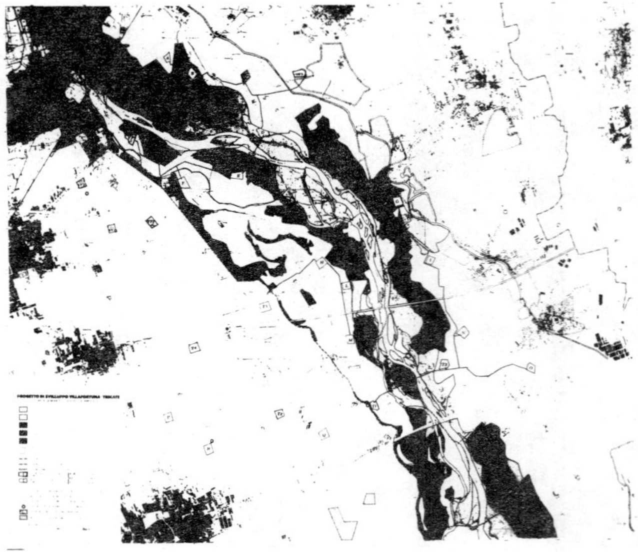

Fig. 6

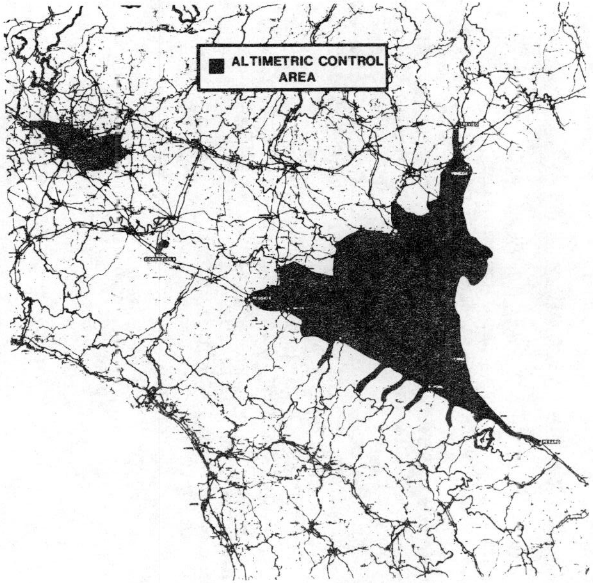

Fig. 7

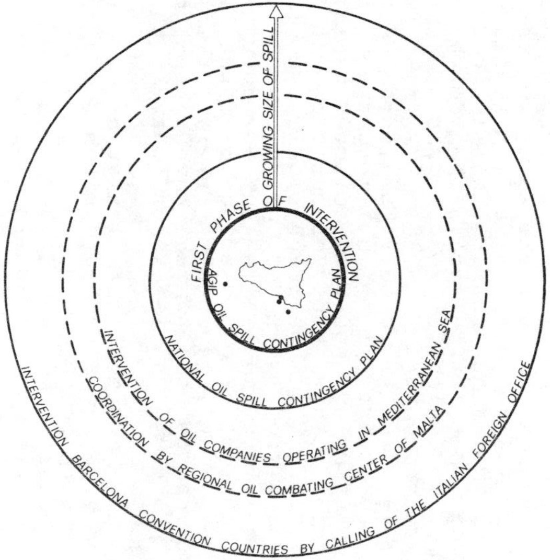

Fig. 8 - Sicily offshore

The offshore esperiences concerned 2 platforms in the Adriatic (PCW-b and Barbara-B, respectively 6 and 60 km form the Ravenna coast): the measurements, repeated 5 times from May 1987 have revealed 1 cm negative altimetry variation on the annual base, that is not, therefore, comparable to that registered in the North Sea (Ekofisk Field, 1984), in which the sea bed was reported to have subsided 4 metres.

In the last few years AGIP has increased the used of provision models parallel to an improvement in measuring techniques. After a geometric discretization treatment of the reservoir, they supply calculation matrix, leaving from a geotechnics, petro-physical type inputs and dynamic behavoir of whatever productive level of the field. The aforesaid cells (up to 5000) express a surface value of the lowering of the soil which corresponds to an elementary field prism, all the variable dynamics of which have been calculated: therefore plotting the iso-value curves to obtain an "iso-subsidence map".

AGIP has adopted the analytic Geertsma model ("nucleus of strain theory") and taken the opportunity to input data in the computer.

In the case of measured subsidence values higher than those foreseen in the initial scenarios of the model, the negative effect can be reduced by adopting slower production rythms thus limiting the stress to which the production formations grains are subjected.

This method is used in AGIP as an efficient prevention factor thanks to its operative elasticity found in its synergic mechanism (model and monitoring feedback). In such a way one works towards a protective worldwide action of land and environment in a general sense. (Fig. 7)

Oil-spill Contingency Plans at sea

The accidental spills of oil at sea certainly represent a big challenge for the technical and organizational skills of oil Companies.

Serious alarm bells were sounded for some well-known oil-spills: 40,000 tons from the ship EXXON VALDEZ in Alaska, March 1989; the sinking of AMOCO CADIZ (1978, more that 200,000 Tons) and the well blowout of IXTOC 1 (1979, 400,000 Tons) faced with which the oil Companies and State Authorities intensified force and work on preventive action together with the development of emergency services techniques.

The oil-spill Contingency Plans, that is the technical-procedural methods necessary in order that an oil Company can be organized and technically ready to combat an accidental loss of oil at sea, have been part of AGIP environmental intervention standards for many years. Two different plans of operative necessity have been identified:

- individually, AGIP for a single oil italian or foreign production area that corresponds to the highest possible risk of an accident, is about 1000 - 1500 Tons of spill;
- for a major oil spill more than 1000 - 1500 Tons.

At the first level this corresponds to a set of the minimum set of eqipment in order to give the personnel the necessary confidence, by means of practice runs: the necessary equipment (booms, skimmers, dispersants) is located in the italian operating bases of Ravenna, Ortona, Siracusa, at the oil terminal of Brass (Nigeria) and on the platform with dock and unloading system for the oil tanker in Bouri (Lybia).

The two Contingency Plans for the Vega (Sicily Channel) and Bouri (Lybia) fields are integrated together due to their contemporaneous presence in the Mediterranean Sea and takes into account the Italian National Contingency Plans so that all the available equipment, in case of necessity, can be mobilized in the most efficient way. (Fig. 8)

4. CONCLUSION

With reference to the above-mentioned, one can deduce the difficulty of putting into practise an environmental protection policy in relation to both the complexity of the operations and the E & P's location activities: in actual fact, these activities go hand in hand with highly densely populated lands and often high natural and landscape values.

It has been necessary for AGIP, therefore, like other oil Companies, th acquire new potential study and technical tools in order to afront ever more impegnative needs.

The object of this effort, used up to now, is to achieve progress in a sector with problems in common also in different fields from the oil industry, therefore contributing to an improvement in the land resources preservation.

AN INTEGRATED EMERGENCY MANAGEMENT SYSTEM FOR SPILLS OF CHEMICAL SUBSTANCES

J.B.NIELSEN and T.GUDMUNDSSON

Danish Hydraulic Institute
Agern Allé 5
DK-2970 Hørsholm
Denmark

H.BACH

Water Quality Institute
Agern Allé 11
DK-2970 Hørsholm
Denmark

Summary

Recent major oil spills into the aquatic environment have once more focussed public attention on the environmental hazards and hazards to human safety which are associated with the production and transportation of hydrocarbons and other chemicals. It is obvious that more should be done to prevent spills of chemical substances. It is unfortunately also obvious that - in an imperfect world - it is impossible to completely avoid spills. Accordingly, it is necessary to establish procedures, tools, and equipment to be applied when spills do occur, in order to minimize the damage.

This paper describes a computerized emergency management system for spills at sea. The system is being developed as a tool to be used in that crucial phase of a spill event, where a few decisions may turn an event into a disaster or a potential disaster into an event of limited consequences.

1. INTRODUCTION

We live in an industrialized world where the production, bulk transport, storage, distribution, and use of hydrocarbons and numerous other chemical substances is an integrated part of our daily life. Many of these chemicals are potentially harmful to the life or health of human beings or to our environment.

The tremendous consequences of a major oil spill in a vulnerable area have been highlighted by the Exxon Valdez disaster and other recent accidents. Less attention is presently focussed on spills of other types of chemicals, although such spills are known to be very frequent and potentially very dangerous. In Canada for instance, where spills are being registered in a national data base, almost 2,000 chemical spill events involving more than 100 different substances were reported during the period 1974-1984 (1).

Accordingly, any strategy aimed at reducing the dangers associated with the use of hydrocarbons and chemicals should concentrate on two aspects:

1 Limit the number and size of spills by introducing prudent preventive actions, i.e. safety regulations, etc.

2 Limit the consequences of those spills which still occur, despite all efforts to prevent them.

This paper concentrates entirely on the second aspect and further more, it concentrates on spills into the sea, i.e.: "How do we manage a spill of a chemical substance into the sea, once it has occurred?"

2. MANAGEMENT AND INFORMATION

Management relies on information. In order to make the correct decisions in a spill emergency situation, management must have access to as much and as reliable information as possible. Among the information of prime importance one can mention:

Spill scenario, e.g.

- Location of spill
- Substance, name and properties
- Amounts (spilt till now and potential further spills)
- Condition of the ship involved (if any)

Safety of humans, e.g.

- People on-board (if ship involved)
- People in the vicinity
- People involved in spill combat

Environmental concerns, e.g.

- Areas/species which are particularly vulnerable
- Properties of possible chemical substances used in spill combat

Forecast data, e.g.

- Weather forecast
- Wave, current and water level forecast
- Likely movements of spill (track, spreading)

- Likely fate of spill (physical, chemical, and biological processes)

Logistics, e.g.

- Short term availability of people and equipment
- Status on possible additional resources for spill combat

"What if" data, e.g.

- What if wind direction changes?
- What if spill amounts are underestimated?

Computer technology has made it possible to get access from one management center (or from a computer on the spill site) to many sources of information and to simulation tools which are able to provide reliable forecasts as well as answers to "what if" type questions. The following sections describe a prototype system in which many of the possibilities of modern informatics are being employed in order to provide the crisis management with one integrated system for information management.

3. THE PROTOTYPE EMERGENCY MANAGEMENT SYSTEM

The "core" of the emergency management system which is being developed by two Danish research institutes is a suite of simulation systems which can be applied for simulation of the track, spreading, and fate of oil and other chemical substances, which are spilt into the sea.

The primary purpose of these simulation systems is to provide **forecasts** of the likely development of the spill. The simulation systems are also able to compute answers to a variety of "what if" type questions.

The simulation systems cover seven relevant **spill pathways** as illustrated in Fig. 1.

The user of the simulation package can select one out of six pre-defined **simulation scenarios**:

1 Floating objects (including disabled ships)
2 Oil spill
3 Patch of chemical other than oil
4 Dissolvers
5 Plumes
6 Objects on bottom

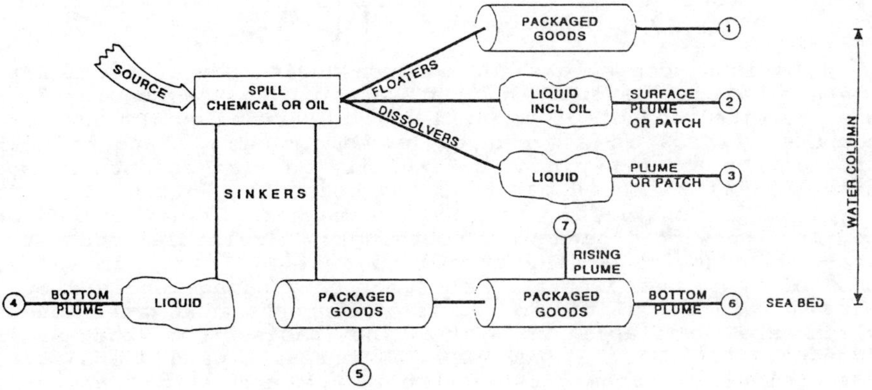

Fig. 1 The Seven Pathways

The six simulation scenarios are covered by a system of nine software modules, as shown in Fig. 2. As an example of the modules invoked for one spill scenario, the modules **highlighted** in Fig. 2 are those invoked for scenario 3, i.e. a patch of a chemical different from oil.

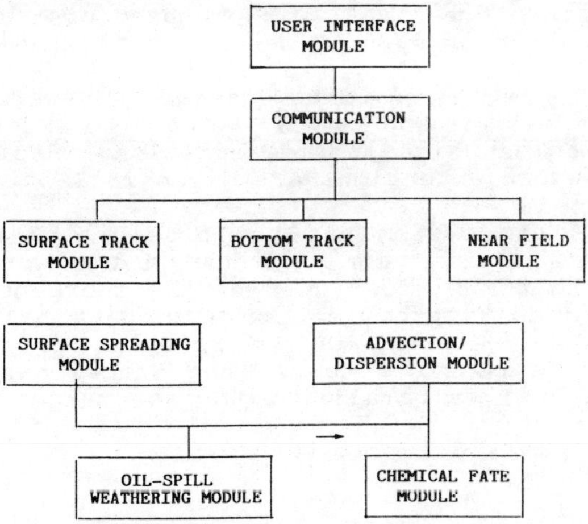

Fig. 2 Main Modules of the Prototype System

Many of the simulation modules shown in Fig. 2 have been developed several years ago by the two institutes and have been applied extensively through the years for oil spill modelling as well as other types of simulations (2),(3).

Time is of course a major concern during the critical first hours of a spill event. This puts a limit to the complexity of the numerical computations which can be carried out on the local computer, which is assumed to be an industry standard work-station. On the other hand **reliability** is also essential. Most major spills at sea occur near the coastline, often (like the Exxon Valdez spill) in confined areas where flow patterns are complex and wind dependent. Accordingly, a reliable forecast of the hydrodynamics of the area is a must. In order to satisfy both of these requirements, the emergency management system is based on pre-calculated forecasts of currents (and water levels) which are available on a routine basis from the Danish Meteorological Institute (DMI). Forecast winds and other meteorological parameters of importance are also transferred from DMI.

Measured data from the area of interest - if available - can often serve as initial conditions for the forecast simulation and validation of the first "warm-up" hours of the simulation results. For that reason, the emergency management system includes facilities for retrieval of real-time data and graphical presentation. It is even possible to compare simulation results (e.g. current speeds) with measured data and thereby obtain an impression of the reliability of a given forecast.

The emergency management system is designed for use on a work-station with a graphical, window based user interface (X-Windows). The system is **open** by design in two important ways:

1 The system is developed using standard software techniques. It is highly modular and it will be easy to extend the system with new modules which provide additional facilities. Some ideas for future extensions are given in Section 5.

2 The user of the system has the possibility to run other software applications in a window on the work-station in parallel with the windows used by the emergency management system. This could provide access to data bases (local or remote, via a network) or to relevant programs. As an example, it is possible to use the Seabel system (running in a DOS window) in parallel with the emergency management system (4).

4. NEW MODULES

Two of the modules which are new in the emergency management system, when compared to previous oil spill models developed by the same institutes are the User Interface Module and the

Chemical Fate Module.

User Interface Module

As described above, the User Interface Module is window based. The basic "entry window" shows a map of the region covered by the particular version of the emergency management system. This Base Map Window also shows the location of any sensors (current meters, etc.) in the area, which provide on-line data to the system. Finally, the Base Map Window contains a number of "push buttons" which activate different other modules, like for instance the simulation modules.

Fig. 3 shows an example of a screen lay-out with only the Base Map Window open.

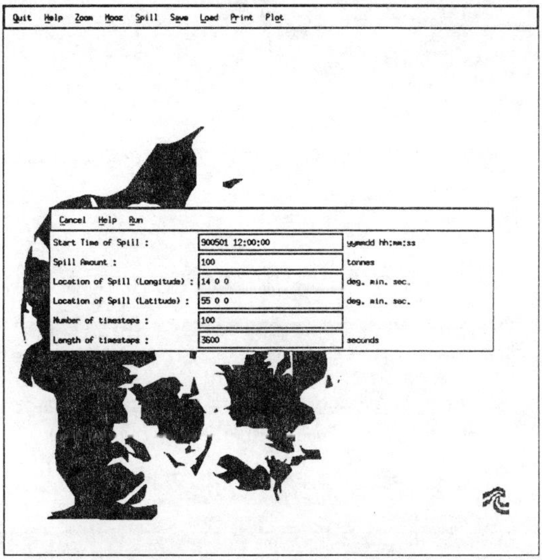

Fig. 3 Base Map Window (Example)

The map of the Base Map Window is also used for presentation of results, e.g. forecast oil spill track and spreading. For this purpose, the base map may be in too large a scale. To provide a better resolution of the results a zoom facility is available.

The push buttons in the Base Map Window activate different other modules like simulation modules (starting with input pages) and presentation modules, which provide a graphical presentation of measured data, simulation results or a combination of both. Fig. 4 shows an example with a timeseries of measured current speeds.

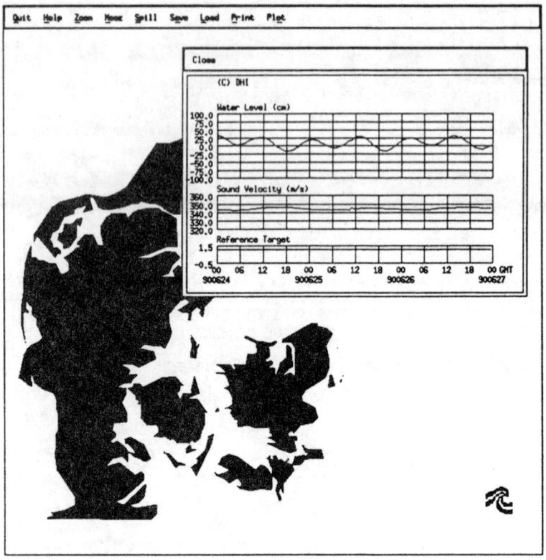

Fig. 4 Presentation Window (Example)

<u>Chemical Fate Module</u>

Another new development is the Chemical Fate Module. This module is applied for simulation of the physical, chemical, and biological processes which transform a spilt chemical substance into one or more new substances. These processes depend on the properties of the chemicals as well as on the characteristics of the receiving waters. Fig. 5 shows the most important processes involved in these transformations.

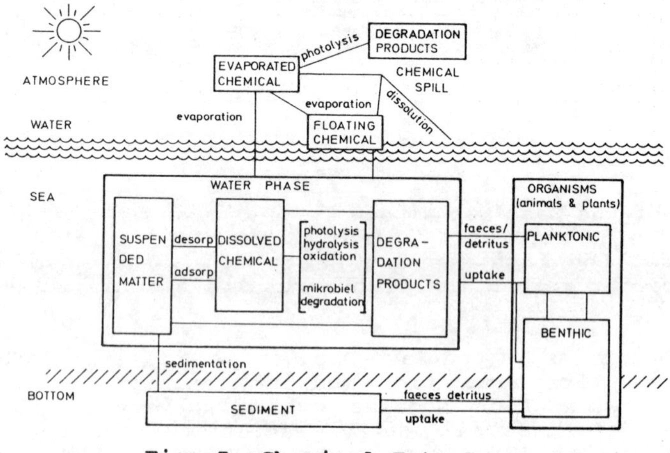

Fig. 5 Chemical Fate Processes

The most important **physical** processes for soluble chemicals are: evaporation, adsorption/desorption, and sedimentation.

The primary chemical and biological transformations are: photolysis, hydrolysis, oxidation, and biodegradation.

In case of non-soluble or slightly soluble chemicals some additional processes have to be taken into consideration. These are: dissolution, emulsification, and dispersion.

One of the main principles of the Chemical Fate Module is to include descriptions of all the known important fate processes for chemicals. The module is structured in a way which allows the user to select the most relevant processes in each specific case.

5. FUTURE EXTENSIONS

A prototype of the emergency management system is presently under development. The prototype will cover the North Sea, the Baltic Sea, and the Straits connecting these two areas. The prototype will be installed at the Danish National Agency of Environmental Protection and thoroughly tested by operators from that agency.

Based upon the test results and practical experience gained through the use of this prototype future modifications and extensions will be planned. Among the possibilities can be mentioned:

- A data base module containing information on equipment and personnel available for spill combat.
- An expert system (or links to an existing system) which provides advice on immediate actions (warnings to the public, evacuation, mobilization of combat resources from other countries, etc.).
- Improved descriptions of the processes (3-D flow, etc.).

ACKNOWLEDGEMENTS

The development of the Integrated Emergency Management System for Spills of Chemical Substances is supported by the Commission of the European Communities, Directorate-General XI and by the Danish National Agency of Environmental Protection. This support is gratefully acknowledged.

REFERENCES

(1) M. Fingas: "Review of Personal Protective Equipment for Spills". Spill Technology Newsletter, Vol. 14 (4), December 1989, Ottawa.

(2) Water Quality Institute and Danish Hydraulic Institute: "Computer Model Forecasting Movements and Weathering of Oil Spills". Report to the European Economic Community, October 1985.

(3) Abbott, M.B., Damsgaard, A., and Rodenhuis, G.S.: "System 21, Jupiter, A Design System for Two-dimensional Nearly Horizontal Flows". J. Hyd. Res., Vol. 11, pp. 1-18.

(4) TNO Division of Technology for Society: "Seabel - A Hazard Identification and Decision Support System for Emergency Response for Chemical Spills at Sea", October 1988, The Hague.

A MULTI-DISCIPLINARY EUROPEAN APPROACH TO DEVELOPING TECHNOLOGIES FOR THE REMOVAL OF OFFSHORE PLATFORMS

R. W. Ebdon Principal Engineer
Advanced Mechanics & Engineering Ltd

R. Surle Senior Project Manager
Comex Services

P. Minardi Project Manager R&D
Tecnomare S.p.A.

SUMMARY

Three companies, AME from the UK, Comex from France and Tecnomare from Italy proposed to EC Directorate General for Energy to perform research and development studies of technology for the removal of offshore platforms. The proposals contained common elements and the technologies proposed for study by each company were complementary. At the suggestion of the EC a co-operation was initiated between the three companies and this paper presents a brief outline of the projects and some results from the co-operative work.

The results reported included those from a review of the possible removal techniques, the types of structures and the market potential for industry. The specific technologies studied by the individual companies were

AME	-	safe and reliable application of explosive shaped charges to the cutting of large steel jacket tubulars and the effect of underwater shock.
Comex	-	establishing the performance of high pressure abrasive water jet cutting by means of extensive testing under hyperbaric conditions.
Tecnomare	-	the design of a diverless underwater cutting system that can be integrated with advanced cutting techniques and provide a commercially viable facility.

1. INTRODUCTION

The development of legislation relating to the removal of redundant offshore platforms requires either partial or complete removal. In either case there will be a growing necessity for safe, reliable methods for cutting structural elements in a manner which minimises costs. Undoubtedly, methods exist currently which enable the removal of small platforms but these are not always reliable and hence can involve additional costs. Also, these methods are not directly applicable to the large platforms in the North Sea and it was to these that the projects reported in this paper were directed. Three companies independently proposed to the EC Director General for Energy to carry out research and development to provide techniques which could be used on large platforms, as well as providing more reliable and cost effective techniques and equipment for removal of small platforms.

In view of the complementary nature of the proposals and some common activities the EC encouraged the companies to co-operate and avoid duplication of effort. In particular this involved review of possible legislation, the identification of the ranges of sizes of platforms to be removed and the potential market for and costs of the removal operations. The co-operation also extended to exchange of information on the conduct of the research on the new techniques and the implication of the results for the removal costs and operations.

AME's project, also supported by the UK Department of Energy, is concerned with the safe and reliable application of explosive shaped charges to the removal of steel jacket structures. It involves a testing programme to evaluate advanced explosive charges, developed through military research, for civilian application offshore on 50D steel up to 100mm thick. The project, started in 1988 and finishing this year, involves extensive analytical developments for the prediction of underwater shock and bubble pulse effects. The findings of these two aspects of the project are being developed into practical recommendations on how explosive shaped charges should be used safely and effectively. Assistance is being supplied to the project by Heerema Engineering Service, who have extensive experience in platform removal and all aspects of offshore construction work, and Defence Technology Enterprises Ltd who were specifically set up to exploit military technology for civilian use.

Comex's project is aimed at establishing the performance of the techniques of Abrasive Water Jet cutting for platform removal operations. Comex had carried out in-house evaluation of high pressure abrasive water jet cutting and identified a further need to perform an assessment of the available techniques to decide on the most promising for the operations of the company. Following this the project is aimed at establishing suitable performance characteristics for the chosen process by an exhaustive series of hyperbaric cutting trials on suitable materials.

The aim of Tecnomare's research project is to design a diverless underwater cutting system that, integrated with advanced cutting techniques and procedures, will reduce the cost of platform removal. Functional requirements have been obtained by a market assessment of platforms to be removed in the North Sea and Mediterranean and by a thorough analysis of removal procedures summarised here As a result of these activities two cutting systems, a multipurpose cutting robot and a Remotely Operated Vehicle have been preliminary identified. These are described in detail in this paper.

The co-operation between the three companies proved to be very effective and undoubtedly enhanced the cost-effectiveness of the projects, summerised results of which are presented below. A more complete report is contained in Ref (1).

2. GENERAL REVIEW

The three companies co-operatively carried out review and assessments on the development of legislation, operational requirements and market potential to provide a background for their indvidual projects.

The review of legislation developed by the IMO may be summarised by stating that existing small platforms in shallow water and all future platforms must be completely removed. For large existing platforms in deep water the onus will be on the operator to prove that removal is not technically feasible or involves extreme costs or risks to the environment or personel if partial removal is to be accepted. It seems likely that the debate will concentrate on the definition of 'extreme cost' as this will determine what measures are taken to make the other aspects acceptable.

A detailed analysis of costs and phases in the removal of platforms was performed by Tecnomare with the objective of evaluating the impact of cutting operations on a complete removal operation. The analysis required the classification of platforms in the North Sea and Mediterranean into five groups and the availability of lifting vessels with specified lift capacity. Generally it was concluded, based on current technology, that cutting operations take about 35-50% of the overall time for platform removal (including time for transport to shore). The cost summary of the platform removal costs according to classification of platform is shown in Table 1 with the classification being shown in Table 2. The cost of overall cutting operations varies from 15% to 30% of the total operational costs. Since a large part of the costs is the high day rates of attendant vessels it is desirable to develop more efficient, and hence inherently less expensive, cutting systems with dedicated tools and procedures which reduce the duration of all cutting operations.

A market assessment was conducted to establish the potential income from the cutting techniques considered. The market assessment was limited to the North Sea and the Mediterranean and assumed a mean platform life of 25 years. Table 3 shows the market

PLATFORM CLASS	REMOVED	LIFTING METHOD	CUTTING SYSTEM	TOTAL REMOVAL COSTS ($)	% OF TOTAL REMOVAL COSTS			
					CUTTING OPERATIONS COSTS	TOTAL OPERATIVE COSTS	SALVAGE YARD COSTS	ENGINEERING
A	TOTALLY	CRANE VESSEL	WATER JET	1,771,000	20	81	14	5
			EXPLOSIVE	1.641,000	15	79	16	5
B	TOTALLY	CRANE VESSEL	WATER JET	2,949,000	25	80	15	5
			EXPLOSIVE	2,710,000	19	78	17	5
C	TOTALLY	HEAVY LIFT VESSEL	WATER JET	6,520,000	15	82	13	5
			EXPLOSIVE	6,794,000	19	83	12	5
		CRANE V. + BUOYANCY TANKS	WATER JET	7,225,000	14	93	2	5
			EXPLOSIVE	7,367,000	13	93	2	5
D	TOTALLY	HEAVY LIFT VESSEL	WATER JET	13,040,000	14	81	14	5
			EXPLOSIVE	12,729,000	12	81	14	5
		CRANE V. + BUOYANCY TANKS	WATER JET	17,076,000	13	93	2	5
			EXPLOSIVE	16,695,000	11	93	2	5
E	PARTIALLY	CRANE VESSEL	WATER JET	12,692,000	19	67	28	5

Table 1 : Cost Summary of Platform Removal Cases

Table 2 : Platform Classification

Class	Water depth (m) from	to	Jacket Mass (t) from	to
A	0	50	200	1000
B	20	75	1000	2000
C	25	75	2000	4000
D	25	75	4000	10000
E	75	190	4000	35000

Table 3 : Market assessment by classes

	North Sea		Mediterranean Sea	
Class	Number	Market (M$)	Number	Market (M$)
A	132	211	60	96
B	42	118	8	22
C	22	145	3	13
D	17	218	0	0
E	34	432	4	51
TOTAL	247	1124	75	182

Table 4 : Removal operations market sharing by year ranges (M$)

	1990 - 1995	1996 - 2000	2001 - 2005	2006 - 2010	After 2010
North Sea	29 - 33	192 - 215	267 - 299	382 - 428	254 - 285
Mediterr. Sea	6 - 7	29 - 33	10 - 21	43 - 48	92 - 103
TOTAL	35 - 40	221 - 248	286 - 320	425 - 476	346 - 388

assessment correlated to the classification in Table 2, and Table 4 shows the potential income for contractors between 1990 and 2010.

A considerable amount of work has been done by the three co-operating companies in preparing the background information summarised above. This information is subject to continuous amendment as conditions change but is an essential starting point for development work in this field.

3. DEVELOPMENT PROJECTS

The following is a very brief outline of the work undertaken by the individual companies to develop further the state-of-the-art in underwater cutting.

The Use of Explosives for Platform Removal (AME)

The use of explosives for the demolition of large structures is a longstanding practice, both in peace and war. Developments from the basic shattering techniques, particularly shaped cutting and piercing charges, have also been in regular use in the oil business for many years. The need for further research and development results from the emphasis on safety and reliability of operation offshore and the reservations about the application of shaped cutting charges to the removal of large structures. The removal of the damaged Beryl A SPM base (2) highlighted difficulties and uncertainties associated with the use of shaped charges, obtaining reliable detonations at a deepwater open-sea site and predicting shock effects. Matters were complicated by a composite steel and concrete base, but the conclusions from that project appeared to be that better procedures and more reliable methods of assessing shock effects were required. The objectives of the project performed by AME were therefore to

- prove the potential of an advanced design of cutting charge in a marine environment,
- investigate the shock and bubble pulse effects on the structure and attendant ships.

The project involved testing and analysis. The testing program was conducted for AME by Defence Technology Enterprises who have organised supply of charges and testing with Royal Armaments Research and Development Establishment (RARDE) and other appropriate establishments. The principal objective of the testing programme was to establish the performance of the optimised linear shaped charges against offshore grade 50D steel. Although there are empirical relationships to predict charge behaviour, and hydrocodes for numerical solution, testing is considered necessary to provide confirmation of expected results.

The linear charge design used in the tests is shown in Figures 1 and 2. This is an advanced optimised charge which ensures that the maximum amount of the explosive energy is imparted to the jet of metal which actually cuts the structural member. It is important that the explosive is detonated uniformly and this is achieved in the RARDE design by mechanical means.

The sequence of detonation is as follows. The booster pellet (1) is detonated by an inserted detonator and then transfers detonation along sheet explosive (2). As the detonation travels along this the longitudinal 'mousetrap' plate (3) is projected downwards to impact sheet explosive (4) uniformly. A linear detonation front then forms in the side sheets of explosive (5) projecting the sloping mousetraps (6) into the back face of the main charge (7). By precise design of the mousetrap angles the detonation front in the main charge can be arranged to arrive at the back of the copper liner (8) as two planes parallel to the liner faces. This detonation mechanism maximises the effect of all the main charge using as much of the available energy as possible to form and propel the liner material. At present the longest charge which has been fired in this way is 300mm. The limitation is the behaviour of the longitudinal mousetrap. However, two charges could be formed back-to-back to produce a 600mm long straight cut.

The testing was conducted in phases starting with tests on flat plates in air and progressed to tests firing through a simulated charge casing and water layer before cutting the target. Following these evaluations, through-water tests on curved plates and a full circle tubular were performed.

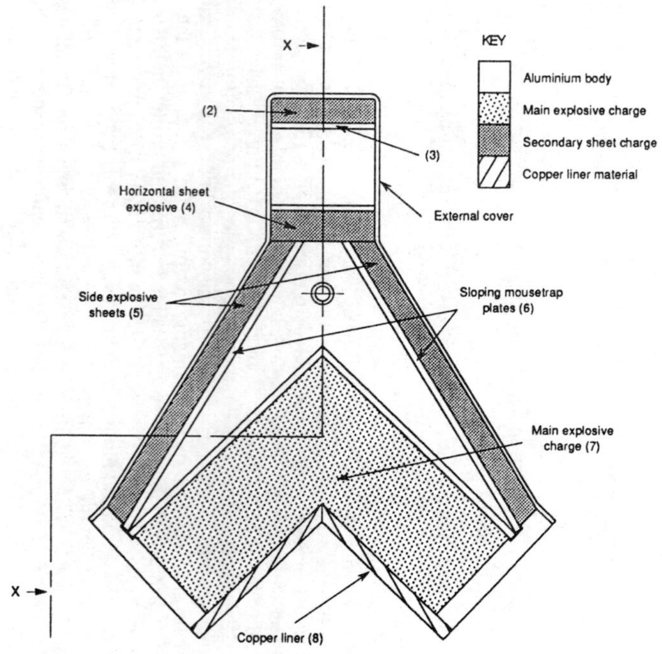

Figure 1 Optimised Shaped Cutting Charge – Second Elevation

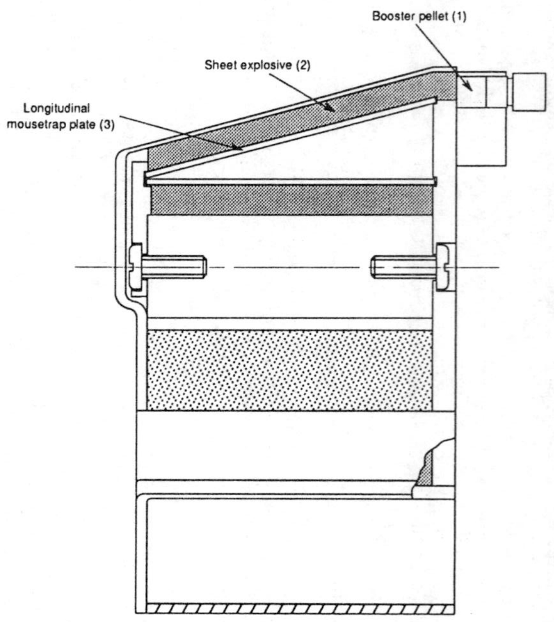

Figure 2 Optimised Shaped Cutting Charge – Part Section

These tests provide the following information.

- the maximum thickness of 50D steel which can be cut with the standard 60mm throat-width charge
- a scaling relationship to predict the size and explosive loading of charges required for changes in steel thickness
- the effects of sandwich cutting of the steel casing, water layer and target for a practical stand-off and performance/size relationship
- the jet behaviour of the charge over a range of stand-offs
- the effects of variable stand-off between straight charges and curved targets
- the interference between end effects of straight charge sections arranged to cut curved steel samples.

The emphasis of the AME project is on the interlinked aspects of safety and reliability of explosive cutting. Having established reliable prediction of charge performance the remaining concerns for reliability centre on the detonation system.

Test Results

The optimised shaped charge of 60mm throat width has successfully made a clean cut of 100mm of 50D steel plate. The second series of tests have evaluated the performance of the charge cutting through the three layers of a simulated steel casing, a water-filled gap and, finally the target material, 50D steel plate.

The primary concern here was to evaluate the likely effect on the jet behaviour of the sandwich of materials of varying density. The influence of the water layer was used to model the fact that installation would be more adaptable if allowance is made for some misfit between the charge casing and the target. Previous work indicated that jet penetration is governed by the square root of the ratio of the densities of the jet and target materials. The expected penetration through water would be 2.8 times that of the equivalent steel target.

Although the results from the full series has yet to be completely analysed it appears that the effect of a steel-water-steel sandwich is greater than would have been expected. However other effects associated with the length of the charges are being eliminated before drawing final conclusions. The tests have been conducted by constructing a steel plate box onto the front of the target plate and filling this with water. This avoids the difficulties associated with firing the charges in a tank and, as the process is so rapid, the effects are expected to be comparable. On the assumption that water density is the controlling factor, ambient pressure should not affect the resultant cut.

The analytical techniques and software being developed by AME utilise the test results and predict the effects of the explosions on the structure and attendant vessels. Thus the project will provide guidance on charge design requirement to satisfy installation tolerances which can be achieved economically, and establish a safe and reliable approach for the use of explosive cutting charges in an offshore environment.

Abrasive Water Jet Cutting for Platform Removal (Comex)

Comex has initiated a development program to evaluate abrasive water jet cutting as a complementary tool to explosives to complete its own operational tool range for use on its worksites. This development will take advantage of Comex's experience in underwater cutting techniques and the latest technological developments in the field of abrasive water jetting.

The current development programme includes the following subjects:

- the development of an abrasive water jet system to a state where it could be used independently or to complement explosive cutting for platform removal;
- the development of operational procedures for the offshore deployment of abrasive water jet to produce cost effective and safe solutions for large platform removal;
- a general study concerning the adaptation of an abrasive water jet equipment to remote controlled systems.

Programme Details and Co-operation

The Comex development programme is spread over two years (1989-90) and covers three areas, in addition to the worked performed in co-operation with AME and Tecnomare

- Development and hyperbaric testing of HP abrasive water jetting system and creation of a parameter data bank.
- Definition of the specification of the deployment systems required for installation, remote control and monitoring of the HP abrasive water jet.
- Basic engineering for the deployment of abrasive water jetting by ROV or by purpose designed dedicated robot.

A general study has been carried out to obtain the latest information about the various manufacturers and operators of relevant equipment. The study has served to evaluate the advantages and disadvantages of each of the methods which are;

- Secondary Injection of dry abrasive at the cutting head level.
- Secondary Injection of a low pressure slurry at the cutting head level.
- Direct injection of HP slurry to the cutting head.

The study assisted in selecting the most appropriate existing method and corresponding equipment or identifying an automatic abrasive metering device prototype in order to test it and adapt it to Comex's specific needs. In the event and for various reasons [1], Comex selected the third of the method listed above.

Comex has finally selected the Diajet 350X2C automatic abrasive metering device designed and built by British Hydraulic and Research Association (BHRA). In order to test it thoroughly it was necessary to evaluate its limits for underwater application and reliability for operational use together with any modifications needed to adapt it to Comex's needs. It's cutting capabilities for platform and deep pipeline removal purposes will be evaluated.

Hyperbaric Testing

The testing site was mobilised at Comex Services in Marseilles for tests from mid-December 1989 to mid-March 1990. Figure 3 shows the equipment layout.

The primary aim of the testing was to investigate the effects of the parameters which contribute to the success or failure of abrasive water jet cutting operations. These are shown in Table 5. The purpose of the tests is to reduce this parameter list in order to establish basic procedures for future on-site cutting operations.

Test Results

The preliminary conclusions from the test results are summarised below.

Abrasive grit efficiency

Three of the four different abrasives tested show little difference in cutting efficiency. Abrasives tested : Garnett, J-Blast, Sicated sand and Rugos.

Abrasive grain size

Fine grain size does not produce reliable and good cutting performance. Medium size grain is recommended.

Abrasive flow rate

The variation of abrasive flow rate is better optimised around 10% to 15% of the total jet flow.

Water depth

No significant variation of the cutting efficiency is encountered between water depth of 50 to 400 m. The cutting rate is not affected.

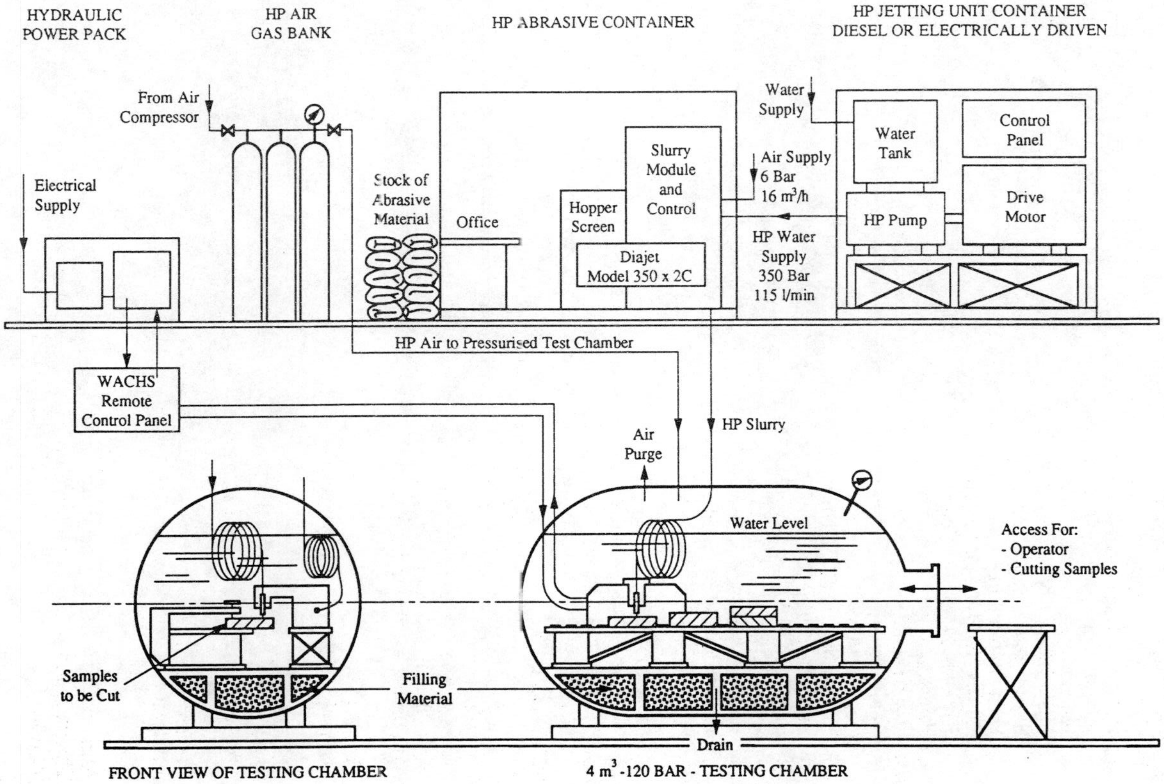

Figure 3 Equipment Arrangement for Water Jet Cutting Trials

Table 5 : Parameters Concerned in Abrasive Water Jet Testing

	INFLUENCING PARAMETERS (VARIABLE OR CONSTANT)
JETTING NOZZLE	
- Nozzle wear	- Nature of abrasive
- Manufacturing material	- Nature of material to cut
- Stand off distance	
- Size of nozzle entry	
NATURE OF ABRASIVE	
- Sand	- Efficiency
- Copper slag	- Cost
- Garnett etc	- Availability
	- Quality
NATURE OF MATERIAL TO CUT	
- Steel	- Thickness
- Steel and Concrete	- Grade of material
- Multiwall material	- Cutting speed
- Armoured hoses	- Kerf width
	- Abrasive feed
CUTTING DEPTH	
- 0 to 200m (platform repair or	- Ambient pressure
removal)	- Stand off distance
- Below 200m (structures and pipe	- Slurry proportion
repairs)	- Water flow
	- HP water pressure
	- Hydrokinetic power

Quality of the cut

Fairly good, even for cutting samples like flexible pipes. Not so good on a reinforced telephone pole featuring some large and hard granite aggregates mixed into a good quality concrete.

Slanted cut

These are perfectly feasible on 25mm and 50mm steel plates at 20mm and 400m deep.

Nozzle life

The nozzle life for a standard tungsten carbide nozzle is still quite acceptable and should be somewhere between 10 to 20 hours at least. This point is still to be confirmed.

So far the hyperbaric cutting tests have shown the potential of the method even at great depth and the application for offshore platform removal looks quite promising. At shallow water depths (between 50m and 150m), the zone of most fixed offshore platforms, the potential should be even greater. The system works with a minimum maintenance (spot checks have to be made regularly for reliability and safety purposes). The abrasive consumption is quite reasonable for underwater use (8 - 11 l/min). Cutting tests with a 200 m long umbilical are scheduled in order to test out the system close to operational conditions. More tests on sandwich materials using a larger nozzle size are still to be done.

Remotely Controlled Cutting Module (Tecnomare)

As a result of the initial analysis by Tecnomare, three cutting operation procedures were identified which require the improvement of the present technologies or the development of new ones. These operations are:

- main pile cutting 5 m below the mudline from inside
- skirt pile cutting 5 m below the mudline from inside
- leg/bracing cutting from inside at intermediate water level

In particular, considering the trend to avoid the utilization of divers in marine operations at great water depth for safety reasons, the opportunity arises to carry out these operations with equipment guided from a surface support structure or vessel. Thus there is a requirement for the development of a dedicated system capable of autonomously deploying the equipment and performing cutting operations.

For these purposes two systems are proposed:

- Multipurpose Cutting Robot (autonomous or ROV assisted) to deploy and activate the equipment required to cut skirt piles from the internal side.
- ROV to perform leg and bracing cutting.

The development of a multipurpose cutting robot is aimed at the integration of cutting tool, dredging pump and centring device in a single module to reduce the number of trips necessary to perform skirt pile cutting from inside. Positioning and entry operations on skirt pile can be autonomously performed by integrating the positioning system in the support frame of the cutting module The utilization of a ROV capable of assisting and guiding the passive module during the positioning could be considered a possible alternative.

Figure 4 shows a possible system. In order to perform the necessary tasks (1) the system will have the following attributes:

The support frame has a wedge shape and is adjustable at different skirt diameters (from 1m up to 2m) with an overall height of about 1.5-2m containing the integrated system.

There must be at least 3 thrusters, to control the displacement in the horizontal plane. Two black and white TV cameras are required to allow monitoring of operations. An electrical power distribution module and a hydraulic auxiliary power pack will be installed in the frame to activate the dredge pump, cutting nozzles and centring arms.

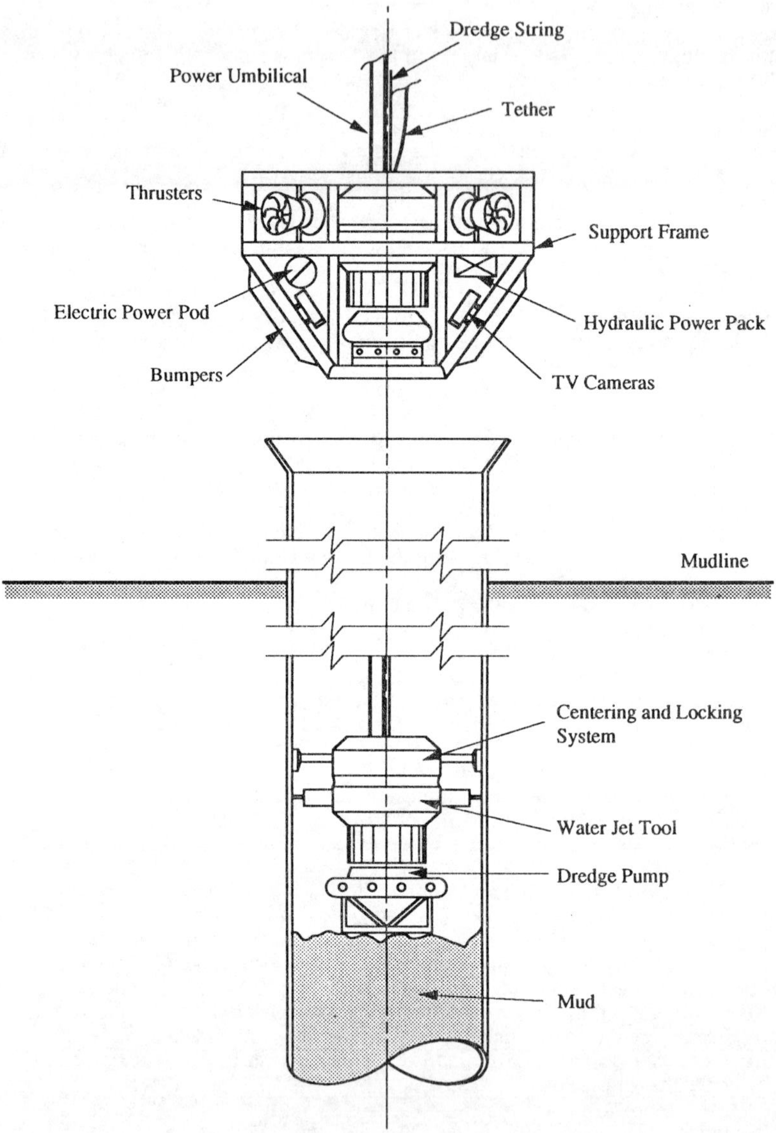

Figure 4 Lowering Multi - Purpose Cutting Robot In Pile Sleeve

The maximum diameter of the integrated system will be 0.8m. The centring arms and the cutting nozzles will be equipped with mechanical adapters, (allowing the same cutting device to be utilized with different skirt pile diameters), and hydraulic actuators for final positioning. For water jet cutting nozzles have to be continuously held at 5 mm from the skirt wall by means of a sliding spacer.

The rotation speed of the cutting nozzles along the pile circumference must allow the complete cutting with respect to steel thickness, for instance:

thickness = 20 mm	cutting rate = 7 cm/min
thickness = 60 mm	cutting rate = 2 cm/min

If explosive shaped charges are utilized the multipurpose robot will be modified for the operation. The cutting site will be cleaned with steel brushes and an inflatable device used to adapt the charge to the wall.

Considering that piles can be 30-35 m high, the integrated cutting equipment could need to be lowered by as much as 40m inside the skirt before reaching the cutting level. The dredging pump is electrically powered and its delivery head must be enough to push the mixture of mud and water at least out of the pile mouth.

It must be pointed out that the integrated cutting equipment could be adapted for main (leg) pile cutting from inside.

The study includes the evalution of the use if an ROV for cutting operations applied to legs and braces.

To cut legs and bracings at intermediate water levels, a ROV fitted with a manipulator arm featuring high accuracy and dexterity, if possible without the necessity of a docking module, seems to be the most promising solution. A preliminary demonstration of the feasibility and of the effectiveness of the approach of using an ROV, enhanced by some kind of automation to increase the accuracy and reliability of the operations, has been given by full scale laboratory tests on the Supervisory Control Telemanipulator developed by Tecnomare during a three year research project.

4. CONCLUDING REMARKS

The co-operative spirit shown by the three Companies has undoubtedly contributed to the breadth and quality of the results from the projects funded by the EC. Decommissioning and removal of offshore structures is a very wide field and the scope for development and innovation is by no means exhausted by the projects reported here. It is the intention of the three Companies to continue their technical activities to provide safe, reliable and economically viable tasks for future offshore removal operations. The success of the joint tasks within the project reported here has had the effect of encouraging the Companies to co-operate in planning future work which will be to the benefit of the European offshore industry at large.

REFERENCES

(1) Ebdon, R.W., Surle, R. and Minardi, P. Removal of Offshore Structures, Proc. of Int. Offshore Decommisioning. Platform Removal and Marine Salvage Exhibition and Conference., Aberdeen, March 1990

(2) Brown, A.R., and Peacock. D.A. The Removal of Abandoned Beryl Field Pipelines and Structures. Proc Offshore Europe 87.

THE NEED OF RATIONAL CRITERIA FOR FIRE SAFE DESIGN OF OFFSHORE STRUCTURES

By R. CAZZULO, C. MURGIA, F. ZILIOTTO

REGISTRO ITALIANO NAVALE
Via Corsica 12
16128 Genova GE - Italy

Summary

The paper outlines the present European situation regarding the criteria and the applicable rules for the design of offshore fire protection, and provides an analysis of the benefits which can be obtained from the study of new reliability based design criteria.

The goals of a project [1], aimed at defining a proposal for European guidelines, are set out, reviewing the themes and concepts which most urgently need to be finalized and harmonized for an effective and reliable fire protection design of the topsides of the offshore platforms.

1. INTRODUCTION

Explosions and fire are among the serious risks which jeopardize human life and the safety of the structures and systems during offshore operations. The analyses of the main accidents on board fixed and mobile offshore units reveal the social and economic importance of an effective design and construction of the platform's fire protection.

The complex scenario consisting of governmental and non-governmental bodies which operate in each European country on the basis of their own rules, compels the engineers to guarantee the design conformity using different criteria, not always in terms of the structures and systems to be protected.

The laws governing the design of fire protection are really often the result of a combination of naval design criteria and industrial onshore rules. The design developed on the basis of such criteria is not a fully rational way to ensure global reliability and the economy of the system.

Moreover, fire protection methods in the first generation of offshore installations were not engineered in themselves, but were specified on the basis of the concept "better to err on the safe side". This involves, even today, penalties in weight and costs, which do not imply an increase in safety.

A rationalization of the design principles, able to eliminate unnecessary requirements and redundant criteria, appears to be necessary, looking at the topsides as a system to be characterized and analyzed under the effects of a dynamic event such as fire.

The introduction of new design parameters and the availability of new materials may lead to an optimization of the layout, construction and protection systems of the topsides, in particular through new specially developed reliability based techniques.

2. OFFSHORE FIRE PROTECTION DESIGN IN EUROPE: THE SCENARIO

Interested Parties

The design and construction of fire protection systems for offshore structures and plants is subject, in each European country, to the approval and control of many bodies, both for mobile and fixed platforms. Designers and companies have, out of necessity, to take into consideration at least the following interested parties:

- Competent Ministry-ies
- Technical and Inspection bodies
- Maritime Authorities
- Classification Societies

Table I shows the present situation for six European Countries, indicating the main parties involved.

Country	Ministry	Technical and Inspection Body	National Class. Society
Italy	-Min. Indus. C.& A. -Min. Interno -Min. Marina Mer.	UNMIG, VVFF C.PP.,	REGISTRO ITALIANO NAVALE RINA
UK	UK Department of: - Energy (DOE) - Transport (DOT)	DOE, DOT	LLOYD'S REGISTER - LR
Norway	Royal Ministry of -Labour/Loc. Gov. -Petroleum/Energy	NPD, NMD	DET NORSKE VERITAS - DNV
Germany	Oberbergamt (OBA)	OBA, Bergamt Meppen	GERMANISCHER LLOYD - GL
Denmark	Ministry of Energy	-Energistyrelsen -Dan.En. Agency	Various DNV, BV,LR,GL,ABS
Netherlands	Ministry of Economic Affairs	State Supervision of Mines (SSM)	"

Table I - Main interested parties

As far as the mobile drilling (and/or production) units are concerned, the situation is particularly hard, since, in order to operate in different territorial waters, the design compliance with all the relevant national laws and rules in force has to be ensured.

Laws, Rules and Standards

The design of the systems for fire protection of the topsides is developed in accordance with:

- Laws and Decrees
- Technical and Inspection bodies Rules
- International Rules and Maritime Conventions
- Classification Rules
- Design Standards

The general laws concerning the safety of the mining installations rarely contain specific technical requirements for the fire protection systems, but appoint the competent technical bodies: the rules issued by such bodies are one of the main bases of the project. While on the subject, the different situation of fixed and mobile units is to be outlined: the inspection bodies of the two leading European countries, i.e. Norway and UK, have issued detailed and complete rules both for fixed and mobile units, whereas other countries, among which Italy, have not entirely defined the safety criteria for the fire protection of the fixed units (see next paragraph).

As far as the mobile units are concerned, the international rules and the classification rules are almost completely aligned: the first are codified in the IMO (International Maritime Organization) MODU CODE, which, as regards the fire protection requirements, is mainly based on the international SOLAS Convention; the second include and integrate both the CODE requirements, and the SOLAS regulations.

Moreover, the laws and the inspection bodies regulations often refer to different design standards and national unifications as regards primary technical aspects (e.g. standard fire tests and the relevant passive fire protection qualification of the materials); they represent another critical basis for the design. Table II shows the main laws, rules and standards in force in the six above mentioned countries.

Country	General Laws/Decree	Applicable Laws and Rules	Class. Society Rules	Recog. Stds.
Italy	DPR 24/5/79 no. 886	-DPR 14/11/7 no. 1154 -Draft Regulation (art.49 DPR 886)	RINA MODU Rules	CNR/ UNI CEI
UK	Mineral Working O/shore Inst. Act, 1971	DOE Offshore Installations Guidance on Design and Construction	-LR Class. R.s for Mob.O/S Units -LR Class. R.s for O/S Structures	BS
Norway	Royal Decree 25/8/67	-NPD Regulation for Mob.Dr.Plat. -NPD Reg. for Prod. & Aux. Systems	-DNV Cl. Rules for MODU -DNV Rules for Fixed Offshore Installations	NS NFPA IEC
Germany	Bundesberggesetz, 1980	GL Rules ---›	GL Rules for Constr. and Inspection of Offshore Install.s	DIN
Denmark	Act 292 1981 Ord 592 1981	Indicated in Order 592	Various	
Netherlands	Mining Reg. 1964	Mining Regulations Continental Shelf	Various	

Table II - Laws, Rules and Standards

The Italian Scenario

The fire protection criteria for offshore mobile units are relatively well defined by Italian Laws and decrees (see table II): they are accurately specified for units which are similar to ships, if they are self-propelled, whereas they are not exhaustive for non-self propelled platforms like many jack ups.

The requirements for this kind of rig, classifiable as "floating" according to the Italian definition, are not indicated either by the decree D.P.R. No. 886, nor by the decree D.P.R. No. 1154 (Rules for the Safety of the navigation and of the human life at sea) to which the first refers.

For the jack ups, in any case, the IMO requirements are applicable on the basis of the classification rules, like that issued by RINA.

The situation of fire protection design for fixed platforms is illustrative of the difficulties connected with the presence of the many technical bodies involved and heterogeneous rules to be harmonized. In the decree D.P.R. No. 886 issued in 1979 the definition of the main criteria for the fire protection of fixed platforms was foreseen by the following year.

For this purpose, a Committee was established, consisting of representatives of the technical bodies of the Ministries for Industry, the Interior and the Merchant Marine.

Defining the draft rules regarding the above mentioned criteria, an alignment with the IMO criteria for mobile units was also introduced, on the analogy of the rules issued by the other leading countries in offshore operations.

With RINA as effective consultant, an integration of some requirements of the Italian laws with the IMO requirements was defined with regard to passive fire protection; the IMO requirements related to active fire protection were almost completely accepted. The result is the "Draft regulation concerning the main criteria for the fire protection of the fixed platforms and similar fixed structures" issued in 1988 which, though it still has to be officially ratified, is considered to be the design reference.

3. SPECIFIC RULES: BENEFITS OF NEW DESIGN CRITERIA

The existence of heterogeneous requirements in the above mentioned rules and regulations, in particular as far as layout and passive fire protection are concerned, justifies the need for a rationalization of the design criteria.

The main benefits which can be obtained through the development of new design criteria for the prevention and protection against the risks of explosion and fire are shown below:

- INCREASE OF THE TOPSIDE SAFETY AND OF THE EFFECTIVENESS OF THE PROTECTION SYSTEMS
- HARMONIZATION OF THE APPLICABLE RULES
- COST OPTIMIZATION AT A FIXED SAFETY LEVEL

The actual criteria used for the layout definition, structural design and fitting out of the topsides are based on two questionable approaches:

- classification of the topsides in "safe" and "non-safe" areas
- definition of the protection requirements (in particular for the passive fire protection) taking into account only the interface between two zones/rooms, without considering the kind and method of propagation of the event.

The latest offshore disasters show that the fatal event to be avoided is really a highly dynamic process consisting of a hydrocarbon leak, explosion and fire, which is able to quickly transform also the areas considered safe into hazardous areas.

To improve the safety on board offshore units, by defining a rational design basis, is the target and the main benefit. To achieve this, it is necessary to look at the topsides as a composed system and to operate in two main directions:

- to study and define the layout and system management defining the following items:
  - specification of the hazard sources
  - layout plants÷quarters
  - operating conditions (open, closed and semienclosed spaces, ventilation systems)
  - area classification
  - safety procedures

- to study and optimize the system on the basis of:
  - a definition of the different kinds of fire, and of the properties of conventional and non conventional materials
  - an evaluation of the safety level of the composed system [monitoring/active+passive fire protection/safety shut downs] of each protected areas, in order to define the probabilities of extinguishing and non propagation of the event
  - simulation of the event propagation
  - simulation of the structural behaviour in presence of thermic stress, to collapse
  - development of a stochastic model able to assess the probabilities of development and propagation of the events
  - optimization of the system layout, used materials and protection methods, through reliability based tools.

A harmonization of the rules aimed at offshore applications and agreed upon by the experts in the different sectors (companies, designers, constructors, classification societies) and the administrations appears to be the most desirable target for all parties involved.

To optimize the fire protection design also in order to obtain a reduction in weight of the topsides at a fixed level of safety, could make a considerable reduction in cost possible in particular for deep water operations. The definition of the most effective combination between active and passive protection, the calibration of safety factors and the introduction of new materials such as composites, can be considered the main goals to be pursued in the above mentioned phases of the study.

4. THE OFFSHORE FIRE PROTECTION SAFE DESIGN: THE GOALS OF A RATIONAL APPROACH

The phases of study and optimization indicated in the previous paragraph are briefly described in the present chapter: these phases are to be considered necessary for a rational approach to the safe design of the "topsides system".

Layout of the systems and safety management

Classification of the materials

The classification of the materials used for the structures and to fit out the topsides is the first essential step to obtain all the necessary input both for the deterministic and stochastic models described in the following paragraphs.

The behaviour of the materials to fire is classified on the basis of:

- endurance limits for fire resistance
- heat transmission characteristics
- toxic gas/smoke emission
- creep properties
- elastic and structural collapse properties.

The main aspects of this classification is the kind of materials to be considered: steel structures and panels, insulated and non-insulated, are not the only items to be examined, also alternative materials such as composites are classified.

Fire and explosion sources/AFP methods

The identification of the ignition (IS) and fuel sources (FS) of the fire are fundamental for the definition of the topside layout. The analysis is to be foreseen for drilling, processing and storage units, and should be able to define, for each possible IS/FS combination, the probability of the event (explosion and/or fire), and the kind of fire (liquid hydrocarbons, gas etc.). Moreover for the different types of arrangements, a complete classification of the active protection (AFP) methods used, as well as of the detection systems (smoke, heat, flame, gas) are necessary.

Segregation conditions

Among the preliminary classifications necessary for an effective simulation and reliability based analysis, is a careful definition of the boundary conditions which is also fundamental for the characterization of the areas and rooms to be protected. Thus it is necessary to define univocally:

- open spaces
- closed spaces
- semienclosed spaces
- naturally ventilated spaces
- mechanically ventilated spaces
- pressurized rooms

Layout of the topsides

Before defining the classification of the safety areas, a global analysis of the topsides feature devoted to the mutual arrangement plant/plant and plant/quarters is necessary.

For the different kinds of service of the platform (drilling, process, storage) and of hydrocarbon (gas, oil) the layout plans used at the present time are to be classified, taking into account, in particular, the mutual position of the living quarters and plants.

Definition of Hazardous Areas

The recent offshore accidents causing an high number of victims, among which the Piper Alpha disaster, have shown, as a result of various inquiries, the need for a different conception of the so-called "hazardous areas classification".

Up to now, design standards and international rules have considered, hazardous areas (Class 1 and 2) those where an explosive gas/air mixture is likely to occur only under normal operating conditions, and regulate the choice of electrical components and machinery to be installed in these areas on the basis of this classification. The goal should be to formulate a dynamic definition of the risk conditions that a plant or component can create, taking into account both its "normal" and "abnormal" operating conditions: as a consequence of "abnormal" operating conditions, vast uncontrolled hydrocarbon leaks can extend the hazardous area to all the unit. Only in this way is it possible to formulate accurate hypotheses about the probable risks to which the living quarters may be subjected, and to define "what´s safe" in the layout, outfitting and protection of the topsides [7].

Safety procedures

The control of safety is not adequately dealt with in the safety offshore rules, and the matter strongly depends on the policy of the oil companies [3][6].

The goal should be to produce a critical review of both the methods for dealing with and controlling the safety of the personnel and plants, and of the emergency procedures, in order to reduce the likelihood of:

- ineffective or insufficient checks and maintenance plans
- incomplete or not optimized emergency procedures
- human error due to insufficient or improper training
- lack of supervision of the safety officers (need for audits and periodical courses).

Maintenance programs and schedules; gas alarm, fire,

alarm, uncontrolled blow out and unit abandonment procedures; training courses and their length; safety audits for officers should be optimized and indicated in the main requirements in the form of adequate guidelines.

Tools for the optimization of the system

Effectiveness of the integrated protection system

The goal is a probabilistic evaluation of the capability of the detection and protection systems to avoid, fight and keep the event within a defined area.

In particular, the following aims are to be pursued:

- probability of gas detection and effective alarm and shut down
- probability of fire detection and effective alarm and shut down
- assessment of the reliability of the AFP Systems for the kind and intensity of fire
- assessment of the reliability of the PFP (passive fire protection) system of the area/room for the defined exposure time and intensity of fire to its complete quenching.

The study may be performed by making use of existing tools for the risk analysis of on-land and offshore production plans, such as the HAZOP techniques, fault tree definition, etc. [2][4][5].

Deterministic models for dynamic propagation

A first deterministic model to be considered is a simple simulation of the event's propagation in time and space. It should be able to take into account the different arrangements of the system and the boundary conditions: outlet possibilities of the blast, ventilation conditions, location and size of equipment and obstacles, properties of materials and matching of AFP and PFP. The model should make it possible to define the direct fire exposure and the temperature in any area/structure/location of the topsides, as processes varying with time.

A second deterministic model is to be developed for the elastoplastic and creep thermal analysis of the structures, able to define their resistance capacity to fire, evaluating the limit state conditions for the applied thermal stresses, and the degradation of the material's physical properties in time.

The model should be able to define the percentage of material degradation and the thermal stresses in any area/structure/location of the topsides, to the local limit state conditions or the collapse of the whole system.

Probabilistic assessment of the system's behaviour

A global probabilistic approach aimed at the analysis of the system could be the main tool for a rational study of the problem. Deterministic tools alone are unable, in fact, to really assess the effect of a fire on topside structures and fittings, resulting from an explosion/ignition or ignition only. The propagation process appears strongly dependent on a stochastic chain: coexistence of ignition and fuel sources, stochastic behaviour of the AFP and PFP systems, deterioration of materials and structures, stochastic nature of the heat transmission´s phenomena etc.

The objective should be to assess the effects of fire in terms of propagation (temperature) or structural deterioration, as a stochastic function varying in time and space, starting from defined initial conditions, in order to determine both the reliability of subsystems and the overall reliability of the system. The assessment should take into account the different layout arrangements, the choice of materials and the classified protection methods.

Therefore it is necessary to define a succession of events to be examined, starting from the event limited to a selected area and finishing with the fatal event involving the whole platform. No tools are presently available for this kind of analysis. However, the problem can be fully and rationally undertaken by taking advantage of system reliability based techniques, in conjunction with the developed simplified deterministic tools.

Reliability based optimization

Using the specially developed reliability based tools, it is possible to carry out a global optimization of the system, varying one or more of the key-parameters outlined in the previous paragraphs. For each single area of the platform (drilling, process, storage, living quarters) and for the topsides in their entirely, it should be possible, in this way, to identify the arrangement able to guarantee the highest level of safety or different arrangements with the same level of safety. In the latter case, the definition of a "target" safety level may make it possible to choose from the different arrangements able to guarantee such a level, on the basis of other parameters such as cost and weight. As far as the choice of materials is concerned, structures built of different raw materials but able to ensure the same level of safety could be defined as "equivalent" on the basis of new equivalence criteria, not only dependent on the results of a standard fire test.

5. CONCLUSION

In the light of the complex European scenario presented here, this paper has tried to stimulate a new approach for the "design of safety" of offshore platforms.

A proposal for harmonized European guidelines could represent the best answer to the present need for effective safety criteria, internationally recognized and aimed at offshore applications. To look at the topsides as a whole system subject to the risk of explosion and/or fire, to assess the reliability of this system in terms of its different arrangements; to optimize the choices as regards layout, protection methods and materials, those are the main concepts and ways of thinking which have been highlighted and which are necessary for the development of rational guidelines.

Moreover, the need for a critical review of the checking, maintenance and emergency procedures and training of the personnel is to be emphasized.

REFERENCES

[1] REGISTRO ITALIANO NAVALE (1990): Optimized fire safety of offshore structures, BRITE/EURAM "OFSOS" Project
[2] BELLO G., GENTILE G.: Offshore safety criteria and risk analysis methods, 3 ASI
[3] ARNOLD K.E., KOSZELA P.J., VILES J.C. (1989): Improving safety of production operations in the U.S., OCS - Proc OTC 89
[4] OREDA Offshore Reliability Handbook
[5] WOAD Worldwide Offshore Accident Database
[6] HOPE B., JOHANNESSEN P.A. (1983): Safety management in offshore development projects, Tanum-Norli Oslo
[7] VICTORY G. (1989): When is 'safe' not safe? MER, June 89
[8] MER (1989): Piper Alpha Enquiry Report

IN SITU MEASUREMENT AND NUMERICAL MODELLING OF THE RESERVOIR COMPACTION AND OF THE INDUCED SURFACE SUBSIDENCE

M.J. BOUTECA (+), D. FOURMAINTRAUX (++), Y. MEIMON (+)

(+) Institut Français du Pétrole
(++) Elf-Aquitaine (Production)

Summary

In the first part of the paper we describe and analyze the reservoir compaction measurements. In the second part we describe and analyze most of the methods which are used to measure ground displacement. The discussion of both parts is summarized through tables 1 to 3. The third part of the paper deals with the numerical modelization methodology. In conclusion we indicate which measurements we consider as most relevant and we give some guidelines for their implementation.

1. INTRODUCTION

When producing oil/gas from reservoirs, the pore pressure decreases. The pore pressure variation induces a deformation of the reservoir rock -compaction- which in turn induces a deformation of the ground surface, the so-called subsidence. Though surface subsidence is the most noticeable phenomenon, its amplitude is much less than the amplitude of rock compaction. In the Ekofisk area for instance [11], the estimated rate of compaction in 1987 was 39.4 cm per year while the estimated rate of subsidence was 28.7 cm per year.

The consequences of the compaction may be beneficial since it may significantly contribute to the oil production in the reservoirs. However when the rate of compaction is too high, rock deformation induces fracturing, shearing and casing collapse within the payzone, the underburden and the overburden.

The effects of the surface subsidence are essentially negative. Onshore, the vertical and the horizontal displacements may distort roads, railways and other civil engineering structures. They may also change the slope of pipes and channels. In Ingelwood, they have been suspected of causing the rupture of the Baldwin Hills dam. On the coast, part of the land could be invaded by sea water especially in sensitive areas like the polders in The Netherlands. On the Bolivar Coast where the subsidence bowl is $450 km^2$ large, dams had to be built to protect the shore. In the offshore environment the Ekofisk subsidence has been so great that the platforms had to be jacketed for security reasons during the summer of 1988.

From this short description of the consequences of both reservoir compaction and surface subsidence, it is clear that a monitoring system should be installed. Since many factors can influence the measurement accuracy, as it will be shown later on, it seems necessary to have a continuous follow-up of the evolution in order to obtain the trend, especially for subsidence. As far as the measuring systems are concerned, one should split the involved principles into two categories: a) the rock compaction is measured in a well and hence one must use logging devices; b) the subsidence is a ground surface movement and hence direct measurement methods can be used.

2. RESERVOIR COMPACTION MEASUREMENT

The methods that have been used for reservoir compaction measurement can be listed under two groups: a) conventional logging tools; b) logging devices specially designed for compaction monitoring.

Conventional Logging

The compaction may be estimated by comparing two logging runs in the same well at different times, or comparing results from surveys in initial and side-tracked wells [11]. In the following we indicate the physical principle involved in the log devices and give the name of the logging tool based on this principle used in the Ekofisk field [11,17,21]:

- natural radioactivity detection (gamma ray)
- induced neutron: one may detect either the thermal neutrons -CNL- or the gamma ray emission due to the capture of thermal neutrons -TDT-.
- compton effect -FDC-

The main drawbacks of these methods are:

- the measured values must be corrected for surface displacement -i.e. subsidence
- the depth of the reference level is not known accurately enough. On the Ekofisk field [11] the accuracy -60 cm before 1985- was of the same order of magnitude as the annual rate of compaction
- one must have good target zones with distinct formation signature.

Although these methods are fairly inaccurate they may be used as a complement to the specific logs hereafter developped.

Specific Logging

Three main problems have to be dealt with when trying to measure the reservoir compaction:

a) the depth positioning: The position of the tool is often estimated through the length of the cable between the probe and the surface. The longer the length, the more inaccurate the measurement. As a matter of fact, the elasticity of the cable produces twists and oscillations. To correct this effect several methods have been proposed. In Allen's probe [1,2], two small wheels are marked with five magnets each. These magnets trigger a switch each time they cross it. A vertical accelerometer is mounted on the FSMT Schlumberger tool;

b) the tool calibration: the compaction measurements are made during a long period of time. The values obtained from several runs must be compared thus implying that tool calibration is necessary for a given tool -instrumental drift calibration-. Moreover the tools may be changed during the timelife of the field production. In Groningen [18] two 1000 m deep calibration wells have been drilled. On the Ekofisk field -offshore environment- the calibration is realized in a riser by means of an invar rod;

c) the observation wells: As underlined by de Loos [18] the most important question when measuring compaction is: how many wells are required? N.A.M. on the Groningen field has been using the delineation wells as observation wells. Their distribution gives a good idea of the whole compaction of the reservoir. Local variations induced by fluid movements in production/injection wells do not interfere with the measurement in those observation wells. In the Ekofisk field a sub-vertical well has been drilled and largely cored. Logs have also been run in the uncased hole in the overburden to calibrate a subsidence simulation model.

The first specific method ever used has been precision casing joint detection -for instance on the Wilmington fields [2]. On the Groningen field two main reasons led N.A.M. to drop this method: a) it is much more sensitive to irregular movements of the probe than the FSMT type of measurement -see next paragraph; b) the recorded signal is complex and could not always be interpreted especially in the case of VAM casings. The method has thus not been used on the Ekofisk field. Another method to be used is the detection of radioactive markers. Basically, radioactive markers inserted in bullets are positioned in the formation with a given spacing fixed by the logging tool geometry itself. For the FSMT tool, the accepted marker distance is 10.5 m ± 1 m. If the distance between two successive sets of markers is greater, it must be carefully monitored by recording the cable movement. In the Ekofisk field, a test has been run using a coil tubing: the raw data quality is much better, but the calibration procedure of the tool has to be reworked. The marker placement requires a good knowledge of porosity and mechanical properties of the rock to select charge sizes. According to Menghini [11] in order to obtain a good signal from the Gamma Ray detector, the penetration depth should be about 20 cm. In the case of open hole completion, the quality of the bore hole wall must be checked using calipers. In cased holes the markers can only be fired in the producing zone. Due to operational constraints it is recommended to have a deviation of less than 30^0. The gun which is currently used is designed for a 5" casing. A 2" gun is available but to our knowledge it has not yet been used. From the experience gained on the Ekofisk field, Menghini [11] concludes:

- optimum placement of the markers is essential for a good GR signal detection
- cable friction along the tubing causes signal deterioration
- the best accuracy -standard deviation less than 10 mm per 10.5 m spacing- has been obtained in non-deviated wells. Note that the worse standard deviation was 325 mm per 10.5 m spacing.

Comments and recommendations -see table 1

Reservoir compaction induces surface subsidence. Monitoring its evolution is thus useful in order to control the surface subsidence. From the large experience gained in the Groningen and Ekofisk fields the best method for measuring the reservoir compaction appears to be RadioActive Markers -RAM- detection. If one also wants to follow the overburden evolution, the decision must be made very early since RAM must be fired in place before casing. The required accuracy implies that any cable friction must be avoided. Hence good quality measurements are expected only for sub-vertical wells. Using delineation wells as observation wells is a good and practical solution in on-shore fields. Finally we have to keep in mind that time duration of the compaction phenomenon implies implementing calibration devices.

3. GROUND DISPLACEMENT MEASUREMENT

Measurements referenced to sea level

In the offshore environment, measuring sea level variation to infer sea floor movement has often been considered more relevant. Thus values must be corrected for sea level variations and referenced to a theoretical level: Lowest Astronomical Tide -LAT-, Mean Sea Level -MSL-, or Average Sea Level -ASL-. Hence any measurement should be corrected for: tide effects, wave effects and water elevation due to the storms or atmospheric pressure variations.

air-gap

The air-gap -distance between sea level and cellar deck- is measured using a graduated weighted tape or radar wave. From 1980 to 1987 in the Ekofisk field [14] the estimated subsidence rate from radar wave data was 44 cm per year while the standard deviation on the

Method	Accuracy	Dimension	continuous recording	Comments
conventional logging CNL, TDT, FDC	60 cm	1D	NO	- must be corrected for surface displacement -subsidence- - needs distinct formation signature
RadioActive Markers (RAM)	10mm/10,5m	1D	NO	- observation wells must be vertical or sub-vertical - the probe must be calibrated

Table 1 : Reservoir compaction measurement

measurement was 20 cm.

bathymetry

The device used is a sonar. The propagation time, when recorded, is converted to depth value using the sound velocity in the sea water. This velocity [20] is a function of water temperature, salinity and density. It also changes diurnally, seasonally and with random or periodic influences such as the tidal stream or rainfall. Finally it is also a function of the water depth which is the required parameter. Complex calibration procedures can however be used. On the Ekofisk field (Fig. 1) the 1986 bathymetry agrees with the 1985 map within an uncertainty of

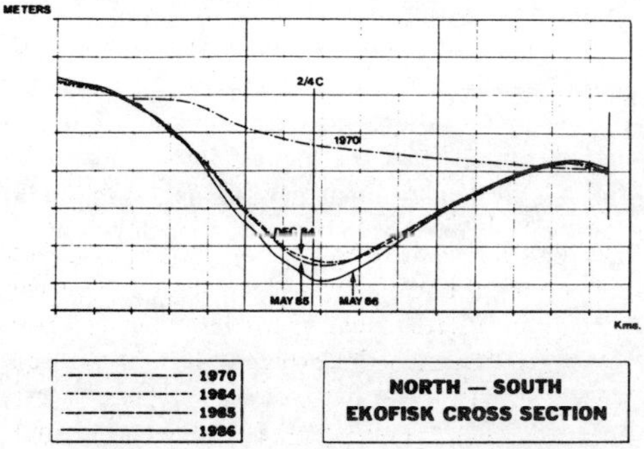

Figure 1 : Cross section of 1970-1986 Ekofisk Bathymetry data after [14]

± 15 cm on the fringe areas [14].

Sea bed Pressure Gauges -SPG

Pressure gauges are installed on selected platforms, near the mudline on a lower platform brace [13]. The relative subsidence between two platforms can be computed by comparing the observed pressure on one platform to a similar value of the other platform. The theoretical uncertainty on the differential water level was 9 mm [13]. We must still be careful since the electronic drift is not yet very well known. One should also keep in mind the complexity of the procedures which are involved while processing the raw data [13]. The experience gained in this type of measurement is not large enough to conclude. It is obviously very interesting due to the fact that the measurement procedure is a continuous one.

Comments and recommendations - see table 2

Air gap measurements are local measurements on one platform and their accuracy is poor. The basic idea seems simple and attracting but turns out difficult to handle.

The bathymetric mapping is a global method. Although the accuracy is poor the method is of great interest since this is the only way to obtain the shape of the so-called subsidence bowl. The accuracy could be improved by systematically mapping the same profiles and referencing the profiles on well defined benchmarks. We expect some improvements in sea floor mapping using side scan sonar [15] and systems like the SHARPS -Sonic High Accuracy Ranging and Positioning System- recently developped [9]. They allow fast and accurate three-dimensional acoustical mapping of underwater sites and would give detailed information on sea floor shape and movement: continuous or discontinuous deformation including steps and mini-faults.

Method	Accuracy	Dimension	continuous recording	Comments
radar	20 cm	1D	YES	
bathymetry	15 cm	2D	NO	- gives information on the subsidence bowl - could be improved
Sea Pressure Gauge SPG	9 mm	1D	YES	- drift unknown

Table 2 : Measurements referenced to the sea level

<u>Ground displacement measurement</u>

Using the sea level as a reference level -see the previous section- only allows determination of vertical displacements. But it is of the highest interest to follow both horizontal and vertical displacements. Horizontal displacement may damage well equipment due to shearing effects. In the Wilmington field for instance, the measured horizontal displacement between 1937 and 1966 was 3.66 m while the vertical displacement was 8.8 m. The techniques which are described in this section give information on both horizontal and vertical displacements.

Levelling

In the Ekofisk area the obtained accuracy is $\pm$ 5 mm $\pm 5.10^{-6}$. In the presence of haze the measurement cannot be carried. It thus seems of some interest to use laser devices in levelling.

Very high accuracy levelling have been carried out in France by IGN -Institut Géographique National. The accuracy for the selected method -motorized levelling- was 0.25 mm per km [7]

In the Lacq field (Fig. 2) the levelling profiles are referenced to the 1967 survey. The 1987 and 1989 surveys were performed using the above-mentioned method. The subsidence bowl was 4 to 5 cm deep in 1979 and 5 to 6 cm deep in 1989. Earlier surveys performed in 1887 along the railway show that the fringes of the bowl have not experienced displacement. As can be seen from Figure (3) the ground displacement is strongly related to the gas pressure drop in the reservoir.

Global Positioning System - GPS

For a detailed description of the procedure one should refer to Mes [12] and Leick [22], for further information one could consult Willis [16]. The GPS is a satellite supported system. Satellites are transmitting signals at two frequencies L1 and L2 -L1: f= 1575.42 MHz, λ=19 cm; L2: f= 1227.60 MHz, λ= 24 cm-. The frequencies are modulated by two frequencies giving two

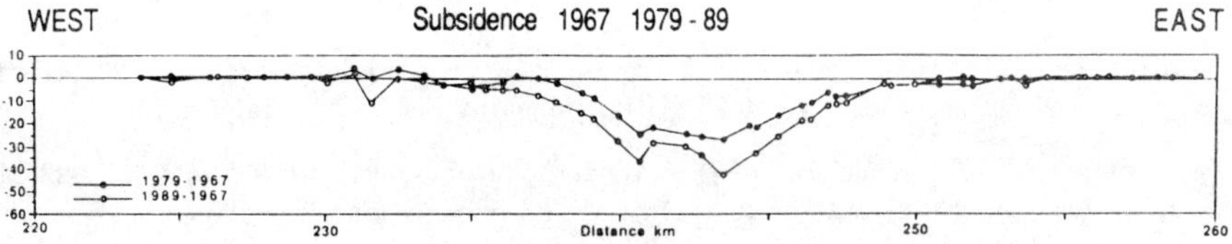

Figure 2 : Lacq field levelling 1967-1989

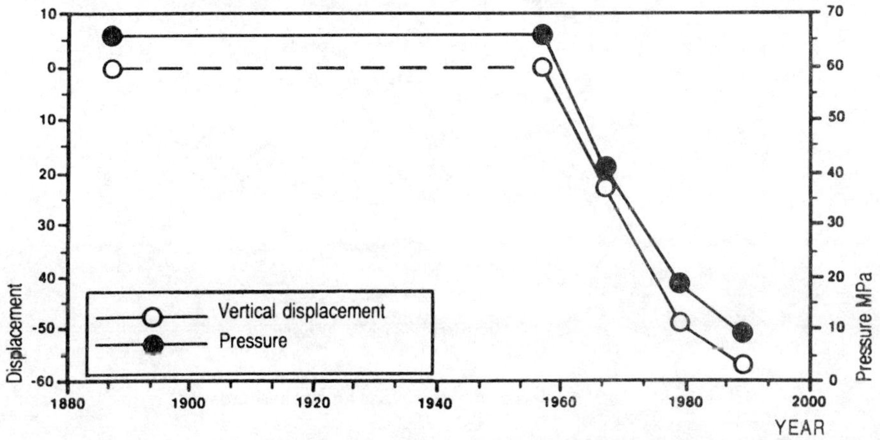

Figure 3 : Subsidence and pressure drop on the Lacq field

codes:

- a precise one -P code- where the carrier frequency is modulated with a 10.23 MHz frequency -λ= 30 m- leading to 10.23 MBPS -MegaBit Per Second- information;

- a coarse one -C/A code- where the carrier frequency is modulated with a 1.023 MHz frequency -λ= 300 m- leading to a 1.023 MBPS information.

The codes give information on the satellite position - the ephemeresis. To obtain an accurate position one must measure and interprete the carrier signal phase.

On-shore Coulon et al [5] obtained a 2 ppm (2.10^{-6}) repeatability and 1 ppm standard deviation.

<u>Comments and recommendations -see Table 3</u>

Both methods are fairly accurate. The levelling method is basically very simple but the measurement has to be performed many times between successive stations. The main drawback however stems from the fact that it can only be performed along roads. Hence, as shown on Figure (4) large areas above oil/gas fields cannot be surveyed, for instance marshes, polders or lakes.

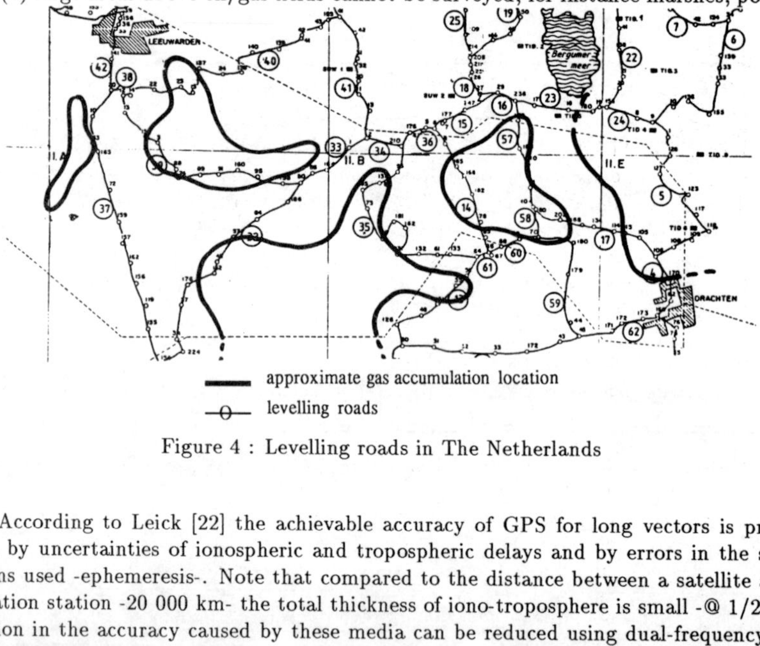

Figure 4 : Levelling roads in The Netherlands

According to Leick [22] the achievable accuracy of GPS for long vectors is primarily limited by uncertainties of ionospheric and tropospheric delays and by errors in the satellite positions used -ephemeresis-. Note that compared to the distance between a satellite and the observation station -20 000 km- the total thickness of iono-troposphere is small -@ 1/20-. The limitation in the accuracy caused by these media can be reduced using dual-frequency observations and water vapor radiometer measurements. The theoretical limit for GPS surveying is expected to be 0.01 ppm using orbital relaxation to reduce ephemeresis errors [22]. A more practical improvement would consist in using permanent GPS receivers associated with remote control facilities for data acquisition and transmission. The marginal cost of each measurement could drastically decrease while data could be obtained at higher frequencies.

4. NUMERICAL MODELLING

As shown above, land subsidence resulting from fluid withdrawal has proven to be a significant problem under a variety of reservoir conditions. Displacement in both the vertical

Method	Accuracy (1)	Dimension	continuous recording	Comments
Levelling		3D	NO	
Ekofisk field	5 mm±5ppm			
HPML	0.25mm/km			High Precision Motorized Levelling
Global Positioning System GPS	2ppm	3D	NO	- Horizontal movements should be computed leading to 3D maps

(1) the accuracy refers to the baseline length -distance between two stations

Table 3 : Ground Displacement Measurements

and horizontal planes may be significant in formations which are underconsolidated, clay-rich, or wherein geochemical processes affect the structural integrity of the rock. Subsidence may also be significant within consolidated reservoirs with relatively high porosity and large thickness. Several mathematical models may be used to compute the reservoir compaction and the induced surface subsidence. The discrete element method is based upon a discretization of the structure into blocks (rigid or deformable) for which the interface behavior is modelled. In the case of rigid blocks, only an accurate description of the geometry (faults, layers) and a good modelling of each block interface is required. The exact efficiency of the method must be evaluated for this type of problem. More often used, the displacement-based finite element method is able to represent the structure behavior [19] especially when materials behave inelasticly. Its application will be developed hereafter.

Modelling overview

In order to determine the subsidence - ground displacement and reservoir compaction- resulting from hydrocarbon production, it is clearly necessary to:

define the geological structure to be modelled

This includes stratigraphic descriptions, geological maps with the locations of structural elements which could be associated with localized deformations in the reservoir and in the overburden, well logs and core data. The need of a rather accurate prediction of displacements due to the reservoir production requires careful description of the geological structure: generally, this implies definition of some variations of the site geometry from a supposed reference description.

define correctly in situ stresses

The displacements due to the subsidence are not reversible because the rock behavior is non-linear. Then, the knowledge of the in situ initial stresses is fundamental for calculations which will use plastic constitutive relations for geomaterial behaviour. This is obviously a complex task requiring the cooperation of geologists and rock mechanicians because the calculation needs not only the stresses level but also their direction. As a rule, the stresses could also be ascertained by analysis and interpretation of well logs (density, neutron, resistivity, sonic velocity, gamma ray and caliper), pressure records from hydraulic fracturing jobs and from leak-off tests performed during drilling operations.

use adequate constitutive equations for the rock behavior

A constitutive equation (called the rheological model) must be established for each mechanical sub-unit of the structure, which must adequately describe each geomaterial behaviour under the expected loading. Depending on the geomaterial type (sand/sandstone, clay/claystone/chalk) an elastoplastic or an elastoviscoplastic model should be used. The simplest elastoplastic model is based on a Mohr-Coulomb yield surface and requires four parameters, two elastic and two plastic, of which the determination needs both logs and shear triaxial tests. A more sophisticated model (called VISPLA) for elastoviscoplastic behavior of geomaterials has been developed [3]. It is based on Perzina's theory of viscoplasticity and a modified Cambridge theory on the critical state of soils. It also includes a damage law for viscous contacts of clay particles that allow the description of progressive damage in the material due to micro-cracks developing during the loading process (for example, creep up to failure). It is useful when the evaluation of long term stability of a critical layer, for which sufficient information exists (cores essentially), must be performed. In this case, three sets of parameters are needed. Seven elastoplastic parameters must be derived from logs and triaxial tests conducted at low loading rate to avoid the influence of the material viscosity. Two viscous parameters can be obtained from creep triaxial tests or/and shear triaxial tests at different loading rates. Finally, two damage parameters can be derived from a laboratory creep test conducted up to failure. Of course, this type of model should be used only for critical layers identified by confirmed experts.

use a mathematical model

The rheological models used in the analysis must be implemented in a finite element program and then, several analyses may be carried out. The simplest consists in applying to the structure the depletion obtained from a reservoir engineering analysis in the way of a volumetric loading varying in the reservoir elements. The mechanical effect is an increase of the effective stresses that induces displacements in the structure: calculations of this type are shown hereafter. Of course, results will depend on the effective stress law used, which is, for rocks, generally different from Terzaghi's principle for soils [8].

For special problems in which the consolidation of the overburden during the production period is to be considered, a coupled mechanical analysis must be performed. Here, the general equilibrium equation is coupled with the pore pressure diffusion equation and, when necessary, with a heat equation (Fourier's law) in the framework of the thermoporoelastoplasticity for saturated porous materials [6].

A more complex analysis could be a coupling of the mechanical model with a reservoir engineering model in an iterative process [4]: at a given time, the reservoir analysis gives first the pore pressure field in the reservoir domain and the mechanical analysis returns the change in the reservoir source terms. This in turn leads to a change in the pore pressure field computation thus generating an iterative procedure.

Finite element calculations

The finite element calculations are performed using a system of programs (called FONDOF) especially designed for geomechanical problems [10]. It includes various interactive pre and post processing facilities. A specific program (called ADELAP) is devoted to the automatic determination of geomaterial parameters for sophisticated models like VISPLA from a laboratory data set, and enables savings in time and gain in efficiency, security and reliability for the user. Two and three-dimensional calculations can be carried out including the modelling of faults by using specific joint elements. Calculations are performed incrementally, step by step.

All quantities -stresses, strains, geomaterial memory variables for instance- are calculated using a classical Gauss integration scheme. Rock properties are provided by the user for each layer and the program interpolates from these data the properties at each Gauss point location in the mesh elements (8-noded for 2-D analysis, 20-noded for 3-D analysis). Mechanical coupling with pore pressure and thermal diffusion will soon be available. More than usual isovalues plots, processing of the results allows one to follow the evolution of global variables (subsidence at the surface, displacements of the reservoir roof or of any interface of layers) and of local key variables as deviatoric stress, mean effective stress, pore pressure, strains and other variable that are important to understand the behavior of the structure. An exemple is given on Figure (5).

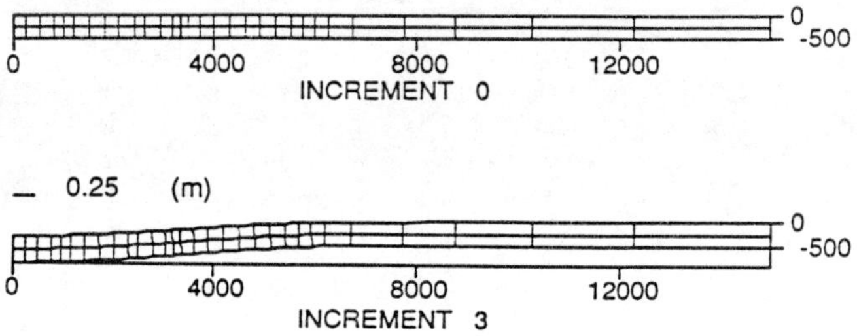

Figure 5 : Deformed mesh at surface level

5. CONCLUSION

The most relevant method for compaction measurement seems to be the detection of radioactive markers -RAM-. The best accuracy that has been obtained was 10 mm per 10.5 m. This accuracy can only be obtained by minimizing cable friction in the tubing. Hence the observation wells must be vertical or sub-vertical wells. For a good description of the reservoir compaction several wells must be equipped. The delineation wells could be good candidates. If one intends to observe the compaction of the overburden, the decision should be made early [11] before casing the well.

Available methods for ground displacement are numerous with a large range of accuracy. Amongst the methods with poor accuracy we favor the bathymetric survey -accuracy ± 15 cm in the Ekofisk area- which gives global information on the shape of the subsidence bowl as opposed to accurate methods in the offshore environment whose measurement devices are located on the platforms. Note that the methodology can be improved and hence the accuracy of the method. The satellite measurements -GPS- with an accuracy of 1 ppm and pressure gauge -SPG- as designed by Phillips with an accuracy of 9 mm should be recommended. One should keep in mind that instrumental drift of the SPG is not very well known and experience still has to be gained. The satellite measurement are being improved and we will use them with a daily recording frequency. Since GPS can be implemented in both on-shore and off-shore environments and since it gives 3D displacements we favor this method.

Onshore, the levelling accuracy is 0.25 mm/km. The ground displacement and/or settlement which is measured is induced by:

- periodic movements such as seasonal movements of the groundwater table, freezing and thawing of surface water, tidal effects near coastlines

- groundwater table movements for water supply purposes
- natural compaction of sediments
- mining including oil and gas production

For the oil producer it is important to determine which part of the ground displacement is truly induced by the oil/gas production. It may thus be very useful to drill shallow observation wells, a few hundred meters deep, as it has been done in Groningen -300 m to 600 m-.

The measurement should be performed continuously as far as possible and the accuracy of the method must be determined by the expected amplitude of the measured phenomenon.

Finally, one should remember that since the ground displacement is always referenced to the initial shape, the accuracy of the final measurement is greatly influenced by the initial shape determination.

REFERENCES

[1] ALLEN, D.R. (1969) - "Collar and Radioactive Bullet Logging for Subsidence Monotoring", SPWLA Trans. Tenth Annual Logging Symposium, paper "G"

[2] ALLEN, D.R. (1981) - "Developments in Precision Casing Joint and Radioactive Bullet Measurements for Compaction Monitoring", SPE 9933, California Regional Meeting

[3] AUBRY, D., KODAISSI, E. and MEIMON, Y.(1985) - "A viscoplastic constitutive equation for clays including a damage law", 5th Int. Conf. on Numerical Methods in Geomechanics.Nagoya, Japan.

[4] BOUTÉCA, M.J. and SARDA, J.P. (1990) - "Fluid Flow in Porous Media and Related Rock Mechanics Problems", European Conference on the Mathematics of Oil Recovery

[5] COULON, B. and CARISTAN, Y. (1989) - "Suivi de mouvements de surface par GPS: étude méthodologique", Rock at Great Depth, vol. 2, pp. 649-655, Balkema

[6] COUSSY, O. (1988) - "Thermomechanics of Saturated Porous Solids in Finite Deformation", European Journal of Mechanics, A/Solids, vol. 8, n. 1, 1-14

[7] KASSER, M. (1984) - "Le Nivellement Général de France, Évolution d'un Grand Réseau de Repères d'Altitudes", Géomètre, No 12

[8] LAURENT, J. and QUETTIER, L. (1989) - "Comportement des milieux poreux consolidés dans le domaine élastique", Rock at Great Depth, vol. 2, pp. 915-921, Balkema

[9] McCANN, A. M. (1990) - "Diving into our past", National Geographic, january

[10] MEIMON, Y., LASSOUDIÈRE, F. and KODAISSI, E. (1987) - "Fondof: a FEM software for the calculation of offshore foundations", Int. Conf. of Offshore Mechanics and Artic Eng. OMAE'87. Computer Book. Houston.

[11] MENGHINI, M.L.(1988) - "Compaction Monitoring in the Ekofisk Area Chalk Fields", OTC 5620

[12] MES, M.J. (1988) - "Accuracy of Satellite Survey Measurements on Offshore Platforms for Monitoring Subsidence", OCEANS'88 Proceedings, Partnership Mar interest, V.4, IEEE Publ.

[13] MES, M.J. (1989) - "Accuracy of Offshore Subsidence Measurements With Seabed Pressure Gauges", OTC 6063

[14] RENTSCH, H.C. and MES, M.J. (1988) - "Measurements of Ekofisk Subsidence", OTC 5619

[15] PITTENGER, R. F. (1990) - "Exploring and Mapping the Sea Floor", National Geographic, January

[16] WILLIS, P. (1989) - "Méthodes de traitement de la phase GPS pour la localisation relative (statique et cinématique). Applications à la géodésie", thèse de doctorat de l'Observatoire de Paris

[17] Log Interpretation Volume I-Principles, Schlumberger document, (1972 Edition)

[18] The analysis of surface subsidence resulting from gas production in the Groningen area, the Netherlands, ed. Nederlandse Aardolie Maatschappij B.V., (1973)

[19] Numerical methods in finite element analysis, BATHE, K.J. and WILSON, E.L. Prentice Hall ed., (1976)

[20] Hydrography for the surveyor and engineer, INGHAM, A.E., Ed. Granada Publishing, (1984) 2nd edition

[21] Fundamentals of well-log interpretation, vol 1. the acquisition of logging data, SERRA, O., ed. Elsevier, (1984)

[22] Satellite Surveying, LEICK, A., ed. Wiley & Sons, (1990)

THE APPLICATION OF PHYSICS IN EXPLORATION AND PRODUCTION OF OIL AND GAS

Quigley T M, Foakes A P, Simmonds S, Tam V H Y

BP Research
Sunbury Research Centre
Chertsey Road
Sunbury on Thames
Middlesex
TW16 4LN
ENGLAND

SUMMARY

The many different operations involved in the exploration and production of petroleum are often complex and require expertise of many scientific and engineering disciplines. The science of physics as applied to the study of subsurface geology, ie geophysics, has traditionally played an important role in the oil industry. Physics also has an established role in reservoir engineering and petroleum engineering. In addition the importance of experimental and theoretical physics in a broader content of oil industry operations is becoming generally appreciated. Physicists, often working in multidisciplinary teams, are active in many upstream and downstream areas. By means of three examples drawn from exploration, production and petrochemicals this paper illustrates the contribution currently being made by the science of physics to the profitable and safe exploration and exploitation of oil and gas.

1. INTRODUCTION

The oil industry has a long tradition of applying the science of physics to the study of subsurface geology. This application is called geophysics and geophysical prospecting is now an essential part of any exploration programme. Physicists are also actively involved in oil and gas production in the fields of petroleum and reservoir engineering. Here they are principally engaged in research and development activities but some are also active in operational areas. As well as these more established areas of activity, in recent years the role of physics in the oil industry has considerably broadened. Physicists are now seen working alongside mathematicians, chemists, engineers and geologists playing their part in helping to understand and hence control the many and often complex natural phenomena involved in the exploration and production of oil and gas, transport and distribution of oil products, and production of petrochemicals.

To help meet the oil industry's increasing need for physics, a number of companies have established groups of physicists in their Research and Development organisations. These groups provide a central core of physics expertise which is available to both support the activities of other scientific and engineering disciplines, and bring physics methods to bear on appropriate operational problems faced by the oil company. Here three examples of the work of such a group are used to illustrate the broad ranging contribution made by physicists. The first example describes how physicists working with geochemists and geologists were able to develop a quantitative understanding of petroleum formation in the subsurface, and hence contribute to exploration risk reduction. The second example describes how physicists and engineers are able to make a significant contribution to the safe production of oil in the offshore environment, by understanding and controlling the consequences of a gas explosion on an oil rig. The final example is drawn from the petrochemicals area where physicists and chemists are working together to improve the catalysts used during the conversion of petroleum feedstocks to bulk chemicals.

2. EXPLORATION - GEOCHEMICAL PROSPECT APPRAISAL

The prime function of petroleum geochemistry is to predict, in advance of costly drilling, the volume of petroleum trapped in an exploration prospect. Physicists, geochemists and geologists now understand the processes reponsible for the creation of commercial accumulations of oil and gas sufficiently well to enable them to derive simple rules and equations to predict the likelihood of a prospect containing petroleum, albeit within an associated precision and risk (1). The procedure is based on three simple equations, the elements of which can either be measured directly or estimated from simple graphical summaries of the results of computer simmulation of the various physical and chemical phenomena involved. Equation (1)

$$M_E = (P_o)\ (PGI)\ (PEE)\ (\rho_{ROCK})\ y\ (AREA) \tag{1}$$

calculates the total mass of petroleum expelled from a source rock. P_o, ρ_{ROCK}, y and (AREA) are, respectively, the source rock's initial petroleum potential (the concentration of organic matter which can be transformed into oil and gas at elevated temperatures), the source rock density, mean thickness , and catchment area. All these can be directly measured for the region of interest by making standard geochemical measurements on source rock samples, and using wireline logs and seismic maps. PGI is the petroleum generation index (the fraction of petroleum-prone organic matter that has been transformed into petroleum). PGI depends upon two factors, the nature of the organic matter in the source rock and its thermal history. The former can either be measured using standard geochemical measurements or inferred from knowledge of the depositional setting of the source rock. The latter can be calculated by coupling an Arrhenius-type kinetic model of petroleum generation (2) to a mathematical model of sedimentary basin thermal history (3). PEE is the petroleum expulsion efficiency (the fraction of petroleum fluids formed in the source rock that has been expelled towards on exploration prospect). PEE can be calculated from consideration of the driving forces of petroleum migration, ie gradients in piezometric water pressure, natural buoyancy of less dense petroluem phases surrounded by denser water, and capillary pressure. By combining these basic forces acting within a fluid-filled compacting sedimentary basin, a mathematical model of petroleum migration can be constructed and hence PEE determined (4). Finally, the values for PGI and PEE predicted by the mathematical models can be checked and calibrated using available geochemical data (1). Figure 1 summarises the subsurface temperature at which most oil and gas are generated and expelled from source rocks.

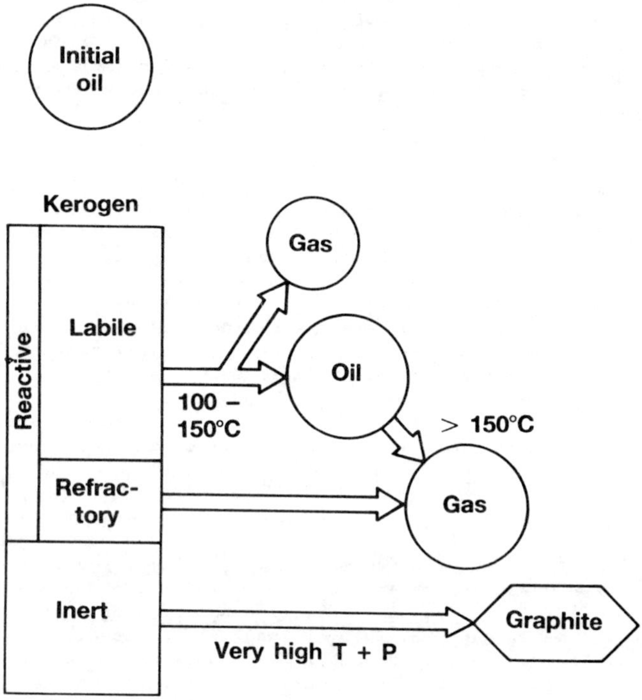

Figure 1: Fate of organic matter in source rocks. Inert kerogen rearranges towards graphitelike structures at very high temperatures (T) and pressures (P). Reactive kerogen is subdivided into a refractory part that yields mainly gas and a labile part that yields mainly oil. Initial oil is the small amount of solvent soluble organic matter normally present in immature source rocks.

Equation (2) determines the volume of petroleum lost (V_L) along the migration pathway:

$$V_L = \phi \ f \ V_D \tag{2}$$

where ϕ is rock porosity, V_D is the volume of rock through which petroleum flows, and f is the residual saturation. ϕ and V_D can be determined using available rock samples, wireline logs and seismic maps. f is difficult to determine accurately but is best evaluated from exploration case history, using results from geologically similar regions where all the variables in equations (1) to (3) except f are known. Some such studies (1) have shown f to lie within the range 1 to 3%.

Equation (3) describes the volume of petroleum charge (V_C) delivered to the exploration prospect in terms of the difference between the expelled volume (V_E) and the lost volume (V_L):

$$V_C = V_E - V_L \qquad (3)$$

V_E is calculated using M_E, together with knowledge of petroluem subsurface phase behaviour and density (4). When lost petroleum volume exceeds expelled volume no petroleum is predicted to reach the prospect. Hence the quantitative understanding described by equations (1) to (3) allows regions, reservoir sequences and prospects to be graded and the relative rankings used to influence exploration policy.

3. PRODUCTION - GAS EXPLOSIONS OFFSHORE

Though the risk of a gas leak and subsequent gas explosion on an offshore platform is small, the threat is real. The consequences of a major gas explosion could be high in terms of life and financial cost. One of the aims of platform designers is to reduce the likelihood of gas leaks, potential ignition sources, and to minimise the consequences of gas explosions through good engineering. Physicists and engineers have been successful in furthering understanding of the consequences of gas explosions thus paving the way towards effective mitigation measures.

When the chemical energy of a flammable gas cloud is released very fast, damaging high pressure waves are generated, and this is a gas explosion. The severity of the gas explosion depends on how fast the chemical energy is released which is determined by the burning rate. The initial flame speed near the time of ignition will be equal to the laminar burning velocity of the fuel mixture, plus a velocity due to the expansion of the hot gases behind the flame front. The laminar flame speeds of gases commonly handled offshore are relatively low (0.37 ms^{-1} for methane, 0.43 ms^{-1} for propane).

The presence of equipment or obstacles inside the gas cloud can cause the flame to accelerate and significantly increase the magnitude of the resultant overpressure. This is due to the interaction of the flame with turbulence. The burning fuel-air mixture forms a flame front moving into the unburnt fuel mixture. This burning of flammable gases causes flow ahead of and away from the flame. In the presence of equipment or obstacles, this flow becomes non-uniform and a significant amount of turbulence is generated. The interaction between combustion and turbulence is complex. It can be qualitatively understood, with much simplification of the physical processes, by imagining the flame front as a continuous sheet which wraps itself around obstacles in its path and becomes wrinkled when agitated by turbulence. Since the burning velocity increases with flame area, the flame accelerates. As the flame moves faster, it generates more intense turbulence which causes the flame to move faster still. This positive feedback process can increase the burning velocity to several hundred m s^{-1}. It is at these high burning velocities that overpressures sufficient to cause structural damage occur.

Further, the magnitude of the overpressure is greatly enhanced by confinement. In the offshore environment, process equipment is contained in modules with ceiling, floor and some walling. As the gas is burnt, heat is released, and the gas expands. The effect of the wall, ceiling and floor is to restrict the expansion of the hot gases, causing a rise in pressure.

The factors described above, namely the effect of obstacles or equipment and venting, are present in all platform designs. The platform designer needs to be able to produce equipment layouts and venting arrangements which are operationally efficient and produce minimal damage to the installation in the event of a gas explosion.

One approach to estimating the consequences of gas explosions is to model the combustion and fluid flow processes described above. BP, jointly with other oil companies, have been sponsoring research on gas explosions at the Christian Michelsen Institute (CMI), Norway for the past ten years. This work has resulted in a numerical fluid dynamics and combustion code, FLACS which stands for FLame ACceleration Simulator. It solves the Navier Stokes equations of fluid motion, couples the combustion with turbulence, and simulates the effect of flame acceleration due to turbulence. Thus theoretically, it could take account of the positive feed back and venting processes. FLACS was specifically intended for use in offshore platform modules. It has been validated by CMI using the results of explosion tests in 1:5 and 1:30 scale models of simple platform layouts. The code is still under development, further details can be found in (5).

A simulated gas explosion in a hypothetical offshore process area, the plan view of which is shown in Figure 2, is used to illustrate an application using FLACS. The area consists of a blast wall and 3 vented walls, with a solid floor and ceiling. The dimensions of the gas filled region are 32 m (along the blast wall) by 28 m and 8.5 m in height. An additional fire wall surrounds an array of control valves on 3 sides from the deck to the ceiling on the north-east side. The effects of 3 types of walling are illustrated below. In order of decreasing vent area, these are : open walls, porous walls and wind walls.

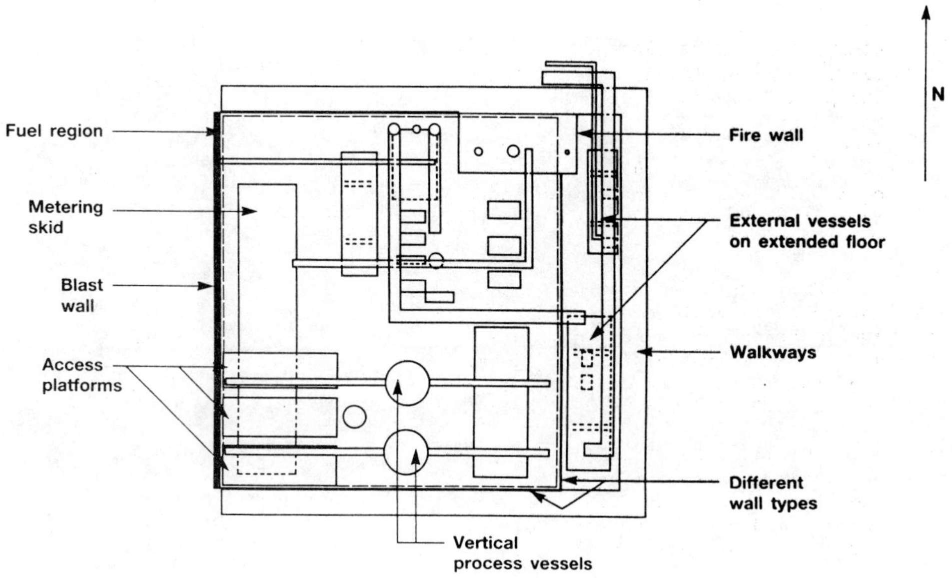

Figure 2: Plan view of equipment layout in a hypothetical offshore process area

The porous walls are assumed to cover the whole vented wall area with an area porosity value of 0.5, representing 50% effective opening area. Wind walls were represented as solid walls with a 1 m air gap at the bottom and 1.3 m at the top. A stoichiometric gas mixture of 75% methane and 25% propane was used in all simulations of this area. The ignition position in all cases was close to the centre of the blast wall.

TABLE 1: Calculated peak overpressures in the process area for the three types of walling.

Wall Type	Peak Overpressure (bar)
Open walls	0.12
Porous walls	0.32
Wind walls	0.58

Peak overpressures calculated at a range of locations are listed in Table 1. As expected, the results showed that open walls greatly reduced the maximum overpressure by effective venting through all 3 walls.

Porous walls presented significant resistance to the flow and increased overpressures considerably. Sensitivity tests were performed varying the porosity value of the porous walls, which showed that overpressures were quite sensitive to the porosity and underlined the importance of an accurate representation of wall types.

Results from the wind wall case showed a pulsating flow from the open gaps alternating between the top and bottom gaps, with corresponding localized high pressure regions close to the vent gaps. The geometry of the wind walls prevented continuous venting and greatly enhanced the maximum overpressure.

There are spatial variation in overpressures within the process area. It is important that the structural and layout design take this into account. Overpressure histories in most parts of the module are very similar. However, for all three wall types higher overpressures arise near the fire wall section surrounding control valves. The wall blocks a section of venting area and narrows the width of the process area available for gas expansion. Similar results have been obtained for large equipment placed close to vented walls.

This example, and other calculations, show that the complex phenomena of gas explosions are not related in a simple fashion to the amount of venting and the average blockage, but depend strongly on the density and particularly the distribution (layout) of equipment and venting. 3-D numerical simulations, suitably calibrated by experimental measurements, are therefore required for the determination of overpressures in the congested geometries common in offshore platforms. 3-D numerical simulations allow the consequences of gas explosions in different equipment and venting layouts to be evaluated, and the design tuned to minimise the consequences of offshore gas explosions.

4. PETROCHEMICALS - HETEROGENEOUS CATALYSIS

Most processes that refine and convert crude oil to bulk chemicals depend on the action of a catalyst to improve both their reactivity and selectivity. One important class of catalysts is that where the catalyst is heterogeneous to the reactants and products, such as the zeolites used in refining, and the solid-state catalysts that are used in many synthesis processes.

Many of these solid state catalysts were discovered by experiment with little fundamental understanding as to why their presence should enhance a chemical reaction. Yet there are some fundamental questions about how molecules interact with the surface of a solid and how their interaction with each other is different on the surface as opposed to the gas or liquid phase. To answer these questions it is important to understand not only the processes associated with the surface, but also the electronic (band) structure of the catalyst itself.

These solid state catalysts often are not simple compounds. Instead, they are a complex mixture of metals and oxides with possibly many impurities associated with both the surface and the bulk. To appreciate fully how this complex structure causes the catalytic action and, indeed, to identify correctly from first principles those components that are crucial to this catalytic action, would require a calculation of almost impossible detail.

However, this detail is not necessary to aid the chemist's search for improved catalysts. In many situations, a simple model needing little mathematical sophistication but a great deal of physical insight can reveal the underlying physics onto which such detail must fit. A good example is the way the surface defects of a metal-oxide, which can act as active sites for a catalysed reaction, increase in number when the oxide is in contact with a bulk metal (6).

In many such catalysts, the oxide can be considered to be a thin layer over some bulk metal. The band diagram for such a system is shown in Figure 3a. The bands in the oxide bend due to the presence of the metal as the metal/metal-oxide system forms a Schottky junction. However, the thickness of the oxide is typically less than the natural length of this junction.

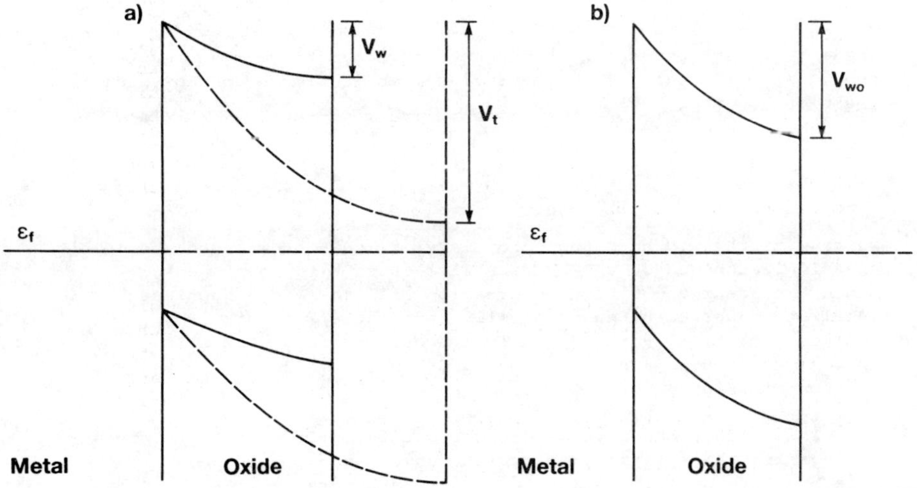

Figure 3: Electronic (band) structure for metal - oxide junction; a) before vacancy formation, b) after vacancy formation.

The main defects in the surface structure of a typical metal-oxide are oxygen vacancies. The creation of these vacancies in the surface of the oxide can be represented thus:

$$O^{2-}_{surface} \Rightarrow \frac{1}{2} O_2 + 2\ e^- \qquad (4)$$

If the metal is within the quantum tunnelling distance to the surface, the bulk metal can accept the two electrons left in the surface when an oxygen is removed. With the band structure of Figure 3a, it is energetically favourable for the vacancies to be formed via equation (4) because the electrons release energy as they tunnel to the Fermi level of the metal, ϵ_f. Once this occurs, a residual positive charge is left in the oxide. This charge causes an electrostatic potential that modifies the energy levels of the electronic structure in the oxide. This potential V is governed by Gauss's law:

$$\nabla^2 V = - \frac{N\ q}{\epsilon\ \epsilon_0} \qquad (5)$$

where N is the concentration of charges q. For constant N, this gives a simple quadratic dependence of V on the distance from the interface between the metal and the oxide. If we define the ratio R as:

$$R = \frac{V_t - V_w}{V_t} \qquad (6)$$

where V_t is the potential at the natural Schottky barrier length and V_w is the potential at the surface, it is simple to show from equation (5) that:

$$R = 1 - \frac{eNw^2}{2\epsilon_0 \epsilon_s V_t} \qquad (7)$$

where w is the width of the oxide layer. The energy advantage for creating vacancies on the surface will cause N to increase and hence R to decrease. Vacancy creation will therefore continue until it ceases to be energetically favourable, ie where V_w reaches a critical value V_{w0}. This usually occurs when R is approximately 0.5. The electron bands in the oxide will have bent as shown in Figure 3b.

This effect can give a 10^3 increase in surface vacancies in a 30 Angstrom thick layer of a typical oxide (eg ϵ_s = 15 and V_t = 0.5eV). In many solid state catalysts, an increase in reactivity of this order is seen. The best example is in the methanol synthesis catalyst, which is a copper/zinc-oxide combination.

Note that nothing has been said about selectivity. This is because this seems to depend on the detailed structure of the particular sites on the surface. However, the simple insight of the above Junction Effect Theory has revealed a major cause of the increase in reactivity seen in metal-oxide based catalysts when some bulk metal is present. Research is now directed at using this understanding to modify existing catalysts and to design new improved catalysts.

5. CONCLUSION

The oil industry has faced and overcome many technical problems in its long history. The technical challenges of the future are likely to be at least as difficult as those of the past; physics and physicists will play a role in their solution. Perhaps the greatest contribution of physics will arise where physicists work together with chemists, geologists and engineers adopting a multidisciplinary approach to technical problem solving.

6. REFERENCES

(1) Mackenzie, A S and Quigley T M, Principals of Geochemical Prospect Appraisal, AAPG Bulletin, Vol 72 (1988), P339 - 415.

(2) Quigley, T M and Mackenzie, A S, The Temperatures of Oil and Gas Formation in the Sub-Surface, Nature Vol 333, No 6173 (1988), P549 - 552.

(3) Guidish, T M, Kendall, G C St C, Lerche I, Toth, D L and Yarzab, R F, Bain Evaluation using Burial History Calculation: an Overview, AAPG Bulletin Vol 69 (1985), P92 - 105.

(4) England, W A, Mackenzie, A S, Mann, D M and Quigley, T M, The Movement and Entrapment of Petroleum Fluids in the Subsurface, Journal of the Geological Society, London, Vol 144 (1987), P327 - 347.

(5) Hjertager, B.H., Computer Simulation of Turbulent Reactive Gas Dynamics, Modelling, Identification and Control, Vol 5, No 4 (1985), P221.

(6) Frost, J C, Junction Effect Interactions in Methanol Synthesis Catalysts, Nature, Vol 334, No 6183 (1988), P577 - 580.

ABANDONMENT AND REMOVAL OF STEEL PLATFORMS

J.M. MARTIN BOURGON
Repsol Exploracion, S.A.

M. MOREU – A. MORON
Seaplace Iberia, S.A.

1. INTRODUCTION

The project has been an approach to the analysis of the problems associated with the removal of offshore steel platforms and then to study feasible solutions from a technical and economical points of view for a series of three platform types.

Platform Type	Weight (M. Ton)		Water Depth	Jacket Height
	Jacket (1)	Topsides (2)	(metres)	(metres)
A	850	1,100	60	66
B	8,500	8,500	106	117
C	14,000	14,000	140	155

(1) MTO of jacket plus removed pile weight.
(2) Design weight.

The aim of this communication is to present some conclusions of the study related to the effective cost of the removal operations.

Costs of independent topsides and jacket removal operations were obtained for each platform type using the following methods:

By crane lifting, hanging from barges, floating and demolition. In addition a mixed method was considered; this method removes the topsides by crane lifting and the jacket hanging from barges.

Finally an integration of costs of topsides and jacket removal was got for the most relevant cases, for which the results are shown below:

2. BASIS OF THE COST STUDY

Many of the removal operations have not been previously performed or there is little experience, so there are obviously some uncertainties.

An estimation of the weather downtime has been made for each removal method but its cost has not been included.

Cost have been obtained as an addition of labour, equipment and marine spread costs.

The marine spread cost has been derived directly from the time schedules using medium daily rates. For studies of sensibility low and high rates which are 30% below and above have been applied.

3. PLATFORM REMOVAL INTEGRATED COST CONCLUSIONS

The most significant total removal cost results obtained for the three types of platforms are summarized hereinafter in table nº 1.

		Type B		Type C	
Removal Method	Type A	Total	Partial	Total	Partial
All cases	4000/5700 (30%)	11000/20000 (45%)	6000/15500 (60%)	18000/31000 (40%)	12500/19500 (35%)
Cranes	5000/5600 (10%)	19000/20000 (5%)	13200/15500 (15%)	27000/31000 (15%)	18000/19500 (10%)
Mixed	4500/5700 (20%)	13000/17000 (25%)	11000/12500 (12%)	18000	14000
Barges	4000/4700 (15%)	11000	8800	-----	-----
Floating	4000	-----	7000	-----	13000
Demolition	-----	-----	6200	-----	14000

Table 1 : Integrated Total Cost Ranges for Different Removal Methods. (Thousand Dollars)

Numbers enclosed in parenthesis represent the % variation between maximum and minimum values.

Notes.– Partial removal for platform type A is not feasible due to its water depth.

- Cranes Both topsides and jacket removed by crane lifting.
- Mixed The topsides removed by crane and the jacket hanging from barges(s).
- Barges Both topsides and jacket removed by a barge method.
- Floating The jacket is refloated after topsides removal by other method. Only partial removal of jacket "B" and "C" are feasible.
- Demolition Similar to the preceding case but the jacket is toppled once the topsides have been removed.

In general for the three types of platforms the methods of removal by cranes are the most expensive ones, and those with mixed marine spread are more expensive than any of the barge methods. Finally, when the operation is allowed and feasible the floating and the demolition methods are cheaper than the rest.

Cranes > Mixed > Barges > Floating or Demolition

The table shows in its first row the cost range of removal of each type of platform. Those values indicate that a careful selection of the removal method must be done because savings of 30 to 60% could be achieved using the most appropriate method.

MECHANICS AND NEW MATERIALS

Session Chairmen : J. BOSIO, ELF (F)
T. TORP, Statoil (N)

POSEIDON: The Multiphase Production Theoretical Approach becomes Reality

A.LAFAILLE

TOTAL Compagnie Française des Pétroles
TOUR TOTAL - CEDEX 47 - 92069 PARIS LA DEFENSE
FRANCE

Summary

In January 1984, TOTAL CFP, IFP and STATOIL launched their joint R&D work on the promising Multiphase Boosting and Transportation Production Concept by creating the POSEIDON Project. The application of this concept to covering long distances between the wellheads and the process center could cut field development costs considerably, but requires the mastery of multiphase pumping and long distance multiphase transportation techniques. All the major problems to be tackled were listed from the outset of the Project; some, it was possible to solve temporarily by extrapolation of the techniques used for short distances, such as those already applied to satellite subsea wells; however, in most of the cases, a theoretical approach was deemed necessary, given the extremely innovative nature of the concept. Such an approach has given birth to new techniques and tools which are among the most promising to meet the demands of the specific constraints of the concept.

1. INTRODUCTION

In the world of Oil Production, the 1980s will always be remembered as the era of new production techniques for offshore hydrocarbons, the fruit of Oil Operators' efforts to tackle the new challenges of the coming century. Most of the giant offshore fields, those which made the good old days at the beginning of the offshore exploitation, have now been discovered and will be depleted before long. Today, more and more discoveries are being made of smaller structures under deeper seas: this is probably one of the biggest technical challenges the Oil Industry has ever faced, heightened by the present oil price projections which do not favour traditional stand-alone developments. European countries have to continue to reduce their dependence on imported hydrocarbons and help maintain oil prices at a "reasonable" level, which means the economical development of the so-called "marginal" fields.

In order to make a profit out of developing these resources, the Oil Industry have already investigated many possibilities: two major trends emerge:

- no modification of the conventional production lay-out where the wellheads and all the process facilities are located on the surface. In this case, the combination of slight and progressive improvements of conventional well-mastered production techniques, together with reduced cost-dependence of the support on the sea depth, offer an increased range of applications. For instance, tension leg platforms can be used in deeper waters than conventional platforms, and the reduction of topside weight will be beneficial to any concept. Although useful on the short-term basis, this way does not offer any vast potential in the long term regarding water depth and field size;

- new production concepts, perfectly adapted to future needs. These demand long and tedious industrial validation, but have the greatest immediate yield and future potential.

Any industrial sector owes to its own R&D a fast and profitable growth. This is very much the case for the Oil Industry as most of the recent improvements and innovative tools are the outcome of long and expensive R&D programmes.

Today in most offshore provinces, major changes in marginal-field economics are to be expected as a result of a new technological breakthrough: the multiphase production concept. The potential of this concept was perceived as early as 1984 by TOTAL CFP, IFP and STATOIL who formed the POSEIDON Project, a joint R&D effort, backed by EEC funding.

The aim of the Project was to list the problems and offer solutions that would ultimately allow the multiphase production concept to become a base-case field development option as viable as any other, for small fields at first, and for any deep offshore field later on. The multiphase technique, already commonly employed to link close wells to the main process center, can obviously yield large savings when considered for satellite fields lying medium to long distances away with higher flow rates and higher pressures. The problems which were not apparent to Operators for short distance applications then become predominant:

- the distance that can actually be covered by the effluents with the available well head flowing pressure, is limited to avoid jeopardizing overall recovery; longer distances imply that energy must be added to the flow,
- multiphase flow in long lines generates numerous phenomena, unknown to single phase lines, and demands innovative know-how on the part of the Operator.

Early in 1984, the first conceptual studies showed the need for a theoretical approach to most of the problems listed; no improvement or adaptation of existing tools was capable of the necessary performances. The key points where R&D efforts had to be focussed were identified as follows:

- multiphase flow, complex behaviour which calls for a very well-adapted multiphase line design to avoid the risk of not even being able to start up the line or operate it safely,

- multiphase boosting that offers all the required qualities of reliability and efficiency, whatever the effluent composition or flow pattern may be,
- subsea motorization, well adapted to the concept constraints,
- transmission of power over long distances, and subsequent remote control,
- specific aspects of subsea stations where multiphase equipment is to be installed.

In the following chapters, we describe how a theoretical approach has played a decisive role in the development of the major new POSEIDON tools.

2. MULTIPHASE FLOW

Multiphase flow: just two words to qualify the many various complex phenomena that occur inside a long line along which hydrocarbons are flowing in thermodynamic equilibrium; one can imagine for example a length of pipe where stratified flow is established: nothing will change as long as the gas and liquid speeds stay more or less equal - a purely theoretical situation -; if the gas speed increases, waves will form on the liquid surface, just like the sea; they will grow until they touch the upper part of the pipe, thus forming short liquid slugs. These short slugs will gradually collect together and form long, to very long slugs, due to coalescence, decreasing temperature along the pipe path causing gas condensation, and seabed profile that may favour liquid accumulation in the lower length of pipe. Such a simple theoretical case demonstrate how stable stratified flow can only be considered an exceptional situation.

Two major elements have been identified as key points:

- a thorough knowledge of the flow pattern for a suitable sizing of the installations is essential; in particular, a fine estimation of the slugs maximum length will avoid having huge slug catchers, and save money on receiving equipment on the host platform,
- the permanently changing effluent composition, in obedience to physio-chemical laws according to temperature and pressure gradients along the pipe, causing problems if not well-monitored.

A mathematical model able to predict the flow pattern must integrate both mechanical and physio-chemical parameters, the second being much better known than the first. On the basis of the famous OLGA model, the POSEIDON Partners have developed two computer programs:

- RETIS (REal TIme Simulator) that includes models for simulation of pumps, valves and pigs, and an improved description of fluids, in addition to OLGA features,
- PIPSS (PIPeline Simulator Station) which consists of two main parts: a predictive and a real-time simulator; the first is to predict the future development of a flow process; the second is run in parallel with an actual process and periodically fed with fresh values from a data acquisition system; as such it can be used for supervision purposes.

Both models can be used during the design phase of new multiphase lines when simulation of various operational conditions is likely to highlight operating problems that may have an impact on the overall design; as an example, start up of a line where low points are full of liquid may prove to be impossible without additional boosting, whereas boosting would normally not be required during normal operation.

Regarding the specific problem of hydrates formation in multiphase flow, studies have led to the formulation of a new inhibitor that prevents hydrates accumulating and forming large plugs. The hydrates remain in the form of small, light crystals, and are carried along in the line by the flow. In other domains of effluents physical chemistry, such as erosion/corrosion and wax deposition, theoretical studies have provided a better understanding and hence mastery of the phenomena.

3. PUMPING SYSTEM

To add energy to a multiphase mixture was probably the main challenge of the Project: in 1984, there was no such thing as a multiphase pump; the main constraints the pump was expected to withstand were listed as follow:

- flow rate range: 15 000 up to 200 000 bbl/d liquid+gas at pump inlet conditions,
- pressure rise: as high as possible, and not less than 15 bars,
- ability to face any effluent composition and flow pattern safely and efficiently, e.g. all GLR from pure liquid to pure gas, acid and abrasives, severe slug flow,
- all the above being guaranteed for a mean time of 2 years in subsea running conditions.

Such demands could obviously not be satisfied by merely adapting an existing product: there was no pump which could keep up its performance when facing a certain amount of gas, nor any compressor that would not burst if too much liquid was throughput. Hydrojet pumps were also ruled out, due to low efficiency and the need for an additional line. A new concept had to be invented, which was able to treat both gas and liquid molecules. Tentative answers to the above items were put forward:

- there should not be the slightest clearance between moving and fixed parts inside the pump, as there is a risk that this would be enlarged by the crude effluent abrasive effect,
- need for a fully open system, well-adapted to effluent density and sharp flow regime variations; the full wellstream should be able to flow through a stopped pump,
- large flow rate capacity without the drawback of a huge machine,
- simple and reliable mechanical concept favouring marinization.

A new helico-axial concept, of the dynamic type, was invented and is now covered by some 30 patents; one pump consists of some 10 to 15 stages according to the pressure rise requirement; each stage is made of one rotating impeller wheel whose blades communicate cinetic energy to each molecule, followed by one rectifier element where molecules will hit the fixed blades thus converting cinetic energy into pressure. Each stage being capable of an average 2 bars, this concept can deliver up to 30 bars at GLR 10. In such a pump, the pressure rise is not dependant on any mechanical clearance, it is a fully open system and the most simple one, as a single isostatic shaft can support the 15 stages. Moreover, the flow rate being the result of the effluent speed, high flow rates would be easily obtainable with a relatively small pump body.

In fact, this design was already well known and used, in particular for the multiphase downhole pumps jointly developed by IFP and TOTAL in the 70s. An immediate extension of the range of this system could well allow wellhead applications provided the in situ GLR is not more than 3 (e.g. around 75% of gas). So what is so new about POSEIDON then ? Based on the previous state of the art, a theoretical approach of the various phenomena occuring inside one pump stage, provided a clearer understanding and helped to develop an accurate mathematical model; this model was then used to improve the blades profile, and thus the hydraulic canal to enhance tolerance to increasing gas fractions while avoiding loss of efficiency through excessive centrifugal effect. This procedure led to the design of a completely new blade profile, with outstanding results, such as the ability to add energy to dry gas (GLR infinite).

Late in 1986, the POSEIDON Partners decided to give shape to these encouraging results, by launching the design and construction of the first industrial pump prototype; "industrial" meant a machine closer to the final product than the numerous "guinea-pigs" that had previously helped to correlate the model: as such it was designed from scratch as a subsea machine. For the bench tests, the machine had been voluntarily limited to low pressure rises (20 bars) and moderate flow rates (25 000 to 40 000 bbl/d) in order to handle with the maximum test loop capacity. At this stage, the decisive role played by a well-instrumented multiphase loop that can achieve accurate data acquisition and results must be mentionned:

Such a loop demands a certain financial investment, with no obvious potential return at first; however, this is the only way to test the expectations produced by years of theory and calculations. The multi-million Solaize test loop made it possible to simulate various flowing situations i.e. long or short gas and liquid slugs, very low inlet pressure, all effluent compositions from pure liquid to pure gas, all the relevant parameters being monitored and recorded in real time.

Hydraulic and endurance tests were performed in 1989 and confirmed the validity of all the above choices:

- excellent tolerance of harsh slug flow,
- hydraulic efficiency practically constant from pure liquid to pure gas,
- still a good pressure rise without any overheating when handling long slugs of pure gas,

- excellent mechanical behaviour at all régimes, in spite of a quite low mechanical efficiency due mainly to the decision to build a marinized machine,
- targeted performances easily reached at 2 000 rpm less than expected.

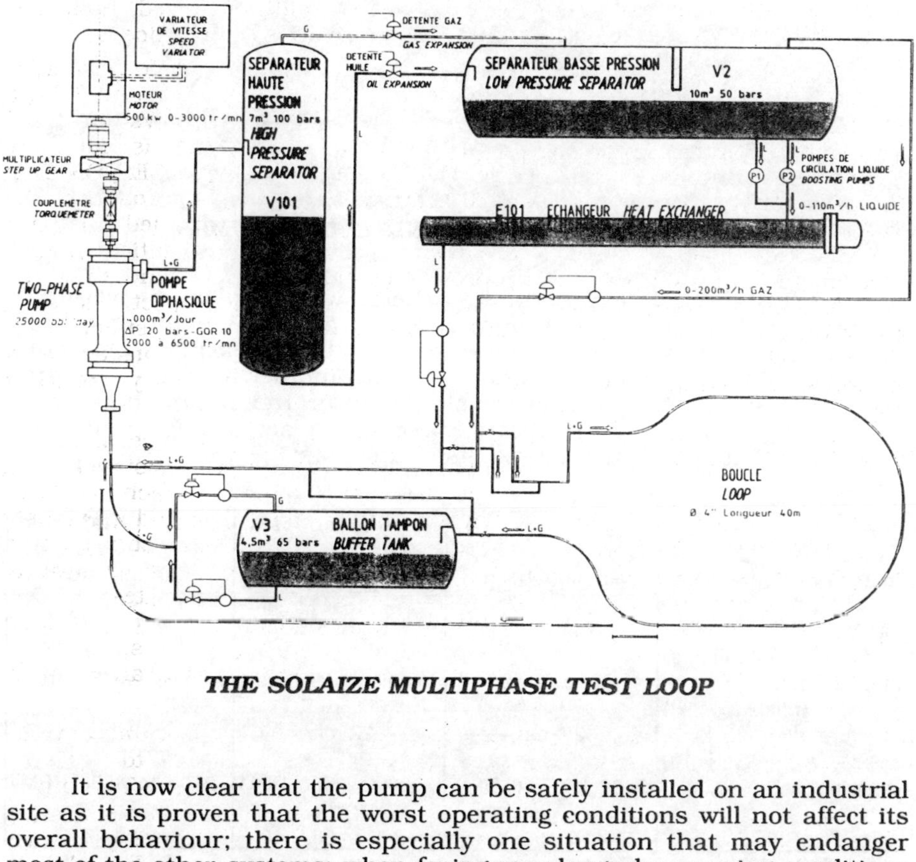

THE SOLAIZE MULTIPHASE TEST LOOP

It is now clear that the pump can be safely installed on an industrial site as it is proven that the worst operating conditions will not affect its overall behaviour; there is especially one situation that may endanger most of the <u>other systems</u>: when facing very long, dry running conditions - a situation that must be considered during the design of any multiphase installation - some pumps will totally "collapse" e.g. will no longer deliver the required pressure rise or flow rate, leading to the following:

- the system blocks unless some liquid is re-injected,
- the pump overheats and seizes up.

In such a situation, it is probably not sufficient to rely on any kind of liquid reserve (that will be empty sooner or later), or on the artificial injection of sea water which may cause more problems than it actually solves.

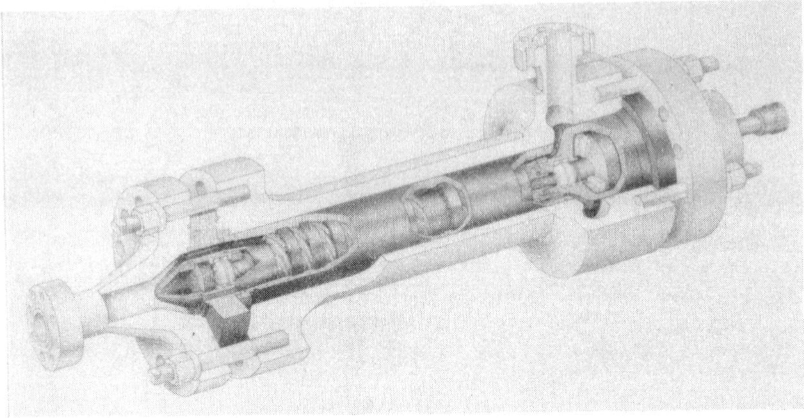

THE POSEIDON P300 MULTIPHASE PUMP

The POSEIDON pump, unlike others, will keep on delivering a sufficient pressure rise to the pure gas, and allow the effluent to keep circulating in the line, even at lower speeds, thus releasing some liquid from the wells to re-prime the whole system. Another interesting feature lies in the ability to face severe slug flow at the pump inlet while maintaining flow rate and pressure rise at an acceptable levels. The following table summarizes the results of the tests:

	P300 (design)	Results of tests	P302 (obtainable)
Total flow rate (inlet conditions)	25 000 bbl/d	35 000 bbl/d	130 000 bbl/d
Gas fraction (inlet conditions)	0 to 90 %	0 to 95 %	0 to 95 %
Inlet pressure (Mini - Maxi)	10 to 40 bar	minimum 5 bars	minimum 5 bars
Pressure rise (GLR 10)	20 bar	22 bar	32 bar
Rotation speed	3000 to 6500 RPM	2000 to 5500 RPM	2500 to 6600 RPM
Hydraulic Efficiency	40 to 50 %	42 % at GLR 10	around 45 %

As a conclusion we can say that the POSEIDON theoretical approach has permitted the roto-dynamic concept to deliver a pressure rise to pure gas, which was far from certain at the beginning of the venture.

4. SUBSEA DRIVING UNIT

It is now commonly accepted that hydraulic motor reliability is offset by their very short range for subsea pumping applications, which is why preference was given, from the outset to electricity: in 1984, the

POSEIDON Partners decided to launch the full-scale design and construction of a subsea electrical motor intended for long-term subsea testing offshore Norway. This motor had to withstand the following constraints:

- power delivered on the shaft: up to 2 MW,
- rotating speed: variable from 0 to 8 000 rpm at nominal power,
- 2 years MTBF.

Electricity likes salt water whether it comes from the sea or from the effluent. Given the fact that a dynamic interface with sea or effluent is unavoidable, the first motor thus received a reinforced insulation at the stator windings; after 4 000 running hours subsea, a short circuit occured inside the motor stator, due to the presence of salt water in the lube oil. This proved that a standard motor equipped with a reinforced insulation system is not sufficient to guarantee very long trouble-free running periods.

From that point, theoretical studies, supported by laboratory experiments, were launched to discover the real phenomena which occur when salt water confronts common insulating materials. These studies are still ongoing today, and some promising solutions are showing up at last.

The conclusion is that it was a mistake to under-estimate the apparently simple problem of running an electrical motor under unusual conditions: unexpected problems are likely to arise and consequently can hardly find satisfying solutions with available technology.

5. POWER TRANSMISSION

Another topic where the POSEIDON theoretical approach is the only valid one: the potential of multiphase boosting and transportation lies in the ability of the technology to cover longer distances thus reaching outlying satellite fields. This ability also depends on the efficient mastery of supplying power to and controlling a remote pump-motor group. In the case of an electrical drive, the power transmission problem will evidently be solved in a much more efficient way than for the hydraulic option, but again some unknown phenomena have to be solved beforehand. Actually the electric motor requires a three-phase variable frequency, alternate current; one way to transport the current could be that commonly employed by Electrical Supplying Companies worldwide, i.e. one separate cable per phase, each cable being installed at a certain distance from the others; this would not be cost-effective for subsea cables; thus the three phases have to be integrated into a subsea umbilical that will also shelter hydraulic ports and optical fibres. Installing three phases together inside the same cable raises fresh difficulties that none of the existing modelling programmes have managed to solve properly, and again a purely theoretical approach to the problem proved the only way to find a valid tool design. The following mechanical analogy can help understand which general phenomena can be expected; as this analogy suggests, a very long 3-phase cable actually behaves like a spring attached to a mass (motor) and excited by large amplitude mechanical vibrations on both sides.The vibrations generate stationary waves in the spring, which if not well mastered, can lead to incorrect driving of the mass and even spring failure.

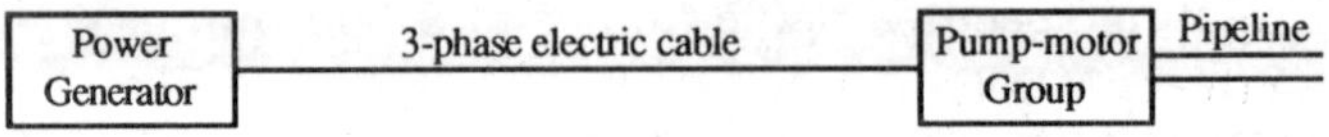

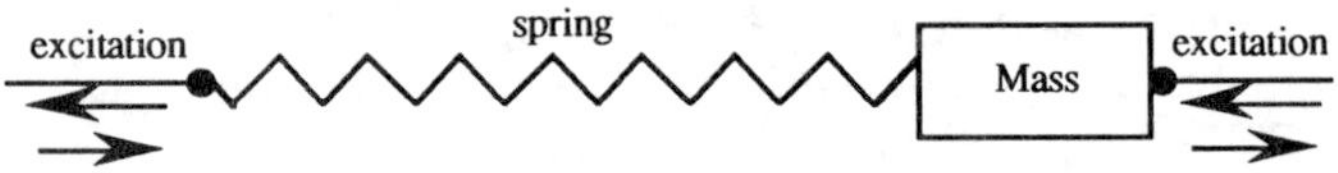

Here the comparison stops as an efficient modelization of the chain cannot be limited to one spring and one mass: the selected architecture consists of the following basic equipment from the control center to the motor:

- an AC current generator (rotating group or static converter),
- a step-up electrical transformer (high voltage required in the cable),
- the three-phase subsea cable,
- a subsea step-down electrical transformer (low to medium voltage required for the motor),
- the subsea electric motor.

This chain has to be optimized with regard to the various running parameters in both stabilized and transient régimes.

Each electrical component can be simulated by an equivalent, made of capacitors, inductances and resistances, the corresponding mechanical analogy being spring, mass and friction; the final model will consist of several groups of the above basic components. The finer the model, the longer the calculations; a good model is an intelligent compromise between calculations which are not too long and results which are not too wrong. The above approach has given birth to "ElPoLD" (Electrical Power transmission over Long Distances). This completely new tool, designed to be operated from a personal computer, can integrate any component present in the electrical chain and simulate various situations such as start-up, shut-down, torque variations, frequency variation, short circuit, failure of any piece of equipment, etc; for each case the main parameters are calculated, and a final compilation of results determines the installation sizing. As cable data and frequency are considered as parameters, the cable can easily be optimized with sufficient guarantees of reliability. Some calculations performed on the basis of theoretical hypotheses have shown that costly precautions may be called for, especially when high frequency values, generated by static frequency converters are considered.

6. CONCLUSION

From the above, one can conclude that theoretical studies are the key to development of new tools for efficient production of marginal fields. When first industrial applications are considered, the knowledge and operating expertise from the oil operator will be the link between the new technology and its successful exploitation. Of course no Operator would take the risk of launching a full-scale development based on such a revolutionary technology without prior qualification and validation of the basic tools involved. This is probably the only limitation of the theoretical approach of which results are requested more than any others to provide strong references before being accepted for good.

THE UNMANNED PLATFORMS MANAGEMENT IN NORTH ADRIATIC SEA

M. GUIDA and S. TONELLI

Production Dpt.
AGIP S.p.A.
S. Donato Milanese
Italy

Summary

The increasing number of platforms installed in the North Adriatic Sea has obliged AGIP - since the first years of '70ies - to increase its engagements toward the control and limitation of operating costs, safeguarding however the safety criteria of men and plants and the level of gas erogation.

In the report are briefly described the main characteristics of process plants, safety criteria and logistics of AGIP gas platforms in Italy; the esperience in production and maintenance management of these platforms is also illustrated.

1. INTRODUCTION

AGIP started gas fields'exploitation in the Northern Area of Adriatic Sea, that is through "RAVENNA MARE" and "PUNTA MARINA" fields, at the beginning of the '60ies.

In these fields were installed about ten single well platforms and one gathering and exploitation platform; gas was then conveyed to the onshore gas treatment plant before being let in the national gas distribution network.

In the following years, always in the northern area of the Adriatic sea, in the zone before the litoral from ANCONA to CASALBORSETTI (north of Ravenna) other gas fields were discovered and put into production and many other platforms were installed.

At present in Northern Area of Adriatic Sea, about 40 platforms of various types (4/6/8 legs) are installed; gas is conveyed onshore by the sealines network and is gathered and treated in gas treatment plants distributed along about 200 km of coast.

Having such a large number of offshore platforms distributed along such a wide area, the limitation of the operating costs is really important; another fundamental aspect to be considered for the management of these platforms is the safety of people and plants.

Increasing of safety and reduction of costs have therefore been the aims that have justified the development of a PLANTS CONTROL COMPUTERIZED SYSTEM (ANDROMEDONE DUE) and of a PLANTS MAINTENANCE MANAGEMENT COMPUTERIZED SYSTEM (SIMSO).

2. ORGANIZATION DEVELOPMENTS

Until about half of the '70ies all platforms were usually manned by operation teams, operating during the 24 hours/day, that locally executed all the operations necessary for the production, control and safety of plants and personnel.

During those years, the continuous development of technology started to make available informatics thechnologies, as well as control and regulation equipments ever more reliable; AGIP, therefore started to take into consideration the idea of applying these technologies to unmanned platform management.

The first experimental application of the above was put into operation in 1975, at the CASALBORSETTI treating plant, for the unmanning management of its platforms.

The success of this experimentation allowed to employ the platform operating team only during daily hours, that is 8h/24h; the platforms operations and safety was so ensured from the CONTROL ROOM of the onshore gas treating plant during 16h/day.

At the end of the '70ies a computerized control system named "ANDROMEDONE" located in the District Headquarters in MARINA DI RAVENNA, went into operation for the management of the offshore platforms.

The positive accumulated experience as well as the increasing production led AGIP to realise at the end of the '80ies a NEW ambitious PLANTS CONTROL COMPUTERIZED SYISTEM, named "ANDROMEDONE DUE", modular and able to control all offshore platforms as well as all the relevant onshore treating plants.

All onshore and offshore plants parameters and signals arrive to the control and supervision system.

By means of telecontrol it was then possible to start a new platforms managing system, leaving the continuous manned system (8h/24h) only on a few platforms, the MASTERS, selected either according to their baricentric position in respect of the other platforms, or to their condition of gas gathering platforms.

Continuing the Company policy of increasing the automation levels, it was then progressively left the continuous manned operating system per MASTERS platforms, arriving to the present operating system where all platforms are manned in "TURNOVER".

In the present management, therefore, all platforms have become "UNMANNED".

Having eliminated the continuous manned operating system that had since then guaranteed the running and maintenance of the platforms plants, it was necessary to maintain an efficiency degree of the plants, such as to grant the regular running of the same.

At the end of the '70ies the investigation about the number and the frequency of the spot interventions made by the Maintenance Service for the resetting of the equipment breakdowns and the increased aero-shipping traffic for maintenance personnel and materials transportation, brought to the decision of passing from maintenance of the platform equipments "upon breakdown" to a "preventive/inspective" maintenance of the same. At about the half of the '80ies AGIP put into operation the PLANTS MAINTENANCE COMPUTERIZED SYSTEM (SIMSO).

To such purpose, all plants components installed in the platforms have been accurately examinated and registered, and have been established the relevant maintenance operations and operating controls to be regularly executed, in order to grant a certain unmanned period of time without breakdowns, that could prevent the regular running of the plants.

Such operations of preventive/inspective maintenance, selected and detailed through informatics' technologies have made it possibile the definition of well detailed working programs.

At present, are in operation three offshore teams of 6/7 maintenance specialists each (mechanical, electric, instrumentation), that cyclically perform such programs in each platform.

The present platforms unmanned cycle, suggested by the experience and the return data so far acquired, is of 48 days.

Such cycle, however is not to be considered as consolidated purpose, but as intermediate one, with the possibility of increase, depending on the plants technology evolution as well as on the maintenance techniques developments.

3. PLATFORM'S PLANTS MAIN CHARACTERISTICS

As already specified, all platforms have been designed to operate without personnel on board (unmanned).

We are now herebelow examining the main characteristics of the gas platforms plants installed in the Northern area of the Adriatic Sea:

3.1 Gas Process

Gas extracted from the wells (whose well heads are placed on the cellar deck) is sent first to separators (placed in the process module on the main deck) where formation water is separated and then through outlet collector and relevant sealine to the onshore gas dehidration plant.

To outlet gas, after having measured its flow rate, is added diethilene glycole before the expansion valve.

The separated field water is then sent to the treating plant.

The fire plant safety is granted by an automatic depressurizing system that conveyes the gas contained inside the intercepted equipment to the vent.

3.2 Electric power generation

Electric power is produced by two main electric generators, gas fed, for operation on normal conditions, and by an electric generator, Diesel oil fed, for operation on emergency conditions.

They are placed in a protection cabinet in which are installed gas detectors, smoke detectors, fire detectors and relevant firefighting system.

On normal conditions, only one of the two main generators is working, while the second one is in stand-by.

In case of main generators breakdown, or on emergency conditions, the main electric services are granted by the emergency electric generator which starts automatically.

Besides, it is installed a series of electric batteries designed to power supply the main electric services for 6 hours at least, in case of breakdown of the main electric generators and of the emergency one.

3.3 Safety Systems

The following safety systems are installed to grant safety to personnel onboard, enviromental defense and plants:

3.3.1 Gas detectors system

Gas detectors are installed on air inlets and inside the rooms where flammable gas can accumulate due to incidental situations (main electric generators, emergency generators, living quarters, well head area, process module, etc.).

Said detectors are set for two thresholds of intervention; the intervention on the first threshold brings into action pre-alarm signal, while the intervention on the second threshold brings into action an alarm signal and, where foreseen, the block system.

3.3.2 Fire detector system

Platforms are equipped with a suitable fire detector system, in order to bring immediately into action the fire fighting system and/or the block system. The following fire detectors systems are foreseen:

- **Heat detectors (fusible heads)**
  the system foresees the use of fusible heads made of a constant fusion point (70 C) metal alloy, set on an air compressed piping system, and located in the well head zone and on the gas separators in the production module.
  In case of fire originated by gas leak, the alloy melts and depressurizes the fusible heads system, bringing into action the plant general block through depressurizing of the relevant equipment there installed, and making easier a quick fire fighting.

- **Flame detectors**
  they are the combined type UV/IR and located inside closed rooms (main electric generators cabinet, emergency electric generator cabinet, electric room, instrumentation room, etc.).
  In each room are installed two detectors, each of them connected to a separate loop : the first loop is reserved to the room itself (it is the room's specific loop) and causes the alarm only; the second loop constitutes the general confirmation loop which brings into action the platform block and the Halon fire fighting system in the interested room.
  HALON automatic fire fighting systems are foreseen for the protection of said rooms.
  Each room is protected with separate independent system.
  The HALON system intervention is brought into action as follows:
  . automatically, through the flame detectors systems
  . manually, through a push-button placed inside the room near the Halon bottle or a quick-release device placed outside the room.
  The intervention of the Halon system is signalled to the onshore control room by means of telemeasures signals and (when the platform is manned) in the platform chief's office.

- **Smoke detectors**
  Smoke detectors are placed inside all the closed rooms of the platform (living quarters rooms, main electric generators cabinets, emergency generator cabinet, electric room, instrumentation room, air compressors room, etc.).
  In each of the living quarters rooms it's foreseen one loop only, connected in logic of intervention one by one: that is in case of intervention of one detector, the relevant detector alarm system is activated.
  In all the other industrial cabinets and rooms two loops are foreseen, connected in logic of intervention two by two, for the intervention of their relevant alarm systems.

3.3.3 Platform automation system

When a CAUSE is produced by a manual or automatic action in connection with gas, flame or smoke detectors, fusible heads or pneumatic pressure switches, it produces as EFFECT a lighting signal and/or an acustic signal and/or operating drive on the production plants, on the electric plants, on the conditioning plants and on the HALON fire fighting system.

4. PLANTS CONTROL COMPUTERIZED SYSTEM (ANDROMEDONE DUE)

4.1 Architecture

ANDROMEDONE DUE is a very powerful tool for control and supervision of the NORTH ADRIATIC SEA offshore and onshore gas plants; its main tasks are: gas field management, production management, data filing and support to maintenance management.

To perform these tasks, ANDROMEDONE DUE manages a series of process computers located in the onshore production plants; these computers allow the control of the onshore treatment plants as well as of the related platforms plants.

Depending on control strategy, the production plants can be telecontrolled both from the control room of the treatment plant and from the ANDROMEDONE DUE operating center.

4.2 Dimensions and performance

The system can receive and process:

- 32.000 digital signals
- 32.000 analog signals
- 4.000 alarms per each telecontrolled plant
- 2.000 synoptic displays.

The system has the capability to manage up to 75 large size platforms and 12 treatment or storage large dimension centers. Other characteristics are:

- Computer user-oriented system's configuration capability by means of particularly simplified operation of fill in the blanks.
- Synchronous and homogeneus configuration between the process computers in the peripheral plants and the ANDROMEDONE DUE.

This powerful real time system allows to collect, process and display all the above mentioned data in less than 2 seconds.

4.3 Computer system

The above characteristics in data processing capacity are allowed by 32 bits minicomputers with Real Time Operating System.

The Hardware layout was designed for the maximum in redundancy; computers have therefore been doubled, as well as memories, boot disks, and dual port access working disks.

The two systems operate in hot stand by and automatically exchange roles of Master and Slave; peripherals such as modems, video controllers, printers and various terminals are automatically assigned to the master by a switching unit. Connections to computers in the peripheral plants are performed by 2 GHz data network radio links, fully dedicated to telemetring.

Data lines have high speed characteristics (9600 Bit/sec) and work with packet switching protocols according the international standard CCITT X-25. The communication software was developed using the OSI standard.

The ANDROMEDONE DUE control room is equipped with 4 operators desks. Each has two color graphics video monitors and a touch-sensitive keyboard driven by a custom man machine interface developed for final user needs.

A color video projector shows graphically plants'measures and outlines on a large screen in the control room.

This application software is fully modular and uses an access protected data bank. Due to flexibility and modularity of the software and hardware components ANDROMEDONE DUE is an "OPEN SYSTEM" easy to maintain and to expand.

The application software has been developed by the engineering department of AGIP through the coordination of a team of specialists.

4.4 Process control

Through computers in the onshore treatment centers and through data network all plants'parameters and signals arrive at ANDROMEDONE DUE operating center.

Here data are handled, filed and shown to the operating center operators in form of graphics video display pages. System manager can define and modify during operation: input data characteristics, computing algorithms, as well as the video displays and the print format.

The man-machine interface allows operators to easy handle the large amount of information available. The operators interface the system through user oriented functional keyboards.

The system provides the automatic selection of possible operation lighting a led on the keys enabled for that condition.

Powerful filtering and correlation tools reduce the number of alarms shown to the operators.

The application software masks alarms due to consequential events permitting the operator to focus the primary event.

The ANDROMEDONE DUE system manager can, according to the funtional needs, assign the supervision of a specific onshore treatment plant with its related platforms to any operator desk.

When an onshore treating plant is loaded "UNDER SUPERVISION" the ANDROMEDONE DUE operator can access information coming from the onshore plant and its related platforms, but is non expected to acknowledge the rising alarms or input operation commands.

When an onshore treating plant is loaded "UNDER CONTROL" the operator must acknowledge alarms and has full operative control over the plant and over all the joined platforms.

4.5 Production management

Automatic routines resolve any possible discrepancy, due to instrumentation disalignment and/or different gas flow characteristics, between quantities measured according to fiscal procedure at national gas network inlet and the sum of measures carried out on the single production string.

The production data, corrected by this procedure, are filed in a relational data base, together with other data gathered directly or manually input by the operators. All other data processing refer to these files in the data base.

Daily and monthly production report are automatically built and sent, through the data network, to the onshore treatment centers and to the production department information system at headquarters in Milan.

4.6 Field management support

The organization of production parameters into a data bank allows the management of production, optimizing the field's exploitation.

The gas field can, in fact be processed and compared with estimated depletion models. In this way real curves of the well's performance obtained, properly reprocessed by experts permit to calculate production capacity of wells.

The reservoir manager can handle any possible request for an increase in production (first and second emergency over-production), keeping in mind the phisical and operating constraints of the fields.

The analysis of the production history will also allow an improvement in knowledge of the field's dynamic behaviour so that any possible anomaly can be transmitted to the reservoir engineering department at company headquarter.

4.7 Maintenance management support

A continuos statistical analysis of the field instrumentation, of the transmission system and of the control system is possible over all the information coming from production units.

Periodic or on demand congruence verification procedures based on current flow data from production network's nodes help to discover possible malfunctioning in the field instrumentation.

Automatic procedures permit the correct functioning analysis on the complete chain of analogical to digital conversion.

Instrument calibration procedures also allow the setting up of pressure and flow trasmitters by comparing their value with that of standard instruments.

These modules furnish to maintenance staff interesting data for planned or rapid intervention.

5. PLANTS MAINTENANCE MANAGEMENT COMPUTERIZED SYSTEM (SIMSO)

5.1 Scope

As said before, the large number of equipments and components to be maintened on the platform plants, the large number of platforms, as well as the large number of maintenance breakdown work orders (RdL) necessary to have high level of plants running, as well as safety standards for working personnel and plants led AGIP to create a PLANTS MAINTENANCE MANAGEMENT COMPUTERIZED SYSTEM (SIMSO).

In fact, any maintenance intervention, since it depends on aero-shipping traffic (overcrafts and helicopters) which do not always satisfy all working personnel transportation needs, and also depends on atmospheric and sea conditions (in case of simultaneous fog and rough sea it is not possible to organize any maintenance intervention to platform plants), presents great operating and organizing difficulties.

5.2 SIMSO main purposes

The main purposes that the maintenance service intends to reach - by means of this computerized system - are:

- To optimize the organization of periodic works of preventive and inspection maintenance through the analysis of works to be executed and of the resources availability (specialized personnel, materials, means of transportation).
- To reduce maintenance breakdown work orders (RdL), which presents a considerable waste of technical and human resources, in favour of planning and rationalization of the same.
- To reduce waiting time for the maintenance breakdown work orders (RdL), through a planning of the same, based on a timely control of specialized personnel, materials and means of transportation availability.
- To individualize plants critical components through the examination of their availability and reliability, as well as equipment failures and equipment stops analysis, to determine improved maintenance interventions.
- To analyze the aging and decay process of components technical efficiency.
- To administrate - in accordance with the most advanced management criteria - the maintenance materials warehouse, obtaining all the information on the materials availability in real time, as well as utilizing specifical procedures for store provisions analysis and their automatic restoring.
- To manage - in accordance with the most advanced practice - the aero-shipping means of transportation for the various movements of maintenance specialized personnel and materials.
- To immediately know up to date maintenance costs for manpower, materials and means of transportation.

Finally we can say that SIMSO is in a position of satisfying these requests:

- To MANAGE the scheduling of preventive/inspective maintenance, as well as of maintenance breakdown "work orders" (RdL).
- To WORK OUT balance sheets about all the interventions carried out on the equipment components, as well as on the detected anomalies and relevant causes.

5.3 SIMSO Data Structure

SIMSO uses a sophisticated system to describe all the equipment components subject to scheduled maintenance.

SIMSO utilizes the following main data filing system containing all the technical, codified and support information for the scheduled maintenance management :

- PLANTS
- EQUIPMENT COMPONENTS
- PREVENTIVE/INSPECTIVE OPERATIONS
- PREVENTIVE/INSPECTIVE OPERATIONS TEXTS
- EQUIPMENT BREAKDOWN WORK ORDERS (RdL)
- RECORD OF ALL THE INTERVENTIONS EFFECTED
- RECORD OF ALL THE OPERATING FAILURES DETECTED

All data files are logically interconnected and proper automatic functions control the validity and the correspondence of input data.

5.4 SIMSO Duties

5.4.1 Technical data administration

The system is able to:

- to supply all technical characteristics of plants and relevant equipment components
- to describe preventive maintenance (I.M.) and inspective maintenance (V.F. and V.T.) interventions, specifying operating frequency, specialized technicians requested and estimated work duration
  (preventive maintenance interventions "I.M." foresee the total substitution of components or of exhausted parts of the same;
  inspective maintenance interventions include carrying out of operating checks "V.F." - which consist in checking the equipment components functionality-and of calibration checks "V.T." on components)
- to describe the work to be carried out, the relevant necessary equipment, the foreseen spare parts and the means of transportation.

For all the data files are foreseen the following operations:

- to insert data
- to change data
- to cancel data
- to question data
- to print data

5.4.2 Maintenance breakdown "work orders" (RdL)

This function manages all the maintenance breakdown work orders.

5.4.3 Work schedule management

This function consists of:

- automatic supply of scheduled maintenance interventions (VF, VT, IM) on the basis of established frequencies;
  the scheduled maintenance interventions are established:
  for each plant;
  for each maintenance specialized team;
  for each components operations to be carried out;
  for the duration in which the operation are to be carried out;

- scheduling of maintenance breakdown "work orders" (RdL);
- printing out the final work programme and the preparation of all the technical documents necessary to specialized personnel to carry out the maintenance intervention works;
- printing out the spare maintenance works schedule, to be carried out in case of unexpected works conditions (i.e. for delay in maintenance interventions to be carried out);
- inserting the scheduled works (totaly or partially) executed, in order to up-to-date the system;
- automatic system memorizing of all the carried out maintenance interventions;
- inserting of technical notes by specilized personnel;
- automatic system memorizing of the components'anomalies and relevant causes.

5.4.4 Historical data files management

This function allows to enter into historical data files of the maintenance interventions carried out (VF, VT, IM, RdL) and of the operating failures detected. Through an advanced informatic question language, it is possible to obtain historical balance sheets, according to the most diversified aggregations and forms.

The most common balance sheets are:

- per equipments components
- per plants.

In order to permit a right interpretation of the historical data, the operating principles adopted for intervention works to be executed are memorized, i.e.:

- frequency of intervention works;
- estimated works'duration and relevant execution real time;
- real total duration time employed (transfer time plus working time);
- number of specialized personnel.

The analysis of historical data allows:

- to evaluate the interventions principles trend, in order to adopt the most efficient maintenance policies, with the purpose of decreasing the interventions' frequency, assuring in any case a high level of plants running and granting safety standards for working personnel and plants;
- to carry out the equipment components stop times analysis, in particular the examination of the stop time frequency trend, as well as the repairs times per each type of failure, per each component and per each plant.

This function can also supply the following indications:

- to find out the components with more maintenance interventions;
- to find out the most unavailable component, due to repairs;
- to find out the most frequent failure type per component;
- to issue a comparate analysis among manufacturers of the same component type;
- to issue a comparison on the operation of same components installed on different plants.

ACKNOWLEDGEMENTS

The autors wish to acknoledge the assistance of the following individuals for their contributions to the ultimate preparation of this paper: Messrs. S. Boschi and L. Mori (AGIP Engineering Dpt.), G. Sbalbi (AGIP Production Dpt.).

PLATINE - RESEARCH PROJECT ON UNMANNED OFFSHORE FIELD

Pierre LEFEVRE - ELF AQUITAINE (PRODUCTION)

SUMMARY

PLATINE is a research project initiated in 1986 by ELF AQUITAINE, aiming at specifying an optimized unmanned central production platform, remote controlled, related to an oil or gas field, and visited yearly (or half-yearly) for inspection and heavy maintenance of the installations.

Intermediate objectives aimed at developing and testing pilots on manned platforms (existing or new), these pilots being related to systems or equipment and designed for unmanned platforms.

Several objectives are already reached but the ambitious initial objectives have been retargeted to a more achievable one.

INITIAL OBJECTIVES

PLATINE is a research project initiated in 1986 by ELF AQUITAINE, aiming at specifying an optimized unmanned central production platform, remote controlled, related to an oil or gas field, and yearly (or half-yearly) visited for maintenance and inspection of the installations.

Intermediate objectives aimed at developing and testing pilots on manned platforms (existing or new), these pilots being related to systems or equipment and designed for unmanned platforms.

The project is characterized by a multidisciplinary approach on two main concepts :

- a systematic review of all current techniques and processes in view of an unmanned operation and a production optimization.
- a search for new "oil technology" (either existing in other industries or completely new).

The stakes of the project are :

- Improving the profitability of field developments, particularly by reducing operating costs (personnel and logistics) and increasing the productivity (in optimizing the process control and the reliability of the installations).
- Allowing development of marginal fields.
- Improving working conditions and safety for the personnel.

An initial economical study has proved the potential interest of the research project. Also it has shown that the expected benefits of production optimization could be as high as those related to reduction of operating costs.

METHODOLOGY

For each system or equipment, four phases are sequentially considered : 1 - Evaluation of users'needs, state-of-the-art and reliability study, 2 - Developing and testing a prototype (or demonstration model) when needed, 3 - On site pilot application 4 - PLATINE specifications or recommendations.

RESULTS

The results of PLATINE - 1st phase - are related to the intermediate objectives (i.e. evaluation, prototypes and on site pilots for systems and equipment).

Some of these objectives were reached in 1989, most of the others will be reached in 1990 and 1991.

Several pilots have been developed, installed on site and tested satisfactorily so far. The most significant pilot is the automatic gas-lifted well start-up system.

The results are presented below in accordance with the project organization : well activation - productions systems - sensors - rotating machinery - maintenance - process control systems - safety and surveillance.

A - WELL ACTIVATION

Two well activation modes have been selected : pumping and gas-lift.

1 - Gas-lift

The most critical phase of gas-lift is well start-up ; in addition, a good distribution of available gas in the activated wells generates an increase of oil output. So, two thems have been investigated : well automatic start-up and production optimization.

The unique injection point gas-lift process was selected in order to minimize well work-overs or servicing. It necessitates high pressure gas.

a - Automatic gas-lift start-up
A pilot application has been developed, aiming at controlling the injected gas quantity and the behaviour of the well. After having tested a prototype on a well, the pilot was developed and installed on an offshore oil field in Gabon.
In successive steps, the aimed well output is reached with a minimum injected gas flow-rate and the maximum production choke opening.
The pilot includes an injected gas flow-rate control valve (electrically actuated), an adjustable well head production choke, sensors, a fault-tolerant PLC* and a specific software.
The results are very satisfactory ; the field oil output has been significantly increased and the high pressure gas consumption decreased by 25 per cent.

b - Gas-lifted production optimization
In order to optimize the gas distribution in gas-lifted oil wells on a field, a pilot system is being developed ; it is based upon the calculation of distribution set points under various exploitation constraints. The optimum is reevaluated when a significant variation of parameters (related to compressors, wells or process) is detected.
The pilot includes pressure, flow-rate and temperature sensors, automatic test separator, a control system including PLC'S and a specific software. An off-line version has been worked out.
It is foreseen that the pilot will be installed in Gabon early 91.

2 - Well pumping

Among the various types of well pumping, the electric submersible pump and the long stroke pump have been selected for PLATINE, when gas-lift cannot be considered (low GOR, high specific gravity, high viscosity).

* PLC : programmable logic controller

a - Long stroke pumping
A reliability study of existing units has shown up the unreliable components. In cooperation with a manufacturer, new components have been selected.
An existing unit has been equipped with these components and tested on site.
Further to these tests, a PLATINE unit is being specified.

b - Electric submersible pump
The existing automation means have been assessed and the study concluded that the best pump control system is the one constituted by two bottom hole pressure sensors (annular space level measurement) and a variable speed controller.

B - PRODUCTION SYSTEMS

1 - Well safety system

This system aims at improving the reliability of the conventional well safety system based on pneumatic logic control by using a modular electronic system. This system concerns tubing safety valve, second master valve, wing valve and annulus safety valve.

Being fault-tolerant (triplicated PLC's), this system carries out safety sequences, leak tests of safety equipment ; it also controls the hydraulic actuation system of tubing safety valve and monitors some parameters.

After a successful prototype test, a pilot has been developed and installed on a platform in Gabon (october 89). It works satisfactorily (no failure in nine months).

2 - Gas treatment

Two cases have been considered :

- produced gas treatment in order to meet transportation specifications (gas dehydration)
- associated gas treatment dedicated to self-consumption (fuel gas, gas-lift)

a - Gas dehydration
A study case, based on a 5,000,000 m^3/d sweet gas output and a specified water dew point of -20°C at 120 bar, has been evaluated. Two processes have been selected : triethylene glycol absorption (with column and regeneration system) and solid bed (molecular sieves without regeneration gas recompression).
A comprehensive comparative study has been performed ; it includes an analysis of process options, a reliability study, basic engineering, cost and load estimation.

b - Gasoline recovery of associated gas
The external refrigeration process was selected. Two study cases have been considered : 100,000 Nm^3/d and 1,000,000 Nm^3/d at 15 barg and a hydrocarbon dew point of -10°C.
A study including process evaluation, basic engineering, reliability study, cost and load estimation has been performed.

3 - Injection water treatment

Water injection aims at maintaining reservoir pressure and must not entail formation plugging, nor damage well equipment by corrosion. The following functions are needed : antifouling chlorination, coarse filtration (100 to 150 microns), bacterial treatment, fine filtration (3 to 8 microns), deoxygenation, safety filtration (100 to 150 microns).

For each function, existing or new processes and techniques have been reviewed. After a preliminary study, a prototype has been designed, built and installed on a test site in TOULON. The test resulted in working out the specification of most of the pilot elements.

Antifouling : an electro-chlorination unit, simplified and more reliable, has been designed, satisfactorily tested and specified for PLATINE.

Coarse filtration : a multihydrocyclone is specified (entirely static, without regeneration).

Bacterial treatment : an ultraviolet process has been specified.

Fine filtration : solid bed filtration was selected and tested with different materials. The garnet/anthracite bi-layer process has been specified (3 microns without adjuvant injection, a backwash regeneration is needed).

Deoxygenation : the selected process was the catalytic one - hydrogenation in presence of a catalyst, hydrogen being provided by the electrochlorinator. Some additional tests are needed before issuing pilot specification.

Safety filtration : a variable-geometry filter (with quick backwash regeneration) is specified.

In addition, sensors have been selected, tested and specified, such as those related to filtration control, chlorine content ; a patented robot-aided bacteria detector has been designed and satisfactorily tested.

A comprehensive pilot engineering study has been completed, including distributed architecture of the control system and active redundancy of the equipment.

4 - Disposal water deoiling

The offshore water disposal is bound by various increasingly strict regulations. For this project, hydrocarbon content in water has been specified at 40 ppm.

After a preliminary evaluation of existing and new processes, a prototype water deoiling system has been designed and built ; it is being tested and is made up of a hydrocyclone separator with a non emulsifying feed pump, a specific coalescer (using oleophilic membrane), a discharge tank with pump, and analyzers. Prototype tests (laboratory and on site) will be achieved in 1990.

A pilot application - including sump caisson with disperser and oleophilic belt skimming - is planned in Congo in 1991.

5 - Oil separation - treatment

A preliminary study investigated the operational needs or problems and evaluated the state-of-the-art. Some preliminary tests were performed such as on site test of hydrocyclone used for oil dewatering (promising results). This study concluded that conventional separators could be able to meet PLATINE specifications but that geometry was a key criterion (specifically for foaming and emulsion). Then a feasibility study has been performed on a specific field ; a simplified process is proposed.

6 - Automated pigging

This system aims at automatically pigging sea-lines between satellite and central production platforms of oil fields. It is specified for a 10 in-pipe (900 ASA), two piggings per week with reinforced foam pigs. Design and detail engineering studies have been carried out and resulted in the following patented system : a pig storage system with coiled flexible strip in a container, a pig introduction and presentation system (barrel and pushrod jack), a conventional pig launcher (door replaced by a valve), a vertical pig receiver with a used pig recovery tank.

C - SENSORS

An on site survey showed that 60 % of sensors faults were due to human interventions, 30 % to the packaging and 10 % to the measurement cell. PLATINE actions aim at increasing the equipment reliability and the overall availability. Studies, surveys and on site tests gave the following results.

The equipment reliability can be improved in the short term by :

- choosing adequate measurement processes (such as differential pressure or nuclear absorption for gas-liquid level, conductivity and nuclear absorption for oil-water level)
- transferring electronics in technical room (but presently this does not exist for all categories of sensors ; a potential solution is the optical fiber sensor)
- putting an end to use of pneumatic instrumentation (for instance a valve actuation study recommends electrics for control and hydraulics for isolation).

The overall availability can be improved by use of analog sensors (instead of on/off ones), common use of control and safety sensors (a risk analysis study is being performed) and an optimal architecture (for instance, reduction of the number of sensors by correlation between measurements).

Optics : Optics can be considered as a potential solution for long-term improvement of equipment reliability. Optical sensors can be made passive, intrinsically safe, electro-magnetic compatible, easily lightning protected and would reduce the overall installation cost. A survey showed that the technology exists but manufacturers need incentives from users to develop industrial products. Some industrial safety sensors already exist and are being tested on site such as explosive gas detectors and flame detectors.

Bottom hole sensor (pressure and temperature)

Existing bottom hole sensors have been tested, but the results showed that their reliability was not compatible with PLATINE objectives - the main factor being the poor behaviour of electronic components at temperatures above 80-90°C. As a result of this, optics potentiality was evaluated. A feasibility study concluded to the selection of the white light interferometry method : the bottom hole sensor modulates the spectrum emitted by a light source in surface and transmitted by a multimode optical fiber ; the modulated spectrum is transmitted back by the same optical fiber to a surface analyzer.

In cooperation with AGIP, some demonstration models of the patented pressure and temperature sensor have been manufactured and are being tested. A pilot including surface unit, optical cable, connector and sensor is planned to be installed in an AGIP well end of 1991.

D - ROTATING MACHINERY

A preliminary wide survey pointed out that the reliability of an equipment is far lower offshore than onshore ; it also made it clear that a great part of the failures were due to peripheral components (instruments, lubrication, cooling, filtration, ...) ; for pumps and compressors, leakage is the main factor.

It is therefore recommended not to use instrument air anymore and to try to suppress external lubrication ; moreover, reciprocating compressors have to be avoided. Some new technology is required.

Gas-lift compression : It is hoped that we will be provided with a high speed centrifugal compressor (with active magnetic bearings and dry gas seals) directly driven by an asynchronous electric motor (with magnetic bearings). For the centrifugal compressor (suction pres. : 4 bar - discharge pres. : 70 bar - mini flow rate : 150,000 Nm^3/d - 22,000 RPM) a feasibility study is to be launched. For the electric motor, the feasibility study is being done by a manufacturer (extrapolation of existing prototype) ; a new generation rotor has been designed. A pilot development is planned in 1991 for on-site installation in 1992.

Pumping : Different technologies (not commonly used in oil industry) have been selected for various applications.

Oil delivery pumps : The "canned" technology has been selected (no rotating seal, no coupling, no bearing maintenance) ; a pump has been installed on a CAMEROON field.

Water injection pump : A multistage centrifugal pump, with silicium carbide bearings (water lubricated) directly driven by an electric motor, has been selected.

Water lifting pump : a multistage centrifugal pump, with immersed electric motor (pump impellers and rotor being on the same shaft) has been selected.

Moreover, it is aimed to improve the ball bearing reliability of centrifugal pumps. Two pilots related to bearings equipped with automatic lubricators are being tested on site in Angola and Cameroon.

E - MAINTENANCE

Three kinds of action have been carried out : drawing up a list of all platform interventions, evaluating condition based maintenance and selecting equipment.

A list of interventions on our existing platforms has been drawn up allowing for setting up 150 equipment maintenance forms with the present operation frequency and the needs for equipment modifications or new technology versus visit frequency.

Condition based maintenance : A preliminary evaluation pointed out that an on-line condition based maintenance system would increase service life of equipment, allow for degraded operation (wear follow-up providing a more reliable diagnosis) and improve programming of maintenance campaigns. It calls on various techniques which have been evaluated in view of PLATINE (for rotating machinery : vibration and thermodynamic analysis, acoustics, lubrication oil analysis ; thermography for heating detection). A general condition based system has been studied ; a general on-line event generator has been specified (turbines, compressors, generators, pumps) ; coupling of vibration and thermodynamic follow-up and analysis is its main characteristics ; it could be developed by a rotating machinery manufacturer. An off-line pilot expert system relative to fault diagnosis of a compressor has been developed and installed on site.

Electric facilities : A general study relative to generation, transportation and rotative equipment has been carried out on a specific case (power : 10 MW).

For platform-coast distances lower than 50 km, an on-line electric generation with underwater cable transportation is recommended.

Valves : Low pressure sea water oil and gas valves have been selected for endurance testing ; the intermediate results show that 50 % of selected valves must be rejected. Valve actuators on site tests are also being carried out (in particular electric actuators). Prototype electric actuators for well head valves are being designed by manufacturers.

Material corrosion : Two non-corrodable material studies have been carried out : a stainless steel water injection system and a preliminary study related to an aluminium alloy deck. For cathodic platform protection, sacrificial anodes with a specific monitoring system are recommended.

F - PROCESS CONTROL SYSTEMS

The concept of an integrated architecture of real time systems including process control, maintenance, safety and production follow-up of an unmanned platform has been studied.

A functional analysis of requested information streams has been carried out. All the collected data will allow hardware and software architecture to be specified and an optimal instrumentation architecture to be designed.

Two expert systems have been initiated. The first one aims at delivering operational diagnosis of a crude desalting unit ; an off-line demonstration model has been developed, tested on site and validated ; a pilot on-line system has been developed (not tested on site yet).

The second one, more general, fulfills the following functions :

- Fault diagnosis of sensors, process and equipment.
- Automated control of start-up and shut-down.
- Optimal process control, including degraded mode.
- Synthesis of relevant data for transmission to operators.

A demonstration model, related to gas dehydration (triethylene glycol contactor) has been developed, tested on a specific simulator in on-line recreated conditions and validated by production experts. A new method has been used to evaluate diagnosis and fault detection of sensors and equipment.

A pilot process control expert system related to oil treatment is being developed in cooperation with AGIP. Including slug catcher, 3 stage separation, desalting and crude stabilization, it is planned to be installed on site early 1992.

G - SAFETY - SURVEILLANCE

Platform operation safety depends on equipment reliability, general design of the facilities, presence of personnel and surveillance of the facilities.

Many reliability studies have been performed either on systems or on equipment and resulted in : modifications of architecture (taking into account redundancy), equipment simplification or modification (in connection with manufacturers), and selection of more reliable components or equipment.

A unique control and safety system is being evaluated, comparing its reliability with that of two independent systems.

An autonomous surveillance system (independent of safety and control systems) has been evaluated and is being carried out. It aims at identifying, locating and assessing : liquid leakage, marine pollution, human intrusion, shock and explosion. A pilot shock and explosion detection system (with accelerometers and hydrophone) has been developed and will soon be installed on a platform.

R and D COSTS

On 1/06/90, the total PLATINE project expenses amounted to 165 millions French francs. EEC has participated to PLATINE project funding through four successive contracts (DG XVII) ; AGIP Spa is associated to the last two ones.

CONCLUSION

At the end of 1989, the feasibility of an unmanned platform in accordance with the ambitious initial objectives has been reassessed in the light of the test results on different systems. These objectives have been reconsidered and became : a yearly (or half-yearly) maintenance and inspection campaign, and some random 1-day visits per year for curative tasks ; the number of these visits will be economically (capital and operating costs) optimized.

Production optimization also remains an important objective.

SUMMARY OF RESULTS

	PROTOTYPE OR ENGINEERING STUDY	PILOT	SPECIFICATIONS-RECOMMENDATIONS	COMMENTS
GAS-LIFT (AUTOMATIC START-UP)	PROTOTYPE	GABON 1/90 GOOD RESULT		
GAS-LIFT (OPTIMISATION)	PROTOTYPE			PILOT IN GABON EARLY91
LONG STROKE PUMPING	ON SITE TESTS (NEW COMPONENTS)			
ELECTRIC SUBMERSIBLE PUMPING			RECOMMENDATIONS	
WELL SAFETY SYSTEM	PROTOTYPE	GABON 10/89 GOOD RESULT		
GAS DEHYDRATION	BASIC ENGINEERING			TESTS OF COMPONENTS END OF 1990
GASOLINE RECOVERY				ENGINEERING STUDY BEING PERFORMED
INJECTION WATER TREATMENT	PROTOTYPE	SUB-SYSTEMS FRANCE-3/90		BACTERIA DETECTOR PATENT
DISPOSAL WATER DEOILING				PROTOTYPE TESTS BEING PERFORMED
OIL SEPARATION TREATMENT	FEASIBILITY STUDY			
AUTOMATED PIGGING	DESIGN AND DETAIL ENGINEERING			- PATENT - STOP OF THE PROJECT
SENSORS			RECOMMENDATIONS	ON SITE TESTS PERFORMED

	PROTOTYPE OR ENGINEERING STUDY	PILOT	SPECIFICATIONS-RECOMMENDATIONS	COMMENTS
BOTTOM HOLE OPTICAL SENSOR	DEMONSTRATION MODELS			- PATENT - PILOT IN-ITALY (AGIP)in 91
GAS-LIFT COMPRESSION				FEASIBILITY STUDY OF ELECTRIC MOTOR BEING PERFORMED
PUMPS		3 PILOTS (CAMEROON, ANGOLA)		
CONDITION BASED MAINTENANCE	OFF-LINE EXPERT SYSTEM + ON LINE FEASIBILITY STUDY			STOP OF THE PROJECT
ELECTRIC FACILITIES	ENGINEERING STUDIES		RECOMMEND/ SPECIFIC.	
VALVES AND ACTUATORS	ENGINEERING STUDIES ON ACTUATORS	PILOT ON-LINE TESTS	RECOMMENDATIONS	
MATERIAL CORROSION			RECOMMEND.	
PROCESS CONTROL SYSTEMS	DEMONSTRATION MODEL			PILOT IN ITALY (AGIP) EARLY 92
SHOCK AND EXPLOSION DETECTION	PROTOTYPE			PILOT END OF 90

COLUMN STABILIZED PRODUCTION PLATFORMS

André REY-GRANGE (SEAMET INTERNATIONAL - Paris)
Dimitri ARMENIS (ALFAPI - Athens)
Robert L. JACK (NOBLE DENTON - London)

Summary

The described concept has been developed by an association of ALFAPI of Greece, NOBLE DENTON ASSOCIATES of U.K. and SEAMET INTERNATIONAL of France and was supported by an EEC funding granted in 1987. SEAMET INTERNATIONAL (61, rue de la Garenne - 92310 SEVRES - FRANCE) was the project manager. It provides a set of predesigned main components or systems as they are usually involved in Column Stabilized Production Platforms in such a way that they can easily be assembled into a complete platform design, so as to suit any specific offshore Oil and Gas field development strategy. As a demonstration of how the construction set works two complete platform designs are proposed and evaluated in terms of performance at sea. In such an exercice, the authors have attempted to take benefit of the up to date industry experience about construction and operation of platforms of the kind ; special attention has been given to safety features in light of the recent mishaps.

1. INTRODUCTION

Slowly but surely column stabilized production platforms are accepted by offshore oil and gas industry, not only for temporary tests, but more and more as permanent installations. Pioneered with success through converted column stabilized drilling platforms, brand new production platforms of the kind now surface.

To date, we tally, worldwide :
in use or decommissioned Column Stabilized Production Stations :

. North sea : 10 including 4 in 1989/1990.
. Méditerranée : 4 the last one decommissioned in 1986.
. Gulf of Mexico : 1 installed in 1988.
. Brazil : 18 including 1 in 1988.
. India : 1 in 1989

But, unlike drilling platforms which are broadly all alike, production platforms have to face a wide variety of circumstances, in relation with reservoir features, environment, and other factors.

Nevertheless, we observe most proposed designs, including the few ones which have materialized, are taking after drilling platform monotypes ; furthermore, equipment design and layout is left to develop by the potential users.

The described research work aims at rationally meet the requirements, deliberatly departing from given concepts. It encompasses both structural features and equipment arrangement in an attempt to produce optimized integral designs.

Thus, the project concentrates on systems, as part of a whole, recognizing they are usually specific of each given Oil and Gas reservoir.

The end product is an array of proposed standardized systems and associated containing structures, as usually involved in production platforms such as : primary structure, ballast and bilge piping system, anchor lines, riser interface, production processing plant, auxiliaries, servicing rig, living quarters, helideck, etc.

In this effort, modularisation of the systems is attempted in order to provide a better construction cost.

When ultimately, the project offers a sample of fully equipped platform configurations it is merely to illustrate how the systems can be put together, and demonstrate their compatibility. For them, a performance and cost appraisal is provided.

2. DESIGN CRITERIA

At the beginning of the research work a market survey was made to complement in house experience and draw the design criteria. It covered European operators, builders, and regulations.

Also, a survey was made of public data on existing or planned Floating Production Systems as well as major equipment manufacturers.

As a result the following design criteria have been retained for the proceeding :

2.1. Field tapping configuration :

Three typical configurations have been taken into consideration reflecting current industry practices. They are mainly governed by the riser system and the ability to service or not a cluster of wells or a manifold center located under the platform either on sea floor or at some height below the sea surface. They are illustrated by figure 2.1.1. We identify :

Fig. 2.1-a : Well heads off the platform corridor ; independent flexible risers.

Fig. 2.1-b : Well heads in cluster and / or manifold center inside platform corridor ; independent flexible risers. serviceable subsea equipment.

Fig. 2.1-c : Well heads in cluster and / or manifold center inside platform corridor ; stiff riser in bundle and jumper extensions ; serviceable subsea equipment.

Additionally we have worked out, at preliminary stage, a configuration of the future made of :

Fig. 2.1-d : Well heads in cluster and / or manifold center inside the platform corridor ; independent in line stiff risers and jumpers ; serviceable subsea equipment.

2.2. Environment :

- Typical North weather criteria :

. 100 year significant wave height : 16.0 m.
. 100 year maximum wave height : 31.0 m.
. 100 year wind (10 min. sustained) : 76.6 KT.
. 100 year tidal current : 1 KT.

- Water depth 250 and 450 meters.

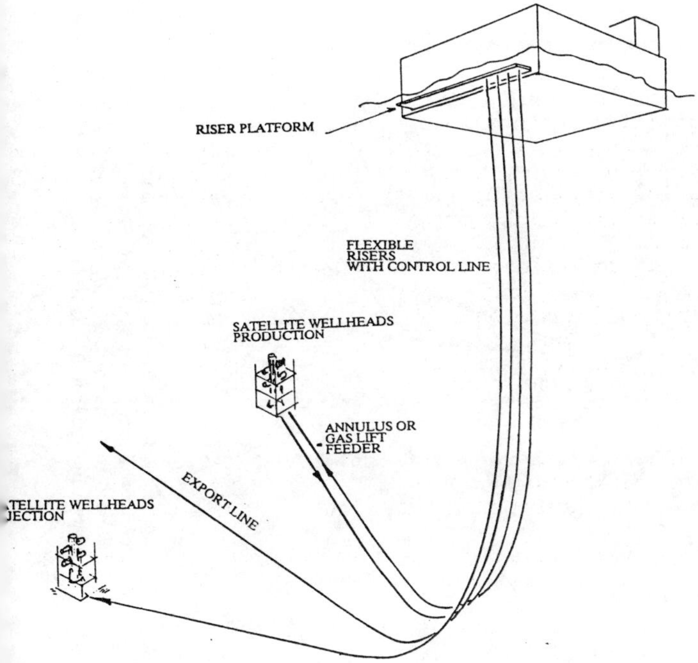

Fig.2.1.a : Well heads off the platform corridor; independent flexible risers.

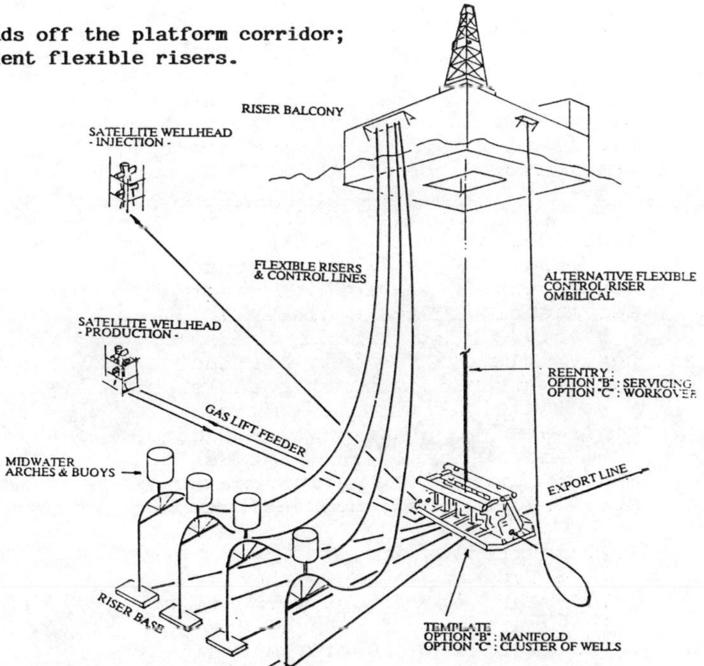

Fig.2.1.b. : Well heads in cluster and/or manifold center inside platform corridor; independent flexible risers. Serviceable subsea equipment.

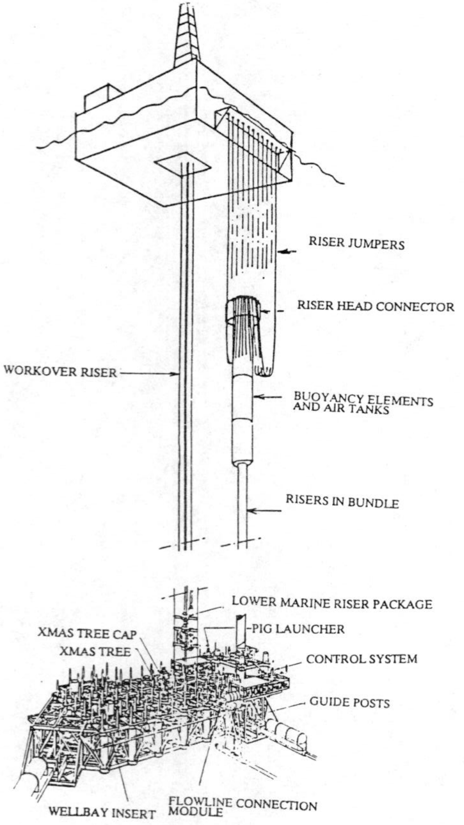

Fig.2.1.c: Well heads in cluster and/or manifold center inside platform corridor; independent in line stiff risers and jumpers; serviceable subsea equipment.

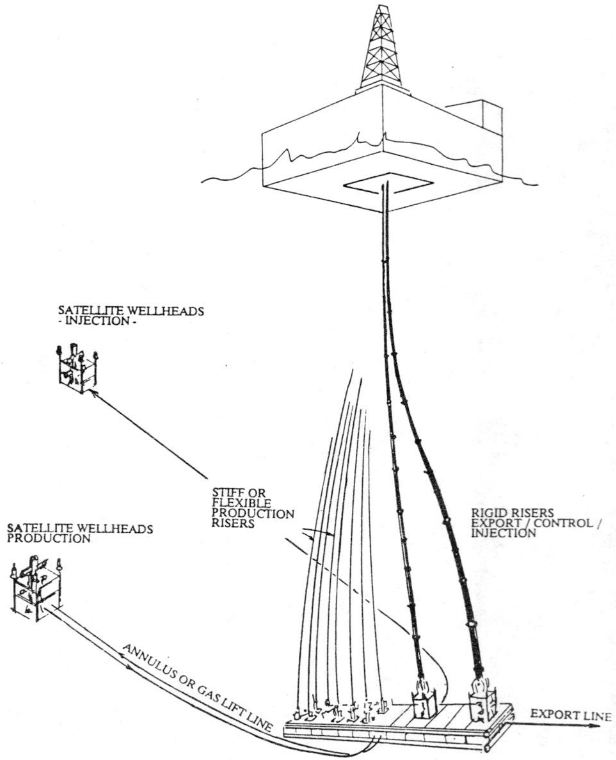

Fig 2.1.d: Well heads in cluster and/or manifold center inside platform corridor; stiff riser in bundle and jumper extensions; serviceable subsea equipment.

These environment criteria are considered the most likely one for future use of the work ; but alternative criteria within reasonable limits can easily be met with minor adjustment.

2.3. Oil and Gas Production rates :

Many combinations have been considered but, for the purpose of research work, and considering the flexibility of the proposed designs, only two typical situations have been selected for the demonstration kits.

Configuration	A	B
bbl/day	60 000	30 000
G.O.R. (m3/m3)	100	50
Separation	3 stages	
Fatal gas	export	flared
Oil export	through sea line	

2.4. Production Plant Auxiliaries :

- Gas injection - Gas turbine powered, consistent with hereabove G.O.R., optional.
- Water injection - 75 000 and 35 000 B.P.D., optional
- Power plant :

Configuration	A	B
Gasturbine generators	3 x 2500 Kw	3 x 1800 Kw
Emergency diesel generator	1800 Kw	1500 Kw

- Flare and burners.
- Crude oil buffer storage : 1000 m3.
- Miscellaneous auxiliaries.

2.5. Servicing rig :

1 - Full size workover rig, 20 000 ft depth capacity with mud system.

2 - Hoist rig for subsea operations.

2.6. Living quarters :

100 and 50 beds.

2.7. Construction facilities :

Assumption is made the Platform will be completely assembled and outfitted inside shipyards, preferably as a single unit built in a drydock or launchway, in order to minimize cost and time.

However care is taken to privilege the " zoning ", a concept in which the job is split in several modules, each one easy to be brought to an advanced stage of completion prior to its mating to the unit as a whole.

As a result, weight and size of prefabricated modules are critical.

After a careful survey of the European shipyards claiming interest for offshore business, we decided to keep the completed platform within the following limits :

- Width : 62.4 m,
- Lightship draft : 6.5 m,

and module weights, not to exceed 440 tonnes.

Such are the main guidelines for the proposed research work : they characterize the objective. Many other secondary design criteria were also taken into consideration as the work progressed at detail level.

3. PROPOSED SYSTEMS

It must be understood such systems and arrangements of equipment must be compatible with each other, and proceed from an anticipated overall platform architecture.

Main features include :

- Column Stabilized Platform configuration :
It provides good behaviour at sea and large deck area in comparison with the displacement.

- Close ring type lower hull : in conjunction with a somewhat like upper hull girder, it provides a compact, sturdy, primary structure, enabling to get rid of the slim bracings which are commonly used on catamaran type platforms and prone to develop problems.

The drawback of this configuration when it comes to transit is not critical since production platforms are likely to remain on the same site for long periods.

- Modular box type upper hull configuration. Optimized for " zoning " concept with regard to the equipment installation and construction procedure.

A double bottom all the way through provides reserve buoyancy, insulation, and makes possible a slick bottom face.

- Standardization is the principal philosophy of the design. Thus all structure has same longitudinal stiffeners spacing, 0.60 m, same web frame spacing, 3.0 m, as well as same type of stiffeners and girders. Main bodies are 6, 9, 12 or 15 meters in size, a multiple of 3 meters frame spacing. In this manner, prefabricated panels with stiffeners can be held ready and easily for the construction of complete vessels whatever be their overall dimensions.

Simplicity aspects are related with the use of flat panels, module prefabricated sections as well as with automatic welding procedures and other specifications commonly used by shipyards.

- Pieces of equipment are unitized as much as feasible ; they are planned with most of piping and instrument systems built-in. As units, they can be completed and tested at manufacturer's plants.

- Decks arrangement saves a good network of sheltered escape routes and care is taken to provide a clean segregation between safe and hazardous areas.

Following main areas have been covered by the research work :

3.1. Primary structure : figure 3.1. illustrates typical body segments of :

. Lower hull,
. Columns,
. Upper hull modules.

Fig.3.1. : Primary structure

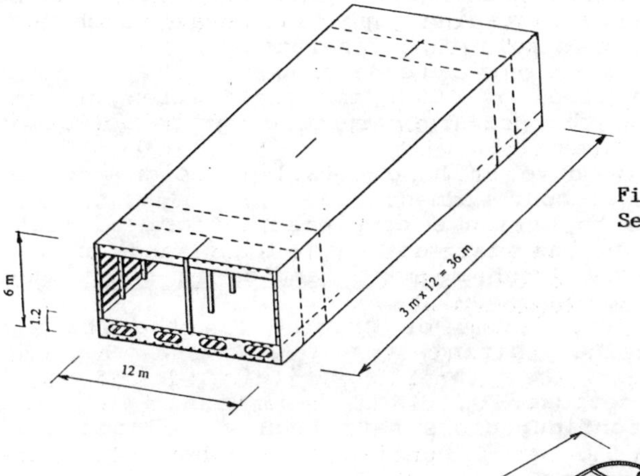

Fig.3.1.1.
Segment of upper hull

Fig.3.1.2.
Segment of column

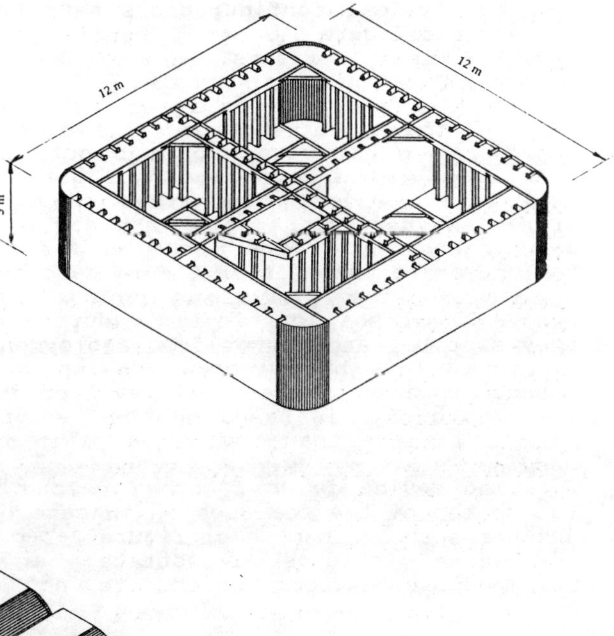

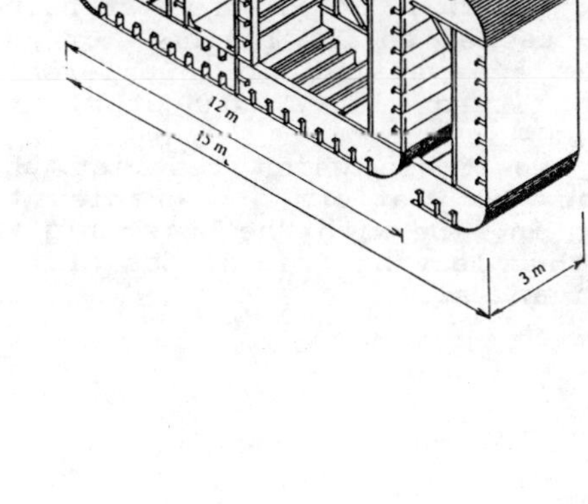

Fig.3.1.3.
Segment of lower hull

Primary structure has been stress analysed, anticipating environmental and fixed loadings as they are likely to occur in complete platforms taking up after the architecture philosophy described in paragraph 4 hereafter.

Thus, typical samplings were developed.

High strength steel of 355 N/mm2 yield strength is used all the way through the primary structure, grade selection to fit every specific area.

As a result we have on hand detailed design and weight assessment of typical body segments.

When it comes to design a complete platform to meet any specific requirement as far as dead loads and space are concerned, we may use such segment samples in body lengths, and heights to follow suit.

Total weight and center of gravity can then be easily computed for preliminary sizing.

3.2 Riser hook up devices.

Two typical configurations have been worked out and are illustrated figure 3.2. :

1 - Flexible risers.

2 - Stiff riser in bundle.

In the configuration 1 two alternatives have been considered : the first one has got the connectors at upper deck level ; the second one has got the connectors at lower hull level. Indeed the market survey shows some operators prefer the connectors up in the air so as to make easier the maintenance. Some other ones prefer them attached to lower hull so as to make easier the release in an emergency.

Our device includes a bridge crane which enables to handle the hoses, particularly during hook up procedure when they must be handed over from deployment vessel to production platform. This bridge crane design is the same whether the connectors are at upper hull level or at lower hull's.

The design is based on the use of mechanical connectors but it is more than obvious a plain flange or union can be used as well.

The device is good both for long flexible lines down to the bottom of the sea such as in case of figure 2-1-a and for jumpers such as in case of figure 2-1-c.

These features illustrate our attempt to produce multipurpose devices which are not bound to a specific concept.

In option 2 the riser system is made of a bundle of stiff lines free to swing on a flex joint ; it is free standing thanks to some buoyancy tanks attached to it. The head of the riser is down into the water at a level out of interference with the platform hull ; it is fitted with a connector, and jumper extensions to the platform.

Our device includes some tensioners which restrict the motion of the riser head versus the platform and enables to shift it from servicing position in line with the hoist rig to the normal operating position, thus leaving free access of the hoist rig to other subsea contrivances.

Fig 3.2. : Riser hook up devices

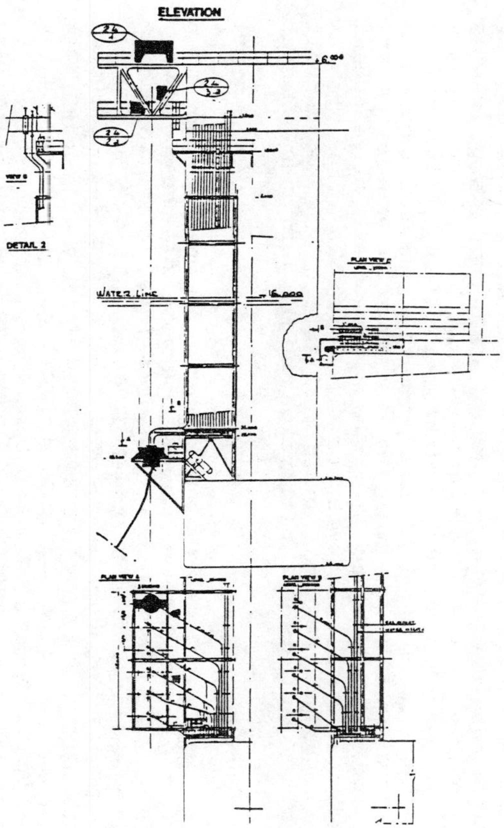

3.2.1. - Flexible Risers

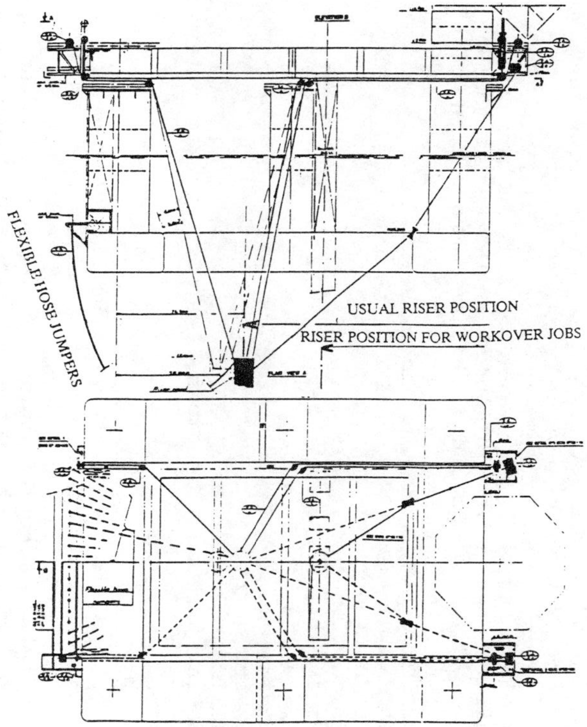

3.2.2. - Stiff Riser in Bundle

With these two basic riser hook up devices we believe we have covered the bulk of industry current trends.

Within a further research effort we plan to work out a design including some stiff individual slim risers extending each well in a cluster straight to the surface in a manner very much like T.L.P.'s.

3.4. Miscellaneous :

Similarly we have developed typical designs of main systems involved in Column Stabilized Production Platforms, namely :

- Workover rig,
- Servicing rig,
- Production processing plant,
- Production auxiliaries,
- Crude oil buffer storage,
- Flare,
- Diving and R.O.V.,
- Living quarters,
- Helideck,
- Ballast and service piping systems,
- Mooring,
- Life saving equipment,
- Fire fighting, etc...

It would be too long to describe the features of all of them within the frame of this paper. We only want to emphasize our designs are modular ; with minor adjustments, they can be incorporated in a complete platform design, provided itself follows a preset concept which is in fact just another system package.

Again, this approach stems from the observation systems included and part of a whole platform are liable to some level of standardization.

Even when the size of equipment has to fit the rate of production, or the depth of wells etc, the basic schemes remain the same.

Instead whole production platforms do not offer such an opportunity, for they have to meet lots of specific requirements calling for more or less complex assemblies and taking into account the many possible field development strategies.

All system designs are available with conceptual drawings, specifications, bill of material, weight and budget cost.

4. OVERALL PLATFORM CONFIGURATIONS AND EQUIPMENT LAYOUTS

The main guidelines of our construction set concept have been described in Chapter 3. The purpose of the following exercice is to illustrate how it works : we offer two models of different sizes, illustrated figure 4.

A rectangular deck shape with six columns was selected for the large unit in order to decrease deck free span as well as to keep a long distance between the living quarters and the hazardous workover and process areas.

For the small vessel a nearly square upper hull has been developed because of the reduced number of equipment, compared with the large unit. Only four columns support deck in this configuration due to the decreased free span.

Configuration A - Principal Characteristics

Length over all (Pontoon length)	78.00 m
Breadth moulded	62.40 m
Length between FWD and AFT Columns center lines	66.00 m
Length between port and STBD Columns center lines	50.40 m
Depth to upper deck	43.20 m
Depth to lower deck	37.20 m
Operating draft	22.20 m
Survival draft	22.20 m
Pontoon Height Moulded	7.20 m
Pontoon width moulded	15.00 m
Large column section	12.0x12.0 m2
Small column section	9.0x9.0 m2
Displacement	35789 .0 m

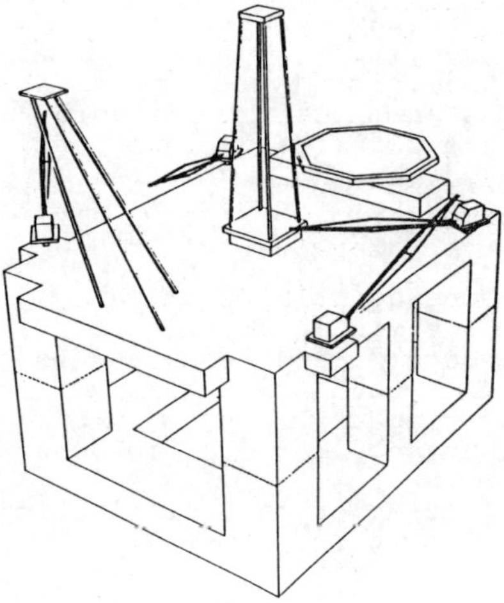

Configuration A

Configuration B - Principal Characteristics

Length over all (Pontoon length)	64.80 m
Breadth moulded	62.40 m
Length between FWD and AFT Columns center lines	52.40 m
Length between port and STBD Columns center lines	50.40 m
Depth to upper deck	40.20 m
Depth to lower deck	34.20 m
Operating draft	19.20 m
Survival draft	19.20 m
Pontoon height moulded	7.20 m
Pontoon width moulded	12.00 m
Large column section	12.0x12.0 m2
Displacement	25364.0 m

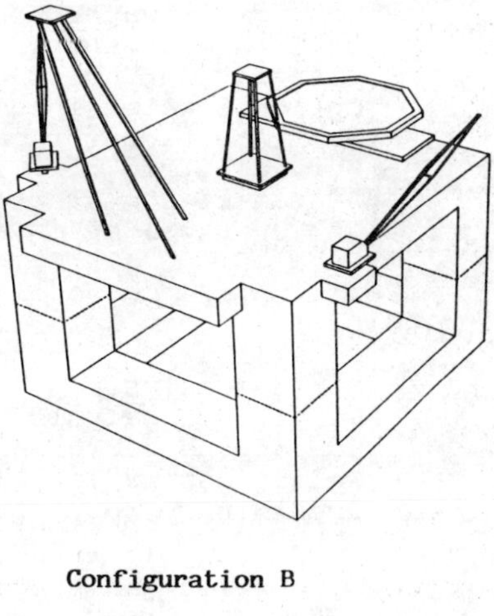

Configuration B

Fig 4 - Overall Platform Configurations

Deck structure in both vessels consists of box and basket type bays. Basket type modules are located in the workover rig and process plant bays where natural ventilation is required, while box types are in the rest deck area. Box type elements as well as deck double bottom provide reserve buoyancy and prevent total vessel loss in case of severe damage.

A single draught aspect for operation as well as for survival condition has been followed. This is a major advantage of the design because both platforms can operate even in the worst 100 years environmental conditions with consequential shut down time to be reduced to zero. Based on this single draught principle, airgap remains constant for all weather conditions and is fixed at 15.0 m.

Both vessels also have been designed to satisfy the stringent rules of N.M.D. and D.N.V. after loss of column buoyancy between pontoons top and deck bottom.

4.1. Monotype Configuration A :

It includes :

- Either flexible risers or stiff risers in bundle ;
- Oil /gas / water separation ;
- Oil and gas export through sea line ;
- Gas flaring during operation transition periods ;
- Water injection ;
- Gas injection (or gas lift) ;
- Workover rig ;
- 100 men living quarters.

Light ship weight is estimated about 15 000 tonnes.

4.2. Monotype configuration B :

It includes :

- Either flexible risers or stiff risers in bundle ;
- Oil /gas / water separation ;
- Oil and gas export through sea lines ;
- Gas flaring during operation transition periods ;
- Optional water injection ;
- Gas injection (or gas lift) ;
- Servicing rig ;
- 50 men living quarters.

Light ship weight is estimated about 10800 tonnes.

For these two models, full evaluation has been made, including :

- General arrangement drawings,
- Stress analysis,
- Stability analysis,
- Seakeeping performances,
- Weight estimation,
- Preliminary cost appraisal.

5. CONCLUSION

The experience and material gathered by the Association in the field of Column Stabilized Production Platforms constitute a valuable asset. Furthermore, the Association developed some unique concepts, which although not original enough to claim patent, are proprietary information.

Among other things, we emphasize the following features :

- The steel structural pattern providing a significant weight saving and a high level of standardization.
- The " zoning " which help to improve productivity in construction and operation.
- A highly professional arrangement of key functions taking after the best of up to date industry experience.

The result of the Research work is an array of predesigned systems, consistent with each other, which can be put together as necessary into an overall platform monotype, to meet any set of specific requirements.

In addition the proposed typical monotype designs, whose main purpose is to show how the construction set works, encompass by themselves many breakthroughs.

Such a tool is prone to generate savings when the case occurs to undertake a new Oil and Gas offshore field development project, using Column Stabilized Production Platforms.

For Operators, it may be particularly useful for preliminary design and cost assessment, then development of basic design and specification with a view to inquire for the construction of a new specific platform, and ultimately let contract to build it.

For builders it may be useful to make preliminary construction magement plans, cost assessment, and get ready to offer an off the shelf design to potential Operators.

It is more than obvious Column Stabilized Production Platforms will be a masterpiece to offshore field developments in the future, while more and more brand new platforms will be involved, as opposed to converted ones.

This Research Work is on time.

DETERMINATION OF IN-SITU STRESSES FOR THE DESIGN OF HYDRAULIC FRACTURES: A THREE-YEARS STUDY

J.P. SARDA, P.J. PERREAU and M. BOUTECA
Institut Français du Pétrole, Rueil-Malmaison, France

P. CHARLEZ
Total-CFP, Paris, France

J.L. DETIENNE
Elf Aquitaine, Pau, France

Summary

The design of hydraulic fractures for oil and gas wells stimulation requires a determination of the minimum horizontal stress in the reservoir and in the adjacent layers. Existing and new methods have been used and compared to interpret 13 microfracturing and prefracturing tests in deep wells. The new methods include a numerical analysis of the fall-off curve, an analysis of the flow-back curve, an inversion of the whole pressure curve. The bases of the microfracturing technique have been analysed by means of an experiment on a large size block. Measurements on cores for stress determination (ASR, DSCA, core discing analysis) have been completed and analysed. Some recommendations are given for the completion and interpretation of stress tests. The minimum horizontal in-situ stress appears to depend on the reservoir pressure.

1. INTRODUCTION

Hydraulic fracturing is the main stimulation technique in petroleum industry. It consists in the development and propping of a very large fracture which afterwards constitutes a high permeability area in the reservoir formation. It is generally admitted that this fracture is mainly of a tensile type: this means that it propagates perpendicular to the least principal in-situ stress. It is also well-known that at depths of interest for oil production the hydraulic fractures are vertical. Consequently, in deep rocks the least principal stress appears to be horizontal and the fracture propagates perpendicular to the least horizontal stress.

Mine-back experiments have shown that the variations of this stress with depth have a major influence on the vertical development of the fracture. A low horizontal stress favors the vertical propagation whereas a high horizontal stress acts as a mechanical barrier. This is illustrated (Fig. 1) by computer results (1). Three stress contrasts between the overburden layer and the formation are presented: 0, 2MPa, 10MPa. Such simulations show that for practical purposes the stress contrast should be evaluated with a 1MPa precision. A wrong evaluation can be very detrimental to the operation: risk of sand-out, possible rupture of the water-bearing zone.

Methods for the determination of stresses in deep holes have been studied for three years. The analysis includes: evaluation of existing methods, test of new methods. The research has been conducted along three ways: interpretation of microfracturing and prefracturing tests carried out on wells, interpretation of hydraulic fracturing tests carried out on a large model-block, interpretation of measurements made on cores extracted from the wells.

2. FIELD TESTS

Microfracturing, prefracturing

Starting from the mechanics of tensile fractures it is natural to look for a relation between downhole pressure at the fracture level and minimum horizontal stress at the same level.

This relation can be written as:

$$p(t) = \sigma_h + \epsilon(t)$$

$\epsilon(t)$ is positive during fracture propagation, positive then eventually negative after shut-in and pressure-decline caused by fluid filtration through fracture faces (fall-off test), positive then negative if the fluid is produced at constant rate (flow-back test). It is assumed that, when the fracture closes up, then $\epsilon(t)$ is equal to zero and the pressure p_{cl} is equal to σ_h. The aim of a stress evaluation method is to determine this point.

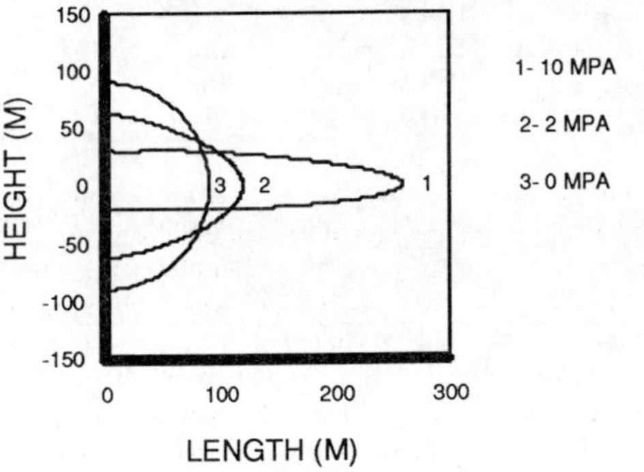

Fig. 1 - Fracture Height versus Stress Contrast

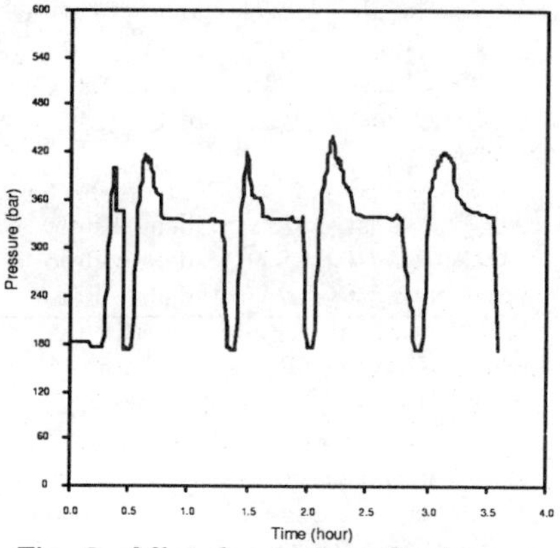

Fig. 2 - Microfracturing Cycles

A first option is to maintain $\epsilon(t)$ as low as possible. This is the idea behind the microfracturing test: low flow rate and low-viscosity fluid to maintain low pressure drop in the fracture. However in the prefracturing tests, carried out in the reservoir, the fluid, flow rate and volume are comparable to those of the actual hydraulic fracturing operation. So $\epsilon(t)$ can be large and has to be evaluated in the propagation, fall-off or flow-back phases.

13 downhole and surface pressure measurements have been completed on 8 wells located on 5 different fields. On 3 of these wells the measurements were performed in the overburden and at the reservoir level. Are indicated in table I: the type of test (microfracturing or prefracturing), the type of pressure decline on which interpretation was performed (fall-off or flow-back), the complementary measurements on cores, the borehole imaging logs. The results are summarized by the value and eventually the direction of the in-situ minimum horizontal stress. Various completion techniques have been tested to implement the microfracturing tests: straddle-packer and injection in open hole, packer in a casing and injection in open hole, injection through a perforated casing.

Fall-off tests

Fall-off test is the pressure-decline test most widely used for interpretation. Actual measurements and modelization show that the main features are closely related to fluid filtration. If fluid filtration is very low, one gets pressure cycles (Fig. 2) on which remarquable points can be easily identified: pressure linearity-limit, maximum or fracturing pressure, ISIP (Instantaneous Shut-in Pressure). If fluid fitration is higher (Fig. 3), the ISIP cannot be clearly identified. In all cases the closure pressure p_{cl} does not appear on the curve and therefore has to be determined through an analysis.

In the overburden the pressure decrease rate is often very low and this pressure will not

Well	rock	depth (m)	test (frac)	pres. test	cores meas.	logs	well pres. (bar)	min. stress (bar)	stress dir. °
A	shale	2145	micro	fo-fb	ASR		220	340	
A	sndst	2160	idem	fo	ASR		225	258	
B	shale	1889	idem	fo	ASR	EVA	178	355	N70E
B	sndst	1917	idem	fo			190	232	
C	shale	1282	idem	fo	ASR, DSCA	FMS	135	180-195	N45E
C	shale	1296	idem	fo	discing	BHTV	136	176-188	id.
C	limst	1328	idem	fo	ASR, DSCA	FMS	139	160	N45E
C	limst	1335	idem	fo	discing	BHTV	140	160	
D	limst	4549	idem	fo	ASR, DSCA discing	FMS	557	1294 ?	
E	limst	1831	pre	fo			178	197	
F	sndst	1898	idem	fb-fo			160	185-200	
G	sndst	1899	idem	fb-fo			160	195-202	
H	sndst	3253	idem	fo-fb			318	410	

Table 1 : Field tests and results (sndst: sandstone, limst: limestone, fo: fall-off, fb: flow-back)

reach p_{cl} during the test. So the only available data is the ISIP and the difference $(ISIP - p_{cl})$ calls for an evaluation. Numerical computations of ISIP have been made (Fig. 4). Due to fracture toughness, this difference can reach 1MPa if the fluid is a brine, and be higher if the fluid is a viscous one, so brine is highly recommended for microfrac tests in impermeable layers.

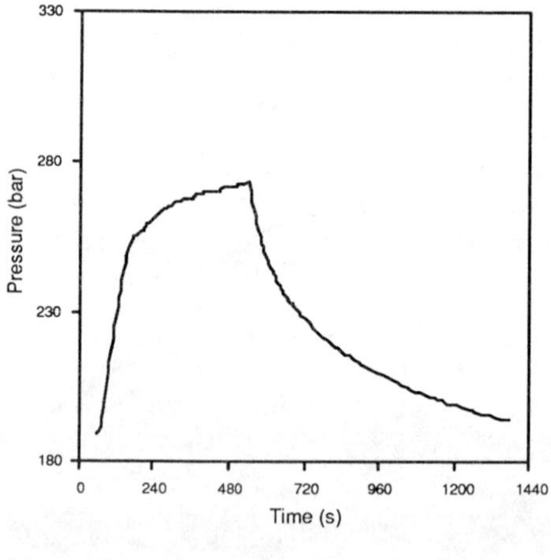

Fig. 3 - Fall-off in reservoir layer

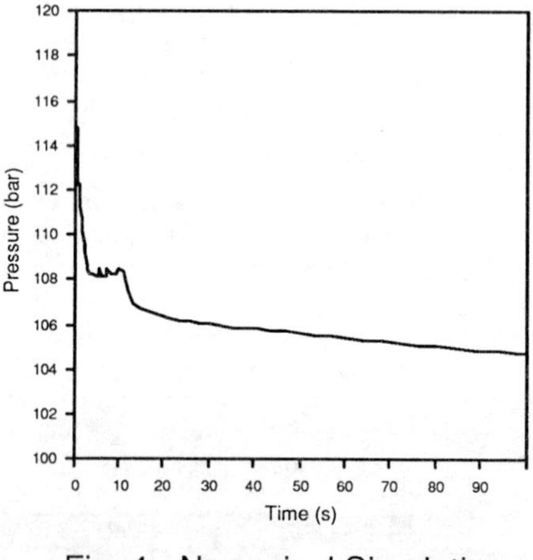

Fig. 4 - Numerical Simulation of Fall-off in the Overburden

When pressure decreases down to and under the closure pressure, particularly in reservoir formations, the analysis depends on the pressure domain considered. If the fracture is still open, p_{cl} is analysed as the lower limit, for pressure, of the regime of fluid filtration from the fracture to the formation: this is typically the Nolte analysis, or the $(p, \sqrt{t})$ analysis, pressure decreases from ISIP to p_{cl}. As soon as the fracture is supposed to be closed, p_{cl} is analysed as the upper limit of a matrix filtration regime, pressure decreases from p_{cl} to the formation pressure. This is typically a reservoir problem and p_{cl} is set at the beginning of the linear part of the Horner-plot (Fig. 5).

This analysis has been applied to cases A, B, C, F, G, H. On well B (Fig. 2), five pressure decline phases lead to very close values of the closure pressure. On well C completion problems explain the scattered values at levels 1282 and 1296: i.e. fluid filtration at the packer level in these open hole tests. On well D, a careful fluid balance analysis based on digital pressure recordings, supports the hypothesis of a hydraulic fracture. However this fracture would be of limited extent and there is no evidence of a vertical fracture on the FMS log. Consequently the proposed stress value cannot be ascertained. On wells F and G, corresponding to reasonably comparable well conditions, the stress values obtained from fall-off interpretation are quite similar (195 and 200 bar).

Flow-back test

This type of test takes place immediately after the propagation phase of the fracture. At this time injection is stopped and production at constant rate is established. There is a fast pressure relief, which, in actual industrial situations, eventually constitutes an advantage. An interpretation is necessary to identify the closure pressure.

The particular interpretation proposed is based on the analysis of flow regimes in a fracture first producing, then closed. As long as the fracture is still open, three flow regimes are possible: linear flow, bilinear flow, pseudo-linear flow. The corresponding pressure laws are linear functions of t^α, α values being respectively 1/2, 1/4, 1/2. However the actual slopes should differ from the theoretical ones since the fracture is closing during the test. After fracture closes-up the possible flow regimes are pseudo-radial regime and pressure relief in the tubing string, and those regimes should be observed.

Taking reservoir pressure as the reference pressure and shut-in time as the reference time, the representation $(log(P - P_0), log(t - t_{si}))$ shows, at the end of the time interval, a linear part with a slope ranging from 0.8 to 0.95. This is identified as a capacity effect eventually backed up by a small production, and this means that the fracture is closed.

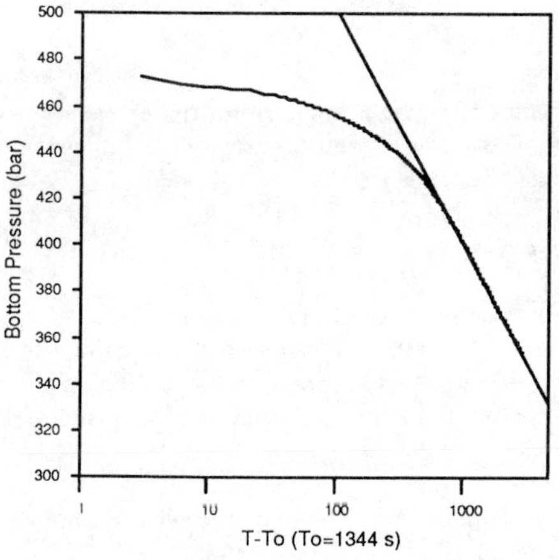

Fig. 5 - Well H. Fall-off after Prefrac.

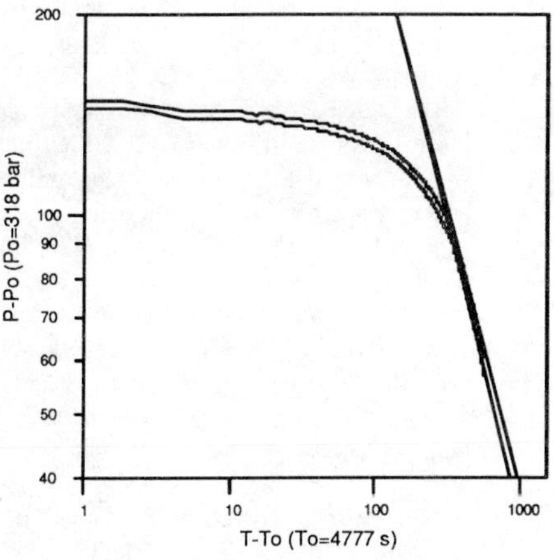

Fig. 6 - Well H. Flow-back

This is quite clear on well H (Fig. 6). The curve shows an angular point: just before this point the slope is 1/2, after this point the slope of the linear part is 0.95. So that the closure pressure is defined very precisely.

On this well the interpretations of the flow-back and fall-off (Fig. 5) tests lead to nearly identical values of the closure pressure. On wells F and G there are small differences between the results of both methods. Finally, although based on very different well tests, both methods lead to very comparable values of the closure pressure, and this is in favour of the mechanical soundness of the notion of closure pressure.

Borehole imaging logs

Borehole imaging logs were systematically run on well C, then, a detailed analysis of the results has been conducted. On this well the best images are provided by the FMS tool, and they show two types of wellbore ruptures: tensile ruptures induced by hydraulic fracturing in limestones and shales, shear break-outs in shales. The directions corresponding to tensile fractures and break-outs are almost perpendicular. These directions lead to the same interpretation: they indicate a minor horizontal stress oriented N45E.

3. TESTS ON A LARGE SIZE BLOCK

To check the efficiency of hydraulic fracturing tests for the determination of stresses, a series of experiments has been carried out on a large size block (1.8 x 1.8 m). A cutting machine was used to draw a block at the bottom of a quarry (Fig. 7). Then the four vertical faces of the block are loaded by means of large rubber flat jacks. A vertical borehole is drilled in the center of the block and a straddle packer (72 mm diameter) equipped with a pressure transducer allows to pressurize a 48 cm long hydraulic chamber.

To initiate a vertical fracture a couple of parallel flat jacks is pressurized and some fluid is injected into the block. The result, checked using an impression packer, is a vertical fracture parallel to the applied loading. Given this initial fracture, a number of fracture reopening experiments are performed under various conditions: several stress states, several flow rates, two different fluids (water and viscous oil). The results are summarized below.

First of all, the characteristic pressures (peak pressure, propagation pressure, ISIP) do not depend on the major horizontal stress σ_H. The peak pressure, in particular, is not at all dependent on a non-fractured elastic state of stress, as it is often admitted: it contains no information on σ_H.

On the contrary, the pressures, particularly the peak pressure and the propagation pressure, are very sensitive to the minor horizontal stress σ_h. The propagation pressure is always much higher than the minor stress which intervenes as an additive constant. The propagation threshold (propagation pressure at zero minor stress) depends on the flow rate (Fig. 8): 19 bar for a 8.5 cm^3/mn flow rate, 22 bar for a 17.5 cm^3/mn flow-rate. The ISIP evaluated through a log-log treatment also overestimates σ_h but to a smaller degree than P_p. Generally, ISIP is not so well correlated to σ_h as P_p.

After this series of reopening tests, the block was extracted and sliced in order to make out a drawing of the fracture. The result is very consistent: the fracture is pseudo-parallel to a block face, it is well contained towards the block upper face, which corresponds to the stressed zone. The final average fracture dimensions are: maximal height 1.5 m, maximal length 0.9 m, mean height 1.3 m.

Interpretation: inversion of the pressure curve

Interpretation of hydraulic fracturing in terms of minor stress depends therefore essentially on the knowledge of the pressure loss in the fracture, and on the appraisal of the rupture process at the tip of the fracture. A new interpretation method, based on mathematical inversion, has been proposed recently (3). Assuming a model for the fracturing process (for example

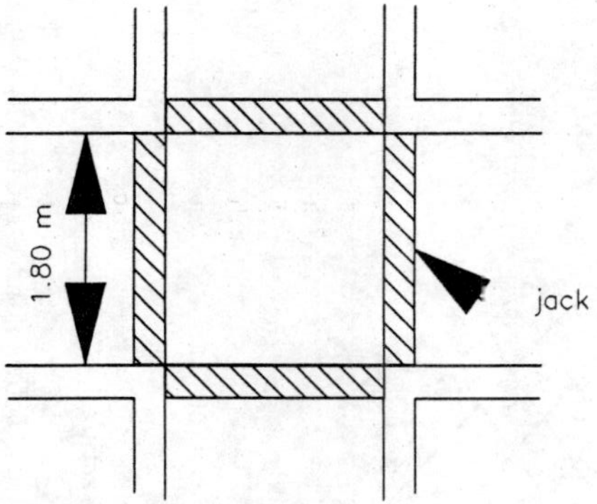

Fig. 7: Block and stresses.

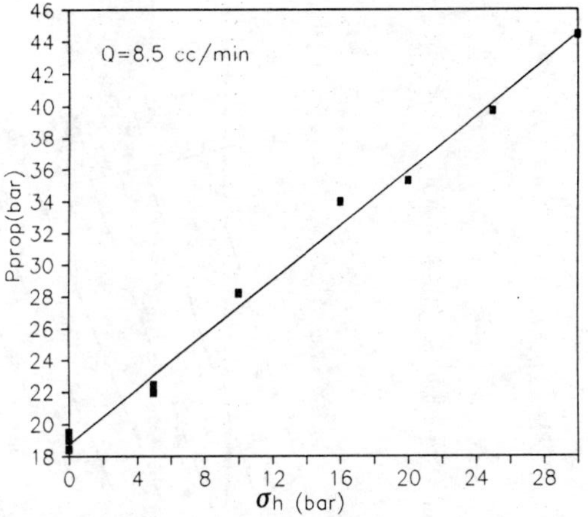

Fig. 8: Propagation pressure vs σ_h

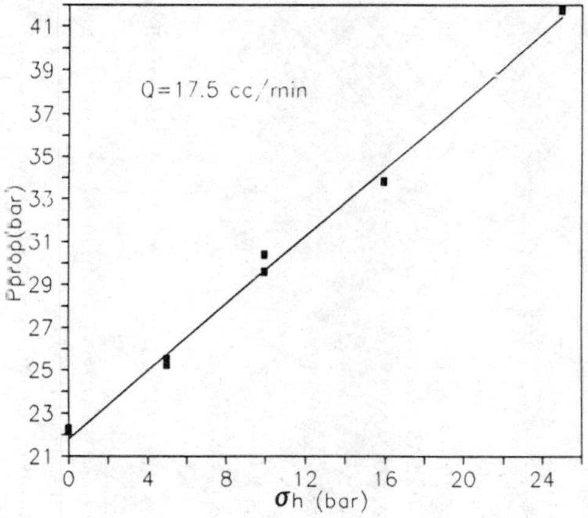

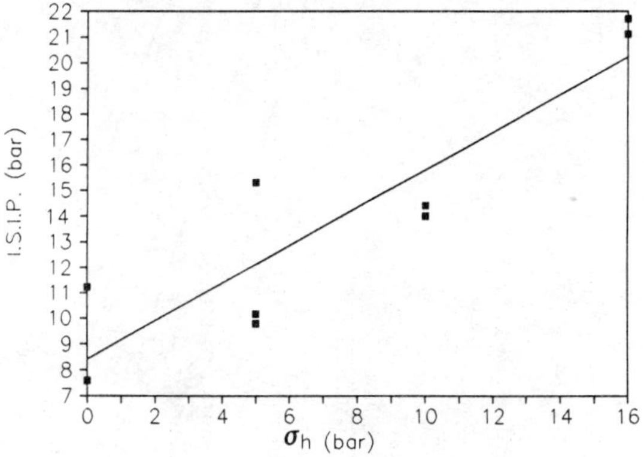

Fig. 9: ISIP vs σ_h

Results	Propagation pressure (bar)	Length (m)	Height (m)
experimental	29.5	0.9	1.5
computed	27.4	0.7	1.33

Table 2 : Tests on the model-block: measured and computed results

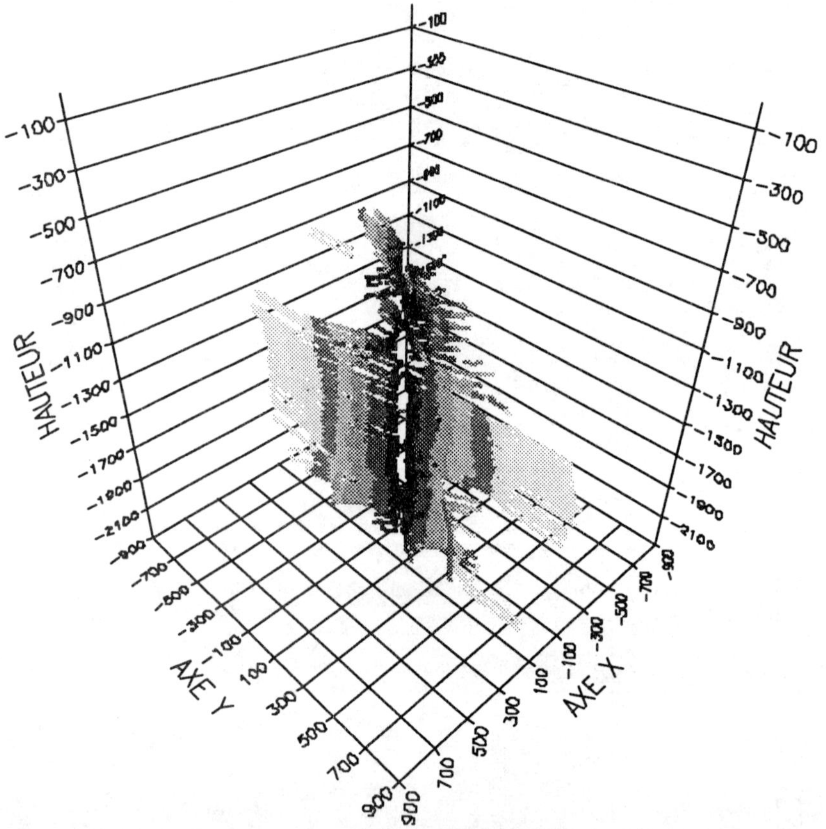

Figure 10 Geometry of the fracture

Perkins-Kern, Geertsma-De Klerk...), the method consists in determining a set of unknown parameters for which the actual pressure fits the calculated one. Minor stress and fluid loss coefficient are the two main parameters, this is in fact an optimisation process.

A tentative inversion was performed using the FRANK model (1). For the considered example the comparison between experimental and theoretical results are summarized in table 4.

The computed values are obtained considering a 20 $bar\sqrt{m}$ fracture toughness. This very high and apparently unrealistic value can be explained for example on the basis of the apparent cohesion concept (5). It can also be noted that, at this very low flow rates, the rupture surface developed is not a unique fracture but a bunch of nearly parallel fractures (see Fig. 10), so the actual length of the fracture front is much longer than the theoretical length and the apparent fracture toughness is higher than the actual one.

However the pressure loss in hydraulic fractures is underestimated in most cases. As a good evaluation of the minor in-situ stress is strongly dependent on the choice of the pressure loss law, an integration of more complex fluid mechanics concepts such as turbulency or tortuosity seems to be a good scope for research in the future.

4. MEASUREMENTS ON CORES

As mentioned in table 1, measurements and analyses have been achieved on cores extracted from wells A, B, C, D.

The principal characteristics of the strain recovery measurements are (4): core placed in a thermoregulated oil-bath, deformation of the core measured using 6 horizontal transducers and one vertical transducer. For differential strain curve analysis, the deformations of the six faces of a cube under isotropic loading have been measured using 30 transducers (5 on each face). For core discing analysis the thicknesses of the discs are measured and their shape is characterized.

Usually the rock behavior is associated with opening microcracks during ASR, with closing microcracks in DSCA experiments. For ASR, stresses are supposed to be proportional to terminal deformation. For DSCA, the actual stresses are supposed to be at the limit between the non-linear part and the linear part of the stress-strain curve. Stress criteria and the shapes and sizes of the discs are used to evaluate the values and directions of the stresses which provoke discing.

On well B the direction of the hydraulic fracture was visible on a core taken after fracturing. This direction coincides with the direction of maximum stress obtained from the interpretation of ASR measurements. The values of minimal, intermediate and maximum stresses provided by the interpretation of DSCA measurements are reported on table 3 for wells C and D.

The method also provides an orientation of these stresses. The minimum stress value on

Well	depth (m)	minimum stress (bar)	intermediate	maximum
Well C	1330	171	218	289
Well D	4549	977	1182 (vertical)	1328

Table 3 : Stress values from DSCA measurements

Well	depth	ASR	DSCA	core discing analysis
Well D	4549	N90E	N70E	N15E

Table 4 : Direction of the minimum stress: 3 results.

well C is in fact confirmed by microfracturing measurements (160 bar, see table 1). However, on the same well, the orientations of horizontal stresses clearly indicated by the borehole imaging analysis are only partially confirmed by ASR and DSCA analyses. On both methods, there is a tendency to exchange the directions of the maximum and minimum stresses. Rock fabric and natural fractures seem to take part in this result. On well D three directions are obtained for the minimal stress when the three methods are used. They are indicated in table 4. The only possible conclusion is an indication of the quadrant containing this stress direction.

In brief, if good orders of magnitude for the stresses can be reached through measurements on cores (especially by the DSCA method) the stress-direction problem requires further research work. The coring process should be taken into account as well as the fabric and natural fissures of the rock.

5- STRESS VALUES

On these wells the stresses are larger in the impermeable layers than in the reservoir or permeable layers. This tendency would not be necessarily true in other basins. The stress contrast between overburden and reservoir is high on wells A and B, located on depleted reservoirs. The stress contrast is low on well C, where the limestone layer is not produced. So that reservoir pressure seems to influence the minimal horizontal in-situ stress. This influence can be schematically explained while assuming an oedometric-type behavior for the reservoir rock (6).

6- CONCLUSIONS

In various microfracturing and prefracturing tests the analyses of the fall-off and flow-back tests lead to comparable values of the fracture closure pressure considered as the minimal in-situ stress.

Tests on a large size model-block confirm the influence of this minimal in-situ stress and its tight link with the reopening and propagation pressures.

Orders of magnitude of the in-situ stresses are accessible by measurements on cores, especially DSCA measurements. However for stress directions, work is still to be done to elucidate the possible influence of rock fabric and natural fractures.

The minimal horizontal stress appears to depend on the reservoir pressure.

Acknowledgements: This work has been performed on behalf of ARTEP (Institut Français du Pétrole, Total-CFP, Elf Aquitaine). It has been partially funded by EEC (DG 17 and DG 12).

REFERENCES

(1) BOUTECA, M.J., (1988). Hydraulic fracturing model based on a 3D closed form: tests, analysis of fracture geometry and containment. SPE/PE, Nov. 445-454.

(2) ECONOMIDES, M.J., NOLTE, K.G., (1987). Reservoir stimulation.

(3) CHARLEZ, Ph., HERAIL, R., DESPAX, D., (1988). Détermination de paramètres de fracturation hydraulique par inversion des courbes de pression. Revue de l'Institut Français du Pétrole. Vol. 43, n⁰ 1.

(4) PERREAU, P.J., HEUGAS, O., SANTARELLI, F.J., (1989). In-situ stresses evaluated from measurements on core samples. European geothermal update. Kluwer academic publishers.

(5) CHARLEZ, Ph., HERAIL, R., DESPAX, D., PERREAU, P.J., (1989) Simulation d'une fracture hydraulique sur un bloc de grandes dimensions. Rock at great depth, Balkema publishers, Vol. 2, pp 983-989.

(6) BOUTECA, M., SARDA, J.P., (1990) Fluid flow in porous media and related rock mechanics problems. 2nd European Conference on the Mathematics of Oil Recovery. Arles, Sept 11-14.

THE FORGED NODES APPLICATION
FOR OFF-SHORE STEEL JACKETS

F. Nicolussi, L. Pellizzari
Tecnomare S.p.A., Venezia, Italy
W. Storesund
Veritec A/S, Hovik, Norway

ABSTRACT

The paper describes the main results obtained in a research project managed and sponsored by the Norwegian oil Company Norsk Agip (with the technical support of the parent company Agip S.p.A.) in cooperation with Saga Petroleum A.S., which proposed, in the recent past, a new configuration of nodes for off-shore steel structures /1/.

In these new nodes, the forged sections constitute smooth brace-chord transitions allowing the shifting of welds outside the hot-spot area and the reduction of the stress concentration.

The principal effect is a significant improvement of the fatigue life compared to the one of the traditional welded nodes.

The main topics of the project discussed in this paper are:

- the new design philosophy and criteria for the applications to off-shore steel jackets.
- the evaluation of the static and fatigue strength as obtained on the base of laboratory tests and Finite Element Analyses;

The results obtained in the project are very encouraging and a significant impact of the new component is expected on off-shore structures as light weight jackets and compliant steel towers.

INTRODUCTION

As oil and gas exploration and production are moving into increasingly deeper waters, fatigue is becoming a more dominant factor in platform design. In conventional steel platform design the welded tubular joints are the critical components with regard to fatigue behaviour.

The traditional joint is combining two generally negative factors as far as the fatigue properties are concerned:

- a sharp transition with high stress concentration factors between chord and brace,
- a weld located in the area with the highest stress concentrations.

The idea underlying the forged node concept is consequently to reduce the stress concentration and to remove the weld from the critical zone.

THE FORGED NODE

Forged nodes (see Fig. 1) are composed of:

- forged segments which constitute the chord-brace transitions: each element includes, in a single piece, the entire stub and the portion of the chord shell adjacent to the stub-chord intersection;

- a rolled steel element (chord) in which the different forged segments are inserted and then welded.

Each forged segment forms the brace to chord transition in a single piece and consequently the welding operations are avoided in the "hot-spot" area. This transition has a smooth profile and a gradual decrease of the thickness from the chord to the stub, the final thickness being reached only at the end of the stub.
In the chord to stub intersection the plug is absent and the chord has then a hole at each brace connection.
The *advantages* of the forged nodes over the traditional welded ones are (see Fig. 2):

- the Stress Concentration Factors (SCF) are lower for forged nodes than for welded nodes as a result of the more regular transition between the brace and the chord,
- the forged node has a significantly better fatigue behaviour than the traditional nodes as the welds are moved from the position where the maximum stresses are reached to positions where they are significantly reduced.

An *additional advantage* of the forged nodes over the traditional ones lie in the welding process: for the new nodes automatic procedures are applicable to all the shop welds. This not only reduces the cost of the welding operations but also produces quality welds. The brace to stub welds, which are performed in the assembly yard, may be carried out by double-sided processes with additional advantages in the fatigue strength of the nodes. This is generally possible due to the presence of the holes in the brace to chord intersections, which allow the welding operations to be done inside the stubs.
Some *geometric limitations* must be satisfied by the new nodes, due to: fabrication procedure, characteristics of the single forged segment and overall node configuration. The main limitations on the geometry of the single forged segment are expressed as follows:

- angle between brace and chord: alfa = 40° - 90°
- brace dia. to brace thk. ratio: d/t = 20 - 28
- brace to chord thickness ratio: t/T = 0.45 - 0.60

The other limitations are related to the overall geometry of the nodes: the cutting of the forged segments must be performed outside the smooth transition between brace and chord in order to limit the SCF in the welds and to have a sufficient space for the execution of the welding operations. This restriction affects the gap between two adjacent braces in the plane of the chord axis and limits the brace diameters as a function of the chord diameter and of the angle between two adjacent braces along the chord circumference.
The *fabrication* of the forged segments is performed utilizing special tools which are produced for each family of identical elements. The time schedule and the fabrication costs are significantly sensitive to the number of different tools needed and, consequently, efforts are requested in order to limit the number of families of forged segments.
The *use* of the forged nodes is recommended for substructures in mean to high water depths, where the nodes are sensitive to fatigue effects. The efficient and economical application of these nodes to off-shore platforms requires, in addition, the development of simple structural configurations and the

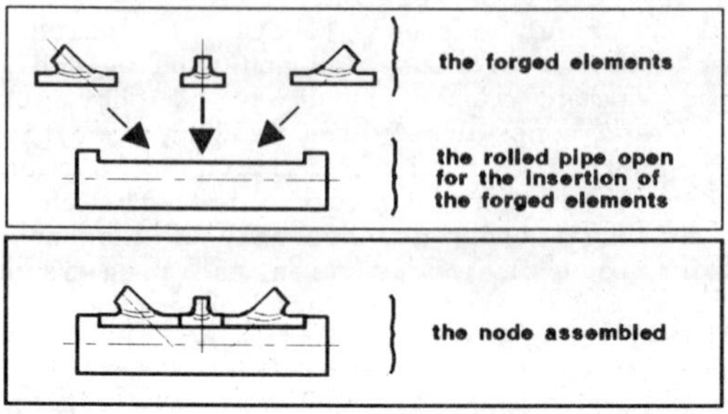

fig.1: definition of the forged nodes

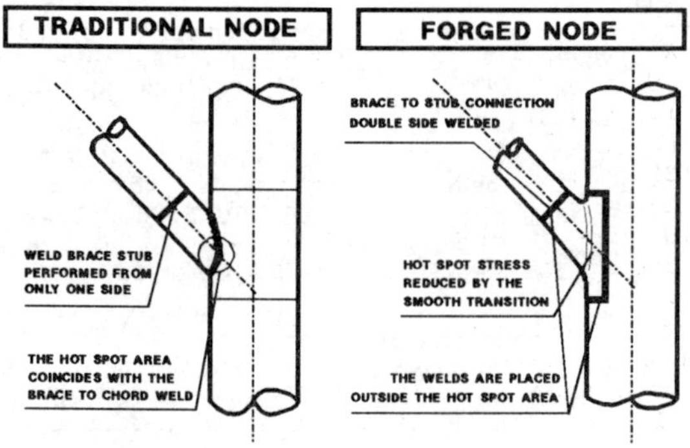

fig. 2: comparison between traditional
and forged nodes

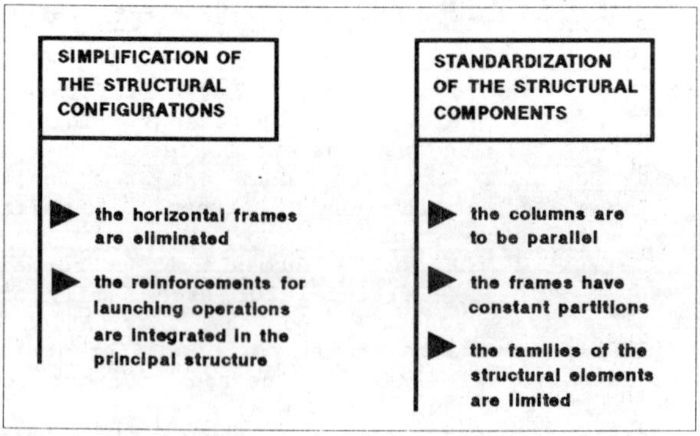

fig.3: requirements for the efficient
application of the forged nodes

achievement of a high degree of standardization of the main structural components (Fig. 3).

THE STATIC STRENGTH OF THE FORGED NODES

The static strength of the forged nodes was evaluated on the base of a laboratory test and of non-linear Finite Element Analyses (FEA).

The laboratory test was performed with the aim of calibrating the results of the FEA and the object of the calculations was the evaluation of the influence of different parameters on the strength of the nodes.

The computer analyses were carried out by use of the MARC computer programme: a four-noded thickshell element was selected to model the node and a combined non-linear behaviour of material and geometry was adopted.

The main results achieved are as follows:

- *calibration*: the non-linear analysis over-evaluates the strength of a K forged node loaded by balanced actions applied to the two braces of a 7% with respect to the results of a laboratory test;
- *thickness of the forged segment of a K node*: an increase of 33% of the segment thickness induces an increase of the static strength of the node of a 31%;
- *thickness of the forged segment of an X node*: an increase of 43% of the segments thickness corresponds to an increase of the static strength of the node of a 55%;
- *yield stress*: an increase of the yield stress of a 25% induces an increase of a 20% of the static strength of an X node.

Particular attention was dedicated to the comprehension of the effect of the complex (multiplane) node geometry on the strength of the forged nodes. For this purpose analyses were performed on simple (all the braces in the same plane) and complex forged nodes for axial forces applied to the braces and/or to the chord.

The main results are shown in Fig. 4 and a typical mesh used in the analyses is illustrated in Fig. 5.

THE FATIGUE STRENGTH OF THE FORGED NODES

The Stress Concentration Factor (SCF) is considered to be a parameter of major importance for the analytical evaluation of the fatigue behaviour of a given structure. A series of linear Finite Element Analyses were carried out by use of the SESAM computer programme in order to get a first set of SCF data: eight-noded thickshell elements were selected to model the node and no attempt was made to perform a particular description of the welds. The SCF were then evaluated by taking the ratio between the maximum principal stress and the nominal stress in the member. The linear analyses were carried out as follows:

1. FEA of an Y-Node to compare the theoretical and experimental results in order to calibrate the mesh and the procedure. The thickness of the forged segment was greater than the one of the chord plate.
2. FEA of the node in 1. with reduced thickness in the forged segment and reduced fillet radius in order to evaluate effects of fabrications tolerances.
3. FEA of the node in 1. with increased thickness in the forged segment.

CHORD AXIAL LOAD IN % OF COLUMN STRENGTH	TENSION ON THE BRACES: SIMPLE X NODE	TENSION ON THE BRACES: COMPLEX X NODE	COMPRESSION ON THE BRACES: SIMPLE X NODE	COMPRESSION ON THE BRACES: COMPLEX X NODE
0.	REF. VALUE	-2.1%	REF. VALUE	-2.3%
100.	-6.4%	-10.1%	-8.8%	-11.5%

fig.4: static strength of forged nodes; effect of chord compression

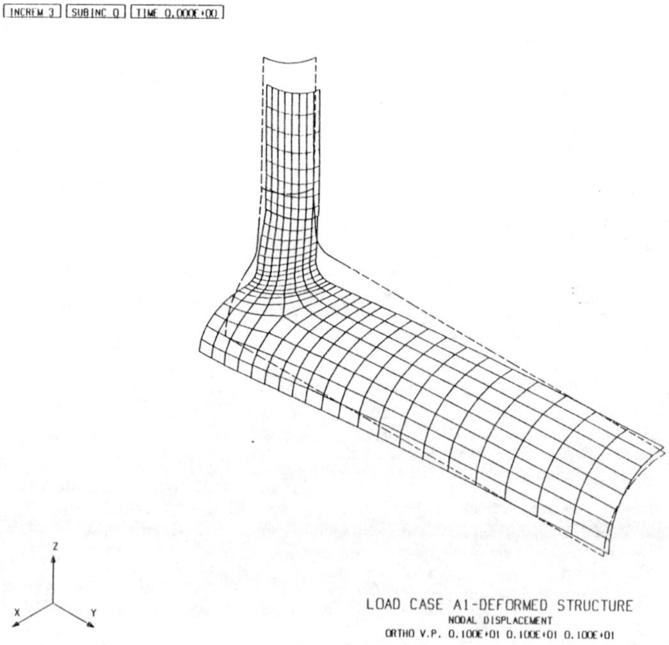

fig.5: typical mesh and failure mode for an X node in compression

4. FEA of a simple (plane) K-Node for two different chord thicknesses.
5. FEA of a complex (multiplane) KTK-Node with geometry as in 4.

The results of items 1,2,3 are summarized in Fig. 6 where only the percentages of difference with respect to the basic case are given. They refer to the maximum values on the three zones of interest for the fatigue analysis: brace weld, chord welds and fillet .

The results of items 4 and 5 are presented in Figures 7 and 8. In this case the calculated SCFs are given, being derived from Finite Element Analyses for the forged configurations and from analytical formulations (/2/, /3/, /4/) for the corresponding welded geometries. A first comparison is given in Fig. 7 between the SCFs of forged and traditional welded nodes, considering two node geometries (Y and K) and two different chord thicknesses. The further comparison between simple and complex forged configurations is summarized in Fig. 8 where only one chord thickness is considered and three nodal behaviour are analyzed (Y, K and Axial Chord), starting anyway from a K geometry for the simple node and from a KTK geometry for the complex one.

The main comments on the results obtained from the tests and the FEA above described are :

- a good tuning has been reached between the laboratory tests and the computer finite element models,
- the forged configurations show low sensitivity to the fabrication tolerances, both for thickness and for overall dimensions,
- the SCF related to the chord weld shows some sensitivity to the change in the chord thickness,
- an improved fatigue strength of forged nodes with respect to ordinary welded nodes is expected due to lower SCFs in the areas containing the welds,
- complex node geometry has no significant effect on the SCF at the brace weld when compared to simple node geometry,
- complex node geometry has an effect on the SCF in the fillet and chord weld area.

A new calculation procedure in now under development in order to face this last problem, in a way that is quite different from the standard off-shore practice: the computer procedure is sketched in Fig. 10 and it could be briefly summarized by the items below:

1. Execution of the jacket structural analyses following the well- established modelling criteria based on simple beam elements.
2. Execution of a substructure partitioning which allows the description of the complex nodes through a certain number of pre-defined "Superelements" representing the forged segments.
3. Execution of a detailed Finite Element Analysis of the selected nodes loaded by the values derived from the global calculations, in order to determine the state of stress in the points of interest.

<u>CONCLUSIONS</u>

- *Static Strength*

Both the test results and the FEA indicate higher static strength for forged nodes than for ordinary welded nodes. This

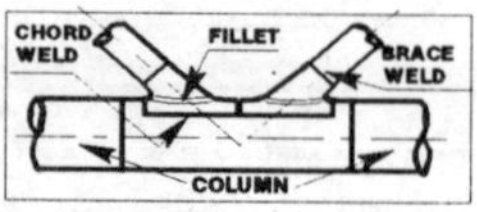

AXIAL FORCE IN THE BRACE

			BRACE WELD	FILLET	CHORD WELD
CALIBRATION	FEA - TEST		-7 %	-5 %	-21 %
FABRICATION TOLERANCES	SEGMENT THK	-6 %	0	+6 %	-2 %
	FILLET RADIUS	-50 %	0	+15 %	+1 %
CHORD	THK	+40 %	-14 %	-25 %	-50 %

fig.6: effect of parameters on the SCF

COLUMN 2800 * 60 CHORD 2800 * 120 *80* BRACES 1400 * 40		FILLET		CHORD WELD		BRACE WELD	
		120	80	120	80	120	80
Y	FORGED	4.49	6.67	3.11	4.57	2.31	3.26
	WELDED			3.89	6.59	3.45	5.16
K	FORGED	1.82	3.03	0.87	1.92	1.68	2.11
	WELDED			1.92	3.38	2.21	2.92

fig.7: comparison of SCF between forged and traditional nodes

COLUMN 2800 * 60 CHORD 2800 * 120 *80* BRACES 1400 * 40 (K) *1100 * 25 (T)*		FILLET		CHORD WELD		BRACE WELD	
		120	80	120	80	120	80
AXIAL CHORD	SIM.	1.75	2.37	1.32	1.83	0.72	0.72
	COM.	2.15		1.97		0.86	
Y	SIM.	4.49	6.67	3.11	4.57	2.31	3.26
	COM.	5.06		3.43		2.36	
K	SIM.	1.82	3.03	0.87	1.92	1.68	2.11
	COM.	2.73		1.36		1.68	

fig.8: comparison of SCF between simple and complex forged nodes

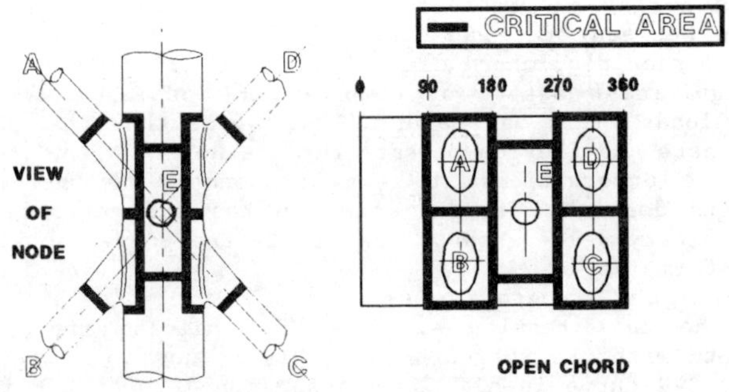

fig.9: critical area of a complex forged node

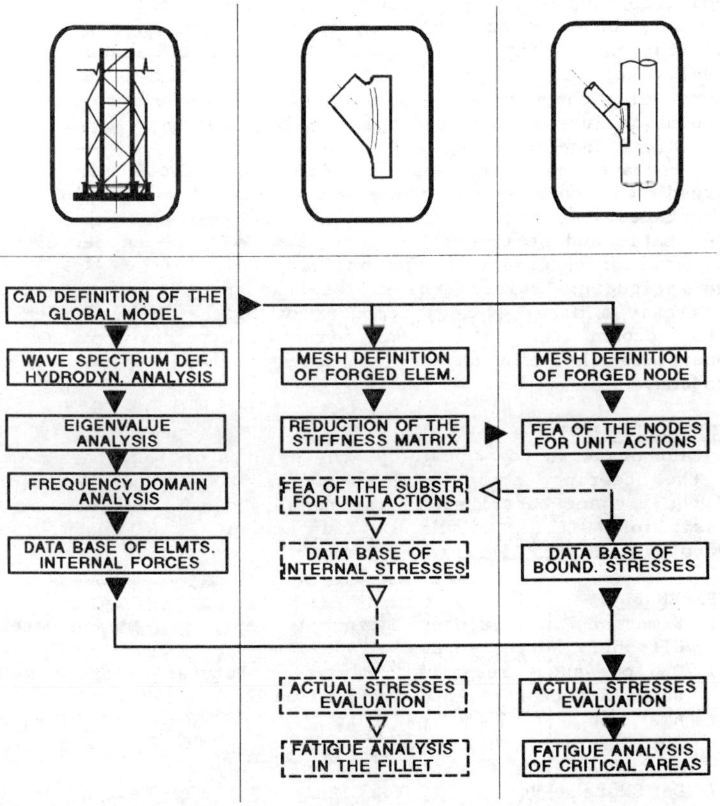

fig.10: flow chart of fatigue analysis procedure

may be due to the increase of brace diameter in the transition fillet between brace and chord giving an increase in the beta value (brace dia./chord dia.).
No significant effect of complex node geometry was found for brace loads. This may be due to the fact that the stubs act as stiffeners and they compensate for the holes in the chord.
On these observations it seems reasonable to base the static strength design of forged nodes on the same procedures in use for ordinary welded nodes, that is by the conventional punching-shear formulae.

. *Fatigue Strength*

Both the test results and the FEA indicate improved fatigue strength with the forged nodes: this is due to a combination of better S-N curve in high stress areas and lower SCF in the weld zones. No significant effect of complex node geometry was found for the brace weld . This may be due to the long stub which moves the brace weld away from the chord.
In the fillet and chord weld area, the complex geometry has a significant influence on the SCF. Also loads in the chord causes high SCF in this area. This is a natural consequence of the redistribution of stresses around the holes in the chord.
On these observations it seems reasonable to base the fatigue calculation of the brace weld on the brace loads only, as usually done for ordinary welded nodes. For the fillet and the chord welds the complex loading of the node should be taken into account as interaction effects are significant.

. *Final Considerations*

The research programme has confirmed the feasibility of the forged node concept and their suitability for off-shore steel structures.
The static and fatigue strength evaluations may be performed by the same procedures used for ordinary welded nodes, except for the fatigue analysis of the fillet area and the chord welds. In this case a more advanced computer procedure is needed to take into account the interaction effects: its developments will constitute the main object of the next phase of the research project.

ACKNOWLEDGEMENT

The authors wish to acknowledge the efforts of the other members of the Steering Committee which contributed to the development of the project described in the paper and particularly: Mr. P. Tassini of AGIP S.p.A., Mr. G. Eide and Mr. P. Simensen of Saga Petroleum, Mr. M. Gibstein of Veritec.

REFERENCES

/1/ Simensen,P.A.,"Future platform concepts for deeper waters", Off-shore Northeren Sea - 1985.

/2/ UEG off-shore research: "Design of Tubular Joints for Off-shore Structures", UEG Publication UR33 - 1985.

/3/ American Petroleum Institute (API): "Recommended Practice for Planning, Design and Constructing Off-shore Platforms", API RP2A, 16th edition - 1986.

/4/ Det Norske Veritas (DNV): "Rules for the Design of Fixed Off-shore Structures", 1977.

EUREKA EU-191 ADVANCED UNDERWATER ROBOTS: THE WIR TECHNOLOGY DEVELOPMENT PROGRAM

W. Prendin, D. Maddalena, A. Terribile
Tecnomare S.P.A., Venezia, Italy
T. Hedge
Ferranti ORE, Great Yarmouth, U.K.

SUMMARY

The Eureka EU-191 project: "Advanced Underwater Robots (AUR) for work/inspection and long range survey", is a joint cooperation effort between Italy and the United Kingdom aimed at developing two advanced underwater robots, namely: the WIR (Work and Inspection Robot), a semi-autonomous tethered vehicle aimed at substituting the divers in performing complex subsea manipulative tasks, and the ARUS (Autonomous Robot for Underwater Survey), an autonomous robot , rated for full ocean depth, capable of carrying out different scientific/industrial missions depending on the payload.

The first (of four) project phase, covering the robot conceptual design and the preliminary design of basic subsystems, has been completed.

This paper, after a general overview of the project, gives a specific description of the key technologies, related to the WIR, which will be developed during phase 2. Particular emphasis is given to the description of technologies such as telemanipulation, positioning, navigation control, working scene geometric modelling.

1. INTRODUCTION

The need to perform tasks in the undersea area has been increasing recently due to the demand for both a better scientific understanding of the complex phenomena relevant to the subsea environment, and for the exploitation of the enormous marine resources.

Until now the development of scientific activities and the exploitation of resources have been carried out by the human beings directly present on site. Safety aspects and operational costs of diving represent a barrier to the extension of man's activities in deeper water or in greater task complexity.

An initial step to solve such problems has been the development of manned submersibles and remotely operated vehicles (ROV). In particular the ROV's seem capable of economically performing simple underwater tasks such as visual inspection, diver assistance, rough cleaning of the structures, etc..More difficult tasks such as non destructive tests (NDT) or maintenance of subsea plants still require the presence of divers, due to the requirements for a high degree of dexterity in obstructed and complex environments.

As a matter of fact it is generally recognized that the present generation of ROV's and relevant telemanipulation systems is rather limited in areas such as man/machine interface, manipulator dexterity and control, access to work site, operational scenario geometric modelling and representation, management of unexpected or unplanned situations.

On the other hand, the field of underwater scientific research typically includes ocean surveys and the use of manned submersibles, requiring extensive on-site involvement of men and equipment and entailing the relevant costs.

The EUREKA EU-191 project aims at overcoming these limitations by a combination of advancement of available technologies together with the integrated application of on-board advanced robotic techniques. This approach allows for effective intelligent control of the systems in a number of operational scenarios with the aim of reducing, and ultimately removing, the need for human involvement in underwater work.

The project goals are the development of two advanced underwater robots namely: a Work and Inspection Robot (WIR), aimed at substituting divers in a wide range of underwater tasks, and an Autonomous Robot for Underwater Surveys (ARUS) able to perform long range deep water survey missions.

The project team is composed of a number of Italian and British organizations with responsibility for the WIR and ARUS being taken respectively by Italy and Great Britain and the system architect being respectively Tecnomare and Ferranti ORE.

The development programme includes four phases, as follows:

* Phase 1: Mission analysis and preliminary design
* Phase 2: Enabling technology research and subsystem development
* Phase 3: Prototype development and testing
* Phase 4: Design finalization.

Phase 1 began in 1988 and, for the WIR vehicle, terminated at the beginning of 1990, while for the ARUS project the end of phase 1 was in mid 1990. The estimated overall project duration is five years and the overall budget is roughly 60 million US dollars.

After a general overview of the two vehicles, this paper aims to describe the key technologies of WIR and their development plans.

Section 2 presents a general description of the two vehicles as per results of project phase 1, section 3 describes the WIR key technologies and section 4 draws the conclusions.

2. VEHICLE GENERAL DESCRIPTION

ARUS

The ARUS vehicle, illustrated in fig. 1, is a free swimming machine capable of travelling submerged to a work site, with a mission dependent payload, performing some tasks and then returning to base.

Fig. 1 The ARUS vehicle.

The planned capabilities are as follows:

* submerged endurance of 2000 km or 240 hours;
* full ocean depth (6000 m) capability;
* transit speed 5 knots;
* construction: single cylindrical pressure vessel made of composite material;

* intelligent data acquisition methods applied to survey sensors;
* high efficiency data storage brought about by the application of expert systems techniques to data reduction;
* on board decision making ability allowing navigation to target location, and performance of task without human intervention;
* standard equipment includes multiple beam side scan sonar, seabed sampling devices and collision avoidance sonar;
* payload capacity 500 kg;
* initial estimates of vehicle size 10 m long and 1.5-2 m diameter.

The navigation function, in conjunction with the Long Range Navigation sensors, will be capable of determining the vehicle position with an accuracy of better than 500 m after a transit phase of 1000 km.

The following tasks have been identified:
* site surveying
* bathymetric survey for charting survey
* pipeline or cable tracking
* sea bed classification
* placement of environmental monitoring packages
* gravity measurements
* search and location of sunk objects.

WIR

This robot will be a tethered semi-autonomous system capable of a wide range of tasks which normally require the intervention of divers due to the required level of dexterity.

The system, illustrated in fig. 2, is based on features typical of the more recent generation of ROV systems:
* a medium size, compact vehicle carrying an inspection "tool kit" and work tools selected and manipulated as needed;
* garage launched to the desired operating depth to reduce damage during launch/recovery operations;
* surface facilities for launch/recovery, power distribution and control.

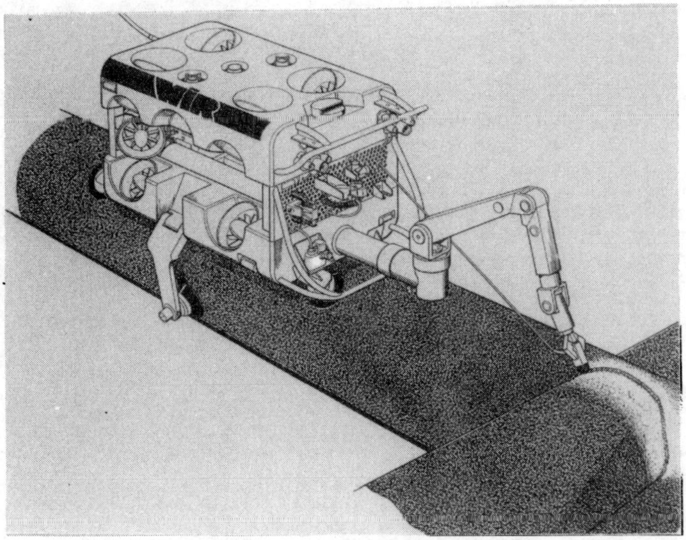

Fig. 2 The WIR vehicle.

Main specifications of the WIR vehicle are:
* Operating depth 300 m (extendible up to 3000 m)

* Dimensions 2050x1440x1260 mm
* Weight 15 kN (in air, neutral in sea water)
* Construction astable tethered vehicle, aluminium open frame, syntactic foam
* Propulsion 2 longitudinal thrusters (260 kg thrust)
  8 transverse thrusters (200 kg thrust)
* Access to narrow areas (e.g. member incidence down to 30°)

The real breakthrough in the WIR system relies on the proper integration and improvement of technologies, operational capabilities and special features considered necessary to replace the divers.

The key areas of development required to achieve this goal are:

* real time vehicle positioning system able to work properly even within an underwater structure, hence allowing a safe navigation to a selected work-site;
* automatic navigation and hovering control capability, to allow access to work-site with the most effective attitude and relative position required to deploy inspection or work tools where needed;
* advanced underwater manipulator system provided with both motion/force control and force reflection in the framework of a supervisory control system;
* work scene geometric modelling system.

The following section provides a description of the WIR key technologies to be developed in phase 2.

3. WIR KEY TECHNOLOGY DEVELOPMENT

Positioning system

Effective positioning systems capable of providing the vehicle position outside and inside the platform would allow safer navigation, less umbilical entanglement risks and shorter work site approaching times; furthermore a system capable of providing relative positioning with respect to the work site would make it possible to automatically keep the vehicle hovering and to make docking operations easier.

To measure the vehicle position the basic idea was not to use the acoustic positioning systems, such as S.B.L., S.S.B.L. or L.B.L., because they need the deployment of a transponder network and possibly (due to both acoustic shading and multipath effects) they do not guarantee the full coverage of all the volume inside a complex platform.

Other positioning techniques were analyzed, making reference to the various operative phases of the WIR robot: the approaching phase in which the vehicle leaves the garage and gets closer to the structure to inspect/maintain, the navigation near or inside the structure, the hovering vehicle phase and the docking phase.

On beginning the project, two approaches were selected as the most viable to solve the positioning problem, namely the sonar and the stereoscopic computer vision; the sonar seems more suitable outside the platform and during approaching phase while video positioning will be considered for close navigation, hovering and docking. Furthermore a large operative overlap exists between the two techniques which will provide a higher reliability level.

As per the sonar positioning system the possible human intervention has been considered; the functions of the navigation sonar will be:

* to determine the vehicle's position with respect to a structure by matching captured images to a previously stored knowledge base;
* to provide the operator with an image of a sufficiently high quality to enable high level control decisions to be made.

Incidentally it is worthwhile noting that, although ARUS and WIR are conceived to have different mission profiles and operate in different environments, some of the image processing and sonar transceiver functions are common to both.

The WIR sonar will be able to work with two resolution modes: a medium resolution mode (in the order of 1 m) featuring a maximum range of 300 m and a update frequency of 4 Hz, and a high resolution mode featuring a resolution in the order of 0.1 m with an update frequency of 2 Hz, but limited to a 50 m range. WIR would use the medium resolution mode to initially determine its position within a structure, and to give the operator a wide field when the vehicle is not in close proximity to any object. The high resolution mode could be initiated when the vehicle approaches an artefact, so that more precise information regarding the relative positions of the vehicle and the artifact could be obtained; furthermore the high definition mode could give the operator a 360 degree field of view when the vehicle is operating in a confined space.

The following processing steps will be applied to the acquired sonar images:

* preprocessing to minimize the effect of noise
* segmentation to divide the image into objects and background
* feature extraction to characterize the different objects
* object classification to distinguish between objects
* world model to determine the vehicle's position relative to the structure.

Since many of the processing algorithms may be performed in parallel, the use of transputer systems or digital signal processing devices (DSP) will be considered.

As per the video positioning system, this will be a computer vision system based on stereoscopic processing algorithms. The basic idea is simple (1): two TV cameras with parallel axes and a known distance between them are used to get two images of the same scene (fig.3). A point in the space is projected in different positions in the two images; by using the equation of fig.3 it is possible to find the coordinates of this point in a frame joint with the cameras. The video digital processor must find the corresponding points by pattern matching between the two images.

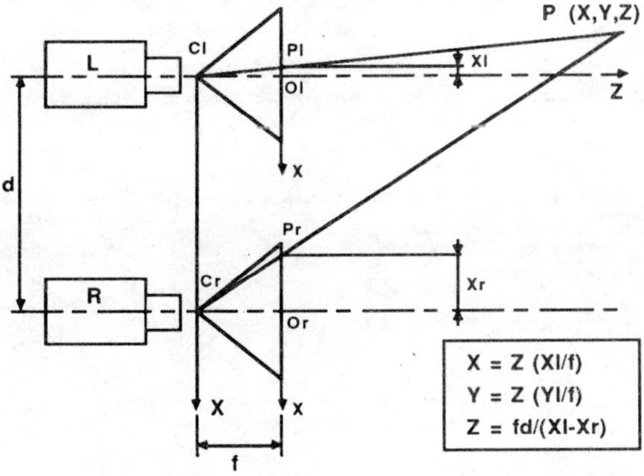

Fig. 3 The basic principle of stereoscopy.

A prototype of a system based on this measuring technique is already available in Tecnomare; such a system is able to provide 3D coordinate measurements of user selected points of the remote working scene imaged by a pair of stereoscopic TV cameras with a rate of 12.5 measurements/sec and an accuracy of ca. 5 mm in a distance range of 2 meters; furthermore the system can track the selected point thus providing coordinate measurements related to the same point even if it is moving. Also this feature is assured by a kind of image processing of a small area around the point being tracked, based on luminance pattern recognition.

This approach can be used for relative vehicle positioning if the measurements with tracking of a point on the scene are combined with measurements coming from attitude sensors, i.e. vehicle pitch, roll, and yaw with reference to an inertial reference frame; this approach is valid as long as the tracking of the same point is possible. If we expect to move along an underwater structure or along the sea bottom a target change is required every time the current one runs out of the visible scene. Since the new target is not known "a priori", it must be chosen by the computer vision system, referring to the position and attitude of the vehicle at the moment of change. This implies that, in the same time instant, both the old and and the new targets must be measured. If the position of the first point is known, this technique allows for an updating of the vehicle absolute position, even if due to the target change, an increase in the absolute positioning error is expected.

An analysis of error propagation has been made through simulation and confirmed by simple experiments of target change conducted in laboratory: the results of these trials showed a propagated error in the simulated vehicle position of only 1% (1 cm per meter of motion) after 14 target changes.

The major improvements relevant to this technology expected in phase 2 regard:

* a better accuracy (also improving the underwater TV cameras);
* the possibility of multi target tracking (to increase measurement reliability);
* automatic selection of most suitable targets (in the available prototype they are user selected).

To achieve these goals a great effort in algorithm and software development will be required along with the availability of a powerful video processor: similarly to the sonar signal processing, transputer and DSP technology will be considered.

Scene geometric modelling

As already pointed out, the WIR robot will be mainly aimed at inspection and maintenance of underwater structures and plants; therefore the working environment will be of a "structured" kind, i.e. "as built" reliable drawings of such structures are normally available In order to exploit such "a priori" information, WIR will be fitted with a subsystem capable of providing a geometric modelling of the working environment; this modelling will be usable for manipulation and navigation purposes and will be based on the automatic fit between a priori geometric knowledge and current measurements.

In detail this subsystem (also called scene reconstruction subsystem) will be a computer vision system in charge of the following tasks:

* position definition of the vehicle during navigation close to the structure;
* exact definition of the docking point during the docking phase;
* definition of the manipulator base position and attitude, when the vehicle is docked
* for all tasks the subsystem will provide a parameter related to the reliability of output data and, in particular as regards docking and manipulation, an estimate of fouling thickness.

These problems have been approached bearing in mind the capability of the aforesaid video positioning system to provide 3D coordinate measurements of the work site: the scene reconstruction subsystem therefore will exploit the video positioning system to get such measurements.

The main functions to be performed by the scene geometric modelling subsystem are:

* image processing for the evaluation of the edges and the features of the image;
* automatization of the choice of the measurement points on the environment;
* interface with the video positioning system to get the measurements;
* best fit to find the most suitable approximation surfaces (by using the a priori knowledge of the working environment);
* rejection strategies to eliminate non consistent measurements.

Key element of this subsystem will be the geometric data base including all the necessary a priori information: such information will be provided by the WIR supervisor computer in due time and format.

Advanced Manipulator System with Supervisory Control

The WIR manipulation subsystem will be required to carry out a wide range of servicing tasks in the fields of inspection, maintenance and repair.

The most outstanding features of the system to be developed are:

* an innovative 7 degrees of freedom manipulator arm, with torque controlled electromechanical joints, lending itself to the implementation of sophisticated motion/force control techniques;
* a telerobot control system architecture according to the philosophy of supervisory control.

The electromechanical joints of the manipulator will integrate modified harmonic drive reduction gears and brushless DC motors in compact, reduced weight, high performance actuation modules. Each of them will be endowed with a resolver for output position measurement and with a sensor of the torque transmitted to the link.

Preliminary laboratory tests on a single joint have confirmed that automatic control of the output torque significantly increases both accuracy and achievable control bandwidth. This allows for the implementation of an accurate and reliable controller of the manipulator motions and of the interaction force with the environment, which is an essential prerequisite for the implementation of higher level control functions.

As for the remote control of the manipulator, purely manual teleoperation of conventional master/slave systems is affected by well known disadvantages: sense of remoteness, partial feedback, need to integrate many information sources and control variables, ...

In order to alleviate these disadvantages and to increase the effectiveness of the system, autonomous functions should be implemented and integrated with teleoperation. For most of the required servicing tasks a standard execution procedure can be devised and their execution can be, at least partially, automated. In fact, although it varies, the working environment is not completely unstructured: typically it is represented by offshore platform nodes or subsea production systems with known geometric features and nominal dimensional data.

On the other hand, autonomous operations, i. e. purely robotic, are not feasible with the present state-of-art technology, especially when decisions and major replanning are required. Therefore the WIR manipulation control system will rely on an "optimal" combination of teleoperation and robotic functions, according to the philosophy of supervisory control. With modern terminology, this approach can be referred to as telerobotics. Its effectiveness has been experimentally demonstrated by Tecnomare with full scale laboratory tests of a telemanipulation system with supervisory control. The tests have confirmed the expectations in terms of time saving, improvement of performances and reduction of operator's fatigue (2).

A more sophisticated control system is going to be implemented now for the WIR development. A hierarchical telerobot control system architecture will be assumed as a reference for its implementation: the NASREM, or NASA/NBS Standard Reference System Model (3). The NASREM has been developed for the application of robots to space, but it is sufficiently generic for robot control systems. It defines a hierarchy of control levels: Object/Task, E(Elementary)-Move, End Effector and Joint, that will be adapted to the specific needs of this application. The hierarchical levels correspond to a set of standard modules and interfaces irrespective of the actual implementation of the different functions (autonomous or teleoperated). This facilitates both integration of software from different sources and evolution through incorporation of newly developed technologies.

Most of the high level functions will be left to the human operator. The automation at the Object/Task level exploits the "a priori" knowledge about the task and the working environment by developing nominal plans that foresee on-line operator's intervention for the adaptation of the plan to the actual working conditions. Nevertheless the operator, at any moment, can abandon the plan and resort to pure teleoperation.

The automation of the elementary manipulator motions, in free space or constrained by the environment, is pursued with a bottom up approach by developing efficient sensor based and model based motion primitives around the motion/force controller of the manipulator joints. This approach makes it possible to deal in real time with environment uncertainties and complexity in typical motion planning problems, such as fine motions in contact with the environment or automatic pathfinding among obstacles.

As for teleoperation, a bilateral control system will be implemented to provide the operator with a feedback of the interaction force with the environment. A suitable teleoperation command device will be developed: a six degrees of freedom hand controller, with motorized joints.

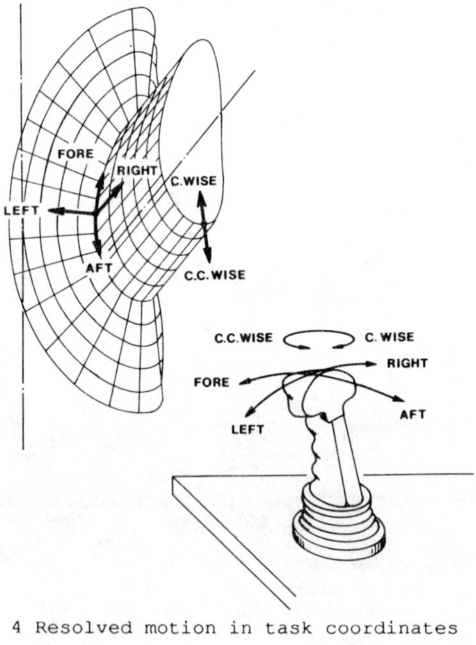

Fig. 4 Resolved motion in task coordinates

According to the supervisory control paradigm, shared and traded telerobotic command modes will be implemented. Sharing of telemanipulative and autonomous input means that only part of the manipulator degrees of freedom are commanded by the operator, the others being managed by the computer on the basis of the geometric model of the working environment or on the basis of the information provided by external sensors. For instance, the operator could command translational motions in a coordinate system of choice, while the computer automatically keeps orientation constant with respect to the environment. The coordinate system can be suitably selected to match the shape of the environment, as shown in fig. 4 in the case of platform nodes (this command mode is referred to as resolved motion rate control in task coordinates).

The sharing level can span an entire range from purely manual to entirely automatic, depending on the task to be executed. In particular, for the execution of specific subtasks the entire control can be temporarily traded to the computer. With respect to the supervisory control system previously tested in the Tecnomare laboratories, the new implementation for WIR will extend the concept of shared command modes to force control, in addition to motion control. In other words, a general sharing scheme will be implemented that makes it possible to specify force directions, in addition to motion directions. They can be either operator controlled (bilateral control) or computer controlled.

Automatic Navigation Control

In order to reach the prescribed working site, the WIR vehicle must be able to perform accurate navigation around the submerged rigs.

The vehicle trajectory will be commanded from the surface control room, either in pure teleoperation by the operator or with computer assistance if a model of the surrounding environment can be provided by the geometric modelling subsystem. Again, according to the supervisory control paradigm, this makes it possible to implement a higher level interface to the operator. Straightforward input command, entered through a joystick, can be converted by the computer into complex trajectories with respect to the environment, e. g. navigation along a platform brace while keeping predefined stand-off distance and attitude. The analogy with the command mode in task coordinates depicted in fig. 4 for telemanipulation is evident.

A fundamental prerequisite for a successful implementation of computer assisted command modes is the availability of an accurate vehicle controller enforcing the execution of the reference trajectory, produced either directly by the operator or by the computer, in spite of the disturbances due to sea currents and waves.

This represents a typical automatic control problem. In order to have access to the worksite and to dock to platform braces, both position and attitude of the WIR vehicle must be controlled, on the basis of the measurements provided respectively by:

* the sonar or the video positioning system, depending on the current mission phase;
* attitude sensors.

Basic issues to be addressed for the implementation of the control system are:

* filtering of position/attitude data, in order to remove measurement noise and to prevent thruster modulation, and estimation of velocity;
* control of a system whose dynamics is nonlinear, coupled, known with uncertainties, in particular as regards the hydrodynamic effects, and subjected to unmeasurable disturbances.

An automatic navigation control system has been preliminarily designed and tested with computer simulations, based on:

* Kalman filters to estimate the vehicle state;
* robust nonlinear control algorithm.

A simplified linear model of vehicle dynamics has been assumed to implement decoupled Kalman filters, for ease of design and reduction of computational burden. The filters also perform on-line estimation of the disturbance forces.

The control law explicitly accounts for the nonlinear dynamics. It is partitioned into two terms:

* a feedforward term, computed on the basis of the available dynamic model, to compensate for the nonlinear and coupled terms of the vehicle dynamics;
* a PD feedback term, to make the control system robust against model errors and disturbances.

Computer simulations have demonstrated the robustness of the implemented control system. Accurate station keeping and relative navigation have been achieved in the presence of moderate speed sea current and hypothesizing substantial uncertainties on the vehicle hydrodynamic and drag coefficients.

4. CONCLUSIONS

The Eureka EU-191 AUR project has been introduced. The development of two different vehicles is foreseen: the ARUS will be an autonomous vehicle devoted to ocean survey tasks, while the WIR will be a semi-autonomous vehicle fitted with all enabling technologies to substitute the divers in inspection/maintenance tasks of underwater structure and plants.

The two vehicles will be developed by a team of companies of U.K. and Italy as follows:

* Italy - Agip, Ansaldo, Enea, Elsag, Riva Calzoni, Saipem, Tecnomare;
* U.K. - British Maritime Technology, Cambridge Consultants, Ferranti ORE, Heriot Watt University, Institute of Oceanographic Sciences, Marconi Underwater Systems, Transfer Technology, United Kingdom Atomic Energy Authority, University of Strathclyde.

Within this consortium Tecnomare and Ferranti are stystem architects respectively of WIR and ARUS.

After a general overview of both vehicles, the paper focused on the advanced technologies to be developed in phase 2 of the WIR project.

Such technologies include:

* sonar positioning system based on sonar image processing and template matching techniques;
* video positioning system based on a stereoscopic computer vision system fitted with tracking capabilities;
* scene geometric modelling based on video image processing and best fitting with "a priori" geometric knowledge of the robotic environment;
* telemanipulation system based on an innovative electromechanical manipulator and a telerobotic control architecture;
* vehicle navigation system featuring state estimation and robust nonlinear control.

Phase 2 is estimated to last about two years and provide as output the subsystems, developed and verified by lab and sea trials, ready for integration.

REFERENCES

(1) Bellin M., Maddalena D., Visentin R., The TV-Trackmeter: an Underwater Non-Contact Position Sensor, Proc. IEEE Symposium on Autonomous Underwater Vehicle Technology, Washington D.C., June 1990.

(2) Maddalena D., Bossi L., Visentin R., The TV-Trackmeter and the Supervisory Controlled Underwater Telemanipulation System, Proc. OTC, Houston, 1990.

(3) NASA/NBS Standard Reference Model for Telerobot Control System Architecture (NASREM), National Bureau of Standards, SS-GSFC-0027.

THE DEVELOPMENT OF A SUBSEA THREE PHASE METERING SYSTEM

T L Dean, Dr E L Dowty and R J J Jiskoot

Texaco Limited, 1 Knightsbridge Green, London, England
Texaco EPTD, Bellaire, Texas, USA
Jiskoot Autocontrol Limited, Tunbridge Wells, England

Summary

Texaco Limited and Jiskoot Autocontrol Limited are developing a subsea three phase metering system based on a conceptual design by and with multiphase metering technical support from Texaco's US research organisation, Exploration and Producing Technology Division (EPTD). Financial support is being provided by the Commission of European Communities, DG-XVII. The project objective is to build and field prove a prototype subsea metering system (SMS) capable of measuring three phase flow subsea to +/- 5%. Texaco Limited in the UK possesses the need for such a system, practical subsea engineering technology and access to suitable production facilities for field trials. Jiskoot Autocontrol recognises the potential of the technology and intends to commercially exploit results of the project under license. Texaco Development Corporation, in support of EPTD, have licensed the design concept to the project. A key element of the system is the Texaco microwave watercut monitor which was also developed by EPTD. Overall project control is by Texaco Limited's UK R&D group which includes Jiskoot Autocontrol personnel with technical support and oversight provided by EPTD. An inter-disciplinary approach is required to bring together theory and oil field practice into prototype equipment suitable for offshore production platform and subsea field trials.

1. INTRODUCTION

In 1981 Texaco started a series of subsea developments to boost Tartan 'A' platform production. Initially, simple hydraulically controlled, near platform, single well satellites were developed. Then in 1985 the sophisticated Highlander subsea oil production template was installed 13 km from Tartan. The Petronella field development followed in 1986 with an offset of 11 km. Work on long offset subsea developments highlighted the high cost of providing metering facilities for flow testing remote subsea wells. Periodic well flow testing is necessary to monitor the performance of wells and reservoirs over time for the following purposes;

- improvement of reservoir models to optimise decisions on new wells and well production rates to maximise ultimate recovery;
- determining the timing of well work overs; and
- identifying when wells become uneconomic to continue producing.

Conventionally a dedicated "test separator" is used to meter individual wells. Fluids from a well are separated into its three component phases, oil, gas and water, in a large vessel and the flow rate of each phase is measured on the respective vessel outlet lines. The same method is currently used for subsea satellite developments by providing a dedicated "test pipeline" from the subsea field to carry a selected well's production to a test separator on the host platform for metering.

The capital cost of these systems rises rapidly with distance. The Highlander well test system includes 13 km of 8" pipeline, a slug catcher and riser at the base of the platform and a dedicated test separator topsides. It was obvious from these developments that a subsea based well test system could allow large capital cost savings in future developments by eliminating the need for conventional test systems (see Fig. 1).

2. BACKGROUND TECHNOLOGY

The earliest related work was initiated in the late 1970's, early 1980's by Texaco's, Texas based research group, Exploration and Producing Technology Division (EPTD). Of particular significance was EPTD's study of water fraction measurement techniques under a project supported by ARAMCO. The objective was to develop a device capable of measuring water/oil fractions of well fluids over a range of 0 to 100% with gas fractions of 10 to 90%. Early work highlighted the limitations of capacitance technology for high water fractions and microwave technology was selected.

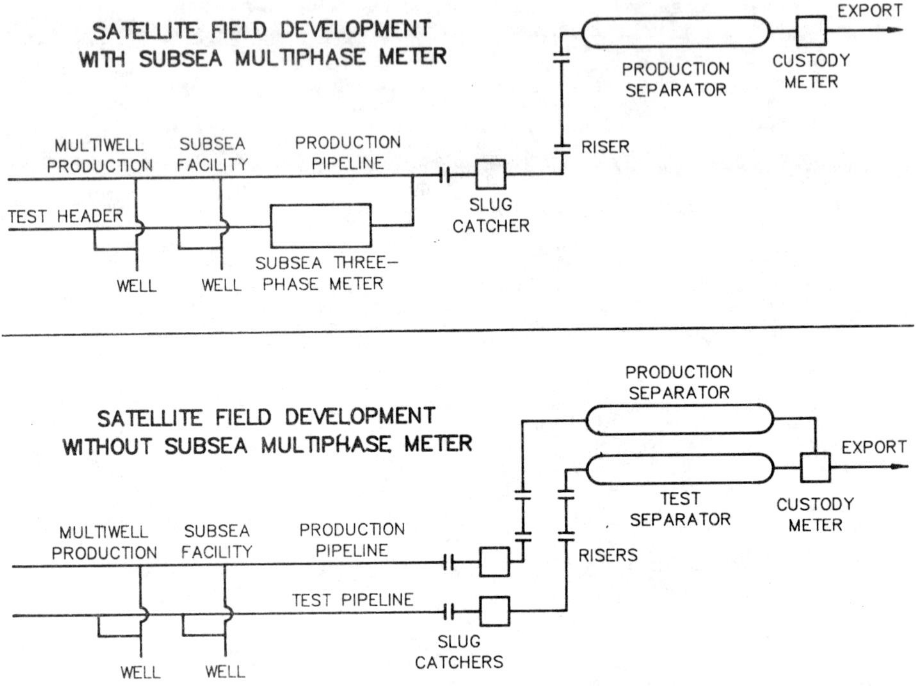

Fig. 1: Subsea developments with and without a subsea metering facility.

The Texaco microwave watercut monitor was developed and refined in following years with the capability of accurately measuring 0 to 100% watercut in both oil continuous and water continuous emulsions. It requires the fluid stream to be essentially gas free and microwave signal strength considerations dictated that it be positioned on a side stream sample line taken from the main production stream. These requirements led to the development of a gravity driven inclined gas/liquid separation system which enables a representative sample of the total liquid flow to be extracted (See Fig 2). Previous work by EPTD on a gas and liquid flow metering system provided the basis for this extraction scheme.

The sampling system can be broken down into two sub-systems, an inclined separator and a side stream sampling loop. The inclined separator stratifies the liquid and gas flows. The liquid, containing some entrained gas, separates by gravity from the bulk gas and flows along the bottom portion of the inclined pipe, while the lighter gas flows in the upper portion of the pipe. The liquid flowing down the inclined pipe has the same characteristics as open channel flow. This highly turbulent flow keeps the oil and water well mixed and allows a representative sample of the liquid to be extracted from near the bottom of the pipe. The extractor probe is positioned so that it is completely immersed in the liquid for the design flow range of the system.

The sidestream separator provides residence time to allow any gas entrained in the sample fluid to be released and returned to the main flow stream. Its physical arrangement provides for the formation of sufficient static pressure head from the separated fluid to create a flow through the microwave watercut monitor. For any flow in the design range, an equilibrium is reached between sample extraction rate, side stream liquid level and flow through the microwave monitor. The system is self regulating and functions without the need for pumps or control valves.

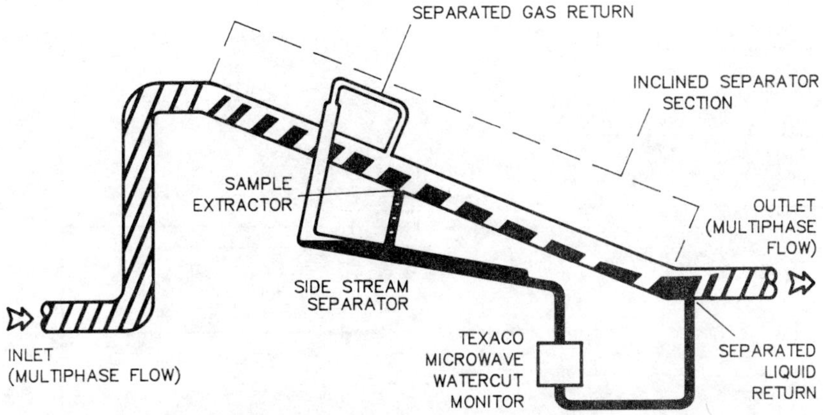

Fig. 2: Inclined sample extraction and separation technique for a multiphase flow line.

These basic principles were then extended to give a three phase metering system (see Fig. 3). Instead of immediately recombining the liquid and gas, separation is enhanced by adding enlarged gas and liquid sections after the incline. A liquid trap is used for the liquid metering run to give a self regulating liquid flow. For any flow rate in the design range of the system, liquid backs up in the inclined separator until sufficient static pressure head is exerted to equal the dynamic head losses through liquid metering run piping. The gas tends to have lower pressure losses and is forced through the gas metering run. Texaco perceived this concept to be suitable for subsea applications as the system required no moving parts and most of the basic technology was commercially available.

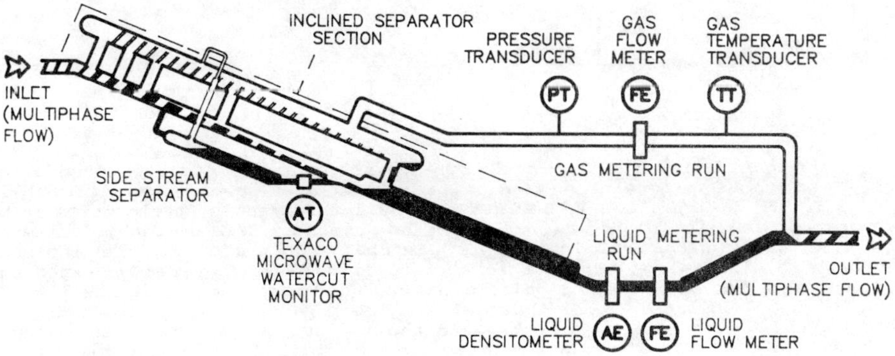

Fig. 3: Early conceptual design for a three phase metering system.

3. INSTRUMENT LABORATORY TRIALS

While the basic instrumentation technology existed, the Texaco microwave watercut monitor was not yet commercially available and few of the instruments had been developed for subsea use. Only basic temperature, pressure and valve position instruments were proven for subsea service and commercially available. It was necessary to draw on industry expertise to extend contemporary technology of the other instruments to give a workable system. Texaco Limited therefore participated with EPTD on a programme of instrument specification, selection and evaluation to confirm the feasibility of developing a Subsea Metering System (SMS) and identify specific instruments to be developed (see Fig. 4).

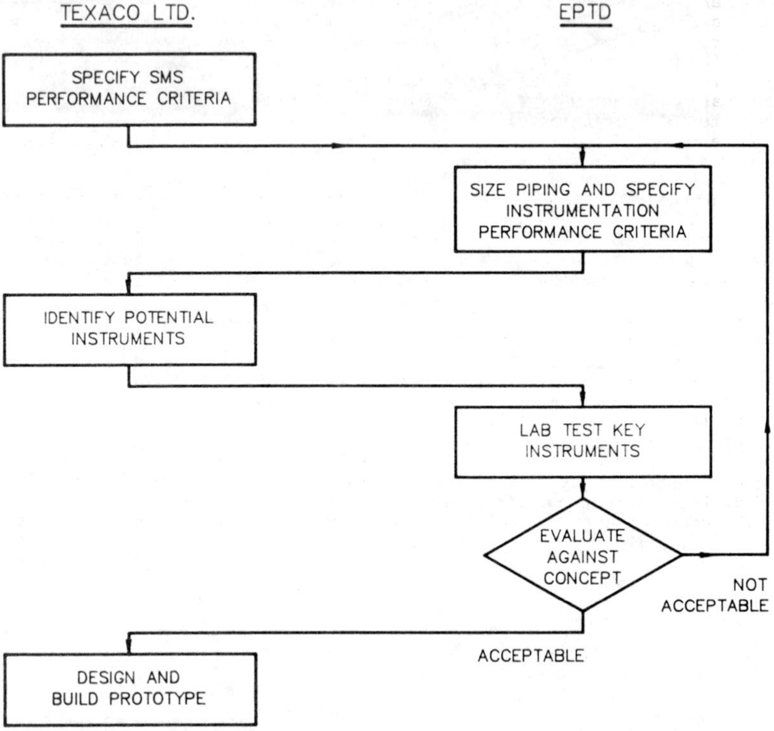

Fig 4: Logic diagram of laboratory trials.

Texaco Ltd prepared design criteria based on maximum subsea production rates and design conditions for the Highlander system as summarised in Table 1. The design criteria defined pipe wall thickness's, fluid mixes to be separated, and liquid and gas volumes to be metered. EPTD used this criteria to size lines and determine slopes, lengths and heads for the inclined separator. They also defined the degree of liquid gas separation expected, ie the degree of entrained liquid in the gas and entrained gas in the liquid. The inclined separator is not intended to achieve perfect separation and there may be as much as 1% liquid in the gas and 15% gas in the liquid by volume. The separation criteria proved to be an important factor governing final selection of flow meters.

Vendors were needed who could supply technically acceptable instrumentation and who were as a minimum willing to modify their instruments to suit subsea conditions. It was assumed that any instrument could be packaged for subsea service by a specialist third party company if necessary providing the instrument vendor was motivated and assisted.

A widespread quotation request was used to identify the most suitable instrument suppliers for the SMS. Instrumentation specifications were drafted in terms of performance requirements and quotations were requested for every type of instrument which could potentially meet the performance specifications. Quotation requests were sent to approximately 50 instrument manufacturers, subsea control system suppliers and stockists as a single package covering all of the instruments required. Some specialised suppliers had the potential to supply only one of the devices while others could offer several of the required instruments.

Parameter	Value
Total Liquids	3,000 to 18,000 BLPD
Water Fraction	0 to 90% by volume
Gas Fraction	10 to 90% by volume
Gas Rate	50 to 490 ACFM
Operating Pressure	100 to 1250 psi
Design Pressure	3640 psi
Overall Accuracy	+/- 5%

Table 1: SMS design criteria

It was decided that some instruments offered by suppliers were sufficiently well proven to be acceptable for the SMS without further evaluation. This included temperature, pressure and differential pressure transducers. "Off the shelf" versions of the remaining instruments were purchased and flow tested in a half scale three phase meter model in EPTD's multiphase flow loop. Table 2 summarises the numbers of instruments evaluated and identifies the vendors ultimately selected to supply instruments for the prototype SMS.

Instrument	Qty Tested	Vendor Selected
Pressure Trans	0	Trans-Instruments
Temperature Trans	0	Bush Beach Engineering
Liquid Densitometer	1	ICI-Tracerco
Gas Meter	4	FSSL/Scheme Eng
Liquid Meter	6	Texaco constriction orifice and Gervase variable orifice
DP Transducer (for liquid meters and liquid level indicator)	0	Caledonia Instr (Solartron)
Watercut Meter	1	Texaco microwave watercut monitor

Table 2: SMS instrument selection

Selecting an acceptable gross liquid flow meter proved more difficult than expected. Acoustic doppler, time of flight and vortex meters were all tested but were found to be unreliable when relatively large amounts of gas were entrained in the liquid. A proprietary constriction orifice meter developed by EPTD was ultimately selected because it is insensitive to entrained gas. A Gervase variable orifice meter was also selected for offshore trials as a possible liquid meter alternative.

Gas meters similarly proved problematic. The majority of gas meters commercially available could not withstand the required system design pressure of 3640 psi. Several of these were tested with the view that further development work could be performed if necessary to meet the required design pressure. Only two manufacturers were capable of supplying gas meters which satisfied the system design pressure. One of these, the FSSL/Scheme gas vortex meter, was ultimately selected because it appeared able to cope with liquid entrained in the gas stream.

The SMS design concept was refined over this period to accommodate the selected instruments and to include improvements in the inlet piping geometry (see Fig. 5).

4. PROTOTYPE DESIGN, FABRICATION AND TOPSIDES TESTING

As an operator of subsea facilities, Texaco deemed it essential to build a prototype SMS and prove it subsea prior to actual use. No operator could risk using an unproven system as a crucial element of a new field development. Due to the high cost of working with equipment subsea, it was also deemed essential first to fully prove all process and metering aspects of the system in "dry" trials with hydrocarbons. Hence it was proposed that the prototype SMS be fully designed and built for subsea service, tested topsides on Texaco's Tartan 'A' platform then tested subsea on Texaco's Highlander template.

The Highlander template is an ideal test bed for subsea trials. It offers a solid level foundation with some protection from marine hazards. Access is

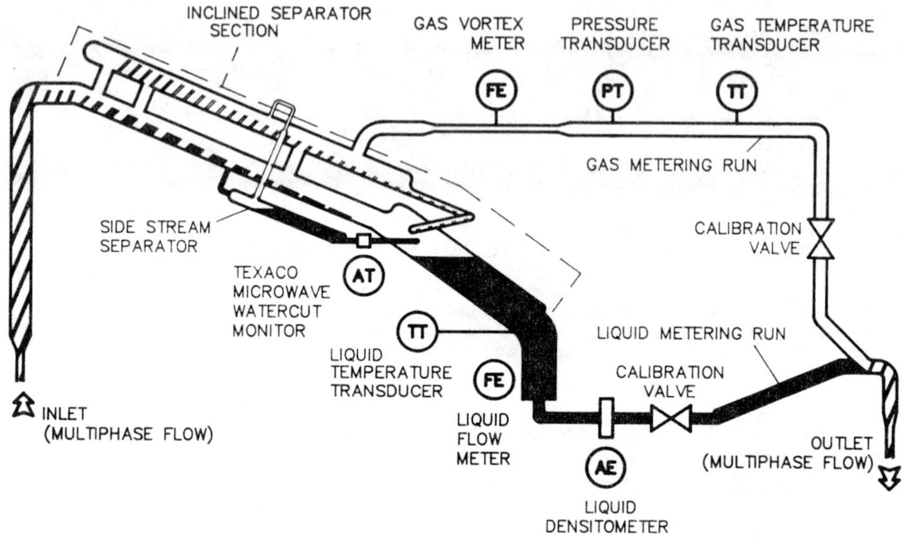

Fig. 5: SMS prototype final conceptual arrangement.

provided to the process fluids through well slot manifold piping connections. The existing FSSL control system can be tapped for electrical power and a spare signal cable is available for relaying back multiplexed data. Existing well slot valves can be used to control flow through the SMS utilising hydraulic control lines from existing control modules and the template well test system facilities can be used to check SMS performance. While the instrument laboratory trials were underway, a programme was mapped out for the design, fabrication and testing of the SMS (see Fig. 6). Financial support was requested from and granted by the EEC Directorate For Energy, DG-XVII. Manufacturers were also invited to participate with the dual objectives of utilising their metering system manufacturing expertise and providing an outlet for commercialising project results. Jiskoot Autocontrol Limited was selected and a Jiskoot Director has been working with Texaco Limited as a member of the project team since September 1988.

Texaco Limited modified the arrangement of the two dimensional conceptual design so that it fit over two adjacent Highlander template well slots. Initial layouts were developed by folding the two dimensional concept into parallel planes to create a more compact three dimensional design to suit the space available.

An engineering contractor, Cameron-Atkins Technology, performed the final detailed piping and structural design work (see Fig. 7). The three dimensional nature of the piping, numerous inclined lines and multiple fittings made it a difficult piping design problem. Many non-standard fittings were required to meet geometrical design considerations. The piping arrangement was subject to careful review by both Texaco Limited and EPTD before a final design could be reached.

The height of the final package exceeds the height of the existing template fence by approximately 5 metres, making it vulnerable to impact by fishing vessel trawl boards. Hence, the frame supporting and enclosing the piping is designed to withstand a 50 tonne lateral impact load. The large changes in pipe size and the complex geometry created pipe stressing problems which were addressed with special pipe supports arranged to selectively allow movement in specific planes.

The selected instruments were purchased independent of a central data acquisition system. Signals from the instruments are transmitted via Tronic

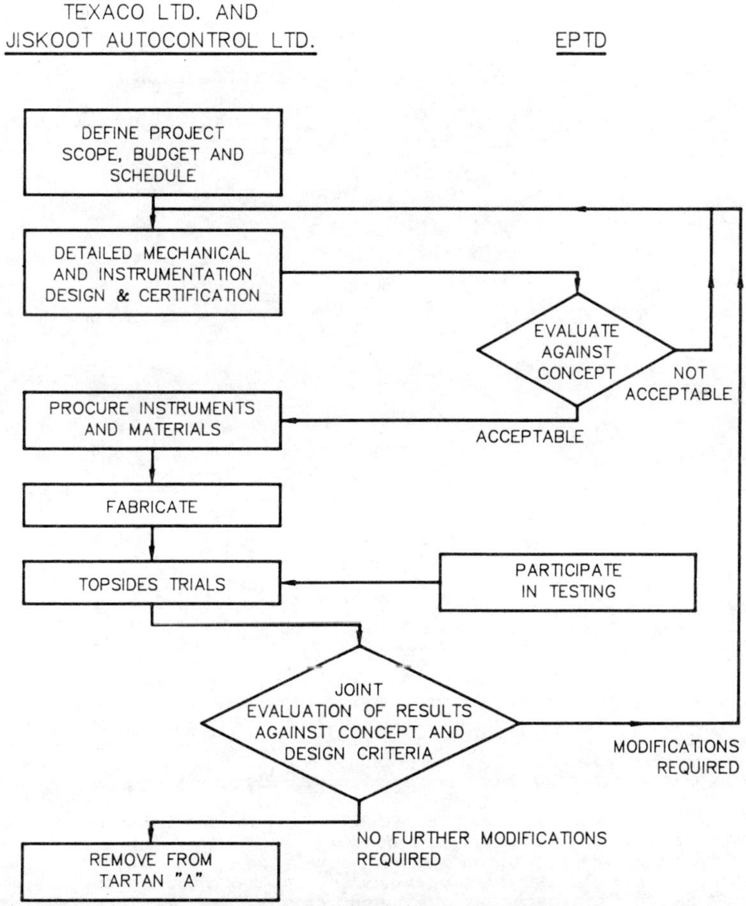

Fig. 6: Logic diagram of prototype fabrication and topsides testing.

subsea maleable pin-to-pin electrical connectors to a subsea data acquisition system developed by N L Shaffer/Liebnitz Lann. The temperature, pressure and differential pressure transducers operate on conventional 24 VDC 4-20 mA current loops. The microwave monitor and vortex meter also use 24 VDC 4-20 mA current loops for readings but require separate power supplies. The data acquisition unit provides a special RS-232 digital signal link for re-calibration of the microwave monitor via a topsides computer. The nuclear densitometer is the only instrument with a digital reading. It reports measurements, discrete radiation counts, with an RS-232 digital signal to minimise loss of accuracy and runs on a separate power supply.

Where possible, instruments were ordered from suppliers inclusive of all design and manufacturing work required to build an acceptable subsea instrument. Fortunately all of the selected instrument manufacturers had some subsea instrumentation experience or teamed with another company which did.

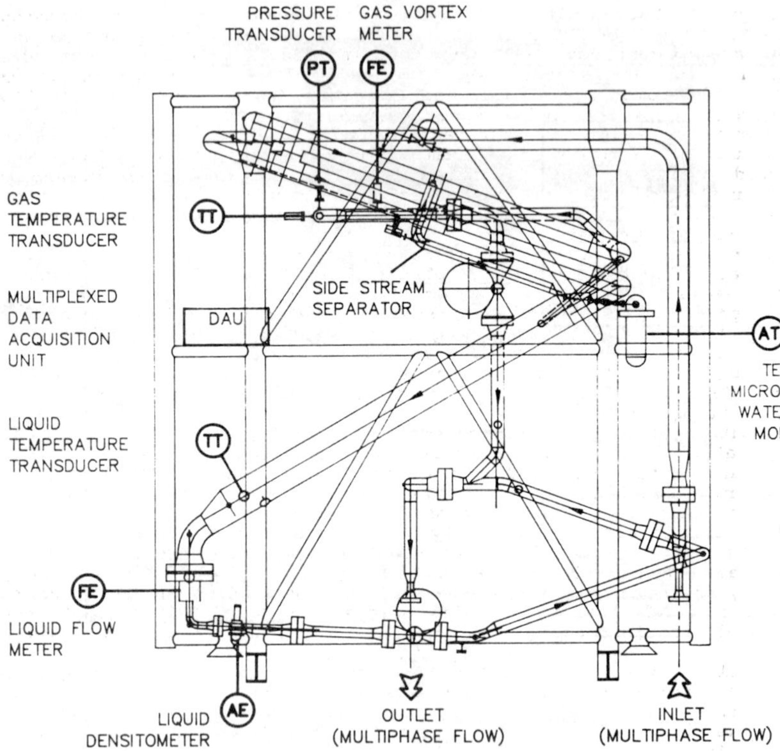

Fig. 7: Side elevation view of view of SMS.

The Texaco Limited project team approved designs and offered advice as required to overcome specific problem areas.

The Texaco microwave monitor was a special case. The high level of the technology dictated that its developers, EPTD, be responsible for assembling and testing the electronics. The instrument however had been used previously only at onshore production facilities and had not been packaged for deployment subsea. N L Shaffer, a Texaco Limited UK contractor, was selected to design and build a suitable subsea enclosure. The design allowed the electronics to be fitted and tested by EPTD in the US. The partly assembled instrument was then returned to the UK for sealing and hydrotesting with a final factory acceptance test performed by EPTD personnel at N L Shaffer's UK works.

All instruments used either subsea or on platforms in the North Sea UK sector must meet certain safety criteria and be approved by an independent certifying authority. Instruments developed for installation subsea are generally designed to meet special requirements and do not have to meet the fire safety criteria required of instruments placed on platforms. Normally all that is required of subsea instruments is that they can safely contain process fluids at design pressures. Topsides instruments on the other hand are normally designed with a flame path to slowly and safety vent any pressure caused by an internal explosion.

The SMS had to satisfy both requirements as it was to be tested topsides on Texaco's Tartan 'A' platform before being tested subsea. The certifying authority was in this case Lloyds Register. Certification of the SMS mechanical design was lengthy but conventional. Certification of its instrumentation and electrical design for the topsides testing was quite difficult due to the rather contradictory requirements.

Testing on the Tartan platform required the development of a methodology to show that each of the SMS subsea instruments was safe for platform use. It was necessary to consider each instrument individually to find a method of certification which was acceptable. For many of the instruments it was necessary to add temporary electrical barriers to the data acquisition system's current loops for the topsides testing.

Throughout the design and fabrication effort practical problems emerged requiring solutions which deviated from the initial conceptual design. When this occurred, details were submitted to EPTD for review to ensure no operational or flow measurement considerations were undermined. Most matters were resolved quickly but some required several cycles before an acceptable compromise could be reached.

The metering system piping and structure was fabricated in the spring of 1989 and installed on the Tartan 'A' platform for trials. It was necessary to strip off several piping spools for the actual lift to keep the weight within the capacity of the platform crane. The metering skid was located on the top skid deck beside the drilling rig. Piping was run from the test header to the SMS and on to the test separator. This allowed wells to be directed through the SMS and test separator in series for testing. Several fire and gas detection instruments were fitted and the platform fire water deluge piping was extended to cover the SMS.

The computer controller was located in the platform control room to best allow coordination with platform operators. Producing wells were flowed individually and in combination from the test header to the SMS and on to the test separator. Initially most of the effort was directed to debugging software and electrical instrumentation interfaces.

Platform trials ran from mid-July to early September 1989. Following this work and a review of the data, key problem areas and potential solutions were identified. Due to the cost of making and testing all of the proposed modifications, further model tests were performed by EPTD in the USA to screen solutions and identify preferred modifications. In March 1990 a final scope of work for modifications and refinements to the original design was defined and revisions to the mechanical design initiated.

The detailed piping and structural designs were revised accordingly and sub-assemblies fabricated onshore. The SMS remained on Tartan throughout and was modified in place. The work was phased to allow testing to continue in July 1990. The specific phases of the continued programme are as follows:

- I Single phase tests of liquid and gas metering equipment
- II Liquid/gas separation
- III Watercut sample extraction, liquid/gas separation and measurement
- IV Total system performance trials

Phase I, "Single phase tests" were completed in August 1990. Some difficulties were experienced controlling all of the variables between the SMS and Tartan test separator. These were of an operational nature and will be overcome by modifying the test procedures. The constriction liquid flow meter performed acceptably while difficulties were experienced with the Ferranti/Scheme gas vortex meter and the Gervase liquid flow meter. Part of the difficulties experienced with the gas meter arose due to attempting to test the meter with single phase gas obtained by throttling high pressure lift gas. The resulting flowing gas temperature was approximately -5 degrees Celsius. This was outside the calibration range of both the SMS and test separator. Subsequent testing on a well with a high gas/oil ratio identified that the gas meter malfunctioned and required servicing by its manufacturer. The results of testing the Gervase liquid flow meter suggested that the differential pressure transmitter was not functioning properly.

Phase II, "Liquid/gas separation" tests are currently underway. Hence no results are available now for this or later stages of the test programme.

The platform trials will be completed in late 1990 and the SMS will be decommissioned from the Tartan 'A' platform. It will be necessary strip the SMS down again to reduce its weight for the platform crane before transferring it to a transport vessel.

5. PLANNED SUBSEA TRIALS

The real proof of the SMS will be the subsea trials. From a project view point however all of the difficult work associated with the basic performance of the SMS, including all of the development work and the vast majority of the fabrication, will be complete by the end of the platform trials. There are important last minute tasks to perform before subsea installation but these are not technically sophisticated nor impact system performance.

The onshore scope of work will include the following tasks:

- Reassemble and leak test piping
- Complete any structural and pipe support modifications required by offshore piping changes
- Fit anodes and bonding straps for cathodic protection
- Touch up paint and subsea markings
- Perform final precommissioning checks

Following the onshore work, the SMS will be installed and tested on the Texaco Highlander subsea template which lies on the sea bed in 130 meters of water. The SMS weighs approximately 50 tonnes and measures 10m X 3m X 10m. It will be installed with a crane by a Diving Support Vessel (DSV) with diver assistance.

The SMS will be formally tested subsea for at least eight weeks with a further period of occasional checks. The intensive test period will establish the functionally of the equipment with follow up checks verifying there are no long term performance faults or problems with the instrumentation.

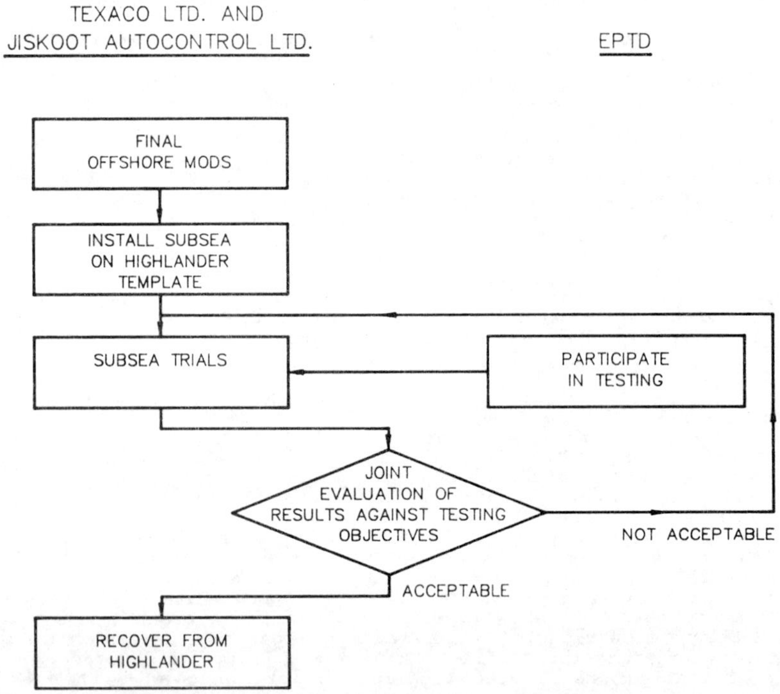

Fig. 8: Logic diagram of subsea trials work.

6. CONCLUSION

The development of a high pressure prototype subsea version of the three phase metering system proposed in EPTD's early conceptual design has required integrating practical subsea engineering expertise and oil field practice from Texaco Limited, multiphase separation and metering expertise from EPTD, metering systems expertise from Jiskoot and the respective expertise of many instrumentation and subsea equipment suppliers.

The project work has been broken down and organised to combine theory, oil field practice and metering systems experience. This was achieved by requiring design approval by both EPTD and Texaco Limited, supported by Jiskoot personnel. Each has access to all aspects of the design and is responsible for covering their respective area of expertise.

On a different plane, the project has drawn on the vast resources and expertise from industry by effectively sieving instrumentation and subsea control system manufacturers to identify those who had both the technology and the will required to develop specific components required by the system. Dozens of experts have contributed to new designs of instruments required by the project through suppliers and sub-suppliers. This work was performed on a strictly commercial basis but that does not diminish the commitment shown or reduce the number of problems that have been overcome. The following is a list of some accomplishments of the project and project suppliers.

- Fabrication and topsides testing of prototype three phase meter
- New high design pressure reduced span subsea pressure transducer
- New subsea temperature transducer
- New subsea gas vortex meter
- New subsea constriction orifice liquid meter
- New subsea variable orifice liquid meter
- New subsea scintillation type nuclear densitometer
- New subsea version of Texaco microwave watercut monitor

REFERENCES

(1) Hatton, G.J., Helms, D.A., Marrelli, J.D., and Durrett, M.G., "A New Microwave-Based Water-Cut Monitor Technology", OTC 6426, OTC 1990.

(2) Abraham, Kurt S. and Terrell, Henry D., "Latest Technology Stresses Smarter, Quicker Operation", World Oil, July 1990.

(3) Hatton, G. J., "Wellhead Metering of Gas/Oil/Water Production Streams", North Sea Flow Metering Workshop, October 1986.

"AN ADVANCED SYSTEM TO PREPARE AND TO FOLLOW DRILLING AND COMPLETION OPERATIONS"

Jacques-Marie COURTEILLE et Philippe BOUTROLLE
ELF AQUITAINE
CSTJF, Avenue Larribau F.64018 PAU CEDEX

ABSTRACT

Drilling & Completion is a technical activity where computing has been considered for a long time as a tool providing help when numerical scientific calculation is needed. It seems logical to anybody to determine by mean of a computer the profile of a well, the pressure losses, the parameters of a kick circulation, etc...

The aim of PINFOR is to extend the scoop of use of the computer to the non-numerical data and to create a general data-base in which all the informations concerning Drilling & Completion operations are stored.

With this data base, different exploitation tools are provided to the user enabling him :

- to seize the data (acquisition),
- to consult them (consultation),
- to select appropriate data in order to satisfy specific need (oriented request),
- to make scientific calculations with these data, either on the rig or in the Drilling & Completion office,
- to build and to send any reporting forms,
- to archive the data in a patrimonial data-base.

In this way, scientific calcultaions softwares become only a part of a general Information System.

Any Drilling & Completion representative can access to the System thru his work-station linked to the computer.

This work-station, available in the same standard of use anywhere in the Company represents the modern tool, used every day by engineers and superintendant to prepare and to follow Drilling & Completion operations with a 1990's tool.

People started to use computer for scientific calculations a few decades ago. Up until now, they were not really familiar with the use of computers for the storage of information and its analysis.

Paper reports are still used in most places with all their inconvenients :

- Length of time for composition,
- Deterioration of the information during transmitting and storage,
- Manual and tedious research for data collection,
- etc...

Modern computerized information systems provide good advantages to the user with :

- a powerful acquisition system with control functions avoiding mistakes,
- an increase of the accuracy and of the exactness in the management of the data,
- a fast and reliable way to get accurate informations when needed,

- a reliable system for archives,
- a "push-button" paper reports system if needed.

That allows the supervisor and the engineer :

- to manage their time in a better way,
- to prepare the operations with more accuracy,
- to take convenient decisions with a greater effectiveness at the time they have to be taken.

The main results will be to get a better quality of the well operations with a saving of time, and money.

PRINCIPLES FOR INSTALLATION & FUNCTIONMENT

The general name of the system is "PINFOR".

The french meaning is "Poste INformatique de FORage" or "Drilling Workstation".

The main workstation is on a HP 9000/835 at the Operational Base looking as a mainframe linked to all the other workstations installed on the different rig sites equipped with a HP 9000/370.

The specific name of the rig application for the acquisition of the data during the operations is PINFOR.

After the operations on a well are over, the PINFOR application is migrated in the permanent patrimonial database whose specific name is "PATRIFOR" (PATRImonial pinFOR).

"CALPIN" is the specific name for scientific calculations module linked to the database (CALculation with PINfor).

Before the beginning of a well operation, the PINFOR application is initiated at the Base. A copy is installed on the rig site.

During all the operations, this application remains on the two sites with the same image except for informations entered from the last transfer.

Acquisition and transfers to the Operational Base can occur at any time.

Acquisition of data is only allowed at the rig site.

Consultation of the data is allowed at any workstation.

By the end of the operations, the PINFOR application is finalized at the Operational Base and stored for ever in PATRIFOR.

This represents the field information system.

The PATRIFOR application can also be transferred to the headquarters in France in a Group patrimonial database.

DIFFERENT USES

The main functions of the system allows :

- to acquire live data in the field,
- to simplify the administrative task of the supervisor and of the engineer,
- to make inquiries in the database,
- to provide daily or statistical reports,
- to run scientific calculations,
- to fill the system with old data.

They consists of :

1 - INITIALIZATION

This function helps to prepare, at the Operational Base only, the PINFOR application which will then be installed on the site machine when drilling begins.

As well as identifying the operations to be carried out, this function enables "pre-entry" of any other information : description of equipment in use, general outlines of the lay-out, types of mud-products and bits in use, estimated progress curve, reference well (all the data), neighbouring wells (deviation data only), and so on...

The application can be transferred to the site in the form of a cassette. A copy of which is kept on the machine of the Base and will receive the data coming from the rig.

2 - ACQUISITION

This function can be used only on the site.

It allows all the relevant drilling data to be entered as the operation is carried out and at anytime of the day. This application is fit to allow the supervisor to "prepare" specific operations such as running casing or cementing, during masked time before operations. Thus he only has to validate them when the operation occurs.

The data acquired are forwarded to the machine at the Base at any choosen frequency. The general intention is that at each connection, everything entered since the previous vacation should be transferred (and not at fixed intervals of 24 hours).

In this way, provided that the discipline of "on-site only acquisition" is respected, the same image of the database is available simultaneously to the rig and the Base, exact whatever time lapse between transfers has been established. Any modifications to be made should then be agreed by telephone and applied on site.

3 - TUBULAR

Designed for the comfort of the user, this facility allows him to get rid of the traditional hand or table electro-mechanic calculator.

As casings (or tubings) arrive on the site, their detailed characteristics (order numbers, lengths,...) are entered once and for all whatever the order.

Thus, the user can later compose on the string, alter it during the descent and transfer the final result into PINFOR the day of the operation.

Obviously, this facility is only available on the rig site.

4 - REPORTING

Composition and automatic transmission for all traditional han-written reports : daily reports, daily telex, mud and cementing reports, end of well reports, maintenance operation report or whatever.

This application was initially intended for use only at the Operational Base, but, could in fact be used on the site at anytime to allow those in charge to edit reports or diagrams whenever it suits them.

5 - FINALIZATION

Upon completion of the operations, this facility allows the

Methods engineer at Base to "end" the PINFOR application by :
- calculating the final drilling assessment data (mud phase records,...),
- entering further informations such as global costing,
- producing the final and official paper editions,
- transferring the whole to the reference data bank at Base : PATRIFOR (certain details such as reduced flowrates noted everyday for safety purpose will be eliminated at this stage).

6 - SCIENTIFIC CALCULATION : CALPIN

We are concerned here with the module available both on the site and at the Operational Base for making scientific calculations.

These can be made either on the basis of data acquired directly or from data coming from PINFOR or PATRIFOR.

In the initial version, six areas of calculation are available (ref. 1) :
- well profile,
- BHA behavior (ORPHEE - ref. 2),
- pressure drop,
- kick control,
- U tube effect,
- file card management.

The second version (coming this year) will extend areas of calculation :
- every day use softwares (lag-time, back-off, nozzles determination, trip speed, overpull allowance, composite drillstring...),
- rock mechanics,
- temperatures while drilling,
- frictions,
- graphical determination of well profile,
- expert system to determine casing : CUVEX,
- expert system to prepare and realize cementing operations : CIMENTEX (ref. 3).

Other algorithms are under testing and validation within the research department and will be included to this module in a further version. Their subjects are :
- expert system to prepare and to run completion program : COMPLEX,
- determination of perforating system performance
- special operation with snubbing or coiled tubing such as cleaning well with foam,
- temperature in a well while producing or injecting,
- fracturing operation program determination,
- fluid speed profile in the annulus while pumping,
- kick control with two phases fluid and with dissolution,
- marine softwares (determination of risers for floating vessel, calculation of anchor lines...).

7 - QUERIES

This facility can be considered as one of the most important tools of the system. It allows the user to make any kind of inquiry into the database concerning any parameters within a single or several different wells.

The data got by these queries can be easily worked with any desk software as OPEN-ACCESS, EXCEL, etc... on a standard desk computer.

This function will be naturally available both on site to help the following of the operations and at the Operational Base where it will be useful in planning future operations, making any statistics, etc...

8 - PRE-FILLING

This function (only Operational Base) is "historical" and provides for the acquisition of old data i.e. concerning wells drilled before the creation of PINFOR.

With an experienced typist such data can be entered at the reasonnable rate of minimum 2-3 wells a day, thus rendering accessible with the same machine all relevant data of previous wells.

The pre-filling has been based on the traditional hand-written well-files to produce this simplified acquisition application. It is available on the machines in the Operational Bases.

9 - MIGRATION

If the pre-filling may be considered as the manual recuperation of former data, MIGRATION is the automatic version for data already existing in computerized form. We refer here to information at present stocked in the various computers used in the Company in the past years (HP 1000, HP 9845, IBM 3090) at the headquarters and in the subsidiary companies. This facility has not yet been realised but will gradually be brougth into being with the subsidiary equipment and according to their needs and particularities.

So, with this system, one can know, at any time the actual status of a well (location, architecture, equipment) and the description of all the operations that have been run on this well (drilling, completion, work-over).

HARDWARE & BASIC SOFTWARE

The System is working on an HP mini-computer type 9000.

At the Operational Base, there is a HP 9000/835, multi-user workstation, working with UNIX (either a single machine or several machines linked together with a local network).

Several people from the drilling/completion department are supposed to work at the same time : method engineer for operation preparation, operation engineer for their following, drilling/completion manager for consulting data or making statistics, secretary for reporting, etc...

The other trades of the Production Division have their own information system built with the same standards and working on the same machine.

So, the PINFOR system can easily communicate with the Reservoir system, the Production system or other and exchange data with them.

This constitute a general information system holding all the production data that we call : INTERSUB (SUBsurface INTERpretation).

On the rig site, a two users machine, a HP 9000/370, is needed.

First user is the drilling/completion supervisor.

Second user is a geologist using the geological information system developped by the exploration people which is installed on the same machine.

This geological system is aimed to collect geological informations. It has the same type of functions that PINFOR and is linked with it. For instance, the description of the lithology can be got directly by PINFOR in the machine when the geologist has entered it in his system.

Basic software used is a commercial database system : EMPRESS/MBUILDER.

The applications that we developed with it represents around 200 screens. These screens appear automaticaly to the user in pre-determined sequences while acquiring or consulting data but one can easily go where one wants to with different menus or automatic selection system.

PRESENT STATUS

The system has been under development for four years and represents around 5 000 day/engineer.

1989 has been the last step for development and validation. The first installation occured in our french subsidiary by the 4th term and was a conclusive test.

- Installation
  We have now started the process of equiping the other field areas, beginning with Europe (The Netherlands, Norway, UK) and starting now in West Africa (Angola, Gabon...).
- Training
  As the system is very powerful and drilling & completion people are not very familiar with the use of a computer, we have developed in the same time, a one week PINFOR training session in order to have informed users on the site.
- Language
  The present version of the Sytem is still in French except the papers reports that can be got in english if needed.
  The whole version will be in english within the end of this year.

ACKNOWLEDGMENTS

The authors thank Société Nationale ELF AQUITAINE for permission to publish this paper.

REFERENCES

1 - R. Fenoul : "Enhanced Drilling Softwares And Integrated Advisor Expert Systems", Paper SPE 19133 presented at the SPE Petroleum Computer Conference, San Antonio, TX, June 25-28, 1989.

2 - M. Birades and D. Gazaniol : "ORPHEE 3D : Original Results on the Directional Behavior of BHA'S With Bent Subs", Paper SPE 19244 presented at the 64th Annual Technical Conference of the SPE, San Antonio, TX, October 8-11, 1989.

3 - G. C. Wolsfelt, C. Roger, and R. Fenoul : "A Cementing Job Preparation Avisor System", Paper SPE 19540 presented at the 64th Annual Technical Conference of the SPE, San Antonio, October 9-11, 1989.

MONITORING OF FLEXIBLE RISERS BY ACOUSTIC EMISSION

ANDRÉ SUGIER,** JOSÉ MALLEN,* PIERRE MARCHAND,** ALAIN MARION,*

*COFLEXIP, 23, AVENUE DE NEUILLY, 75016 PARIS, FRANCE
**INSTITUT FRANÇAIS DU PÉTROLE, BP 311, 92506 RUEIL-MALMAISON, FRANCE

Summary

In floating production systems operators need to rely on flexible risers to link subsea installations to production units. For safety reasons it seemed of particular interest to develop an in situ monitoring system adapted to these dynamic risers. This paper presents the main aspects of the work carried out by Coflexip and I.F.P. on this subject as well as the preliminary results obtained using acoustic emission technique.

Introduction

On both UK and Norwegian sectors of the North Sea, as well as in some other areas such as Brazil or the US gulf of Mexico, there is a clear trend towards an increasing use of Floating Production Systems (FPS) and subsea completions as temporary or permanent development schemes for the production of offshore oil & gaz.

These FPS's rely on dynamic flexible risers to link the subsea equipment to the floating production units (fig.1). For clear safety reasons, it seems of particular interest to develop an in-situ monitoring system adapted to these dynamic risers. This paper presents the main aspects of the work carried out by Coflexip and I.F.P. on the subject of the monitoring of unbonded Coflexip pipes in dynamic applications, as well as the preliminary results obtained using acoustic emission techniques. This work is conducted with the technical assistance of CETIM and CEPM.

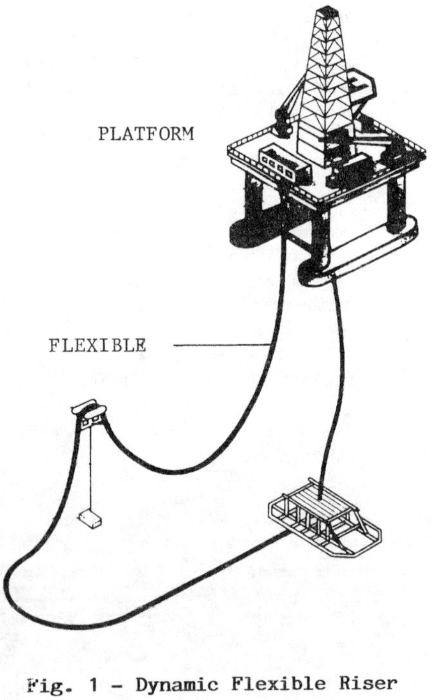

Fig. 1 - Dynamic Flexible Riser

Failure mode analysis

The construction of COFLEXIP flexible steel pipes is a composite arrangement of steel and plastic layers (see fig.2) independent one from another. For multiphase production applications, the pipe structure is composed of plastic sheaths for internal and external leakproofness, an inner interlocked steel carcass preventing the collapse of the internal plastic layer, a zeta layer providing the resistance to internal pressure (hoop stress resistant layer) and armour layers providing the resistance to tensile and torsional loads. For dynamic high pressure riser applications, an additional plastic layer is introduced between the zeta and the armour layers.

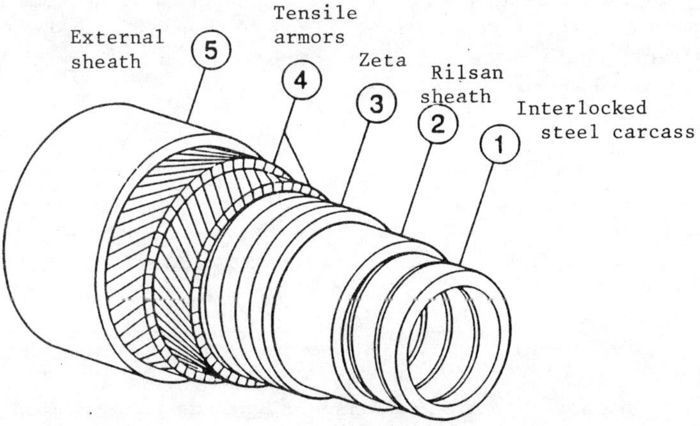

Fig. 2 - Flexible structure

In normal operating conditions, a dynamic flexible riser is exposed to the conveyed fluids at certain pressure and temperature levels. In addition, it is subject to bending variations which, in a unbonded pipe, induce a relative slip between the various layers of the structure. At the level of the armour layers, relative displacements between the wires of one layer and the wires of the other layer may range between 1 and 10 mm typically, which, in the case of poor lubrication between the wires, may lead to seizing.

The most probable cause of a possible failure of a flexible pipe are linked to the following phenomena :

- wear between the armour layers producing a wire cross section reduction finally yielding an ultimate rupture mode associated with fatigue,
- ageing of the thermoplastic material composing the inner pressure layer which ensures the leakproofness of the pipe.

The degradation of the plastic material may ultimately result in the loss of leakproofness of the riser. This would not however result in the mechanical rupture of the steel reinforcing elements of the pipe structure.

On the other hand, the rupture of armour wires is likely to not only yield a loss of pressure containment but also a mechanical failure of the pipe. This type of damage is therefore considered as critical. It was then decided to lay the emphasis on this particular failure mode.

The objective of the work was to develop monitoring techniques able to globally assess the wear phenomenon as well as its evolution towards wire seizing. Acoustic emission has been selected as a promising technique in order to detect whether the flexible riser is in normal operating conditions or, alternatively, to act as an alarm bell system as soon as seizing occurs within the armour layers.

Used as an early warning system, this would allow the operator to take all necessary actions to avoid a catastrophic pipe failure and implement, in a suitable time frame, any corrective step such as riser relubrication or replacement, while increasing safety and reducing potential downtime.

Service life analysis

The analysis of the behaviour of COFLEXIP dynamic flexible risers has shown that the main failure mode was associated with the wear phenomenon between the layers of the tensile armour wires. The decrease in wire thickness generated by wear/friction results in an increase of the static and alternate stresses in the wires which must be assessed from a fatigue point of view.

GOODMAN-SMITH diagrams have been established by IRSID (see fig. 3a) for the steel grades most commonly used by COFLEXIP e.g. FM 15 & FM 35. The FM 35 material, specially treated for H_2S compatibility, displays better mechanical resistance and increased fatigue resistance as compared to these of FM 15 material.

Combined wear and stress increase phenomena are fully analysed by means of a HAIG diagram (see fig. 3b) which displays the material fatigue behaviour as a function of wear-induced wire thickness reduction. From this diagram it is possible to estimate the service life of a dynamic riser based on the knowledge of stresses induced by the pipe internal pressure and its dynamic loadings.

The wear rate is dependent upon the operating conditions of the flexible pipe (internal pressure, relative displacements within the pipe layers induced by the dynamic motion). It depends equally importantly on the lubrication conditions between the wires, in particular whether seizing is taking place or not. It may vary as well once seizing has occured even after the reinstatement of a normal lubrication regime.

A programme studying the influence of these parameters on the wear rate and establishing correlations between wear regimes and acoustic footprints has been initiated in laboratory conditions. It is pursued during bench testing of flexible pipe samples.

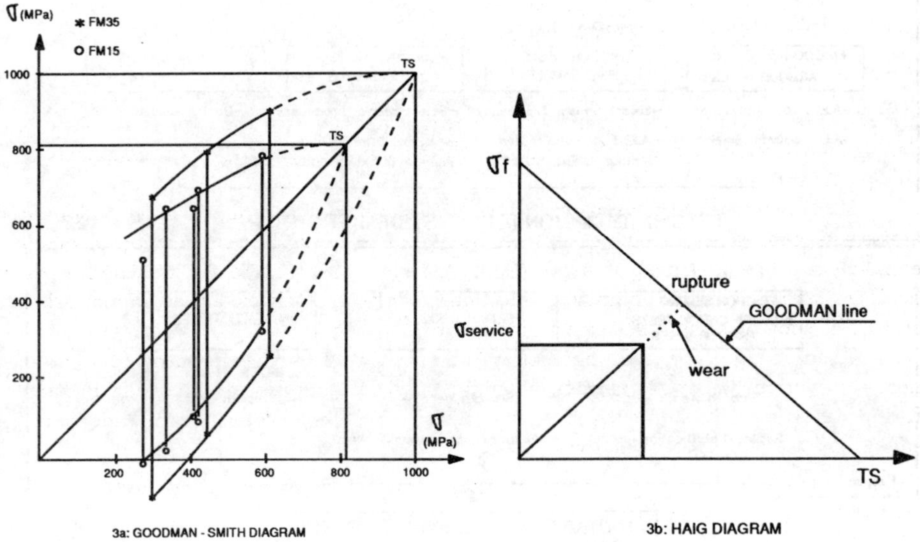

Fig. 3 - Service life analysis

Test programme

It consists of two test phases :

- laboratory tests under various contact pressures and wire displacements in different friction conditions,
- tests of flexible pipe samples (typically 10 to 12 m long) on a suitable test bench.

The laboratory testing conditions will be defined on the basis of expected operating conditions of dynamic risers as well as of measurements obtained from the bench testing of pipe samples (dispacements, loads, contact pressures).

The simulation of seizing incidents generated in laboratory conditions will be reproduced during the pipe sample testing phase.

The table 1 below presents the main steps composing the overal project. This paper will only present a selection of the currently available results from laboratory and bench tests.

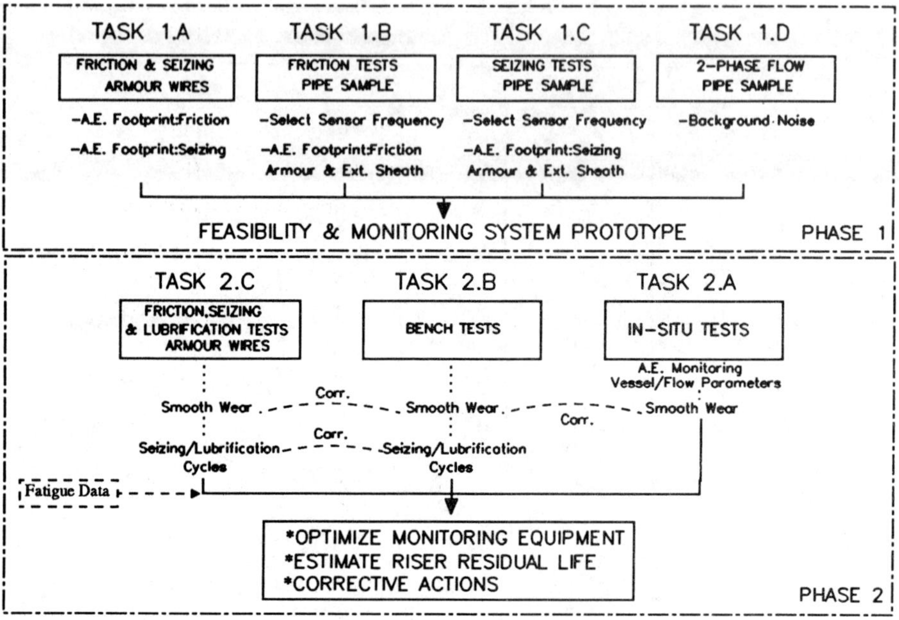

Table 1

Laboratory tests

Laboratory tests have been conducted on armour wires using tribometres the characteristics of which are presented in table 2 below.

The tests have been carried out in I.F.P. laboratories, with collaboration from CETIM concerning the acoustic measurements. They essentially aimed at verifying the technical feasibility of acoustic emission techniques for the pupose of dectecting variations in wearing conditions, seizing and of studying the effect of a wire relubrication after seizing had occured.

I.F.P. possesses various apparatus which allow the study of the wear phenomenon on armour wires under varioux contact pressures, with a range displacement corresponding to the operating conditions generally encountered in dynamic flexible risers.

The acoustic emission (AE) sensors used in the experiment had been previously selected during the first trials conducted on pipe samples, in particular in relation with the noise generated by various multiphase flow regimes under a wide range of gas oil ratios.

The instrumentation used consists of a global monitoring with a continuous follow up of burst and oscillation counting (ring-down counting), coupled with a periodic

Description	Triboflex	M 10 P
No of measurement units	1	10
Range (double amplitude)	0.2 - 12 mm	0.4 - 5.5 mm
Contact orientation	horizontal	vertical
Max. wire sample size	80×20×10 mm	50×15×15×5 mm
Contact load	82.5 N min. 1000 N max.	49 N min. 3140 N max.
Actual contact pressure (6×3 mm wires)	4.1 - 49.4 MPa	2.4 - 155 MPa
Actual contact pressure (14×3.3 mm wires)	0.6 - 7.0 MPa	0.3 - 21.8 MPa
Eriction load measurement		
- type	stress gauge	piezo-elec.
- range	± 1000 N	± 5000 N
- data acquisition	Analog	Digital
Contact temperature	ambiant	ambiant -60°C
Testing medium	air, water	air, water neutral gaz
Continuous wear measurement	possible	possible

Table 2 : Characteristics of tribometers used for wire testing

analysis of the distribution of the EA within the cycle through rms determination. An additional device for individual burst characterization has been introduced to allow for a frequencial analysis of the signals.

In a first approach, only the cumulative counting and the root mean square (RMS) measurements have been annalysed as they seemed to adequately characterize the acoustic emission linked to the wear and seizing phenomena. The other measurements are kept for further processing should the need arise during the course of the investigation.

An excellent correlation has been observed between the cumulative counting of acoustic events and the measured friction load, either in low displacement (1 mm) or in high displacement (4 mm). It has been shown as well, and this is perhaps the most important observation, that the occurence of seizing is well detected (see fig. 4a & 4b) using ring-down counting methods.

Rms measurements of the AE signals have appeared as interesting and reliable for the detection of seizing. Figure 5a shows that in normal wearing conditions, the rms signal is flat. As soon as seizing occurs, it is detected (fig. 5b). The rms signal amplifies dramatically as seizing increases (fig. 5c and fig. 5d).

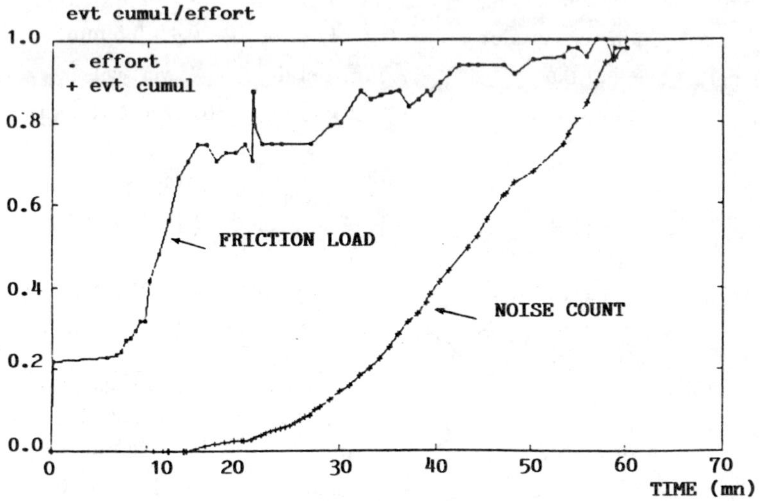

Fig. 4a - Bad lubrication

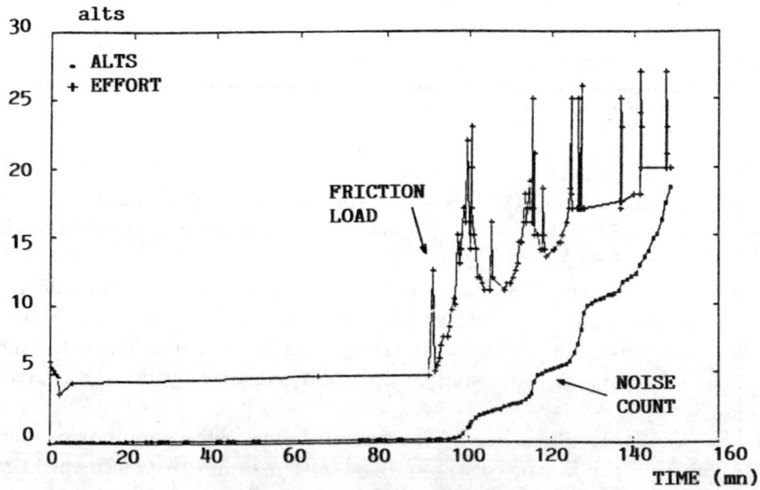

Fig. 4b - Good lubrication

One of the objective of this experiment was to determine whether a wire relubrication could eliminate seizing and whether the efficiency of the method could be detected with AE measurements. Additional laboratory tests have been conducted to investigate the effect of relubrication after an initial seizing occurence. The evolution of

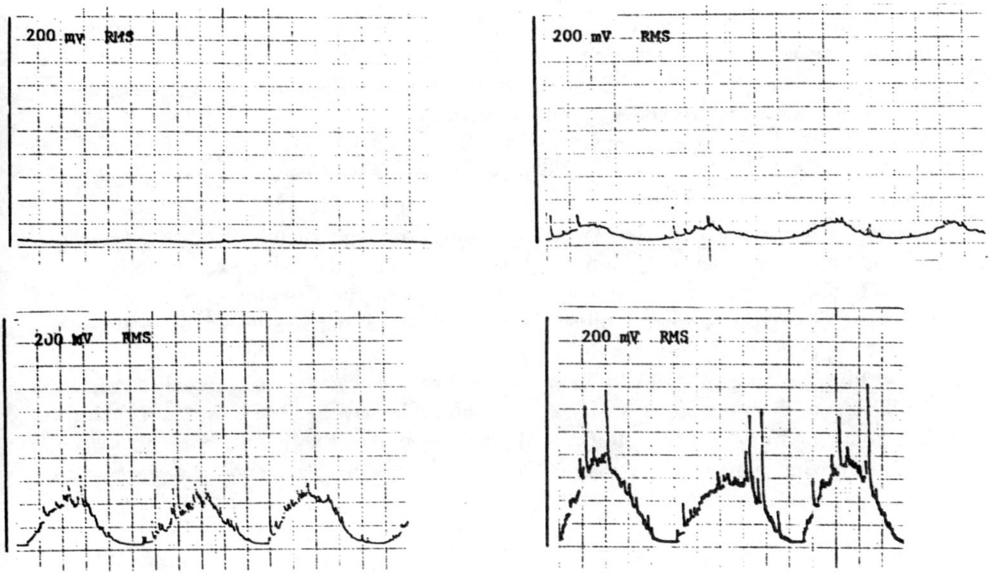

Fig. 5 - RMS signal evolution

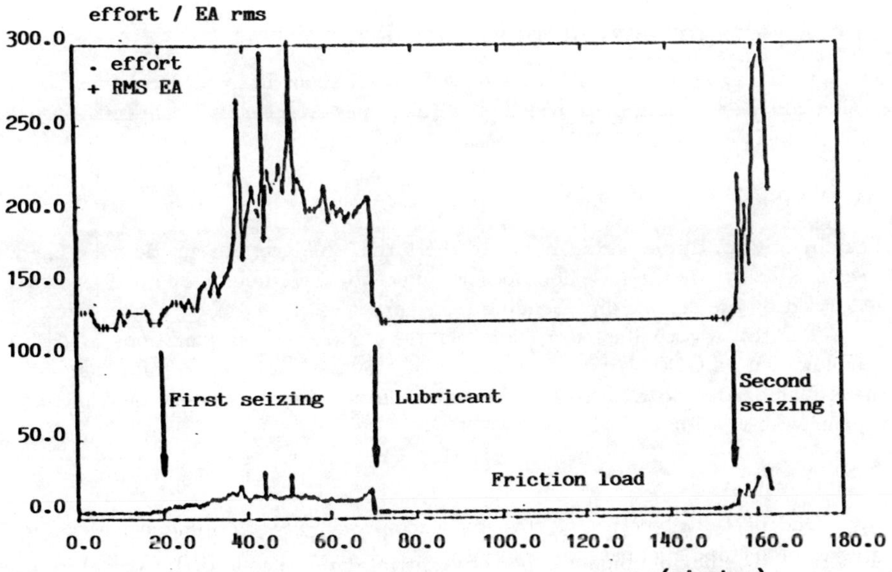

Fig. 6

The acoustic emission has been followed by cumulative counting and rms measurements. The results of the experiment are presented in figure 6. They relate to an accelerated test whereby the appearance of seizing has been obtained while operating at high contact pressure and without lubrication.

The first seizing occurence is clearly detected. After approximately a period of one hour of operation in seizing conditions, a new wire lubrication has been conducted which resulted in a significant decrease in the friction load and the associated acoustic emission, representative of the new friction conditions. Removing the lubricant a second time induces a new seizing phase which is easily detected by the AE sensors.

It appears that in laboratory conditions, acoustic emission is an efficient and reliable technique in order to follow the friction conditions between armour wires and detect the appearance and disappearance of seizing.

In order to establish a full correlation between the acoustic signal, the friction conditions and the resulting wear rates, a comprehensive test programme is set up in well controlled laboratory conditions. This programme is meant to cover most of the operating conditions that a dynamic COFLEXIP flexible riser may encounter in situ, with displacements ranging from 2 to 10 mm and contact pressures ranging typically between 20 and 60 bar.

The acoustic signatures obtained from this parametric study in normal friction conditions or in seizing conditions will then be correlated with measured wear rates. Correlations with measurements obtained from pipe sample testing will also be established.

Bench tests on flexible pipe samples

These tests were conducted in order to verify that acoustic signatures of friction and seizing conditions could be recorded on actual pipes structures. To this respect, three series of tests were conducted, as described here after.

- Multiphase flow circulation

The objective of this experiment was to check that the level of acoustic noise likely to be generated by a multiphase flow does not affect the recording of the acoustic emission generated by the flexible pipe bending under pressure.

The test was conducted in IFP facilities (Solaize). Test conditions consisted of slug flows with a GOR varying between 0 and 1100 Sm^3/m^3. Test results indicate that the AE associated with these types of flows do not disturb the AE associated with armour wire wearing and seizing.

- Friction tests

These bench tests have been carried out over a 6 weeks period in Coflexip facilities (Le Trait). The flexible pipe was not in a critical damage condition (normal friction). The sensors were located at various positions along the pipe sample, representative of different curvature levels, and at one end fitting. Finally, at each one of the positions along the sample, 2 sensors were used (one on the armour wires and one on the external plastic sheath).

The main results of the experiment were as follows :

- The acoustic emission is more important where the pipe curvature is higher which corresponds to the highest armour wire displacements;
- Internal pressure variations affect the acoustic measurement in the high curvature section where high pressures tend to increase the AE signals;
- Strong similarities are observed between the acoustic sensors located on the armour wires and on the external plastic sheath (fig. 7);
- Very little evolution of the characteristics of the acoustic emissions has been observed over the 6 weeks experiment duration. This result is as expected considering the normal friction conditions of the pipe sample.

As a result, acoustic emission seems to be associated with pipe curvature changes and there are strong indications that the monitoring of armour wires can be achieved from the outside of the external plastic of the pipe.

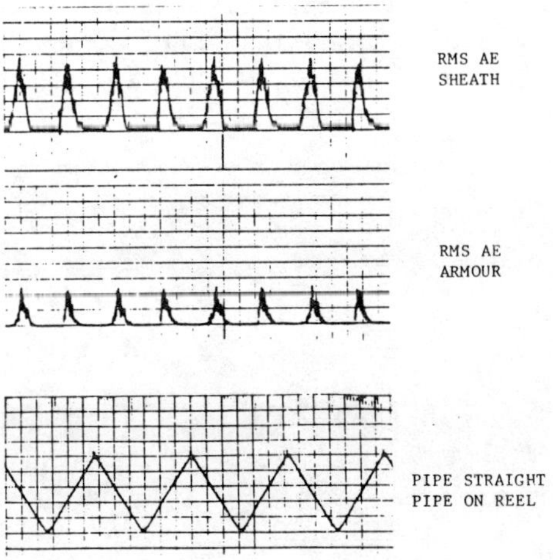

Fig. 7 - Lubricated - 600 cycles

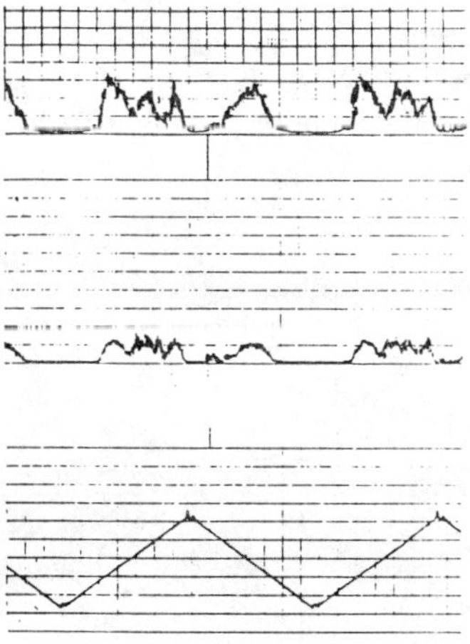
Fig. 8 - Non lubricated - 7400 cycles

- Seizing test

The objective of this test was to confirm that the AE signals measured on the armours and on the sheath are indications of the same physical phenomenon and that the acoustic emission characteristical of the seizing of a flexible pipe sample (the same as the one tested for multiphase flows) can be detected. Drastic bending conditions have been used to accelerate the armour wire seizing development and lubricant has been removed by cleaning several times with a solvant.

A first series of test was started under the conditions :

P = 200 bars (Internal Pressure)
T = 16 tons (Axial Tension)

3600 cycles have been carried out in these conditions. Then the pressure was increased to 350 bars and the tension to 30 tons. Test conditions were characterised by a cycle duration of 32 seconds and a curvature variation of 0.59 m^{-1}.

It was confirmed (fig. 8) that the acoustic emission measured from the outside of the external sheath corresponded to friction between the amour wires and that the seizing occurence is well detected.

Conclusion

The preliminary results clearly demonstrate that the evolution of acoustic emissions relates well to changes between smooth wear and seizing phases, representative of normal and critical operating conditions of dynamic flexible risers.

A attempt at quantifying the phenomena involving wire overstressing due to friction coefficients increases and their evolution will be carried out through laboratory and bench testing. Finally, probabilistic GOODMAN diagrams will be established. in order to take full account, in HAIG'S diagram, of the transient phases associated with overstresses and accelerated wear.

INJECTION OF GLYCOL INTO CO_2-CONTAINING NATURAL GAS IN PIPELINES FOR CORROSION MITIGATION

Glycol/water phase behaviour

by

J.N.J.J. LAMMERS

Koninklijke/Shell-Laboratorium, Amsterdam
(Shell Research B.V.)
Badhuisweg 3, 1031 CM Amsterdam, The Netherlands

Summary

The transportation of wet, CO_2-containing gases in pipelines requires special measures to prevent corrosion of the pipewall. A relatively new option involves the injection of glycol, the presence of which in the water phase mitigates corrosion.

One of the essential parameters determining the corrosion rate is the composition of the condensing glycol/water mixture along the pipeline wall. The paper presents a method for the calculation of this composition. First principles have been applied for phase equilibrium and mass transfer calculations. Natural gas condensate may have a significant effect on glycol and water mass transfer. An approach is presented which enables quantification of this effect.

Calculations have been carried out on the newly designed 66 km long, 36" trunklines between the Troll platform off the Norwegian coast and the offshore processing plant. Results for various sets of representative conditions are presented.

1. INTRODUCTION

Recently, Norske Shell completed the conceptual design of a twin, 66 km long 36" gas pipeline system for the transportation of natural gas from the offshore Troll field to the mainland. Out of several possible transportation modes for the CO_2-containing natural gas, the final choice was made in favour of the "wet gas transportation option". In this option water-saturated gas is allowed in the pipeline and drying of the gas is carried out onshore. Installation of the glycol-treating unit onshore instead of on the platform represents, apart from capital savings, improved operational flexibility.

Transportation of wet, CO_2-containing natural gases in pipelines, however, requires special measures to prevent unacceptable corrosion of the steel pipewall. One of these measures is the injection of glycol into the pipeline. In the presence of glycol, liquids condensing onto the relatively cold pipewall will be relatively rich in glycol. Such water/glycol mixtures have been shown to exhibit reduced corrosion rates compared to pure water (1). The level of corrosion reduction depends, amongst other factors, on the composition of the liquid glycol/water mixture.

In this paper the basis for the calculation of the composition of the glycol/water mixture along the pipeline is outlined. In addition, routes are presented which quantify the hitherto little understood effect of condensate in the pipeline on the composition of the glycol/water condensing at the top (i.e. onto the upper side) of the pipeline. Results are presented for various operation scenarios of the Troll pipelines.

2. BASIS OF CALCULATION

Glycol, injected at the upstream end of the pipeline (usually on a platform), is assumed to form a liquid glycol/water phase in equilibrium with the wet natural gas at the pressure and temperature prevailing at the pipeline inlet. Equilibrium will be promoted during the flow in the vertical riser (the processes taking place in the riser will not be discussed any further here). In the subsequent, more or less horizontal part of the pipeline, the liquid glycol/water will flow along the bottom in case of stratified flow conditions (see below). While flowing further downstream the pipeline, the fluids will cool down, causing, among other things, condensation of glycol/water at the top of the pipeline.

It is conceivable that the top glycol/water condensate is somewhat richer in water than the glycol/water at the bottom of the pipeline. When some glycol/water condenses, the composition of the remaining gas will shift to a somewhat higher relative water content, because the glycol/water condensate is relatively richer in glycol than the gas. As the mass transfer across the bottom liquid-gas interface has a finite rate, the resultant non-equilibrium condition cannot immediately be corrected by mass exchange with the bottom glycol/water. Condensation somewhat further downstream against a cooler top of the pipeline then results in a glycol/water mixture that is somewhat richer in water than the bottom phase.

In the case of a liquid hydrocarbon condensate layer covering the bottom liquid glycol/water, the mass transfer across the bottom liquid-gas interface would be further hampered. Consequently, the water content of the glycol/water condensing against the top of the pipeline would increase as compared with the situation where no such layer is present. This phenomenon will occur under stratified flow conditions; in the annular and slug flow regimes no "top of the line" condensation will occur; moreover, in the slug flow case gas and liquid will resume equilibrium because of intense contact. The assumption of stratified flow therefore represents a most severe situation from the point of view of top of the line condensation.

On the basis of the above concept a proprietary computer program has been written which calculates at specified positions along a natural gas pipeline the composition of the bottom and top glycol/water condensates. The calculation is performed for specified pipeline operating conditions, using first principles for phase equilibrium and mass transfer calculations (Appendix). Input data are the natural gas flow rate, gas average mole weight, pipeline diameter and temperature, pressure and hold-up profile (including glycol/water) along the pipeline as calculated by a multiphase flow computer program. The program also calculates the quantity of glycol needed to achieve a specified water content of the bottom glycol/water mixture at the downstream end of the pipeline given the water content of the inlet gas.

The model describing the condensation process is based on a number of simplifying assumptions (see also Ref. 1):

- The water/glycol mixture condensing at the top of the pipeline is in equilibrium with the vapour phase entering from the previous pipeline section at the local pipe wall temperature. This assumption is conservative with respect to predicted corrosion rates since it leads to underprediction of the glycol concentration of the "top of the line" condensate.
- After condensation, the liquid runs down the wall to mix with the liquid accumulated at the bottom of the line without further exchange of water or glycol with the gas phase.

- The model assumes the rate of evaporation of glycol from the liquids at the pipe bottom into the bulk gas phase and the rate of absorption of water in the glycol mixture at the bottom of the pipe to be completely gas-phase-limited. The interfacial area is calculated from the liquid hold-up assuming stratified smooth flow.
- Vapour-liquid equilibrium calculations for glycol (EG, DEG or TEG) and water are based on the principle of equal fugacities of each component in the gas and liquid phases. The calculation of the gas fugacity coefficients has been generalised via the gas density to be applicable to different types of natural gases.

3. EFFECT OF CONDENSATE ON GLYCOL AND WATER MASS TRANSFER

When hydrocarbon condensate flows as a stratified layer on top of liquid glycol/water along the bottom of the pipeline, the mass transfer between the glycol/water liquid phase and the gas phase will be hampered. This is due to the very low solubility of water as well as glycol in liquid hydrocarbons. In the case of a stagnant layer of condensate, the mass transfer will be fully governed by diffusion and therefore become so low that effectively no mass transfer takes place. Under turbulent conditions in the condensate layer the mass transfer will be increased of course, but expectedly only to a small extent, again due to the very low solubility of glycol and water. Only under such conditions of flow that the glycol/water liquid starts to disperse in the condensate, can direct exposure of glycol/water droplets to the gas phase occur. The mass transfer would then again become gas-phase-limited as in our standard case of full exposure (Appendix).

To estimate the effect of the degree of exposure of bottom liquid glycol/water to the gas phase we have introduced an "exposure coefficient". This coefficient represents the fraction of the liquid phase surface area which is occupied by glycol/water and thus available for mass transfer to and from the gas phase. In order to obtain an estimate for the value of the exposure coefficient it has been investigated to what extent we may expect dispersion to occur.

The following assumptions have been made to estimate the effect of dispersion on exposure:

(a) In case of full dispersion, exposure of the liquid glycol/water will occur in the volumetric ratio of glycol/water: (condensate + glycol/water); this implies an exposure coefficient equal to that ratio.

(b) In case of no dispersion, the exposure coefficient will be zero.

(c) In the intermediate case, exposure will depend on the fraction of glycol/water droplets taken up by the condensate, transported to the liquid/gas interface and exposed for a certain period of time. The exposure coefficient from (a) should then be multiplied by a reduction factor. We do not aim at the moment to present a method determining the value of such factor for the intermediate case.

Criteria for dispersion have been obtained following two different routes: one concerns the breaking-up of two liquid layers into a dispersion, the second route concerns the dispersion of already formed liquid droplets into a second liquid.

1. The breaking-up of two liquid layers has been approached via a criterion which was developed by Ishii and Grolmes (2) for the entrainment of liquid in a gas flow. When applied to our case this criterion implies that entrainment occurs when the drag force F_d exerted by the continuous liquid (hydrocarbon condensate) on the liquid to be dispersed (glycol/water) exceeds the retaining force of the surface tension F_σ. For large

film Reynolds numbers and viscosity number $N_\mu < 1/15$ the criterion reads:

$$I = \frac{\mu_w . v_L}{\sigma} \sqrt{\frac{\rho_c}{\rho_w}} \geq N_\mu^{0.8} \quad \text{(eq. 1)}$$

The viscosity number is defined as:

$$N_\mu = \frac{\mu_w}{\{ \rho_w . \sigma \sqrt{\frac{\sigma}{g.\Delta\rho}} \}^{1/2}} \quad \text{(eq. 2)}$$

The Reynolds number of the film glycol/water is defined as follows:

$$Re_{f,w} = \frac{4\Phi_w . \rho_w}{p_w . \mu_w} \quad \text{(eq. 3)}$$

The wetted perimeter: $p_w = \frac{1}{2} d(\alpha + 2\sin\frac{1}{2}\alpha)$ (eq. 4)

The angle α is calculated with eq.A.2 using the glycol/water hold-up. The latter has been obtained from the total hold-up (from multiphase flow calculations) and the assumption that the ratio of the glycol/water hold-up to the total liquid hold-up equals the ratio of the flows (from multiphase flow calculations and calculation of the bottom glycol/water flow). The glycol/water hold-up is calculated to be a quarter of the total hold-up. The value of this factor appears to vary only by ± 0.05 and is for our approach assumed constant all along the pipeline for the four cases considered. The value for σ has been assumed 0.02 N/m. In Table I four different cases have been considered (see for description further down).

Table I: Criterion for entrainment (2)

Case	μ_w (Pa.s)	v_L (m/s)	σ (N/m)	ρ_c (kg/m^3)	ρ_w (kg/m^3)	I (-)	$N_\mu^{0.8}$ (-)	$I>N_\mu^{0.8}$	$Re_{f,w}$ (-)
I	.002	1.3	.02	650	1025	.052	.023	y	4000
II	.002	1.7	.02	650	1025	.069	.023	y	4300
III	.002	2.0	.02	650	1025	.081	.023	y	4700
IV	.002	2.8	.02	650	1025	.113	.023	y	7400

2. The dispersion of droplets into a second liquid has been approached via a criterion developed by Davies (3) for dispersion of solid particles in liquid pipeline flow. A similar approach was followed by Duckler and Taitel (4) in gas/liquid pipeflow for the transition between intermittent flow and dispersed flow of gas bubbles. Dispersion occurs when the eddy fluctuation force exerted by the liquid exceeds the sedimentation force of the particles.

The sedimentation force: $F_s = \frac{\pi}{6} \delta^3 . \Delta\rho . g . (1-c)^n$ (eq. 5)

The eddy fluctuation force: $F_e = \rho_c . (v')^2 \frac{\pi}{4} \delta^2 / (1+\epsilon . c)^2$ (eq. 6)

The eddy fluctuation velocity v'is given by: $(v')^3 = P.\delta$ (eq. 7)

Extension of this theory to liquid dispersions requires determination of the dispersed phase droplet diameter. The droplet diameter has been calculated according to Hinze (5):

$$\delta_{max} = 0.725 \left(\frac{\sigma}{\rho_L}\right)^{0.6} . P_L^{-0.4} \quad \text{(eq. 8)}$$

The value of d_{max} has been used for the droplet diameter in the various equations. This renders the dispersion criteria conservative. In our case of three-phase flow, P denotes the power dissipated in the two liquids flowing along the bottom of the pipeline as a result of friction with the pipewall. It can be derived that:

$$P_L = \frac{\tau_w . a_L . v_L}{\rho_L . h . \frac{1}{4}\pi d^2} \qquad \text{(eq. 9)}$$

The wall shear stress τ_w may be calculated from:

$$\tau_w = \tfrac{1}{2} f_w . \rho_L . v_L{}^2 \qquad \text{(eq. 10)}$$

Using a wall friction factor calculated according to the Blasius equation: $f_w = 0.08 Re_L{}^{-0.25}$, it follows from eqs. 9 and 10 that

$$P_L = \frac{2}{\pi} \left(\frac{0.08}{Re_L{}^{0.25}}\right) \frac{v_L{}^3 a_L}{hd^2} \qquad \text{(eq. 11)}$$

where the wetted wall area per unit pipelength $a_L = \frac{1}{2} d\alpha$, α is defined and calculated according to the Appendix. The Reynolds number of the liquid layer is defined as:

$$Re_L = \frac{\pi d^2 h v_L \rho_L}{p_L \mu_L} \qquad \text{(eq. 12)}$$

where the wetted perimeter $p_L = \frac{1}{2} d(\alpha + 2\sin\frac{1}{2}\alpha)$. (eq. 13)

Results of calculations are presented in Table II. The value for c is as before 0.25. The internal pipeline diameter is 0.854 m.

Table II: Criterion for dispersion (3)

Case	v_L (m/s)	h (-)	p (m)	Re_L (-)	P_L (W/kg)	α (°)	v' (m/s)	F_e*10^6 (N)	F_s*10^6 (N)	F_e>F_s
I	1.3	.010	.62	7100	0.5	42	.05	3.0	2.2	y
II	1.7	.010	.62	7700	1.2	42	.06	2.2	0.9	y
III	2.0	.009	.59	8600	1.9	40	.06	1.8	0.5	y
IV	2.8	.008	.56	13500	5.0	38	.08	1.2	0.15	y

Both criteria described above indicate the likelihood of glycol/water to be dispersed in the condensate. The exposure coefficient is then determined, according to assumption (a) above, by the phase ratio, which as mentioned before is approximately 0.25.

It should be noted that the two criteria applied above do not account for a homogeneous dispersion. The concentration of glycol/water droplets will therefore probably be greater in the lower part than in the upper part of the condensate layer. The exposure coefficient according to (a) above may therefore be somewhat optimistic, especially in case the above criteria are only just fulfilled. On the other hand, our approach assumes a flat interface between gas and liquid. This is a conservative assumption since in reality the interface will be wavy and moreover, especially in the case of a flow regime close to the stratified-annular flow boundary, the interface will be curved. The actual interfacial area will therefore be appreciably larger than the flat interface assumed in the above approach.

4. RESULTS

Several gas production scenarios have been considered by Norske Shell, of which we selected a limited number to perform our calculations (Table III).

The pipeline temperature and pressure profiles have been calculated using a multiphase flow computer program. Table III shows the pipeline inlet conditions for calculation of the water content of the natural gas, the water content of the aqueous condensate at the end of the pipeline as used in our calculations, and the calculated amount of glycol injected (DEG, 90 %wt, per pipeline).

Table III: Operation scenarios for one Troll 36" pipline used in calculations

Case	Inlet conditions Temp. (°C)	Inlet conditions Press. (bar)	Flow (10^6 stand.m^3/d)	Water end line (%wt)	Injected glycol (kg/s)
I	55	105	27.5	33	1.3
II	55	80	27.5	35	1.4
III	55	105	37.5	33	1.8
IV	55	80	37.5	38	1.6

Results of computer simulations with our computer program are shown in Figs. 1 and 2 for Cases I and III using a value of 0.25 for the exposure coefficient.

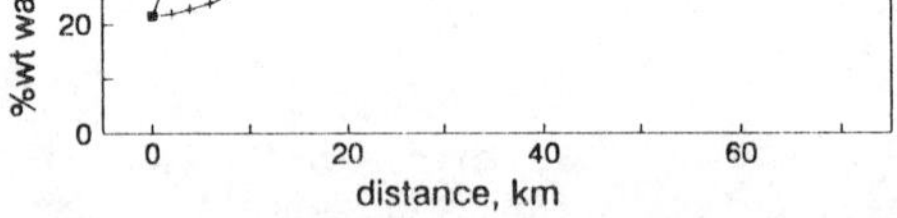

Fig. 1 Composition of bulk and "top of the line" glycol/water condensate along Troll 36" pipeline, Case I, 0.25 exposure coefficient.

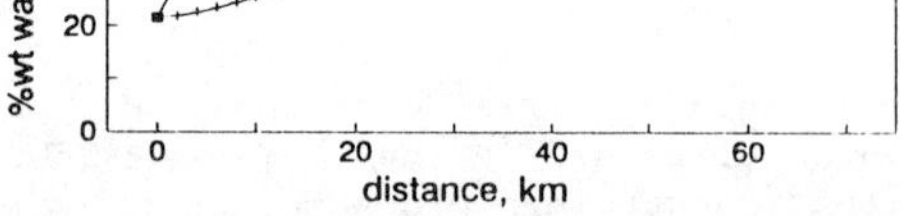

Fig. 2 Composition of bulk and "top of the line" glycol/water condensate along Troll 36" pipeline, Case III, 0.25 exposure coefficient.

The results show a gradual increase in water content of the bulk glycol/water up to the specified water content at the end of the pipeline (Table III), as expected. The composition of the "top of the line" condensate, however, shows a remarkable course, especially between approximately km 8 and 22 for both Cases I and III where the water content exceeds 90 %wt. This must be attributed to the reduced mass transfer between the bulk glycol/water at the bottom of the pipeline and the gas phase due to the presence of a layer of condensate.

The effect of coverage of glycol/water by condensate is demonstrated in Fig. 3 on the basis of Case III by changing the value of the exposure coefficient. Four different values of the exposure coefficient e = 0.1, 0.25, 0.5 and 1 have been used. For e=1, expectedly, the compositions are nearly identical to the "bottom of the line" compositions (compare Fig. 2). For e=0.01 over quite a stretch of pipeline nearly 100 % water condenses at the top of the pipeline.

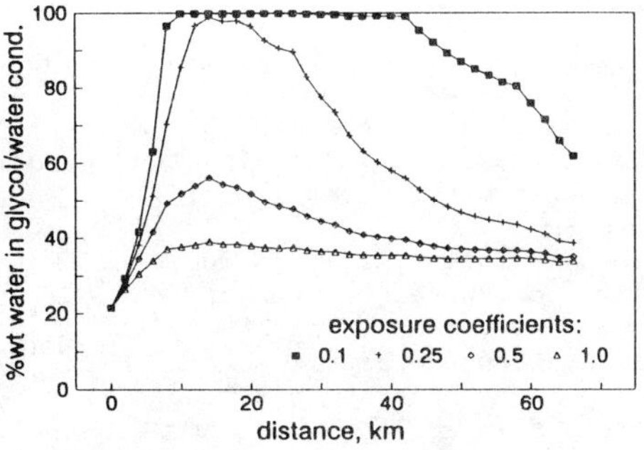

Fig. 3 Composition of "top of the line" glycol/water condensate along Troll 36" pipeline for different values of the exposure coefficient, Case III.

Further sensitivities have been demonstrated in the following:

- Lowering of the pipeline inlet temperature from 55 to 45 °C. The major effect of this change is to be expected from a lower water content of the inlet gas. Injection of the same amount of glycol as in base Case III (1.8 kg/s) indeed results in a much more favourable composition of the glycol/water condensate (Fig. 4).
- Increase of the amount of injected glycol with 50 % (taking 55 °C for inlet temperature as in base Case III). This also leads to a favourable change in glycol/water "top of the line" condensate composition (Fig. 5), though less than in the case of lower inlet temperature.

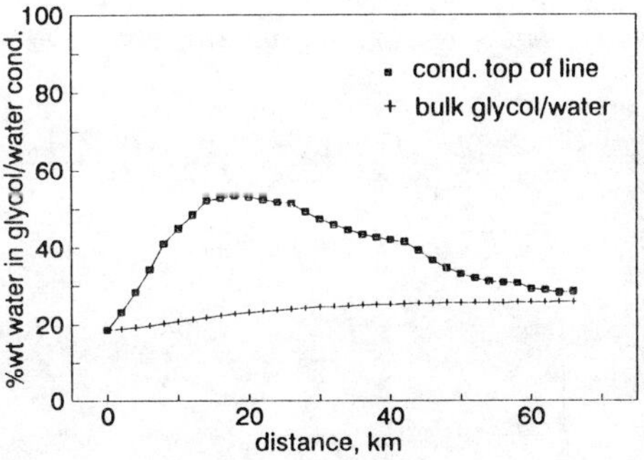

Fig. 4 Composition of bulk and "top of the line" glycol/water condensate along Troll 36" pipeline, Case III, inlet temperature 45 °C, 0.25 exposure coefficient.

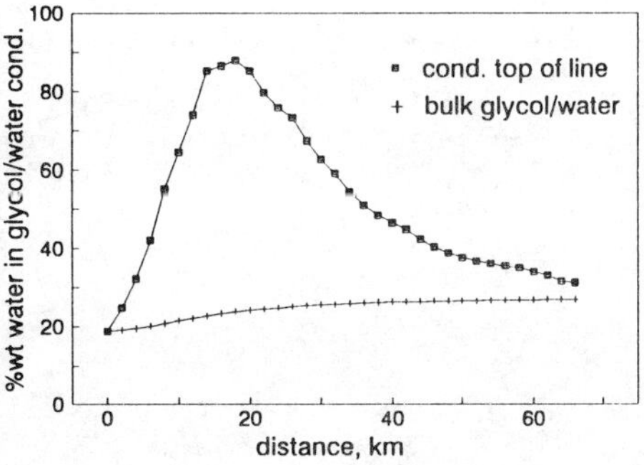

Fig. 5 Composition of bulk and "top of the line" glycol/water condensate along Troll 36" pipeline, Case III, 0.25 exposure coefficient, 50% increase of injected glycol (2.7 kg/s).

4. CONCLUSIONS

A computerised method has been described which enables the calculation of the phase behaviour of glycol in wet CO_2-containing natural gas in pipelines, the composition of the glycol/water condensate in such a natural gas pipeline being a determining factor in the anticipated corrosion behaviour.

The effect of natural gas condensate on the mass transfer of glycol and water between liquid glycol/water at the bottom of the pipeline and the gas phase is approached by means of two dispersion criteria. In the various cases considered for the operation of the Troll natural gas pipeline system, glycol/water is shown to be dispersed in the condensate flow.

Calculation examples indicate that while the water content of the bulk glycol/water towards the end of the pipeline gradually increases, the glycol/water condensing at the top of the line in a section near the beginning of the pipeline has a high (>90 %wt) water content when the effect of hydrocarbon condensate is taken into account. Increasing the amount of injected glycol, but especially lowering of the pipeline inlet temperature reduces the water content of the "top-of-the-line" glycol/ water and therefore would improve corrosion mitigation.

REFERENCES

1. L. van Bodegom, K. van Gelder, J.A.M. Spaninks and M.J.J. Simon Thomas, Corrosion 88, March 21-25, 1988, St.Louis, Missouri, USA.
2. M. Ishii and M.A. Grolmes, AIChE J. 21, 308 (1975).
3. J.T. Davies, Chem. Eng. Sci. 42, 1667 (1987).
4. A.E. Duckler and Y. Taitel in: "Multiphase Science and Technology", Vol.2, (Ed. G.F. Hewitt et al.) Springer (1986).
5. J.O. Hinze, AIChE J. 1, 289 (1955).
6. R.H. Perry and C.H. Chilton, "Chemical Engineers Handbook", 5th ed., McGraw-Hill, New York (1973).
7. D. Ambrose and D.J.Hall, J. Chem. Thermodyn. 13, 61 (1981).
8. J.L. Hales et al., J. Chem. Thermodyn. 13, 591 (1981).
9. M.A. Villamanan, C. Gonzales and H.C. van Ness, J. Chem. Eng. Data 29, 427 (1984).
10. H.C. van Ness and M.M. Abbott, "Classical Thermodynamics of Non-Electrolyte Solutions", McGraw-Hill, New York (1982) pp. 220-231.
11. Dow Chemical Company, "Gas Conditioning Fact Book", Dow, Midland, Michigan (1962).
12. API Technical Data Book, Petrol. Ref. Vol.II, 4th ed. (1983).

LIST OF SYMBOLS

a,A	=	area	(m^2)
c	=	glycol/water fraction on total liquid	(-)
d	=	pipeline inside diameter	(m)
D	=	gas phase diffusivity	(m^2/s)
e	=	exposure coefficient	(-)
f	=	friction factor	(-)
F	=	force	(N)
g	=	acceleration due to gravity	(m/s^2)
h	=	total liquid hold-up (volume fraction liquid)	(-)
I	=	Ishii & Grolmes dimensionless parameter (eq. 1)	(-)
k_{og}	=	gas phase mass transfer coefficient	(m/s)
N_μ	=	viscosity number (eq. 2)	(-)
p	=	total pressure or	(Pa)
		wetted perimeter (eq. 4,12)	(m)
p^*	=	pure component saturated vapour pressure	(Pa)
P	=	power dissipated by the flow per unit mass	(w/kg)
Re	=	Reynolds number (eq. 3,11)	(-)
Sc	=	Schmidt number (μ/ρ/D)	(-)
Sh	=	Sherwood number (eq. A.3)	
v	=	axial velocity	(m/s)
v'	=	eddy fluctuation velocity	(m/s)
x	=	liquid phase mole fraction	(-)
y	=	gas phase mole fraction	(-)

Sub/super scripts

c	=	condensate
e	=	eddy fluctuation
f	=	film
L	=	total liquid (condensate + glycol/water)
max	=	maximum
n	=	exponent for hindered settling (Ref. 3)
s	=	sedimentation
w	=	glycol/water or wall

Greek

α	=	angle of sector in pipeline cross-section covering liquid hold-up	(rad)
γ	=	activity coefficient (liquid phase non-ideality)	(-)
δ	=	dispersed droplet diameter	(m)
ϵ	=	damping factor for eddy turbulence energy taking account of finite concentration dispersed droplets	(-)
$\Delta\rho$	=	density difference glycol/water - condensate	(kg/m^3)
μ	=	dynamic viscosity	(Pa.s)
Π	=	Poynting correction (expressing effect of total pressure on component fugacity in the liquid phase)	(-)
ρ	=	density	(kg/m^3)
σ	=	interfacial tension water/condensate	(N/m)
τ	=	shear stress	(N/m^2)
ϕ	=	fugacity coefficient (expressing gas phase non-ideality)	(-)
Φ	=	volumetric flow rate	(m^3/s)

APPENDIX: Calculation of the composition of glycol/water condensate along a natural gas pipeline on glycol injection

The basics of the calculation model have been described in detail in (1, Appendix). We will restrict ourselves here to the main lines and the enhancements.

Water-saturated gas entering the pipeline is contacted with regenerated glycol of specified residual water content. The two-phase equilibrium mixture then enters the pipeline. Stratified smooth flow is assumed throughout the line.

Component mass balances are set up for glycol and water each and the change in gas phase composition along the pipeline is calculated in a stepwise manner on the basis of the mass transfer rate between gas and bottom glycol/water and (assumed) equilibrium between the gas and the condensate on the pipewall. The pressures and temperatures along the pipeline are obtained from a multiphase flow computer program.

The mass transfer rate is the product of interfacial area and mass transfer coefficient k_{og}. The interfacial area per unit pipe length is calculated from the pipeline liquid hold-up h given by multiphase flow calculations and the pipe geometry, see Fig. A.1. The resulting equations are:

$$A = d * \sin(\alpha/2) \quad \text{(A.1)}$$

$$\alpha = 2\pi * h + \sin(\alpha) \quad \text{(A.2)}$$

The mass transfer coefficient will be gas phase limited for both glycol and water evaporation or absorption because of the infinite mutual miscibility of water and glycol and the high concentrations of each. We first will perform the analysis without condensation of hydrocarbons. In that case, k_{og} can be estimated from the following equation which holds for turbulent flow over a smooth surface (cf. eqn 10-50 Ref. 6).

$$Sh_g = k_{og}\ d/D = 0.023\ Re^{0.8}\ Sc^{0.33} \quad \text{(A.3)}$$

The diffusivities of glycol and water in the gas phase have been estimated from eqn 3-29 in (6).

The phase equilibrium conditions for glycol and water read:

$$\phi_i \cdot y_i \cdot p = x_i \cdot \gamma_i \cdot p^{\circ}_i \cdot \Pi_i \quad \text{(A.4)}$$

Phase equilibrium data (7-12) have been used to generate suitable relations for the parameters in equation (A.4).

For the description of the mass transfer process when hydrocarbons do condense along with the water/glycol mixture we refer to the main text. In this case, for the calculation of the gas-liquid interfacial area, the total hydrocarbon + glycol/water hold-up is applied.

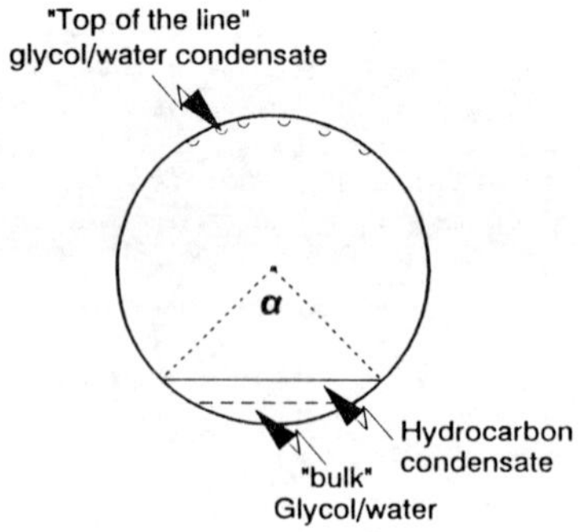

Figure A.1

Schematic pipe cross section showing hold-up = volume fraction hydrocarbon condensate + bulk glycol/water, "top of the line" glycol/water condensate and definition of α.

A NEW INSTRUMENT FOR MEASURING GROSS HEATING VALUE OF NATURAL GAS

D. INGRAIN[1], J.H. ZORIO[1], G. DESERT[2] and L. MONDEIL[2]

[1]DIRECTION DES ETUDES ET TECHNIQUES NOUVELLES
GAZ DE FRANCE (FRANCE)

[2]DIVISION RECHERCHE ET DEVELOPPEMENT EN PRODUCTION
SOCIETE NATIONALE ELF AQUITAINE PRODUCTION (FRANCE)

A new recording calorimeter was developed by ELF-AQUITAINE PRODUCTION RESEARCH CENTER for the measurement of gross heating value of natural gas type fuels (1,2).

This calorimeter is an differential combustion calorimeter to be used in an unprotected environnement, based on the following general principles :

* measurement of the integral of the calorific flow produced by the combustion of the gas ;

* cancelling of the heat balance and heat exchange errors ;

* automatic reference to a standard gas.

This apparatus, controlled by a microcomputer, uses the last developments reached in matters of sensors associated with numerical signal processing.

This calorimeter has now been evaluated and tested by GAZ DE FRANCE RESEARCH AND DEVELOPMENT DIVISION. The calorific values of several synthetic natural gas (34,7 to 44,8 MJ/m^3(n)) were measured with this apparatus and compared to the theoretical gross heating value calculated with the international standard ISO 6976. The test procedure and the results of measurements are described ; the accuracy (0,3%), the repeatability, the time reponse and the interval of calibration with a standard gas are given and conclusions presented.

References :

1 MONDEIL L. J. and ROBERT F. M., MESUCORA PARIS, pp. 79-87, (1985)
2 MONDEIL L. J. and ROBERT F. M., GAS QUALITY, pp. 111-120, (1986)

NEW SELECTIVE WATER CONTROL PROCESSES : MAIN CHARACTERISTICS AND FIELD TESTING.

N. Kohler and A. Zaitoun (Institut Français du Pétrole), J.F. Coste and J.P. Zundel (Elf Aquitaine)

Water breakthrough at producing wells is generally responsible for increasing production costs : problems of corrosion and scale deposits can occur and activation is soon necessary. Also due to the heterogeneity of reservoirs submitted to waterflooding operations, the sweeping efficiency is generally poor and the oil recovery factor limited.

NEW PROCESSES CHARACTERISTICS :

We have recently developed new water control processes based on the injection of single high-molecular-weight water-soluble polymers directly into the producing interval. As compared to treatments using gels or crosslinked systems, these processes exhibit a good selectivity for reducing the relative permeability to water without affecting the relative permeability to oil. Further advantages of water control by aqueous polymer solutions are their selective placement in the watered out layers and the easiness of their chemical destruction if necessary. When set in place into the drainage area of producing wells, these polymers provide a selective pseudo-skin effect and change the production distribution between layers, in favour of unswept ones. A higher draw-down has to be applied after the treatment in order to produce at the same flowrate and for economical reasons the candidate wells have preferably a good potential.

Three original water control processes have been developed so far. Their fields of application as regards the salinity of produced brine and the reservoir temperature are well defined. They are shown to be complementary and particularly adapted to water production control of most matricial sandstone or limestone reservoirs.

FIELD TESTS :

Although the different processes are still in a development stage, a number of field tests have already been accomplished. The poster will present photographs and flow sheets of implementation and treatment results of a representative field test for each process.

Process 1 was proposed to control water production from an underground gas storage well. The existence of a high-permeability streak and an active aquifer induced excessive water production in this well and the neighbouring ones together with a low gas inventory. Since the treatment, the well has functioned as a gas injector during summer and as a producer in winter, showing excellent characteristics for both uses : gas injectivity is as good as prior to the treatment ; gas productivity is also very good with a very reduced water production. Compared to the neighbouring wells, the water/gas ratio from this well is less than half of the ratio of the best of these wells, with a total gas production which is more than double. The treatment is still efficient two years after the treatment.

Process 2 was implemented on an offshore oil producing well with BHT of 95°C and a produced brine salinity of 90g/1 TDS. Before the treatment this well which gross production ranged between 500 and 700 M3/day had to be periodically shut-in due to uneconomical WOR induced by an uncontrolled water influx. Treatment was performed by injecting 1900 M3 of polymer solution with an average concentration of 1.0 kg/M3 and an average injection rate of 29 M3/hour. Real time monitoring of the bottomhole pressure allowed to follow the setting-in place of the polymer into the formation. The total job was completed within 3 days including a previous sea water injectivity test and a small hydrocarbon postflush. After a short shut-in time of 4 days, the well was reopen progressively and no well impairment was detected. One year after the treatment, this well still exhibits interesting production characteristics with WOR low enough to keep the well open

Process 3 was tested on a horizontal well whicn was showing an abrupt increase in water production following an attempt to reduce the detrimental effect of sand on the overall pumping rates. The total job was performed in one day and the well shut-in for five days in order to achieve polymer in-situ swelling by a concomitantly injected specific chemical agent. Well production was reinitiated by increasing progressively the pumping rate. Excellent results were recorded : oil production increased from 2 to 7.2 M3/day and the water production dropped from 12 to 8.5 M3/day. The operation was paid out in 45 days and the water cut is still decreasing 9 months after the treatment.

MARELAB - An advanced offshore laboratory platform for oceanological research and field testing of new underwater technologies

G. Sebastiani
CEOM S.C.p.A.- Palermo, Italy

As it is well-known, R&D related to offshore industry normally requires model testing and often full-scale field testing of prototypes of new underwater equipment and technologies.
It is also known that in Europe exists a high number of land-based model testing basins, but virtually no offshore facilities dedicated to experimental research activity.
Ceom (Centro Oceanologico Mediterraneo), a new research company recently established in Sicily and owned by ESPI (Ente Siciliano Promozione Industriale) (45%), AGIP S.p.A. (45%) and TECNOMARE (10%), active in the field of Marine Science and Technology, is designing an advanced offshore multidisciplinary laboratory platform called MARELAB, to be installed in the Mediterranean Sea and fully dedicated to oceanological research and technological experimentation.
CEOM will have also a land-based Oceanologic Research Centre, with offices, advanced laboratories, computer facilities and general staff.
The main features of this "unique" research facility, called MARELAB, are:

1. Tension-leg platform, capable to be removed and reinstalled in a different site after a period of permanence of one or more years.
2. Fully instrumented, to test and control its own behaviour.
3. First installation site: Sicily Channel, in about 400 m water depth.
4. Equipped with moon pool, handling system, laboratories, computer, data acquisition system, etc. in such a way to allow the simultaneous development of various research and experimentations.
5. Normally manned and capable to host external groups of scientists and engineers.
6. Available for use, on rental basis of laboratories and/or services, by third parties i. e. Scientific Institutions and Industrial Companies, on national and international basis.

Finally, MARELAB could offer the occasion for the development of large international cooperation research projects, among Engineering Companies, Oil Companies and Marine Operators.
Some possible research themes are:

- Full-scale dynamic responce measurement for platform and /or risers
- Equipment and components in-field testing (functional tests, reliability, durability)
- New materials performance (fatigue, corrosion etc.).

MASS BALANCE EVALUATION IN GAS TREATING AND PROCESSING PLANTS BY PC-ORIENTED PROGRAMME PACKAGE

F. Simon (1), Z. Csermely (1) and J. Siklos (2)

Hungarian Oil and Gas Research Institute (1)
H-8201 Veszprém, Pf. 167 Hungary

Trust of Hungarian Oil and Gas Industry (2)
H-1502 Budapest, Pf. 22 Hungary

1. INTRODUCTION

Mass balancing is an everyday problem for engineers both in the process design and the plant management. A plenty of processing units and material streams together with recirculation loops causes the calculation to be time-consuming and tedious. Till now, many computer programmes, useful among others in the hydrocarbon exploitation and processing, have been developed for this purpose. The advent of powerful PC's offers a new dimension to solve the problem.

2. METHOD AND OPERATION

The UNIBAL modelling system and programme package is worked out to evaluate mass and component balances for units and plant processes in the natural gas treating and processing.

Units and processes are represented by three kinds of basic models: 1) Separation; 2) Mixing; 3) Branching. These models consisting of linear balance and performance equations are generated by equation-oriented mode, then the solution of the process model is carried out simultaneously by a matrix method.

The input data are: flowsheet topology, total flowrates and composition of any material streams around the flowsheet, and performance parameters for the units - i.e. yields for components in the outlet streams for separation, ratio of total flowrates at the outlet side for branching - according to degrees-of-freedom of the process. The package is configured to handle maximum 50 units, 99 material streams and 15 components for each cases. The results give, for each stream, the total flowrate, the flowrates of all components and the composition.

3. TECHNICAL FEATURES

The UNIBAL package runs on IBM PC/AT and compatible machines under MS-DOS V3.X and MS-DOS V4.X operation systems. It is a user-friendly tool with menus, on-line HELP, input graphics, forms, and manifold data correction and modification facilities.

A FLOW LOOP TO TEST DRILLING FLUIDS UNDER BOTTOMHOLE CONDITIONS

D. DEGOUY and J. LECOURTIER

Institut Français du Pétrole

The basic properties of drilling fluids are subject to variations during the drilling process, with the combined effects of temperature, pressure, shear and contamination. Drilling fluids however need to maintain specific rheological and filtration properties to properly carry out their functions throughout the operations. The knowledge and the control of their evolutions are therefore of major importance for selecting and optimizing the right fluid formulations.

For this purpose, a fully automated flow loop (10 m length) has been developed for the comprehensive testing of drilling fluids under actual bottomhole conditions at temperatures up to 180°C, pressures up to 50 MPa and shear rates up to 10000 s−1. This innovative loop can be used for both applied and basic research programs. Applied programs aim at evaluating the performances of new drilling fluids formulations or additives under specific bottomhole conditions. Basic research aims to determine the general laws governing the behavior of drilling fluids as a function of bottomhole parameters.

2. MEASUREMENTS

The variations in drilling fluids characteristics are determined:

- Inside the loop during aging. Rheological measurements are performed by means of an HP–HT coaxial cylinder–type rheometer set on a bypass line. Static and dynamic filtration properties are measured in a battery of sintered–metal filters.
- Off the loop after sampling. The evolutions of physicochemical mud properties are characterized.

3. DESCRIPTION

The equipment includes two main parts, namely the fluid circulation and regulation part and the automation and data processing one.

<u>Fluid circulation and regulation</u>

The loop has been designed to have a minimum fluid volume (13 L), thus minimizing the operating cost. The metal constituents of the loop in contact with the fluid are made of Inconel 625 to ensure true chemical inertia, even under highly corrosive conditions of temperature and salinity.

A metal–diaphragm pump circulates the fluid under bottomhole temperature and pressure conditions at flow rates up to 20 l/min. Fluid flow is regulated by means of dampeners set on both sides of the circulation pump and is monitored by a gear–wheel flowmeter. The shearing effect is triggered by a 20 cm length restriction in the main line's diameter.

A double–acting–plunger pump provides the fluid pressurization at the beginning of the test and the regulation of the working pressure throughout the test as well. It is also used to inject salted solutions or solid suspensions into the loop, simulating a possible fluid contamination by formation influxes. Another plungerpump is used to create and maintain differential pressure between the two sides of the filtering area during filtration measurements.

The heating device includes 30 units, each one having its own PID regulation, to ensure effective control of the temperature all around the loop. Insulation is provided by expanded silicone–base materials.

In addition, numerous acquisition points, including 12 diaphragm–type pressure gauges, 50 thermal sensors, 40 switching thresholds for valves and pump control, are distributed all over the test loop for automatic operating.

<u>Automation and data processing</u>

The loop is fully monitored by means of a programmed automation which carries out the following functions:

- Control and regulation of the operating parameters
- Monitoring of all the functional safety sequences
- Real time data acquisition
- Two way communication with a microcomputer unit
- Continuous visualization of all operating components of the loop, on both a synoptic panel and the screen of the microcomputer.

APPLICATIONS OF HIGH PERFORMANCE COMPOSITE TUBES TO DEEPWATER RISERS

C.P.Sparks and P.Odru
Institut Français du Pétrole

1. COMPOSITE TUBES. DESIGN DETAILS AND ADVANTAGES

High performance composite tubes made from carbon fibres, glass fibres and resin have been developed by IFP and Aerospatiale since 1979. The tubes are filament wound with separate structural layers to resist axial and bursting stresses. Particular attention has been paid to the design and test of the steel end pieces. Advantages of composites for offshore applications lie principally in their low weight, high strength, fatigue resistance, and elastic characteristics.

2. TEST PROGRAMMES

During 1979-84, more than sixty 4" dia. tubes samples were tested in tension, pressure, bending, fatigue, wear, ageing. In 1983 three 15 m long tubes were fitted to a drilling riser and tested satisfactorily in the North Sea. Since 1986 several 9" dia. tubes have been built (see photograph) and tested as part of a Joint Industry Programme. Tubes resisted pressures of 105 MPa and 1 million tension cycles of 100+/-75 tonnes.

3. APPLICATIONS

TLP Risers: For deepwater TLPs, composite risers (with steel production tubings) can be operated with top tensions three times lower than for steel. This tension reduction leads to increased payload. Optimized design leads to quasi-zero thermal and pressure induced expansion, allowing tensioners to be eliminated. Axial elasticity results in low dynamic loads on the platform.

Satellite lines of Compliant Risers: Light weight composite satellite lines lead to a reduction in buoyancy requirements. Low thermal and pressure induced expansion characteristics allow lines to be fixed at both extremities without inducing secondary loads on the subsurface buoy.

Kill & Choke Lines: Drilling risers equipped with lightweight composite Kill & Choke lines require reduced quantities of buoyancy material. This leads to a reduction in the riser mass (about 30%) which reduces dynamic loads in the riser in hung off mode.

15 m long 9" dia. tube

4. REFERENCES

(1) Odru,P.,Guichard,J.C.,Levier,J.F., "Drilling Risers for Great Water Depth. Advantage of Mass Reduction by means of Composite Materials" Proceedings of D.O.T. (1985), Sorrento.
(2) Sparks,C.P.,"Lightweight Composite Production Risers for a Deepwater Tension Leg Platform". Proceedings of OMAE (1986), Tokyo.
(3) Falcimaigne J., "High Performance Composite Pipes for a Deepwater Multiline Production Riser". Proceeding of OMAE (1987), Houston.
(4) Odru P., Métivaud G., Sparks C.P., " Lightening Deepwater Offshore Structures by High Performance Composite Tubes". Proceedings of D.O.T. (1987), Monte Carlo.
(5) Sparks C.P., Odru P.,Bono H.,Métivaud G., "Mechanical Testing of High Performance Composite Tubes for TLP Production Risers" OTC (1988), Houston. Paper No. 5797
(6) Sparks C.P., Schmitt J. "Optimized Composite Tubes for Riser Applications" Proceedings of OMAE (1990), Houston.

THOR 2 – TIG HYPERBARIC ORBITAL ROBOT
SECOND GENERATION AUTOMATIC HYPERBARIC WELDING SYSTEM

J. BLIGHT, COMEX Technical Manager, & R. ROUGIER, THOR 2 – Project Manager

COMEX, 36 Bd des Océans, 13275 Marseille Cedex 9, France – Tel 91235000 Fax 91401280

1. OBJECTIVES

The aim of the project is to develop, build and test a fully integrated pipe preparation and welding system: THOR 2, which can be used for advanced diver assisted or future diverless hyperbaric welding.

2. RESULTS

The engineering and development have lead to a manufacturing phase and a complete testing program in workshop, shallow water and offshore conditions of the following components:

- A surface controlled hydraulic pipe clamp which can serve both to remove pipe ovality and to carry different tools for pipe cutting/bevelling or welding.
- A set of tools installed on an orbital carrier either by a manipulating arm, or by a diver to perform pipe machining, joint metrology and welding, all under surface remote control.
- An advanced automatic welding system consisting of 2 orbital heads working simultaneously, ensuring a faster welding cycle, 2 multiprocess (TIG, MIG) power sources, software and hardware controls, demagnetisation system, induction pre–heating.
- A laser diod and camera groove tracking system, interfaced to a data bank of weld parameters for their real time and automatically selected by the weld control computers.
- A malfunction diagnostic assistance and maintenance program based on artificial intelligence.
- All equipment integrated into portable containers and submersible modules linked with umbillicals featuring fiber optics, easily mobilised on a diving support vessel.

3. CONCLUSION

In addition to the features already offered by previous mechanised welding systems:

- improved weld quality
- improved Quality Assurance and documentation of the weld
- improved working conditions of the offshore personnel and the divers.

THOR 2 is the first hyperbaric welding system which can be used in a diverless mode and as such has a major impact on:

- deep water tie–ins and repair work down to 450 meters
- use of ROV and subsea robotics
- welding of special pipes (high grade X70, X80, stainless steel, duplex, clad pipe)
- reduced costs of qualification phases
- automation topside and bottom side of the machining and welding cycles
- improvement of personnel safety.

COMPUTATIONAL TOOLS FOR THE ANALYSIS AND DESIGN OF TLP TETHERS

J.F. McNAMARA and M. LANE

Marine Computation Services International,
3 Buttermilk Walk, Galway, IRELAND.

1. INTRODUCTION

This short communication describes a suite of computational tools for addressing problems associated with all phases of the installation and operation of tension leg tethers. The loss of a tether during the towing phase of the recent Jolliet TLP installation and the subsequent urgent reanalysis of the partially moored platform highlight the importance of this area of offshore engineering analysis and design.

2. SOFTWARE OVERVIEW

The suite of computational tools comprises a range of software codes in both time and frequency domains and two- and three- dimensional reference frames. All the programs embody a general finite element formulation and share a common data input format. They are uniquely capable of accurately modelling TLP tether configurations in the towing, upending and operational phases. All appropriate load terms are incorporated, including hydrodynamic forces due to regular and random waves and currents, buoyancy, gravity (including clump weights) and attached vessel motions. In the three-dimensional codes, noncollinear waves and current, and 3D vessel motions may be considered. Both the 2D and 3D time domain programs incorporate a highly accurate algorithm for the consistent handling of finite section rotations such as occur during tether installation. Both codes also incorporate a seabed contact facility which is crucial to the analysis of the installation phase.

3. EXAMPLE APPLICATIONS

The programs have been applied to a range of test cases adapted from field problems. Particular attention has been devoted to comparisons between time and frequency domain approaches in order to verify the respective accuracy and limits of applicability of each method. This is of importance with respect to verifying the linearised three-dimensional hydrodynamic force computations. The frequency domain method is found to be very efficient in terms of computational speed and the generation of statistical results, and is particularly suitable for parametric studies. On the other hand, final verification using the full nonlinear time domain formulation is also necessary. Particular steps such as the upending and sinking phases must be handled in the time domain due to large nonlinear rigid body structural motions.

TLP TETHER INSTALLATION

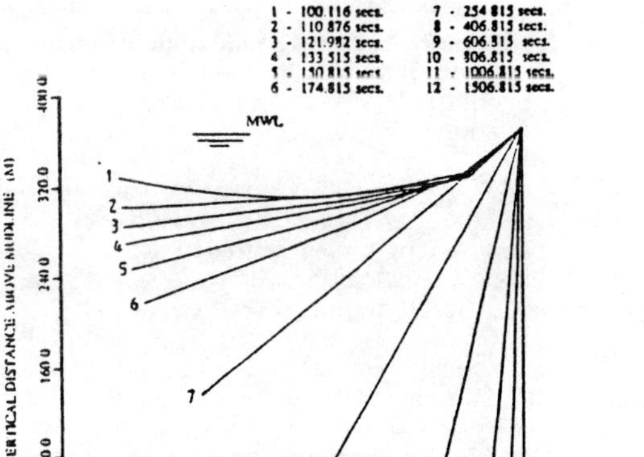

4. CONCLUSION

The overall conclusion is that very adequate and efficient computational tools have been developed for the safe design and analysis of all aspects of TLP tethers.

ULTRASONIC SENSOR FOR MEASURING FREE- AND DISSOLVED GAS IN DRILLING FLUIDS

O.M. VESTAVIK, B.AAS and G.W. HALSEY

ROGALAND RESEARCH INSTITUTE,
P.O.Box 2503, N-4004 STAVANGER, NORWAY

Gas entering the well during drilling operations may often lead to expensive well problems and may in some cases lead to disastrous blowouts. Acoustic interferometry has been investigated as a basis for improved detection of gas entering the well. An acoustic interferometer is an instrument which creates and measures sound resonances in a fluid. Several laboratory prototypes have been built to cover the ultrasonic frequency range 10 kHz to 10 MHz, some which could preform measurements on fluids with temperatures up to 100°C and static pressures up to 400 Bar. The major aim of the study is to investigate possibilities for detecting free and dissolved gas in liquids, specially drilling fluids. The experimental work was performed in the laboratory, however, the methods are adaptable for use downhole in connection with MWD.

The study has shown that the interferometers are effective tools for the detection of free and dissolved gas in the drilling fluids. Free gas is identified through changes of the velocity and attenuation of sound, whereas dissolved gas is identified through changes in the acoustic cavitation threshold.

FREE GAS: The interferometric measurements were very sensitive to the presence of free gas in the liquid. The gas increased both the velocity and attenuation of sound, which is identified through changes of the resonance characteristics of the fluid. The advantage of the interferometer in comparison with other gas detection methods is the ability to detect very small amounts of free gas. The detection level for gas bubbles in light drilling fluids was lower than a volumetric concentration of 0.02%. The detection sensitivity decreased with increasing drilling fluid density due to sound attenuation by the weighting material.

DISSOLVED GAS: Dissolved gas can be detected through a change of the acoustic cavitation threshold. The acoustic cavitation threshold refers to the minimum acoustic pressure amplitude required to produce microscopic bubbles in the fluid, i.e. cavitation bubbles. Acoustic waves with very high amplitude have been utilized to measure the acoustic cavitation threshold, which is found to be dependent on the amount of dissolved gas in the liquid. The onset of cavitation was identified through changes in the acoustic properties of the fluid, and through the presence of subharmonic signals from the cavitation bubbles.

The same ultrasound equipment is suitable for both detection of free and dissolved gas in drilling fluids. The methods are judged to be applicable for downhole use in connection with MWD. The methods are especially useful for early gas detection when drilling in shallow gas zones, and can be used for measurements on oil based as well as water based drilling fluids. If implemented in a downhole measurement tool, the methods would provide improved capability for early kick detection, and therefore reduce the probability for disastrous blowouts.

STRING FAILURE AND REMEDIAL ACTIONS IN ULTRADEEP DRILLING

G. BORRIELLO and M. PRAZZOLI

AGIP Drilling Technologies R&D Department

1. INTRODUCTION

AGIP has been involved, since 1982, in an ultradeep drilling activity in the Po Valley, in Villafortuna–Trecate Field. The wells depth ranges from 6000 m to 6700 m with high pressures (over 1000 bar), high temperatures (up to 180°C), hard and abrasive formations, aggressive fluids. In such a complex technical scenario, an abnormally high occurrence of drill string failure was recorded.

2. CASE HISTORY

The initial step was the failure analysis of the first seven drilled wells, including accurate laboratory tests. A total number of 39 failures was recorded. 13 of which led to fishing jobs which caused 2 side tracks, with an estimated loss of about 6.5 M$. The analyses put in evidence some of the probable causes of the failures such as:

- not adequate steel mechanical properties;
- not adequate API specifications and standards;
- very high fatigue phenomena in the threads;
- not adequate threads profile;
- not adequate components traceability.

The overall distribution of the failures was 77% in the BHA and 23% in D.P.; the 7 5/8" Reg. thread in 9" D.C. has to be considered particularly risky.

3. ACTIONS

It became evident the necessity of increasing the equipment reliability. Since failures were mainly associated to fatigue phenomena induced by vibrations, to prevent problems, a drill string monitoring campaign began immediately, pipes were regularly inspected assuming intervals on the basis of a minimum detactable crack dimension of 0.3 mm. No failures have been verified any more in the monitored wells. A considerable number of fatigue cracks has been detected, confirming the magnitude of the potential danger and the riskiest elements, mainly in the BHA.

Although the monitoring program has achieved the expected results, it is susceptible to many improvements, in order to decrease cost and time loss and to increase its reliability. The normally available NDT equipment, based on electromagnetic sensors, and MPI does not offer sensitivity and reliability. For that reason we are performing some experimental tests and feasibility studies to define new generation equipment. Particular effort is put into the U.S. control of the D.P. body and the upset area; U.S. are also feasible for crack detection in the threads; new sensors will be utilized for automatic control of the pipe body and the internal upset geometry.

The entire activity, to improve the drill string reliability, could be of a limited impact if components are not properly managed, so that an automatic system for pipe traceability is under test. It is based on a temporary code detected by a U.S. probe.

For next future the evaluation of the cumulative stress and the expected fatigue life prediction seem to be possible.

Another aspect to be considered is the drill string and casing wear, which can be qualitatively predicted by mathematical models. Most rejected pipes present severe eccentrical tool joint wear. A considerable increment in pipe wear resistance and reduction of casing wear can be achieved by tool joint laser hardbanding.

4. CONCLUSIONS

Results obtained so far are positive and encouraging, but projects of such a complexity emphasize the need of interdisciplinary integration of various activities and active participation of all the parties involved: users, suppliers and manufacturers.

"Dimensional verification to be carried out at open sea on offshore structures by means of photogrammetry"

Moreno Rampolli

Agip Spa, Offshore Department
Industrial Photogrammetry Group
Cologno Monzese (Mi), Italy

The SER Photo System is an innovative device to carry out offshore dimensional controls and other types of survey on land or on the open sea.
It consists of a special stereo-photogrammetric instruments' assembly, installed on AB412 Agusta helicopter.
It has been conceived and patented by Agip (inventor : Mr G.Bozzolato) and developed by Agusta, with the partial financial support of the European Economic Comunity, under contract TH 15.44/83 Italy .
The system has proved especially attractive because of its ability to achieve high accuracy level and chiefly because it permits high speed dimensional surveying of large-size objects under varying conditions including when in motion.
The possible applications are:

- Dimensional check of off-shore clusters on the following aspects: alignments, parallelism, perpendicularity, general bearing plane, etc.
- Measurement of floating structures
- Survey of damage on ship or petroleum platforms after collision
- Control measurement of real deformation suffered by lattice structures (i.e jackets) during open sea launching
- As-built survey of complex industrial plants, installed on the off-shore platforms one after the other (i.e."wafer" system of construction).

The following surveys have already been carried out:

- Survey of the Garibaldi A platform, fixed structure in Adriatic Sea, off-shore of Ravenna.
- Survey of a supply vessel during the navigation.
- Survey of an gas treating plant in Ravenna.

Although the setup of the S.E.R. was rather complex and costly, its realization has been found convenient because of the exceptional needs claimed by off-shore petroleum engineering.
The solution adopted consists of a pipe transversally fitted to the helicopter, which runs through the passenger cabin. At the two ends of the pipe there are two pod-shaped containers for the photographic equipment (two metric UMK 10/1318 cameras synchronized by an impulse gun) and two range finders. The range finders are fixed to the cameras thus forming two units bound to the pod structures via spring and cushion devices, while the viewing apertures in the pod shells are protected by vanishing doors.
The photograms can be taken both forward and downward, and through all the intermediate positions. The change of trim, which is simply realized by rotating the pipe round its axis by means of an electric actuator, finds a reserve feature in the complication of having a suspension which must respond to the shift in direction of the two photographic units' weight relating to the pods, in which they are contained.
The S.E.R. Heliborne Pivoting Support also includes the control and checking devices for all functions, the monitors which enable the photogrammetrists to suitably target the subject, and to check that the helicopter is stationary when the photograph is taken.
All the instruments are contained in a transportable shelter which can be used as a dark-room in order to develop the photograms.
The setup of the helicopter takes about three people three hours.

DEVELOPMENT AND TESTS OF A SUPERVISORY CONTROLLED TELEMANIPULATOR SYSTEM FOR INDUSTRIAL UNDERWATER APPLICATIONS

P. MINARDI and D. MADDALENA
Tecnomare S. P. A., Venezia - Italy

1. INTRODUCTION

Two basic approaches may be employed to perform telemanipulation tasks:

- master/slave: the operator enters the arm motion commands using a scaled replica (master);
- supervisory control: this approach includes a computer as "intelligent interface" between the operator and the remote arm, translating high level commands such as "fetch the tool A", "clean the weld" in the suitable arm commands.

The outstanding advantages of the supervisory control versus master/slave are demonstrated by laboratory full scale dry and wet tests.

2. SYSTEM DESCRIPTION

The overall system is composed of a console (including processing units) and a skid which is to be eventually mounted on a Remote Operated Vehicle (ROV). The skid includes the remote arm, the tool package and the TV cameras. The console includes the supervisory computer, a graphic system which shows simulations and an enlarged view of the remote scene, a touch panel to select functions and tasks, joystick and master arm. Furthermore a computer vision system (TV-Trackmeter), fitted with stereoscopic underwater TV cameras, provides a geometric modelling of the remote environment.

The main system functions include:

- fully automatic task execution (e.g. cleaning of a node weld with a water-jet tool);
- computer assisted operations (e.g. Non Destructive Test probe motion according to joystick commands with automatic probe alignment along the weld direction, this operation mode is referred to as "resolved motion");
- manual operation (i.e. master/slave).

Since the TV-Trackmeter can provide the coordinates of working scene points even if it is moving, the system is able to perform tasks with a loose docking or a hovering vehicle.

3. TEST RESULTS

The lab tests were carried out simulating a docked vehicle placing the skid in front of a full size structural node, and a hovering vehicle mounting the manipulator base and the TV-Trackmeter cameras on a moving base placed close to a well head mock-up.

The test results showed a faster and safer task execution with the supervisory control: the close visual inspection was carried out in a time two to five times shorter than with master/slave, the optimal water jet cleaning of the weld node was performed automatically, the NDT (Eddy Current) inspection was reliably carried out automatically keeping the probe with the proper alignment (this task is simply impossible with the master/slave approach). Furthermore the moving base trials demonstrated the feasibility of tasks such as valve actuation, insertion of a guide line or a transponder, cutting of a wire rope, due to the automatic base motion compensation implemented using the TV-Trackmeter.

ACKNOWLEDGMENTS

This work was supported by EEC and IMI funding and sponsored by Agip, Ansaldo, Enea, Saipem and Tecnomare.

REFERENCES

(1) MADDALENA D., BOSSI L., VISENTIN R., The TV-Trackmeter and the Supervisory Controlled Underwater Telemanipulation System, OTC 1990.

INFORMATION PROCESSING

Session Chairmen: F. ROCCA, Polytechnic of Milan (I)
L. MATTEINI, AGIP (I)

THE USE OF SEISMIC VELOCITY ANOMALIES IN PREDICTING GAS SATURATED LAYERS

A. Carlini, S. Cornini

Agip S.p.A.
Geophysical Research and Development Department
20120 Milan, Italy

SUMMARY

Local variations in the propagation velocity of seismic waves can be detected by the anomalous fluctuations they induce in the stacking velocity of underlying reflectors. Using such anomalies requires adequate development of new algorithms and thorough integration of the various phases of measurement, interpretation and tomographic inversion of the stacking velocities. This makes possible the estimation of the local velocities as well as the correct physical position of both these and the reflectors used to identify them. The integrated approach was put into practice along a seismic line in the Adriatic Sea, and its application proved that it is a valid method for detecting the presence of gas in a turbidite depositional system.

1. INTRODUCTION

Local variations in the propagation velocity of the acoustic waves are associated with subsurface heterogeneities (eg. changes in facies, salt masses, accumulations of hydrocarbons etc.) and can be detected by the perturbations they cause in the travel times of reflected waves. These travel times perturbations show up as clear anomalous fluctuations in the optimal stacking velocities relating to reflected events underlying the heterogeneities (1).

Part of the EEC sponsored research project - contract no. TH 01.008887 - studied the possibility of using information related to the travel times of reflected waves by measuring, interpreting and inverting the anomalies in the stacking velocities to emphasize variations in the propagation velocities of seismic waves.

In order to measure the stacking velocities, the problem of increasing the resolution of coherence functionals used in processing the seismic data was faced by introducing the concept of the complex trace. The mesurement of the hyperbolic coherence of reflections in relation to trial velocities and times is usually represented in the form of velocity spectra in which the relative maximum areas identify the optimal stacking velocity for that particular reflection time.

In order to interpret these spectra, a process which is usually carried out manually, a semi-automatic picking algorithm has been developed which is able to identify physically and geologically significant coherence peaks obtaining a stacking velocity field time and space consistent.

The use of stacking velocities in estimating the propagation velocity of acustic waves has been studied in great detail in the past, beginning with Dix's first works. The concept of inversion of stacking velocity anomalies is more recent however and its potential has been highlighted by Loinger (2) Rocca (3) and Toldi (4) using the classic tomographic

theory. Following from these studies, a method of tomographic inversion has been developed which is also valid for complex structural situations. This was done by adding a step which determines the propagation geometries through the positioning of common reflecting points.
The ability of such tomographic inversion to show up the presence of gas accumulation was proved using actual seismic data recorded offshore in the Adriatic Sea.

2. THE DATA

The data set used refer to a 2D seismic profile about 10 km long which was acquired in a basin located along the Italian offshore of the Adriatic Sea. The basin is largely filled with resedimented clastic deposits from the Plio-Quaternary age.

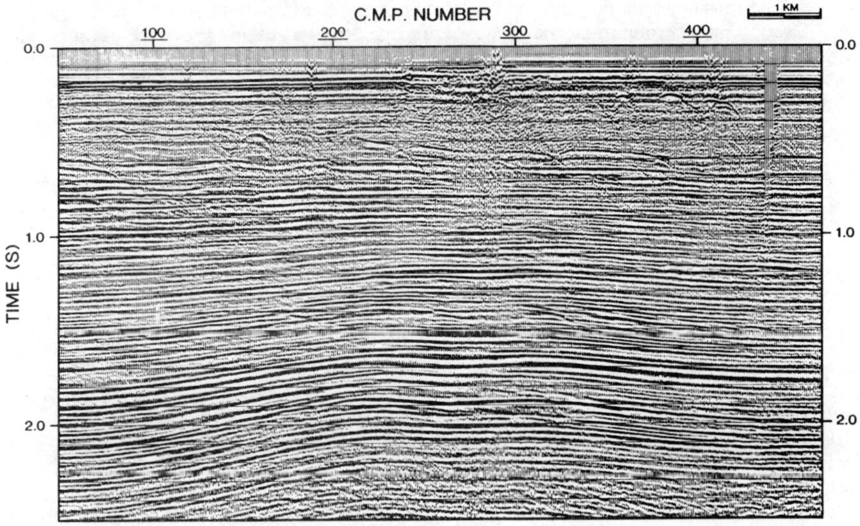

Fig.1 - Stacked seismic section.

The seismic section (Fig.1) shows some interesting characteristics. The most interesting is the weakly deformed anticlinal structure visible for arrival times over 1 second. Some bright spots are clearly present at the top of the structure; these generate zones of low seismic waves propagation velocities causing a false geometrical configuration (time sag). It is belived that these velocity variations are due to the presence of gas-saturated sands and to probable gas seeping upward from them; relatively little can however be said about the exact extension.
This is therefore a suitable case for verifying the results of tomographic inversion in the presence of a subsurface heterogeneity. A well situated along the line in correspondence with the anticlinal structure would, moreover, make these results verifiable.

3. ANALYSIS OF STACKING VELOCITIES

The stacking velocity analyses was one of the first steps in the standard digital processing to be realised and codified. Some of the work in this respect, (5) (6) and (7), date from the end of the 1960's, that is, only a few years after the large scale introduction of digital computers. In these works, most of the theoretical and experimental bases of methodology were established, which have been used ever since without substantial modification.

Stacking velocity analysis actually means the measurement of the hyperbolic coherence of the reflections in relation to trial velocities and times. Various coherence functionals may be used; they are generally represented by semblance and cross-correlation.

The introduction of the complex trace (8) allows the functionals known in the real domain to be redefined in a complex domain, obtaining in this way (particularly with the cross-correlation) coherence measurements which generally allow a considerable resolution increase in determining the stacking velocity values.

The main advantages of using complex functionals are of two basic types:

* the imaginary component, being in quadrature with respect to the real component, has minimum or maximum values at the zero-crossing of the real trace, that is, where the local coherence values, given by real functionals, are low;
* a complex velocity spectrum can be subjected to a phase-gain (that is a controlled rotation of the instantaneous phase).

The phase-gain is a simple technique which allows numerical wave-forms of any origin to be compressed (9). In the present example this is useful for compressing the coherence peaks in the velocity spectra. These peaks are then easier to identify and to interpret. Phase-gain is therefore nothing more than an enhancement factor which multiplies the instantaneous phase increasing its corresponding value in the complex spectrum. This operation is carried out so that the increasing widening of the coherence peaks related to the increase in the trail arrival times and velocities is compensated for.

Fig.s 2 and 3 compare the velocity spectra obtained by applying the semblance and complex cross-correlation functionals in the velocity analysis of the seismic line in correspondence with the CMP 300. In particular, in the complex cross-correlation, a phase gain has also been introduced. This comparison shows the clear increase in resolution and in quality of the velocity analysis which is obtained when the complex trace concept, together with a phase-gain technique, is used. In fact careful analysis shows that complex cross-correlation allows various coherences to be recovered, especially for the most shallow and lowest energy values. It also makes the velocity function's trend more continuous and more easily to interpret. There is also a significant increase in the sharpness of all the coherence peaks and a decrease in background noise due to the application of the phase-gain. No information is lost: the general trend of the velocity function is therefore completely optimised.

The pre-stack data, arranged in common mid-point gathers, can be represented three-dimensionally; CMP coordinates, arrival times and offset between source and receiver. The velocity analysis transforms this volume of seismic data into a coherence volume where the offset coordinates have been substituted by the trial stacking velocities.

Different coherence spectra can be obtained by slicing the volume appropriately: slices with constant CMP (conventional velocity spectra), slices with constant time (time slices) and slices with variable time (continuous velocity analysis).

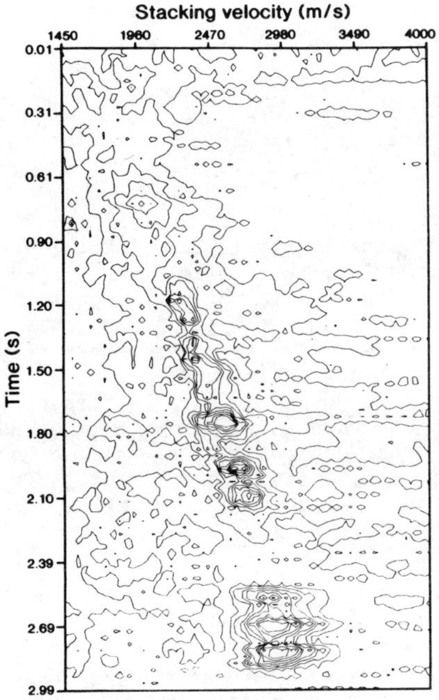

Fig.2 - Velocity spectrum obtained with the semblance.

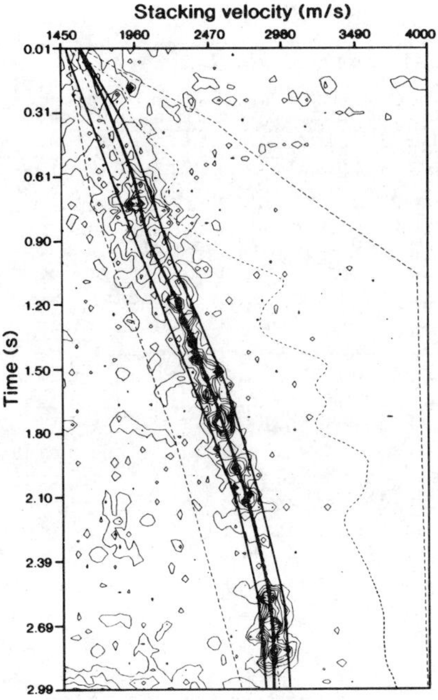

Fig.3 - Velocity spectrum obtained with the complex cross-correlation.

4. INTERPRETATION OF VELOCITY SPECTRA

In order to estimate the optimal stacking velocities along the seismic profile, the coherence volume must be interpreted. Choosing the path which physically and geologically best links the most significant coherence peaks is almost always difficult and uncertain. This is why is usually done manually, since no completely automatic methods are available which can satisfactorily make the kind of decisions a geophysicist does. These decisions depend not only on mathematical criteria, but also on physical and geological constraints, usually called "a priori" information. On the other hand however, there are many velocities to interpret and so this manual phase can become a bottle-neck in the processing. A reasonable compromise is to adopt a mixed system in which automatic interpretation is first guided and then checked and corrected manually.

The basic philosophy considered in the automatic interpretation of spectra is similar to that introduced by Toldi (10). This consists of maximising the line integral computed on the coherence matrix along the velocity function. The algorithm for finding the optimal function is of iterative type.

An inherent problem in the conventional optimisation methods based on the gradients is the possibility that the convergence at the best solution may be blocked by a relative maximum. One way around this inconvenience

is to carry out the first iterations on a smoothed version of the data being examined. In this way the first iterations make the integration trajectory converge towards the area with the highest average coherence, which is the zone of greatest interest. From a certain point onwards, more accurate, or less smoothed, coherence matrices are used, until finally, for the last iterations, the actual data is used.
In the iterative procedure followed by the picking algorithm there are two important principles.
The first is that the interval velocities which correspond to the selected stacking velocities should be physically significant. There are lower limits (usually those of weathered layers or water) and upper limits (the denser rocks) which mean that some of the possible trajectories are excluded, even if these maximise the line integral. These limits can be specified by introducing a velocity corridor and they depend on the "a priori information" about the area in question.
The second principle is the possibility of imposing a maximum frequency with which the optimal stacking velocity can vary in relation to time and space. As well as these two principles other control constraints can be used on the algorithm. It is thus possible to define a stacking velocity corridor so as to delimit the values considered most reasonable, or, establish a starting function for the iterative picking process.
Fig.s 3 and 4 show two examples of automatic interpretation of velocity spectra obtained from the actual seismic data.

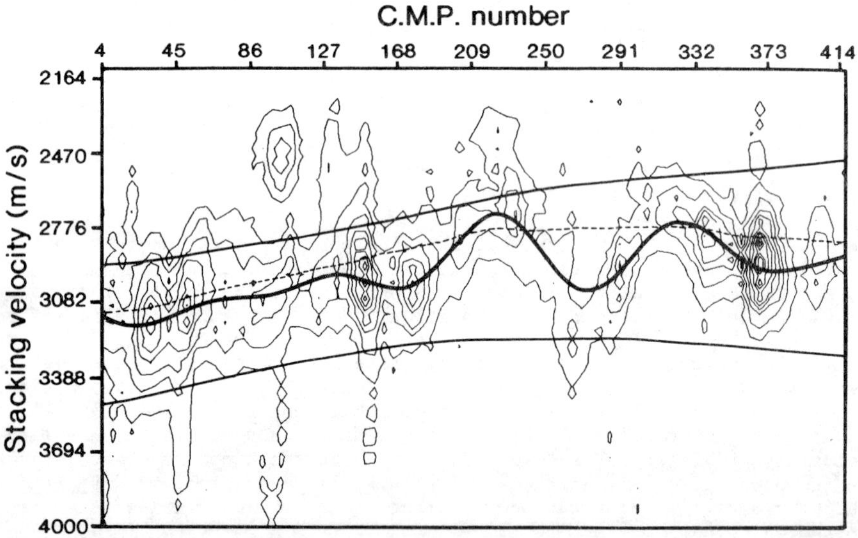

Fig.4 - Continuous velocity spectrum obtained with the complex cross - correlation.

Fig.3 shows a slice of the coherence volume corresponding to the CMP 300, whereas the slice which produces the surface in Fig.4 was made by following a reflecting interface along the line.
Both were obtained with complex cross-correlation and phase-gain and in both cases the continuous bold line indicates the stacking velocity values selected automatically by the picking algorithm. It is clear that

a combined effect of all the physical constraints results in bringing the picking function towards the most reliable solution, avoiding undesirable events and interferences with noise peaks.

For the specific aim of this work, four horizons were selected on the seismic line. These horizons were chosen for having good reflectivity and lateral continuity so that the coherence volume produced by cross-correlation and phase-gain could be sliced appropriately. The corresponding continuous velocity analyses were then automatically interpreted to obtain the stacking velocity values to be used as input in the tomographic inversion.

Fig.5 represents an overall picture of the stacking velocities associated with the four horizons; the zero-offset reflection times of the selected horizons are also marked (broken lines). It can clearly be noted that, for times greater than 1.8 seconds, there are several fluctuations which have the typical appearance of those due to local velocity anomalies associated with subsurface heterogeneities (in this case, the presence of gas).

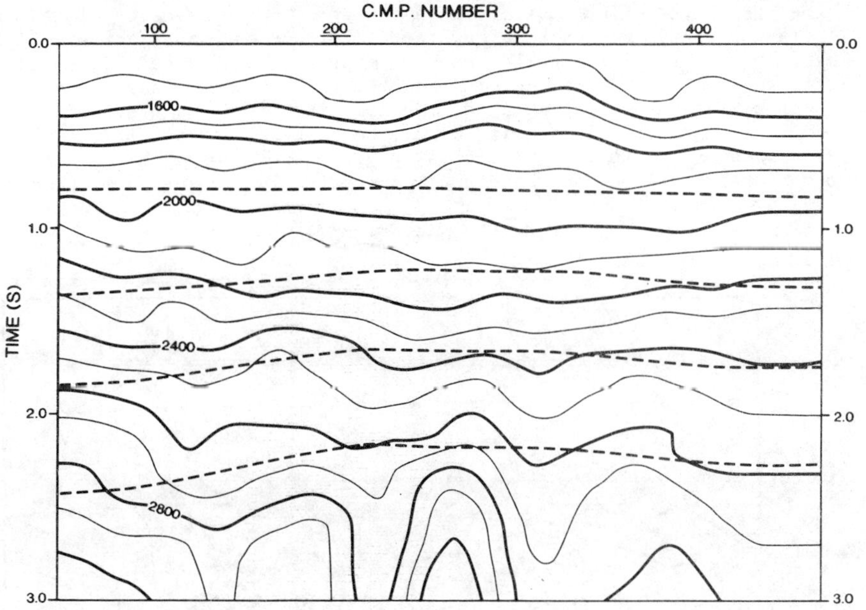

Fig.5 - Stacking velocities field.

5. TOMOGRAPHIC INVERSION

The tomographic inversion for estimating the velocity field of the seismic waves is usually carried out by evaluating the arrival times of reflected signals (travel-times inversion).

This approach is a rather delicate one in that it presupposes the interpretation of the pre-stack data. One satisfactory solution is that of inverting the stacking velocities (11).

This not only has the operational advantage of not requiring any additional operation to the normal processing sequence, but is also

advantaged by the stability of the input data.
Infact the velocity analysis procedure itself provides statistically more significant measurements, in which noise interferences is greatly reduced through space and time averages. The essential steps which form the basis of the tomographic method used are as follows:

* stacking velocities along a certain number of interpreted seismic horizons are noted, and from these an initial model of the depth distribution of seismic propagation velocity (interval velocity) is drawn up,
* the raypaths corresponding to many source-receivers pairs are traced, as a function of both their relative distances and of the position of their midpoint along the seismic line,
* the traced rays are adjusted so as to minimize the dispersion of the respective reflection points at depth.
  This may even mean modifying the position and angle of the reflector segment involved,
* the travel times along the rays are evaluated in relation to the actual velocity model,
* once the travel times have been established, the velocity field is re-parametrized, resolving a system of linear equations obtained by minimizing the difference between the input data and the actual model data for the stacking velocities and the zero-offset times.
* in this way an updated model of the velocity field is produced which may be satisfactory in itself or may be used as a new initial model. Returning again to the second step, an iterative cyclical process begins which can be prolonged until a stable final solution is reached.

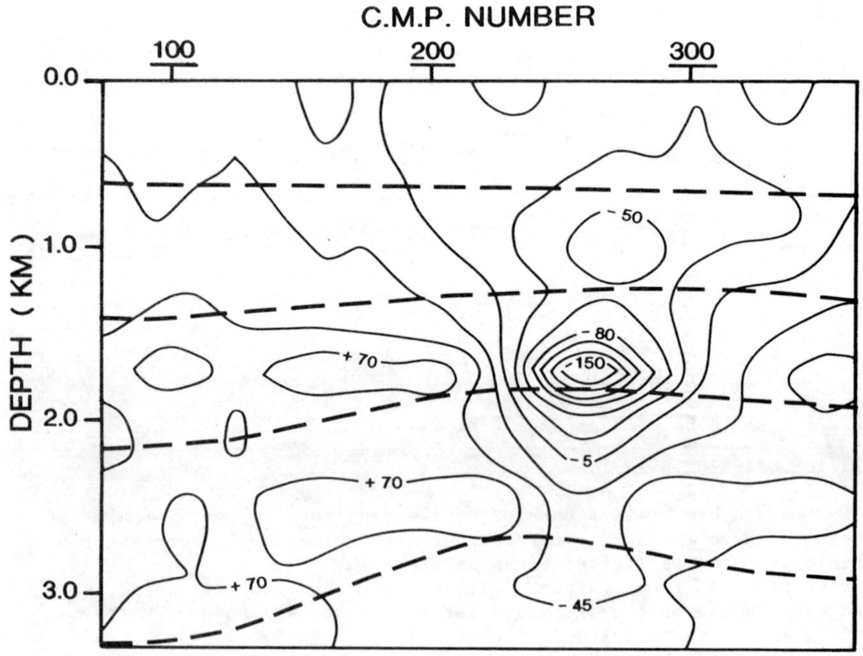

Fig.6 - Field of local variations in the interval velocities.

There are some unusual hypotheses in this approach. Firstly, the velocities and the reflection geometries are parametrized independently. This set up is different from the usual model concept where the layers have constant velocity and their boundaries are always conveniently marked by reflection.
Secondly, reflecting surfaces cannot generally be assumed either flat and continuous. As the raypath is well defined by its reflection point, the reflections can be parametrized as a set of such points.
Thirdly, since the high spatial frequency of the trasmission velocities cannot be inverted on the basis of the travel times, a velocity model structured in "bins" must be excluded, that is, when the velocities change sharply only at the boundaries of the "bins".
It was therefore preferable to define a continuous velocity model as the sum of smooth, two-dimensional basis functions, centered on a grid, of which it is possible to predetermine the resolution (within the physical limits set by the Fresnel zone, obviously).
The tomographic inversion algorithm was therefore applied using, as main input, the stacking velocity values relating to the four horizons selected along the seismic line. Fig. 6 illustrates the field of local variations in the propagation seismic velocities obtained by tomographic inversion.
The four horizons (in depth) are also plotted in fig.6 (broken lines).
The result agrees with the well data available; local variations of low propagation velocity correspond to the clearest "bright spots" on the stacked section particularly in the geological layers where gas is present (basically between horizons 3 and 4).

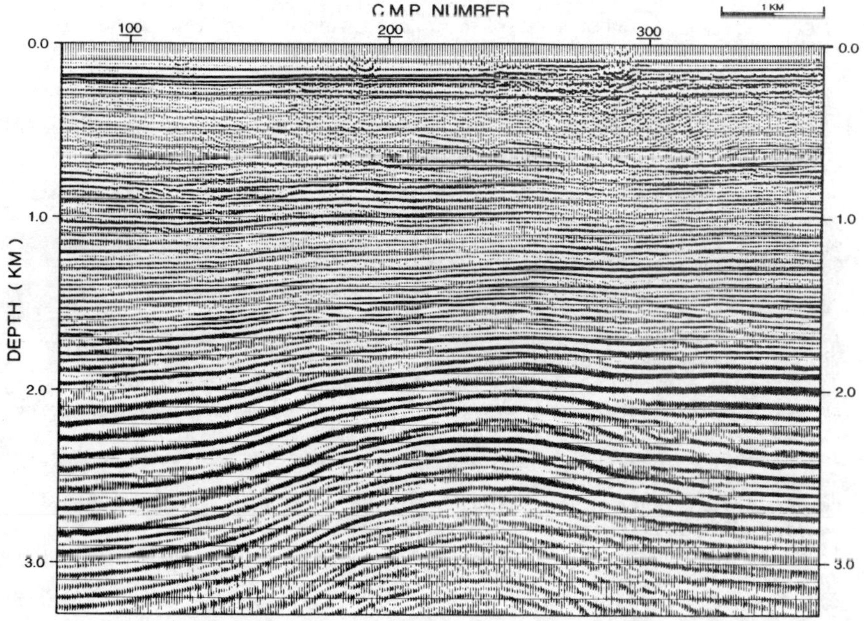

Fig.7 - Depth migrated seismic section.

The velocity decrease in the gas saturated sands, measured by sonic log, is of about 200 m/s. This value is slightly lower than those obtained by tomography, which suggests the possible presence of gas outside the producing levels. This is actually highly probable considering the non-linear relationship between the P-wave velocities and the gas saturation of the rocks for low gas content (12). Tomography therefore appears to be appropriate in detecting the presence of gas even though it is not able to evaluate the quantity because of the similar effects of reservoirs and gas leakages on the P-wave propagation velocity.
Depth migration (Fig. 7) allows the geological likelihood of the velocity model defined by tomography to be evaluated. The result obtained is clearly satisfactory; the geometrical setting at depth of the anticlinal structure is actually realistic and agrees exactly with the regional geological knowledge of the area.

6. CONCLUSIONS

Local variations in the propagation velocity of seismic waves can be detected by the anomalous fluctuations they cause in the stacking velocity of underlying reflectors. Three integrated phases have thus been introduced; velocity analysis, automatic interpretation and tomographic inversion of stacking anomalies. The integrated processing system proposed represents a valid method of detecting the presence of local anomalies in propagation velocities which are induced by subsurface heterogeneities (gas accumulations).

ACKNOWLEDGEMENTS

The authors wish to thank W. Harlan and A. Vesnaver of the Osservatorio Geofisico Sperimentale of Trieste for their valuable contribution in realizing this project.

BIBLIOGRAPHY

1 Al-Chalabi M., (1979), "Velocity determination from seismic reflection data", in: Development in Geophysical Exploration Methods-1, edited by Fitch A.A., 1-68.
2 Loinger E., (1983), "A non linear model for velocity anomalies", Geophysical Prospecting, 31, 98-118.
3 Rocca F., Toldi J., (1982), "Lateral velocity anomalies", SEP 32, 1-23.
4 Toldi J., (1984), "Resolution of interval velocities from stacking velocity anomalies", SEP 38, 89-104.
5 Garrotta R., Michon D., (1967), "Continuous analysis the velocity function and of the move-out corrections", Geophysical Prospecting, 15, 584-597.
6 Schneider W.A., Backus M.M., (1968), "Dynamic correlation analysis", Geophysics, 34, 330-356.
7 Neidel N.S., Taner M.T., (1971), "Semblance and other coherency measures for multichannel data", Geophysics, 36, 482-497.
8 Taner M.T., Koehler F., Sheriff R.E., (1979), "Complex seismic trace analysis", Geophysics, 44, 1041-1063.
9 Sguazzero P., Vesnaver A., (1986), "A comparison between different methods of stacking velocity evaluation", in: Deconvolution and Inversion, edited by Bernabini et al., 267-286.
10 Toldi J, (1989), "Velocity analysis without picking", Geophysics, 54,

191-199.
11 Harlan W., (1989), "Tomographic estimation of seismic velocities from reflected raypaths", SEG 59, 922-924.
12 Toksoz M.N., Cheng C.H., Timur A., (1976), "Velocity of seismic waves in porous rocks", Geophysics, 41, 621-645.

DRILL BIT NOISE AS A SEISMIC SOURCE IN GEOPHYSICAL SURVEYS

G.P. ANGELERI (*), S. PERSOGLIA (**), F. POLETTO (**)
(*) AGIP - P.O.Box 12069 - 20120 Milano, Italy
(**) OGS - P.O.Box 2011 - 34016 Trieste, Italy

Summary

The signal produced by a drilling bit may be used as a source in seismic surface investigations. Two main problems make this method difficult: first, the very high environmental noise level of the drilling area; second, the time incoherent nature of the waves continuously generated by the rotating bit. To get a usable seimic signal, it is necessary to simultaneously register the signal from the drilling bit and from traditional seimic surface signals. By crosscorrelation, the direct or reflected drilling bit signal present in the seismic data is thus collapsed into a spike. This method currently allows the production of reciprocal VSP, but may be further developed for geophysical studies of reservoirs because of its applicability in low cost well-to-well cross-investigations. For testing such a possibility, an experiment was done in which reference signals were recorded together with the main environmental noise of the drilling yard. Multichannel seismic data were collected along a line 1.3 Km long. In the present paper, the most interesting results are presented; in particular, a comparative analysis of different types of reference signal, the application of noise elimination techniques, and the improvements related to the elimination before correlation of the drill reverberations present in the reference signals.

1. INTRODUCTION

The Osservatorio Geofisico Sperimentale of Trieste (OGS) undertook and completed the data collection and processing that formed the research project "Use of Drilling Noise to Generate a Reciprocal VSP", partially funded by the Commission of European Communities, General Direction for Energy. AGIP S.p.A. demonstrated an interest in the objective of the research and participated by setting up the contract. This contract allowed OGS to monitor the signals produced by the drill-bit in the environs of some production wells, with the aim of identifing and separating the various components in all the recorded signals.

The results obtained during the research demonstrate clearly that the signal generated by a rotary drill-bit may be identified and separated from the others produced by the various plants operating in the drilling yard; it may thus be used as a passive seismic source. This result is the basis for determining the location in space of the bit by passive surface listening, for producing reciprocal VSP, and for tomographic investigations in the area surrounding the well using the energy emitted by the rotating bit.

2. PHASES OF THE RESEARCH

The research was divided into three different stages: the planning of the data acquisition phase; the execution of the survey around two production wells; and the processing of the resulting data.

The first required the defining of the parameters of the seismic data acquisition, the types of sensors to place at the top of the drill-string, the locations in the yard of noise sensors, and the positioning of the auxiliary wells in which three component geophones were buried.

The second stage was the construction of interfaces for the sensors, the location and testing of the geophones buried in the auxiliary wells, and the continuous daily acquisition of the signals around production wells.

Finally, the third stage consisted of a non-conventional use of standard programs for seismic processing by the development of original algorithms (e.g. for the identification and removal of coherent noise) and by processing the collected data directly in the field with an intermediate power graphic workstation.

3. DATA ACQUISITION

The bit rotating and breaking the rock may be considered as a compressional vibration source similar to Vibroseis. A fundamental difference consists in the fact that, while in the Vibroseis system it is possible to theoretically define the source signal and control it, in this case the continuously generated bit signal is time incoherent due to the unpredictable coupling between bit and rock, and thus can only be recorded in the field (Pilot Signal). This is very difficult, primarily because the bit is continuosly rotating (and so the start time is not easily definable) and, secondly, due to the very noisy environment in which the signal produced by the bit has to be identified. Engines, generators, rotary drilling rig, mud pumps and human activity in the drill yards are all sources of unwanted, very high energy noise that may almost completely hide the bit signal.

Thus, the key for a successful data collection methodology consists in the ability to correctly record the following information:

- the total earth-filtered wavefield;
- the pilot signal to be used as reference as in the Vibroseis method;
- the environmental noise in order to subsequently filter it out for a signal to noise ratio enhancement.

In this research project by OGS, seismic data were collected using the following parameters:

- number of channels	: 120
- station interval	: 20 m
- offset of first station	: 100 m
- P geophones	: 10 Hz, 12 per group
- geophone pattern	: regular and linear
- S geophones	: 10 Hz, 2 (Hx and Hy components) every 5 stations
- sampling rate	: 2 ms
- listening time	: 24 s
- records for each level	: 20
- distance between levels	: 2 m
- n. of investigated levels	: about 500

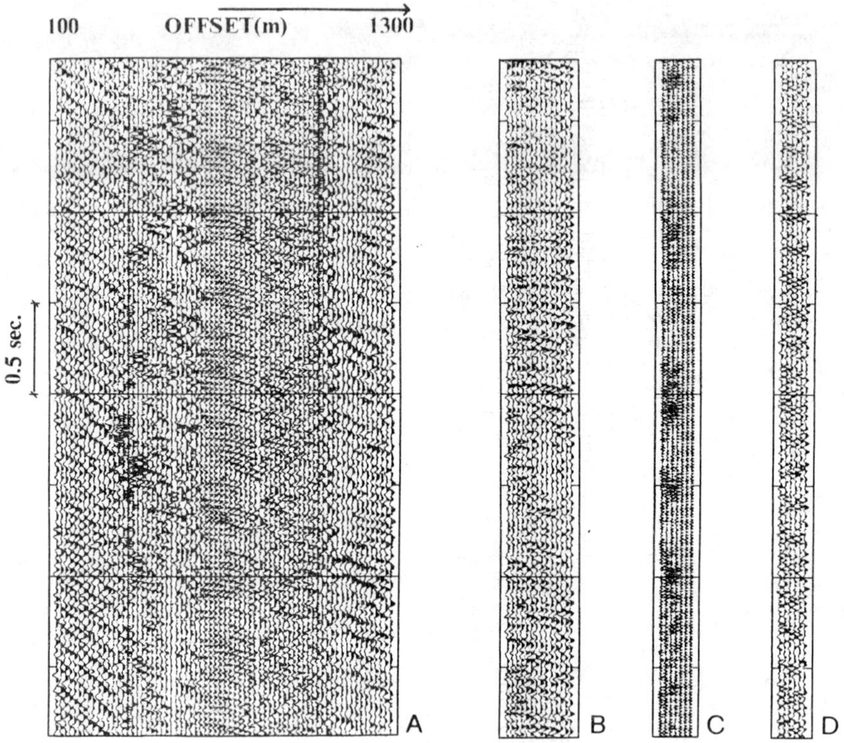

FIGURE 1. Field data: A - seismic line data; B - pilot signals in auxiliary well; C - pressurometer and accelerometer signals; D - drill yard noises.

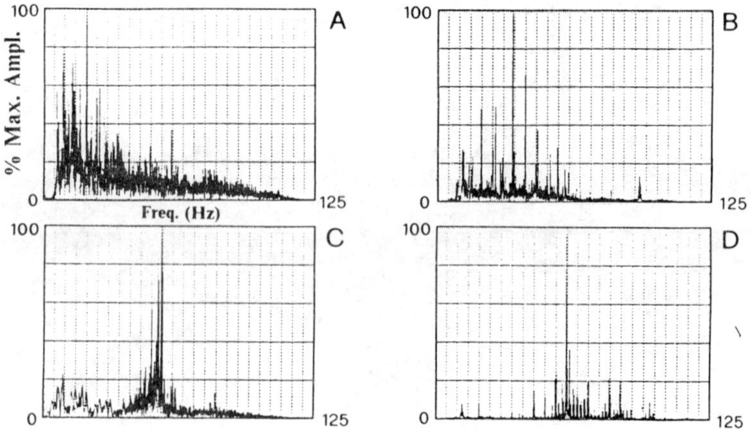

FIGURE 2. Amplitude spectra: A - seismic line data; B - noise (z-component); C - pilot signal (z accelerometer); D - pilot signal (pressurometer).

With regard to the pilot signal, various possibilities were investigated. The basic idea is that the longitudinal vibrations produced by the drill-bit propagate through the drill-string and can thus be detected by an accelerometer placed at its top. To this aim, two accelerometer triads of different technical characteristics were fixed to the swivel using a particular shaped collar. The first triad was composed of three linear, zero mass movement accelerometers produced by Shaevitz Engineering, having a range of 5 g, a natural frequency of 120 Hz, and a linearity of 0.04 %.

The second triad consisted of a triplet of miniaturized accelerometers with an extension-bridge tranducer mechanically coupled to a mass with freedom of movement in the direction of the measurement axis. The range was 5 g, the natural frequency 180 Hz, and the linearity 1%.

Another possible pilot signal envisaged is the variation of the power absorbed by the motor of the rotary table while drilling. Thus, a current shunt was fixed to the primary circuit of the rotary table motor to measure the variations of the current intensity.

A further possibility investigated is that the bit, fracturing the rock, produces pressure waves in the drilling mud that may be sensed by a pressurometer fixed to the rotary hose. This miniaturized sensor, designed by PCB Piezotronic to measure quasi-static or dynamic pressure, consisted of a battery of quartz plates for the conversion of pressure into electrical charge, coupled mechanically and electrically to an accelerometer. The sensibility was very high, the natural frequency 300 kHz and the linearity 1%.

Moreover, two auxiliary wells, 100 meter deep, were drilled with offsets of about 100 m and 400 m from the well head. Vertical geophones were inserted at every 10 m depth, and three component geophones at 0, -50 and -100 m were also buried inside to simulate an antenna oriented towards the bit. The signals collected were used to synthesize another pilot signal.

The environmental noise was sampled at the rig by geophones put in holes near the mud pumps, the generators, the rear support of the tower and the rod store. A geophone was also fixed to one of the supports of the tower to monitor the tower vibrations.

Figure 1 shows an example of the raw data collected in the field while the bit was at 950 meters depth. The related amplitude spectra (Fig. 1.) show large bandwidth frequency distributions for the seismic data, the noise having a common, although lower, frequency band. The spectrum of the pilot signal collected by the accelerometers, excluding a peak at the mains frequency, shows a pattern similar to that of the data of the seismic line.

The signals sampled along the line, together with that of pilot sensors, were recorded for 24 seconds and, in the processing phase, intercorrelated and summed. About 20 records were used for every level. This listening time corresponds roughly to a drilled thickness of 0.5 meter in soft formations and of 1.5 meter in the hard formations of the target zone.

4. DATA PROCESSING

To extract the drill-bit signal from the data recorded at the surface, a processing sequence was adopted with the following main steps:

- crosscorrelation with a pilot signal;

- suppression of environmental noise;

- source deconvolution

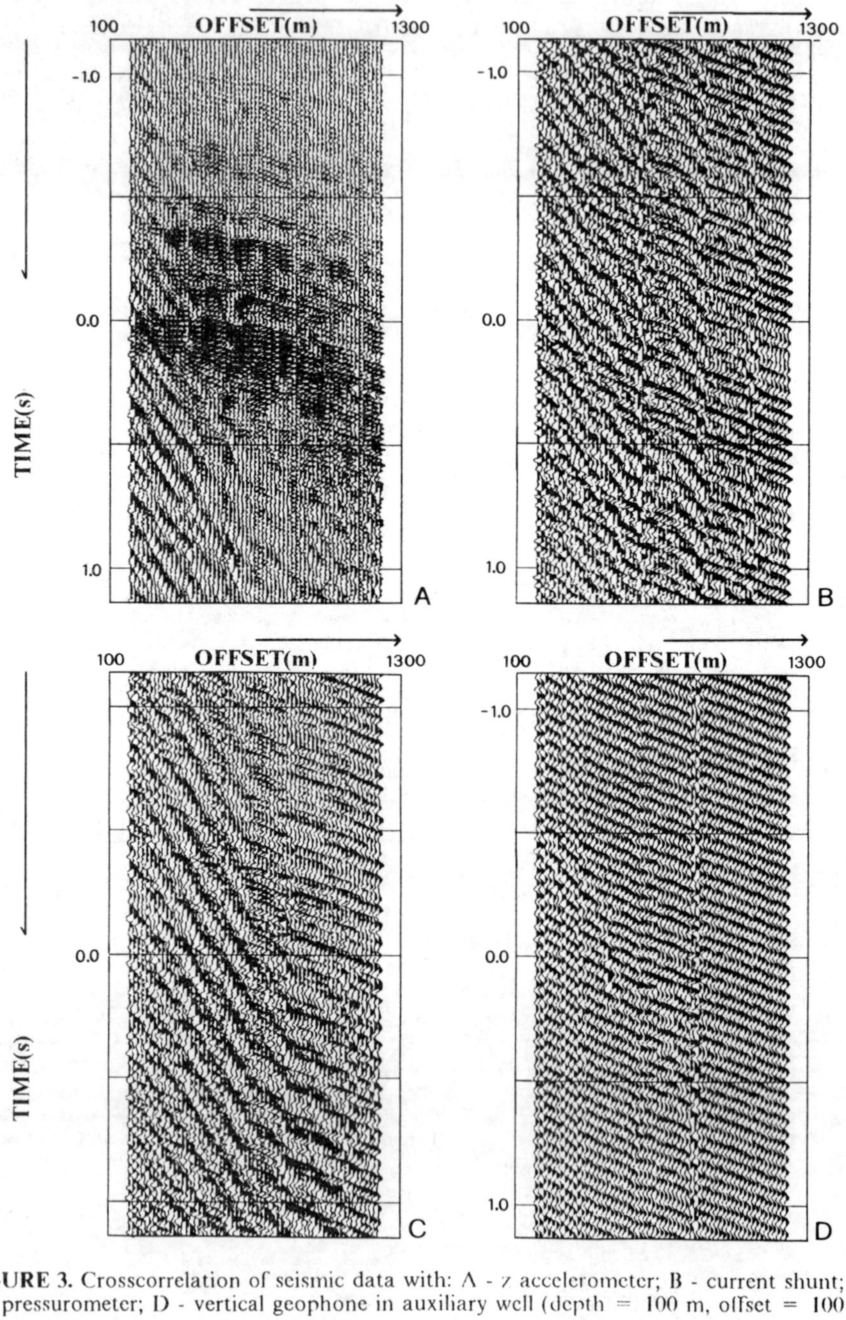

FIGURE 3. Crosscorrelation of seismic data with: A - z accelerometer; B - current shunt; C - pressurometer; D - vertical geophone in auxiliary well (depth = 100 m, offset = 100 m).

Crosscorrelation is used usually to compress the well known and temporally well defined Vibroseis waveforms. The continuosly generated drill-bit signals are, however, collected by several surface sensors; there is, therefore, a strong traveltime path effect in the pilot signal that depends on the physical mechanism of the energy transfer and has to be taken into account.

The accelerometer signals were crosscorrelated with those sampled along the seismic line. The first contain the drilling bit signal together with its reverberations along the complex drill-string system; in the latter, the same signal occurs with their primary and multiple earth reflections.

The lag corresponding to the maximum correlation of the signals is equal to the difference in travel times between the direct arrivals in the earth and in the steel drill-string. All the results obtained using different pilot signals agree with the lags theoretically estimated. Events occur also at negative correlation times with respect to direct arrival times because of the presence of multiples in the pilot signal.

The sensors which achieved the best results, with clearly identifiable hyperbolic first arrivals, at about 0.1 s in Figure 3.A, were the TML accelerometers; the signal to noise ratio was worse for the others.

The delay behaviour is different for the other sensors: in the case of the current shunt (Fig. 3.B), the lag in the pilot signal is comparable with that of the accelerometers. The pressurometer measures a signal that propagates in the mud with a velocity lower than that of the earth. So the lag of the maximum of the correlation occurs at negative correlation times (Fig. 3.C). Finally, there is practically no lag to consider between the pilot signal recorded with the antenna of geophones buried in the listening wells and that of the geophones on the line at the same offset, because the travel paths to them are similar (Fig 3.D). We note that, although the accelerometers generally achieved the best results, at certain depths of the bit, best results were obtained by the pressurometer or by the current shunt.

A strong environmental noise component having a linear trend was always present in the records, together with the useful signal and, in particular, in the geophone traces. The method we developed to reduce it can be defined as an active type, and is based on the availability of recording channels dominated by the noise generated by the equipment operating in the yard. The noise subtraction is performed by mean of a multichannel filter which decomposes the recorded noises into statistically independent signals (orthogonalization).

In the survey, we removed from the data correlated with the pilot signal the averaged correlations between the noise to reduce and the data of the seismic line (Fig. 4). The averaging was done by summing these intercorrelations with different offset along the noise propagation curve, in order to eliminate any appearance of the bit signal in the trace to subtract.

At this stage of the processing, the data are not yet suitable for the computation of a VSP because of the strong presence of long and short period source multiples (Fig. 5.A). The first are due to reflections at the edges of the drill-string; the second are reverberations in the bottom hole assembly (BHA). The latter produce an effect that is stationary as long as the BHA is not modified. The behaviour is different if we consider, instead of the accelerometers, other pilot sensors.

The signal reaches the pressurometer through the mud and a similar reverberation phenomenon takes place. For the current shunt, the energy propagation mechanism is, in general, as for the accelerometers. For a

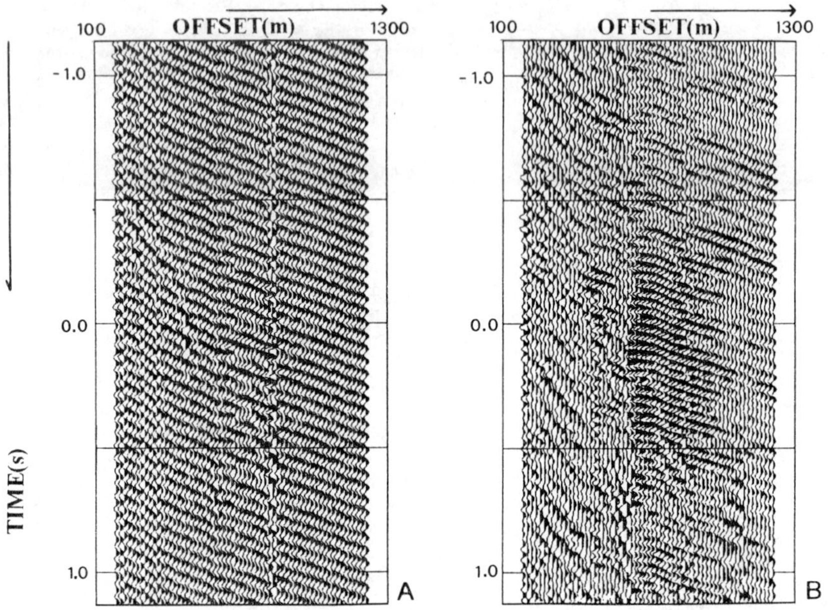

FIGURE 4. Crosscorrelated data: A - before, B - after noise suppression.

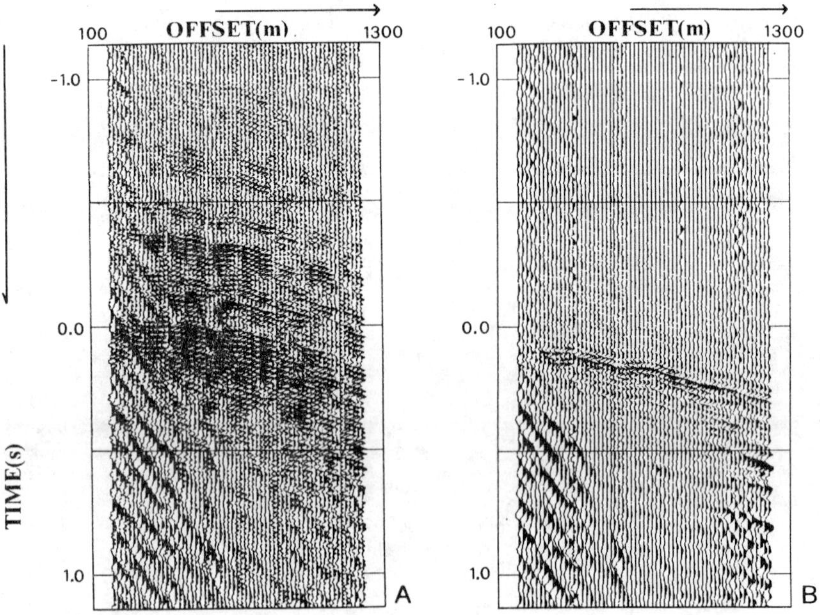

FIGURE 5. Crosscorrelated data: A - before, B - after source deconvolution.

signal sensed by the geophones of the antenna, however, the presence of possible reverberations of the drill-string has to be evalued. Probably, only a neglegible energy fraction of these multiple reflections is transmitted into the surronding rock, and, in this case, the filtering and shaping of the pilot signal is due exclusively to the earth itself.

To remove the multiples, several methods have been adopted. The best results were produced by estimating the source signature and its reverberations in the BHA.

To get this waveform the complex cepstrum of the z component trace of the accelerometer was calculated by averaging more than 150 records covering a significant depth variation of the bit.

Operating over this interval, not only the long period multiple reflections but also the yard noise and the bit signal, assumed incoherent, are averaged. The remaining contribution should thus come only from the stationary reverberations into the lower part of the drill-string. Converting in depth the main period of these short reverberations, we obtain, in fact, the length of the BHA used when the survey was performed. The estimated wavelet is then removed from the data, to which a second, long period, deconvolution is applied (Fig. 5.B).

Having, after the described phases of processing, signals of good quality, geometrical considerations were used to confirm the nature of the direct arrivals. Velocity information exploited by surface seismic and direct measurements in the well, together with the depth reached by the drill-bit, were used to determine a synthetic model in good agreement with the results obtained by processing the real data (Fig. 6). From this point of view, the cinematic behaviour of the first arrival time curves, as the bit moves deeper, is very significant. This is because the location of both the unwanted noise sources and the sensors is fixed during the execution of the survey.

The possibility of really using the first arrival times for the location in space of the drill bit will require in the future a more detailed analysis of the characteristics of the identified signals and of their resolution.

A second interesting result is the possibility of getting real time reciprocal VSP for exploring the layers that have still to be reached by the drill bit. The signals reflected by the strata underlying the levels reached are weaker, and are usually hidden by direct arrivals. The removal of these is one of the aims of VSP processing, which may be directly performed after compensation in the final correlated data of the time delays in the propagation of the drill signals towards the pilot sensors. Promising results have also been reached in this direction.

5. CONCLUSIONS

During the research, many results were obtained. Real data were used and carefully analyzed by means of new processing techniques aimed at the identification and the separation of the signals produced by the drill bit from those due to other drilling equipment operating in the yard.

Using such identified signals, a preliminary reciprocal VSP was produced in which reflected events match analogous reflections appearing in seismic sections derived with standard surface methods are present.

The possibility of determining the first arrival times along the surface seismic line demonstrates the feasibility of using the drill-bit as a source in the study of local velocities around the well by means of tomographic inversion techniques.

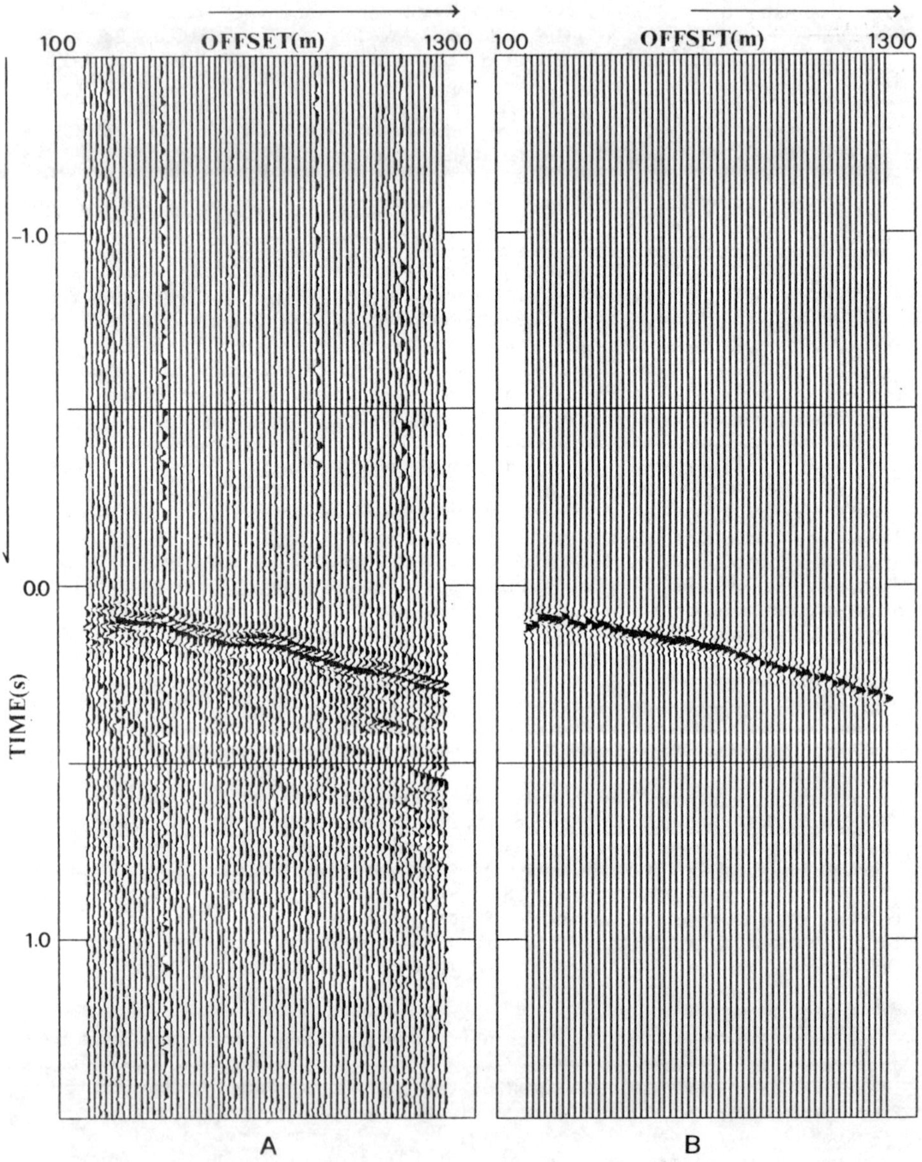

FIGURE 6.
A - drill bit signals after processing phases, including crosscorrelation with vertical accelerometer;
B - synthetic expected signal.

ACKNOWLEDGMENTS

The research has been undertaken under contract n. TH/0113/87 of the Technological Development Project in the Hydrocarbons Field of the Commission of the European Communities, whose support is gratefully acknowledged. AGIP SpA had a significant role in the research in terms of partial financial support, permission for acquiring data around production wells and contributions to the development of the research itself.

The authors wish to thank dr. A. Craglietto (for his basic contributions in the development of computer programs), dr. B. Manca (for the choice and installation of special sensors), mr. G. Dordolo (for the data collection in the field), dr. G. Valenti (for his contribution to the VSP processing of drill-bit data) and, expecially, prof. F. Rocca, project leader, whose ideas were crucial to the research.

REFERENCES

1. Cassano,E. and Rocca,F.;1973:"Multichannel linear filters for optimal rejection of multiple reflections",Geophysics,Vol.38,P.1053-1061.
2. Hardage,B.A.,1983;"Vertical Seismic Profiling. Part A: Principles"; Geophysical Press, London.
3. Kostov,C. and Zanzi,L.;1988;"Analysis of Drill-bit Data: Preliminary Results";Stanford Exploration Project, SEP-57.
4. Lutz,J.,Raynaud,M.,Gstalder,S.,Quichaud,C.,Raynal,J., Muckleroy,J.A.; 1972; "Instantaneous Logging Based on a Dynamic Theory of Drilling"; Journal of Petroleum Technology, JPT-3604.
5. Moore, Preston L.;1986;"Drilling Practices Manual: Second Edition"; PennWell Publishing Company, Tulsa, Oklahoma, USA.
6. Ulrych,T.J.;1971;"Application of Homomorphic Deconvolution to Seismology";Geophysics,v.36,no.4,p.650-660.

HORIZON PROCESSING IN 3-D SEISMIC INTERPRETATION

J.C. Mondt
Koninklijke/Shell Exploratie en Produktie Laboratorium
Volmerlaan 6
2288 GD Rijswijk ZH
The Netherlands

Summary

A 3-D interpretation project starts with the interactive interpretation of a number of selected horizons on a limited grid of seismic lines, known as the control grid. Automatic tracking programs then extend this preliminary interpretation over the whole survey area. Several iterations in which the control grid is extended into difficult data areas may be necessary. Irregularities related to missing data and artefacts caused by the tracking programs in combination with noise interference are removed by filter programs that avoid distorting meaningful geological features. Dip and azimuth of the interpreted horizons are calculated for the delineation of faults and flexures. On the basis of the fault delineations interpreted on the horizons and the fault interpretation on vertical sections, three-dimensional fault planes are constructed. Subsequently, "fault slices" are derived by extracting seismic data along surfaces parallel to the fault planes. These fault slices are used to study the fault throw and the juxtaposition of lithologies across the fault plane. In some cases hydrocarbons can be predicted from seismic attributes.

1. Introduction

The advantages of 3-D over 2-D seismic with respect to better subsurface sampling and improved structural and stratigraphic definition of the subsurface have been widely published (1). A major breakthrough in the interpretation of 3-D data sets has been the development of automatic horizon tracking programs. Beginning with a limited set of interactively interpreted grid lines, these programs extend the interpretation of an horizon to every grid point of a 3-D survey.This makes it possible to apply powerful horizon processing methods to bring out features that might have been seen on individual sections but not correctly interpreted without their areal context .

By taking the spatial derivatives of the interpreted horizon, maps of dip magnitude and dip azimuth can be obtained for delineating faults and flexures that could, for example, represent permeability barriers to fluid flow in the reservoir (2). Subjecting these maps to artificial "illumination" can highlight subtle features that would otherwise go unnoticed. On the basis of the faults delineated on several horizons, together with the interpretation on vertical sections, three-dimensional fault planes can be constructed with increased confidence. Analysis of "fault slices" derived by extracting seismic data parallel to the fault surface on both sides of the fault results in maps describing fault throw, juxtaposition of lithologies or sealing capacity across the fault plane (3). The reflection amplitudes along the horizon provide information about the presence of hydrocarbons.

Section 2 describes horizon interpretation by means of an automatic tracking procedure. Section 3 deals with the calculation of dip and azimuth for delineating faults. Section 4 describes how, once the faults have been determined, fault slicing is applied to study lithological juxtaposition across the fault plane. Amplitude mapping will be discussed in Section 5, which is followed by some conclusions in Section 6.

2. Horizon interpretation

In interpreting 3-D data sets a selected number of horizons must first be interpreted interactively on the workstation for a limited set of seed lines that constitute a so-called "control grid". The required number of seed lines depends on the capability of the automatic tracking program to expand the interpretation of each horizon over the whole survey. In areas that have a poor signal-to-noise ratio or complicated faulting with large offsets, the control grid needs to be quite dense. If faults have either a small offset or fault zones are characterised by relatively low amplitude levels and if at the same time the character of the reflection wavelet is fairly constant within each fault block, only a limited number of seed lines are needed After the interpreted control grid is provided to the automatic tracking program, it expands each selected horizon laterally away from the seed lines. In this way the amount of data that is interpreted interactively can be limited, whereas the information from each trace in the 3-D survey will become available for subsequent analysis. An example with only two seed lines is shown in Fig. 1.

Various methods of automatic tracking can be used, but basically the approach consists of comparing unpicked traces with the seed-line traces and finding, within a pre-defined time window, a similar reflection wavelet . The program then "snaps" to a point on this wavelet, which commonly is taken to be either its maximum, minimum or zero-crossing. As long as similar reflection wavelets are found, the tracking program continues the expansion of the horizon. When similarity in wavelet shape or amplitude disappears, or when the picked point falls outside the set time window, the trace is not picked. This would, for example, occur in the case of a fault characterised by a loss of reflection strength.

The result of the automatic tracking procedure is a picked horizon with gaps due to either faults or a poor signal-to-noise ratio. These gaps need to be examined and additional seed points or changed picking criteria will have to be provided to extend the tracking further. An example of the results of a first round of tracking is shown in Fig. 2. The results for this data set were verified by extensive interactive interpretation of all critical areas. In the example the horizon expansion could be seen directly or monitored from diagnostic displays (Fig.3). It appeared that the tracking program was able to recognise and stop at the fault zones (note the small gaps in the en-echelon fault on the right-hand side). The continuity in the tracked horizon between the two sets of en-echelon faults in the final result indicates the consistency in the expansion from the two seed lines.

Usually the tracking is done in several iterations consisting of tracking using certain quality criteria, inspection of the results, optional interpolation to fill the gaps, calculation of dip and azimuth for quality control, addition of new seed lines and back to tracking, possibly with relaxed quality criteria. Dip and azimuth (Section 3) are particularly useful as a quality check. If the dip and azimuth patterns are very noisy, then the tracking procedure has not been finding the horizon consistently, and re-tracking using different quality criteria is needed. Sometimes it becomes evident that the control grid was not interpreted consistently and loop skips have occurred. After these iterations the final dip (and azimuth) map should clearly indicate a geologically consistent fault pattern (Fig. 4).

After the horizon has been tracked, the remaining gaps can be filled in by straightforward interpolation or triangulation, or by more sophisticated methods developed in image processing. Also, irregularities related to missing data, artefacts caused by the tracking programs or noise interference can be removed by means of filtering. The filter program should, however, avoid removing or distorting meaningful geological features. An artificial example showing the filling in of missing lines is shown in Fig. 5.

3. Fault delineation

Once horizons have been tracked, dip-magnitude and dip-direction maps are made to delineate faults and flexures. For good-quality data a small computation template consisting of four neighbouring traces can be used (Fig. 6), providing maximum resolution of subtle structural variations. For increased stability in the case of noisy data larger templates should be used. The results can be displayed as separate dip (Fig. 7a) and azimuth (Fig. 7b) maps , as overlay maps (Fig. 8a) or combined dip/azimuth maps (Fig. 8b)[3]. On Figs. 7a and 7b the control grid is also shown. The high density of the control grid (every fourth line) was mainly needed for the deeper horizons. Only horizons were interpreted, fault cut-outs were left as blanks. From the displays it can be seen that it is essential to integrate the control-grid interpretation and dip and azimuth information. Fault A in Fig. 8a is not expressed on the azimuth map (Fig. 7b), because the fault plane azimuth equals the horizon azimuth; Fault B is not expressed on the dip magnitude map (Fig. 7a), because the fault plane has only a moderate dip. These plots are particularly useful for determining whether reservoirs are separated or in communication. For example, in the enclosed area of Fig. 8a the continuity of faulting and the direction of fault hade changes rapidly over a very short distance. This discontinuity, an indication of likely fluid communication between the blocks, could easily have been missed on a grid of interpreted in-lines and cross-lines. Another useful display highlighting subtle features is obtained by artificial illumination. An example is shown in Fig. 9.

4. Fault slicing

Once faults have been delineated on several horizons and interpreted on vertical sections through the 3-D data set, fault planes can be constructed. Using palinspastic reconstruction, in which the structural configuration at the time of deposition of each interpreted interval is reconstructed, the integrity of the fault-plane interpretation can be checked. Additionally, the possibility of fault sealing is investigated (4). Fault sealing requires the lateral juxtaposition of reservoir against a sealing lithology across the fault plane. Alternatively the fault plane itself could be sealing, owing to smearing of ductile clays along the fault plane during faulting. The best way to study juxtaposition of lithology and clay smear is by construction of fault slices. Fault slices are extractions of seismic data from the three-dimensional seismic data set along surfaces parallel to the fault. These slices, made on both sides of the fault (footwall and hanging-wall slices), are compared to derive

- a throw map along the fault plane, which could indicate not only dip-slip but also strike-slip components;
- juxtaposition of lithologies and degrees of clay smear using estimated or observed properties of the shale layers between the reservoir intervals;
- maps of amplitude anomalies;
- maps of of undisturbed seismostratigraphy.

Examples of a fault, a fault slice, a throw map, a juxtaposition map and a clay-smear potential map are shown in Figs. 10-14.

5. Amplitude mapping

Especially in young deltaic areas hydrocarbons express themselves quite clearly on seismic. In the case of a sand-shale sequence in which the sands have a lower acoustic impedance than the surrounding shales, hydrocarbons cause a decrease in the acoustic impedance of the reservoir. This shows up as "bright spots" or high-reflectivity areas on horizon amplitude maps. An example is shown in Fig. 15. As can be seen from the contours, the anomalies broadly conform to structure, which in itself is a strong indication of hydrocarbons (horizontal fluid contacts).Time contours are shown in the example of Fig. 15, since a laterally constant overburden velocity can be assumed.

Amplitude anomalies, known to be related to the occurrence of hydrocarbons, can alternatively be used to fine-tune the time-to-depth conversion.

6. Conclusions

To exploit fully the information present in a 3-D data set, relevant horizons should be interpreted on every available trace. This can be done economically by means of control grids and automatic tracking programs. Image processing methods can subsequently be used to enhance the information and to delineate faults and fractures. Using fault slices, the sealing potential across the fault plane through lithologic juxtaposition or clay smear can be estimated. Finally seismic amplitudes can be used to determine the pore fluids.

References

1. Nestvold, E.O.
   The Use of 3-D Seismic in Exploration and Production.
   Energy Industries Council Oil, Gas and Petrochemical Seminar, Baghdad, Iraq, 1989.
2. Dalley, R.M., Gevers, E.C.A., Stampfli, G.M., Davies, D.J., Gastaldi, C.N., Ruijtenberg, P.R. and Vermeer, G.J.D.
   Dip and Azimuth Displays for 3-D Seismic Interpretation.
   First Break 7, 86-95, 1989.
3. Brown, A.R., Edwards, G.S. and Howard, R.E.
   Fault slicing, a new approach to the interpretation of fault detail.
   Geophysics, V. 52, 1319-1327, 1987.
4. Bouvier, J-D., Kaars-Sijpesteijn, C.H., Kluesner, D.F., Onyejekwe, C.C. and Van der Pal, R.C.
   Three-Dimensional Seismic Interpretation and Fault Sealing Investigations, Nun River Field, Nigeria.
   AAPG Bull. V. 73, No. 11, 1397-1414, 1989.

Acknowledgements

The author is indebted to Shell Internationale Petroleum Maatschappij for permission to publish this paper.

The development of the methods discussed in this paper represent the combined efforts of staff in Shell Internationale Petroleum Maatschappij (SIPM), many Shell Operating Companies worldwide, Shell Oil Company (U.S.) and Koninklijke/Shell Exploratie en Produktie Laboratorium (KSEPL).

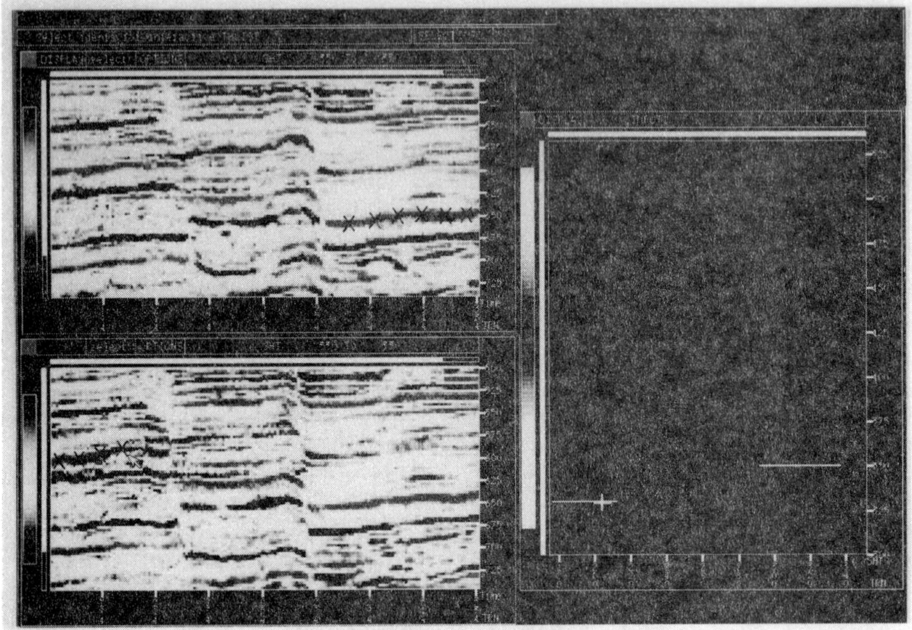

Fig 1: Input control grid

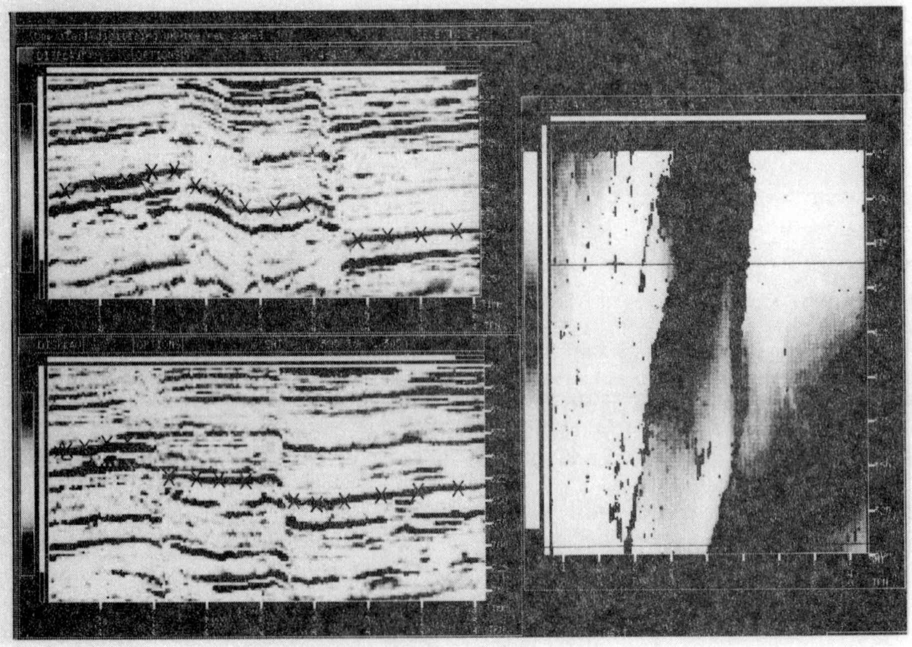

Fig 2: Output horizon time map

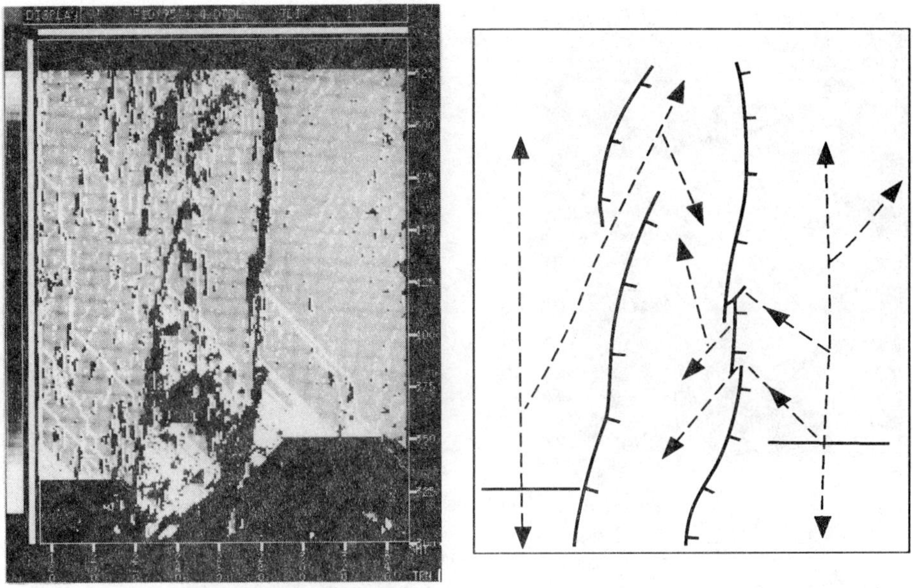

Fig 3: Tracking progress

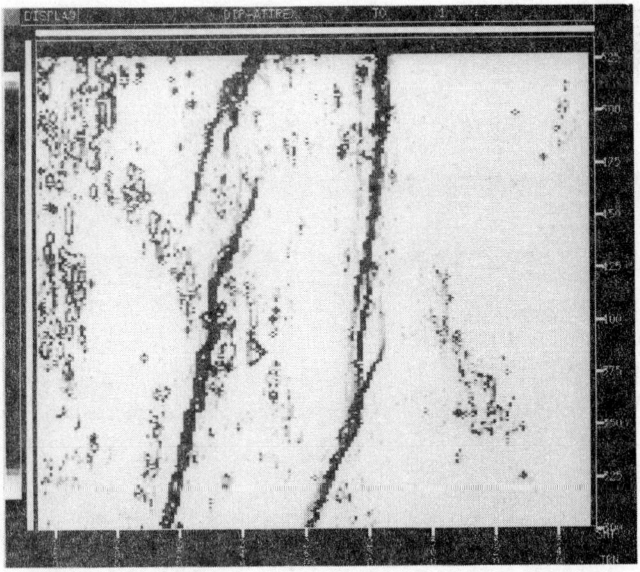

Fig 4: Dip magnitude map, steep dips (black) indicate faults

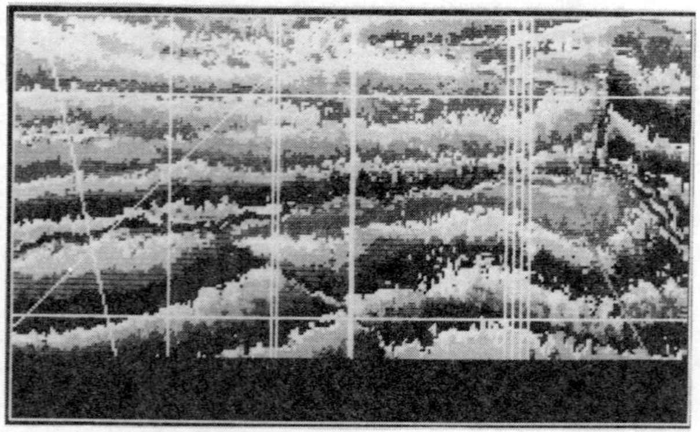

Fig 5a: Input horizon

Fig 5b: Output horizon

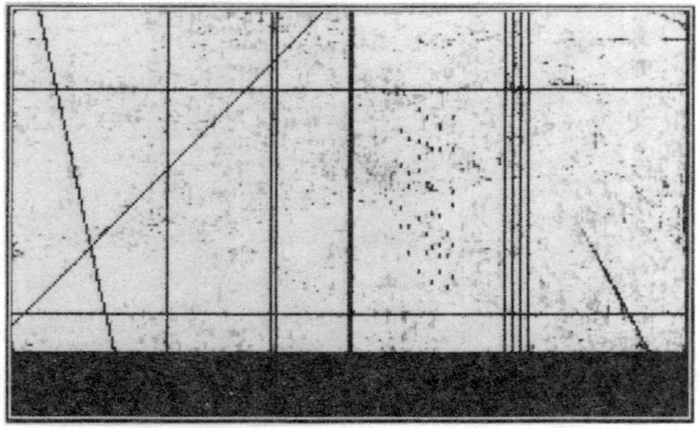

Fig 5c: Diffence horizon

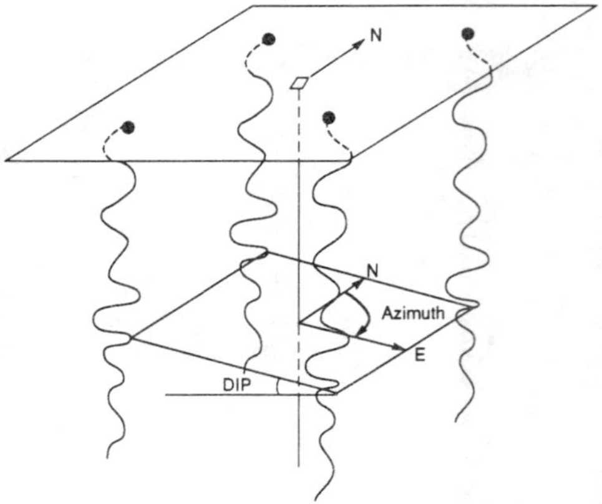

Fig 6: Dip and Azimuth calculation

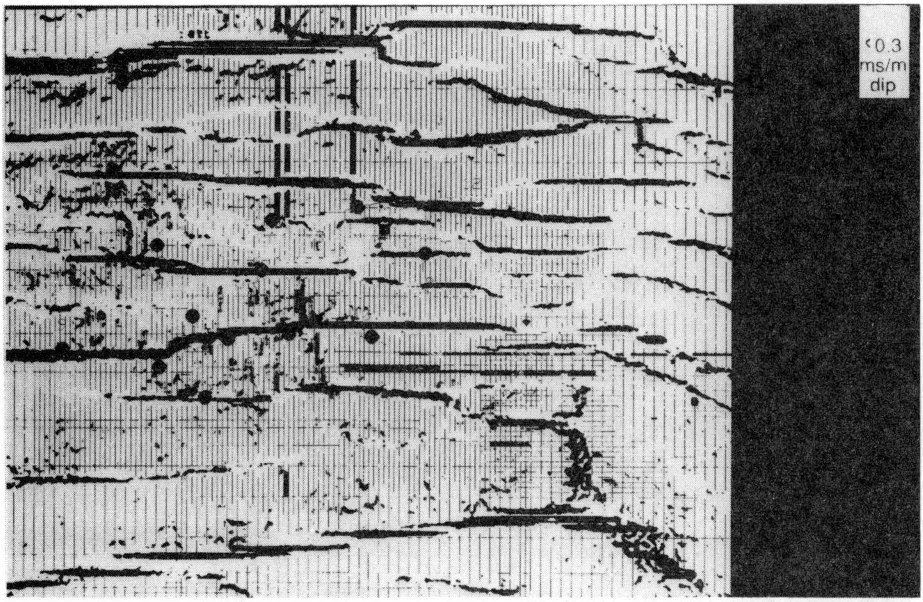

Fig 7a: Fault dips and interpreted control grid

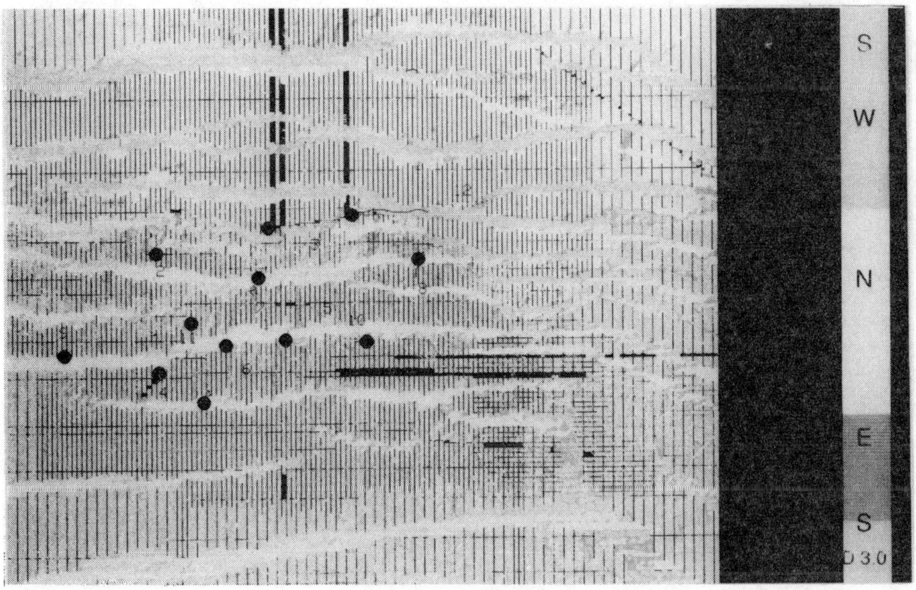

Fig 7b: Azimuth and interpreted control grid

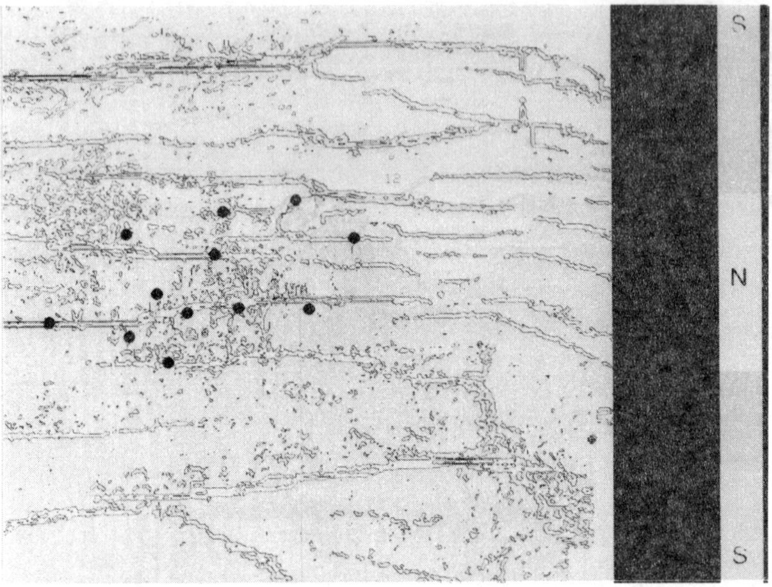

Fig 8a: Azimuth map with contoured Dip-magnitude overlay

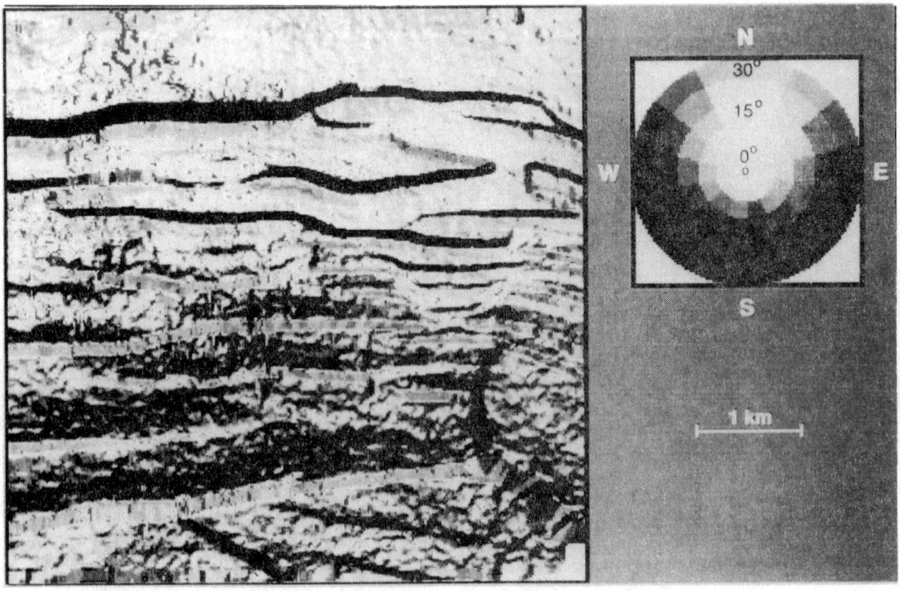

Fig 8b: Combined Dip and Azimuth map

Fig 9: Faults/Flexures highlighted by artificial llumination

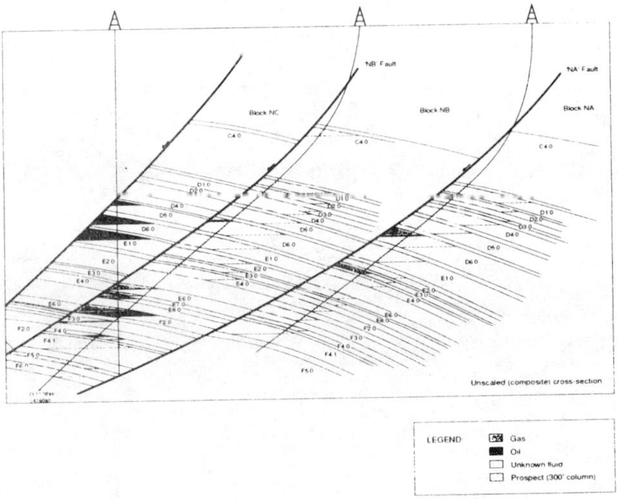

Fig 10: Fault-scooping wells

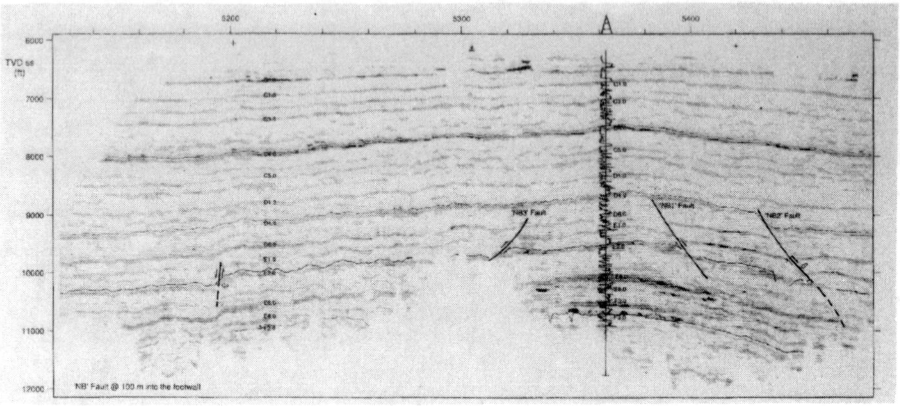

Fig 11: Fault-slice

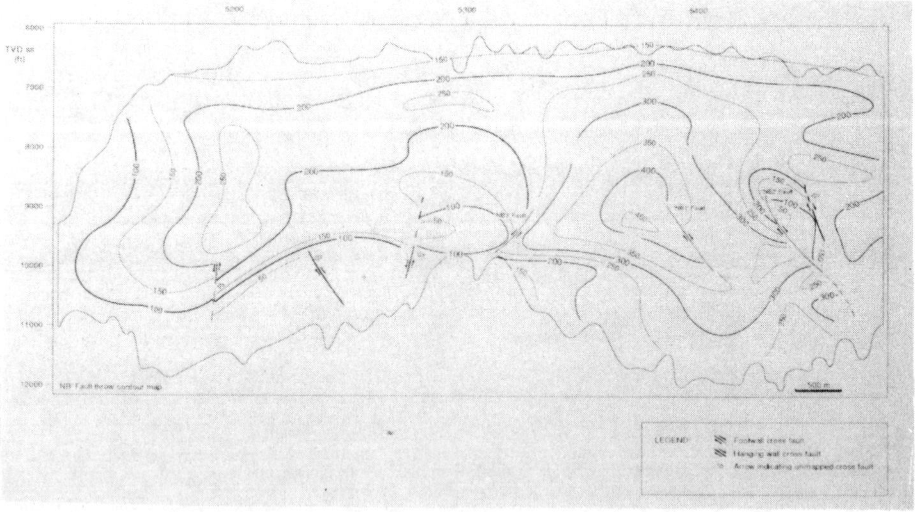

Fig 12: Throw contour map

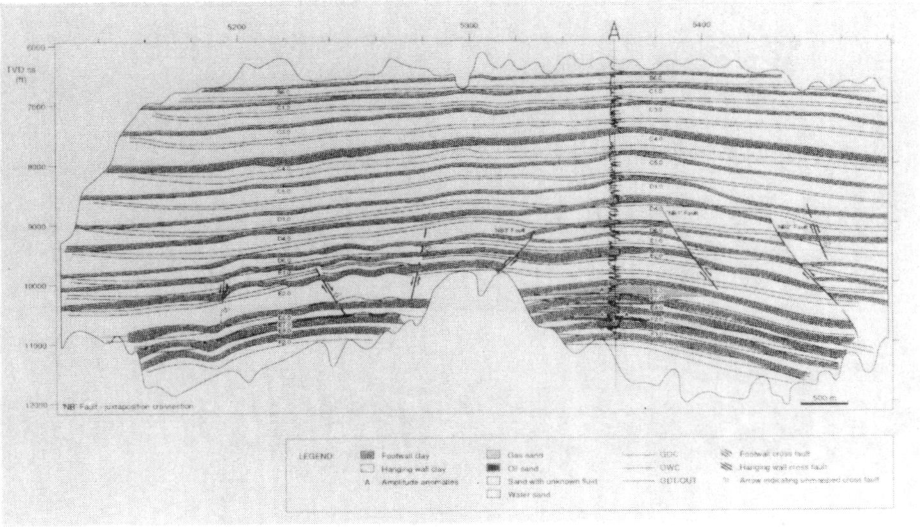

Fig 13: Fault-juxtaposition cross-section

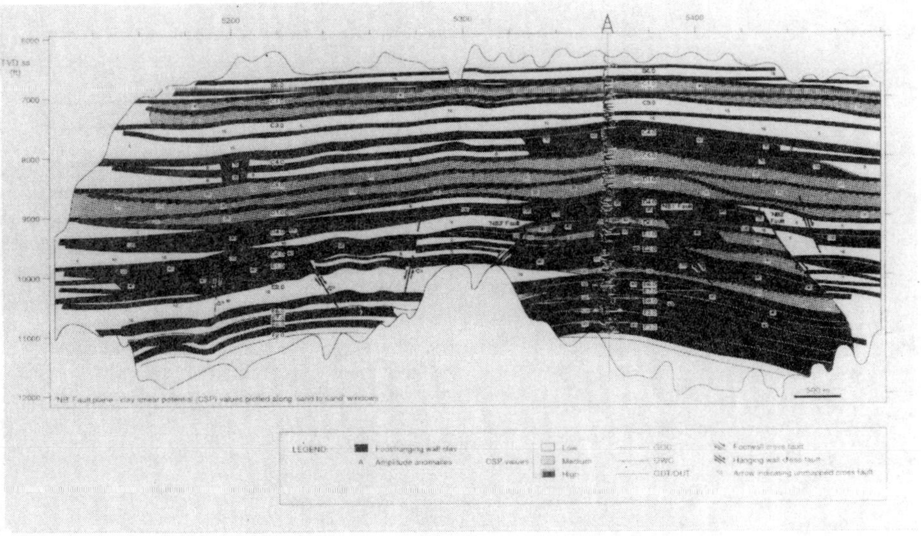

Fig 14: Clay smear potential values

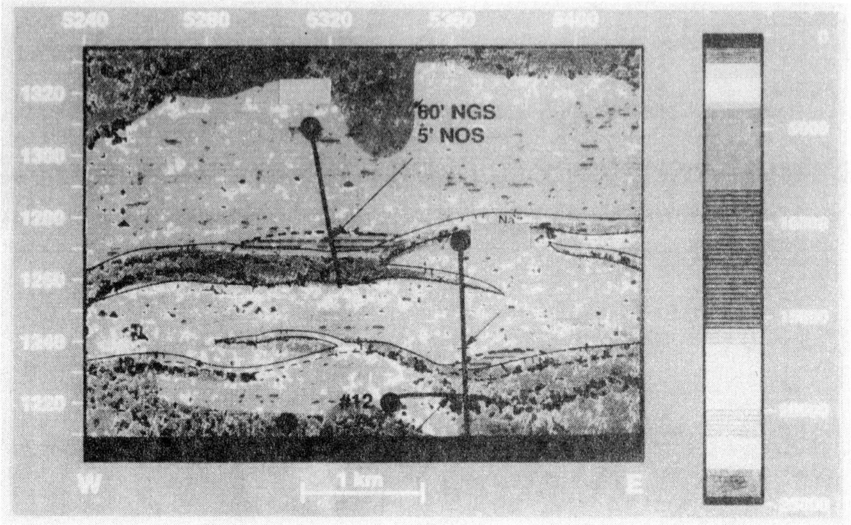

Fig. 15: Amplitude map

MONTE CARLO SIMULATION OF LITHOLOGY AND POROSITY FROM SEISMIC DATA

P.M. Doyen and T.M. Guidish

Western Geophysical, a Division of Western Atlas International
455 London Road, Isleworth, Middx TW7 5AB

SUMMARY

A two-step Monte Carlo procedure is introduced for numerically simulating lithology and porosity cross-sectional models from seismic impedance profiles and well log data. In the first step, the spatial arrangement of sand and shale units is simulated. In the second step, the inter-well lateral distribution of porosity is simulated within the sand compartments. The technique accounts for the fragmentary and imperfect nature of the geophysical information and for the resulting non-uniqueness in the modelling process. Rather than calculating a unique reservoir model, the Monte Carlo method provides a family of alternative geologic images, all of which are consistent with the seismic and log data at hand. The multiplicity of models reflects the uncertainty in the seismic prediction of lithology and porosity, and is used to assess the accuracy of reserve estimates. The reservoir modelling technique is illustrated using a high resolution seismic line and well log data from a Canadian oil field in Alberta. Sand/shale and porosity cross-sectional simulations are generated along the seismic line which intersects four wells. The simulated models reproduce the log-derived lithology and porosity observations at the wells; they are conditioned by acoustic impedance data obtained by inverting the seismic amplitude stack section, and are consistent with the spatial autocorrelation and crosscorrelation structures of the seismic and well data.

1. INTRODUCTION

In heterogeneous formations, spatial variations of petrophysical properties usually cannot be inferred from sparsely distributed well data. After careful data processing, lateral variations of seismic amplitudes can be used to help predict inter-well changes in reservoir properties, such as porosity and lithology.

One of the main challenges in this seismically based prediction process is the averaged and relatively imprecise nature of the seismic information. First, the acoustic parameters which can be derived from band-limited and noise-contaminated amplitude data are inherently nonunique. Second, changes of acoustic parameters, such as seismic impedances, generally cannot be related to a unique physical cause. Instead, variations of acoustic properties integrate the effects of a number of different geological variables, such as porosity, lithology, fluid type and saturation, pressure, temperature and rock microgeometry. In practice, the seismic prediction is restricted to rock properties, such as porosity, lithology, and gas occurrence, that contribute the most to changes in acoustic properties. Even so, in view of the limited information content of seismic data, it is important to quantify the reliability of seismically derived reservoir models and to evaluate how modelling errors affect predictions of reserves and production.

Conventional methods for estimating reservoir properties, such as porosity, from seismic data rely on empirical or regression formulas which are constructed, for example, by cross-plotting seismically derived impedances against porosity measurements in wells. Such approaches treat the data as spatially independent observations and ignore the existence of spatial patterns in the variations of rock properties. Moreover, the reliability of the resulting reservoir models is rarely assessed.

Here we present a two-step Monte Carlo procedure for predicting the spatial variations of lithology and porosity from seismic and log data. This technique, based on the work presented in [1] and [2], accounts for the ambiguous nature of the seismic information. Rather than calculating a unique reservoir model, the Monte Carlo method provides multiple, equally probable models, each of which is consistent with the seismic and log data available. The variability among models reflects the uncertainty in the seismic prediction process and can be used to assess risk in reservoir development.

In Section 2, we briefly describe the Monte Carlo procedure for the simulation of lithology. A similar method applies for the simulation of porosity. In Section 3, the Monte Carlo method is applied in two steps to generate cross-sectional lithology and porosity models of an oil-producing formation along a high resolution seismic line.

2. MONTE CARLO SIMULATION OF LITHOLOGY

Figure 1 schematically illustrates the lithology modelling process which combines seismic and well data. The figure depicts two wells intersecting a seismic section, which is represented by a regular grid of acoustic impedance (Z) observations. In practice, the acoustic impedances are obtained from the inversion of an amplitude stack section, as explained in Section 3. At each pixel $\tilde{x} = (x, z)$ of the seismic grid, a lithologic indicator variable B is defined by

$$B(\tilde{x}) \begin{cases} = 1 \text{ if } \tilde{x} \text{ is in sand} \\ = 0 \text{ if } \tilde{x} \text{ is in shale} \end{cases} \tag{1}$$

This unknown lithology indicator must be inferred at each point $\tilde{x}$ from observations of the variable B in nearby wells and from the knowledge of the local seismic impedance value, $Z(\tilde{x})$, at that point. The advantage of using acoustic impedances in inferring lithology is that the seismic data provide spatially dense information between boreholes. However, the seismic information is limited in that, in general, acoustic impedance is not uniquely related to lithology. That is, the impedance ranges for sands and shales often overlap. For instance, depending on the presence of gas, pore pressure, age, mineralogy and porosity, reservoir sands can exhibit higher or lower impedances than those of adjacent shales. Here, we do not assume that there is a unique correspondence between B and Z. Instead, we model their dependence statistically using a spatial crosscorrelation function.

The Monte Carlo lithology simulation procedure is based on estimating at each pixel of the model the probability p that the pixel corresponds to a sand. This probability, which is calculated from the well and seismic information, is called a conditional probability. It is defined at each pixel $\tilde{x}$ of the cross section as:

$$p(\tilde{x}) = Prob\{B\,(\tilde{x}) = 1|\, B(\tilde{x}_1), ..., B(\tilde{x}_n), Z(\tilde{x})\,\}\,, \tag{2}$$

where, in the conditioning dataset, $B(\tilde{x}_1), ..., B(\tilde{x}_n)$, represent n binary-valued observations of the lithology at spatial locations $\tilde{x}_1, ..., \tilde{x}_n$ in the wells, and $Z(\tilde{x})$ is the seismic impedance at $\tilde{x}$ (See Figure 1). The probability is calculated at each pixel using a Linear-Mean-Square (LMS) estimation procedure. The LMS estimate, $p_{est}(\tilde{x})$, is given by

$$p_{est}(\tilde{x}) = \sum_{i=1}^{n} \omega_i(\tilde{x})\, B(\tilde{x}_i) \; + \; \alpha(\tilde{x})\; Z(\tilde{x}) \; + \; c(\tilde{x})\,, \tag{3}$$

where weights $\omega_1, ..., \omega_n$ and α assigned to the data are determined at each pixel $\tilde{x}$ by minimizing a mean square prediction error criterion. This minimization only requires knowledge of the spatial autocorrelation and crosscorrelation structures of the variables B and Z. It is performed by solving a system of normal equations analogous to a cokriging system [3,4]. In (3), the constant, $c(\tilde{x})$, is determined from the condition that the average proportions of sand and shale in the simulated models be equal to those observed in the wells. Note that the relative magnitude of weight α in (3) depends on the degree of local crosscorrelation existing between seismic impedance and lithology; i.e., the stronger the correlation, the larger the weight value. In practice, this crosscorrelation is determined by comparing the lithology interpreted in wells with impedance data derived from seismic traces which are in the direct vicinity of the wells. The probability estimate p_{est} mixes binary-valued observations with impedance data, corresponding to values outside the interval [0,1]. In practice, prior to performing the LMS estimation of p_{est}, the Z data are transformed so that the new values lie between 0 and 1.

The simulation method involves the sequential estimation of the conditional probability p_{est} at all pixels of the model, or equivalently at all sample points of the seismic impedance profile. The algorithm for generating one sand/shale model can be summarized as follows.

For all pixels $\tilde{x}$ in the cross section:

1. Obtain the LMS estimate $p_{est}(\tilde{x}) \; = \; Prob_{est}\{B(\tilde{x}) = 1 \mid data\,\}$.
2. At $\tilde{x}$, draw a simulated value B_s that is equal to 1 or 0 with probability $p_{est}(\tilde{x})$ and $1 - p_{est}(\tilde{x})$, respectively, and
3. Add the simulated value $B_s(\tilde{x})$ to the conditioning data set of the probability in Step 1 above.

Note that, when the first location $\tilde{x}$ is considered in the simulation grid, the probability p_{est} is calculated only from the observations of B in wells. However, when the ith pixel is selected, the probability estimate is calculated from the $(i-1)$ previously simulated values, in addition to the well observations. The simulation process stops when all pixels of the model have been assigned a sand or shale value. The lithologic model constructed using this process is not unique. In fact, repeated application of the algorithm yields different models by varying the order in which the model pixels are visited and the

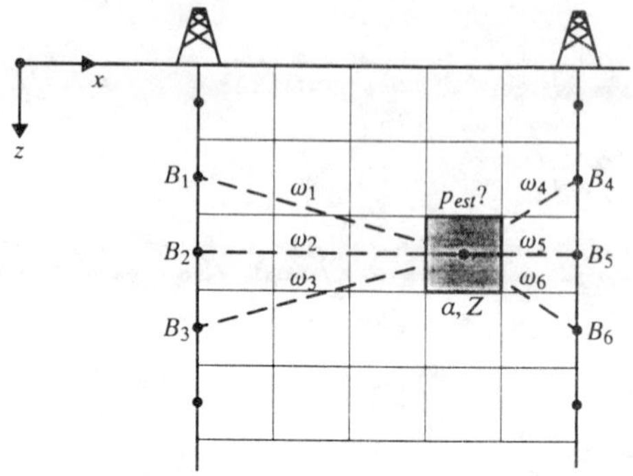

Figure 1. The sand probability, p_{est}, is calculated from the seismic impedance, Z, and from the lithologic data in wells.

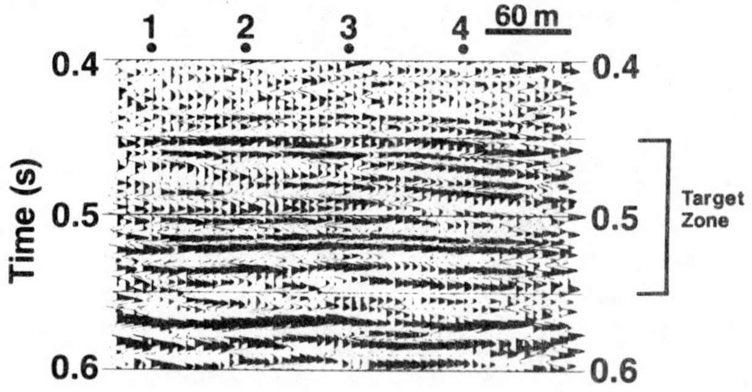

Figure 2. Relative-amplitude-processed stack section intersected by four oil-producing wells.

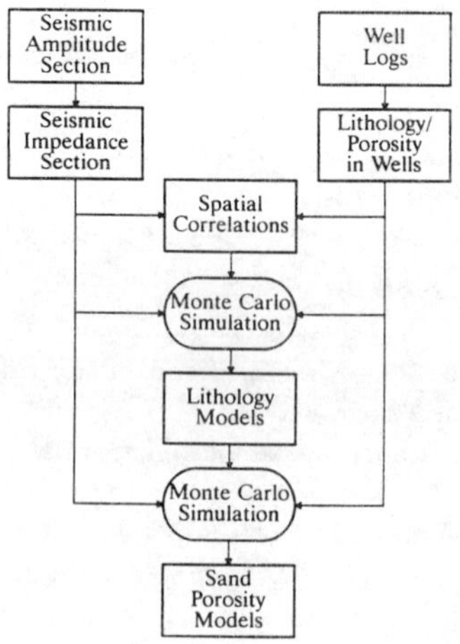

Figure 3. Flow chart of the reservoir modelling process.

sampling of the sand/shale probability distribution. It can be shown that the models simulated using the above algorithm all have the following properties:

- the models reproduce the vertical sand/shale sequences observed in wells,
- the models reproduce the average sand and shale proportions, or equivalently the average net-to-gross ratio inferred from the well data,
- the areal extent and thickness of the simulated sand/shale units are constrained by the spatial autocorrelation structure of the lithologic data, and
- the models are consistent with the seismic data in that they agree with the crosscorrelation between lithology and acoustic impedance.

Finally, note that a similar Monte Carlo procedure can also be applied to generate porosity simulations.

3. CASE STUDY OF A CANADIAN RESERVOIR

The Monte Carlo technique was applied to obtain cross-sectional lithology and porosity models of the Mannville Formation along the high resolution seismic line displayed in Figure 2. The line intersects four closely spaced oil-producing wells numbered 1 to 4. A full suite of interpreted logs was available at each of the wells. The target formation is located between approximately 450ms and 550ms two-way travel time. This time interval corresponds to depths ranging from 480m to 650m. Note the high temporal resolution of the seismic data: the signal bandwidth extends from about 20Hz to more than 200Hz. This broad bandwidth was achieved by using specialized acquisition and processing techniques similar to those discussed in [5]. Note also the complex lateral changes of seismic amplitudes in the seismic window. These changes reflect the variability of the lithology and porosity in the Mannville Formation. The purpose of the Monte Carlo modelling was to delineate lateral lithofacies and porosity changes which could affect the performance of oil recovery operations. In view of the high frequency content of the data, we were interested in obtaining detailed reservoir models with a vertical resolution on the order of 5m, or approximately one quarter of the dominant seismic wavelength.

Figure 3 shows a flowchart of the reservoir modelling process, starting with the processed seismic amplitude data and the interpreted well logs. First, the seismic amplitude data were converted to acoustic impedance estimates by applying a recursive inversion process similar to that described in [6]. Figure 4 shows the estimated impedance section for the target seismic window after conversion from travel time to depth. A low-frequency trend calculated from the sonic and density logs was added to the seismically derived impedances. Note the fine layering of the impedance model which reflects the high frequency content of the data and the complex interfingering of the sands and shales in the Mannville Formation. In order to check the quality of the inversion process, we compared inverted seismic impedances and log-derived impedances as a function of depth at the four wells along the seismic line. For well 1, this comparison is shown in Figure 5 with the two impedance curves sampled at a 2m spacing. Though lacking the fine details

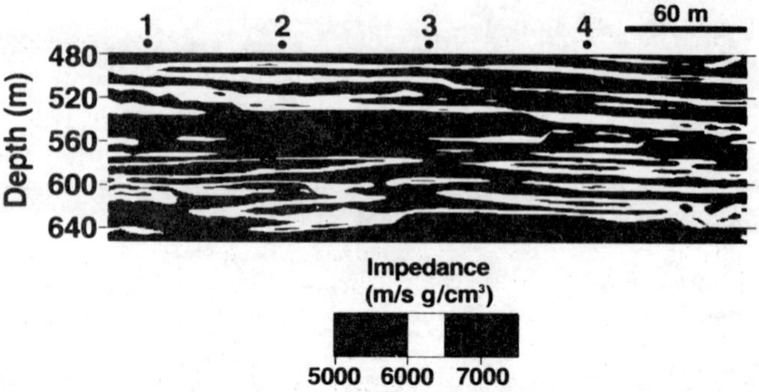

Figure 4. Depth-converted seismic impedance section of the target zone.

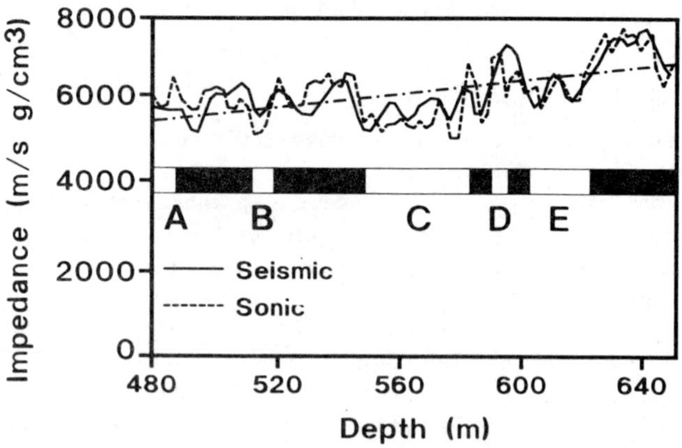

Figure 5. Comparison of sonic impedance and inverted seismic impedance at well 1. The five sand members penetrated by the well are labelled A through E.

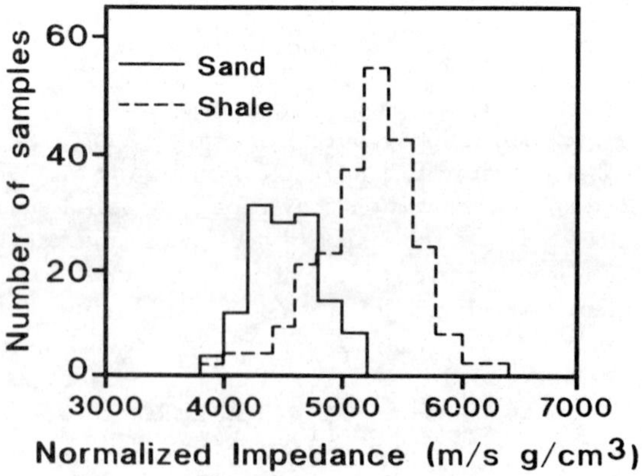

Figure 6. Normalized impedance histograms for sands and shales.

apparent on the impedance log, the seismic impedance function closely approximates the log across the depth interval corresponding to the Mannville Formation.

Figure 5 also displays the sand and shale intervals which were interpreted from the gamma ray log at well 1. This well penetrated five sand members which are labelled A through E. The five sands are oil-saturated with the lower sand being the main producer. Comparing lithology and impedance at well 1, it is apparent that the major sands, such as C and E, tend to have lower acoustic impedances than adjacent shales. However, due to the trend of increasing impedance with depth, a deep sand, such as E, may have an impedance as high as a shallower shale, such as the shale between sands B and C. To remove this feature, which would adversely affect the seismic discrimination of lithology, the linear trend depicted in Figure 5 was subtracted from the impedance depth function at each CDP along the seismic line. Figure 6 displays seismic impedance frequency histograms for sands and shales after removal of the linear depth trend and normalization to a reference impedance of 5000 m/s g/cm^3. These histograms were calculated from the four wells along the seismic line. The two histograms show that sands exhibit lower impedance, on average, than shales. However, there is considerable overlap between the impedance ranges of the two lithologies. The histograms were used to estimate the coefficient of crosscorrelation between the variables B and Z. This coefficient is needed in the calculation of the sand probability p_{est}.

Figure 7 shows four sand/shale cross-sectional models generated during the first simulation step (See flowchart in Figure 3) from the seismic impedance section and from the binary lithologic data at wells 2 to 4. Well 1 was used as a hidden well where the accuracy of the lithology prediction can be evaluated. Note that the black and white bands overlaid at the wells correspond to the log-interpreted sand/shale intervals. The simulations were performed at all sample points of the impedance profile which is composed of 53 traces, each containing 86 depth samples. The spacings between samples in the horizontal and vertical directions are 6m and 2m, respectively. In the Monte Carlo modelling, geometrically anisotropic, exponential correlation models [4] were selected for the variables B and Z.

As indicated in the flowchart, the modelling of lithology was followed by the simulation of porosity within the sand units. Figure 8 shows four sand/shale cross sections with overlays of the sand porosity distributions generated during the second simulation step. The sand porosity simulations were also derived by combining the impedance data and porosity observations in well 2, 3 and 4. The simulated images represent alternative lithology and porosity models which are consistent with the seismic and well log information available. Note in particular that all the models exactly tie at wells 2, 3, and 4. Differences between simulations reflect the uncertainty in describing inter-well reservoir heterogeneities.

Examination of Figures 7 and 8 shows that certain low spatial frequency simulated features consistently appear in all models, while other higher frequency patterns are variable from one simulation to another. The variability in the high spatial frequency details of the models is related to the limited model resolution that can be achieved from the data. For example, comparison of the lithology models in Figure 7 shows that the gross morphology of the sand members labelled C through F in simulation 1 is well

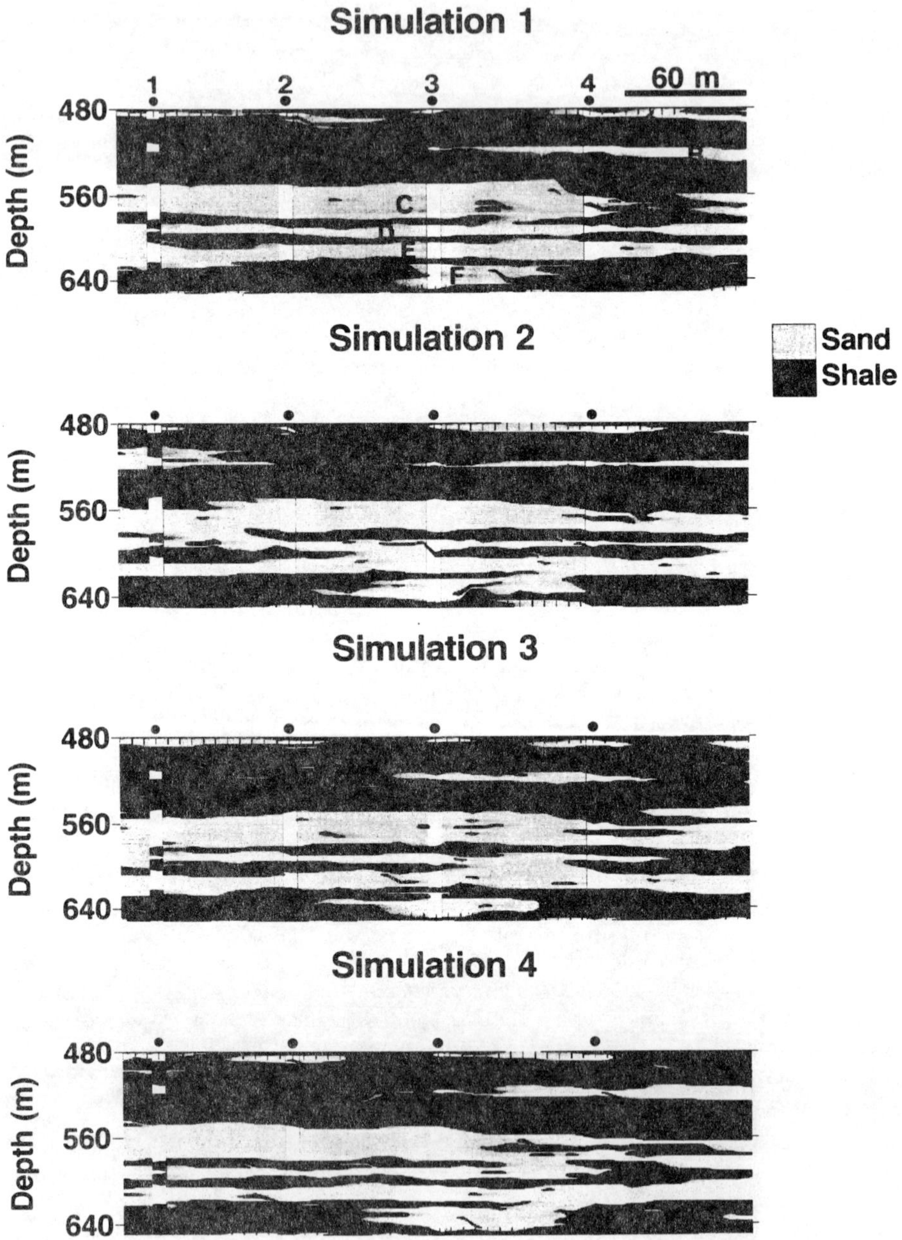

Figure 7. Four lithologic models. The black and white vertical bands at the wells indicate the log-derived lithology. Well 1 is a hidden well in the simulation process.

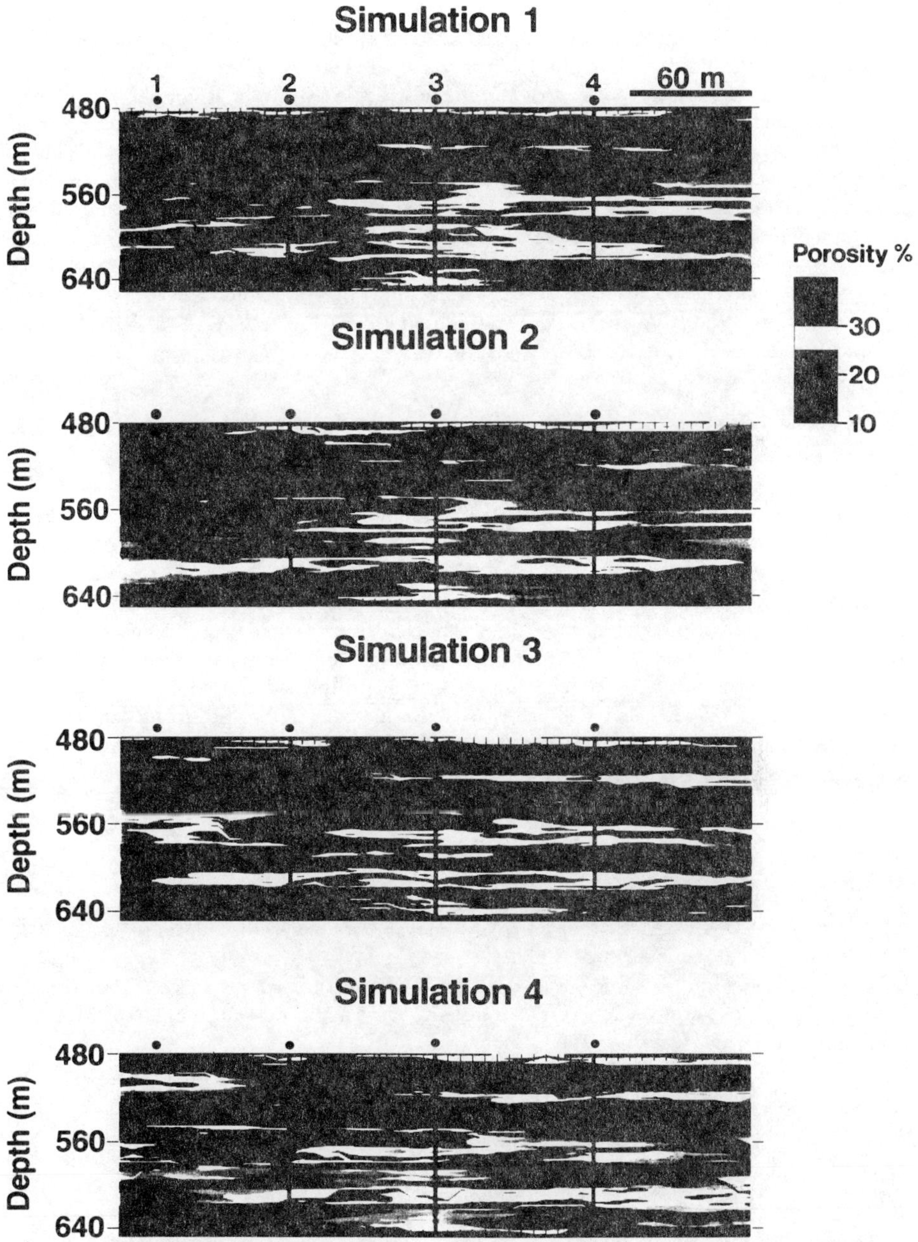

Figure 8. Four lithology and sand porosity models. Well 1 is a hidden well in the simulation process.

constrained. In addition, the position of sands C, D, and E are predicted accurately at the test well 1. By contrast, variations from simulation to simulation in the geometry of sands A and B indicate that the lateral extent of these thin sands is poorly defined. Similarly, for porosity, comparison of the models displayed in Figure 8, shows that the gross contours of the high and low porosity zones are resolved, while the exact position and details of the porosity highs and lows are not well defined spatially. In addition to the level of spatial frequency, the simulation variability is controlled by the quality and proximity of the observations which are used in the modelling. For example, differences between lithology models are greatest in areas of the seismic profile where the impedance values lie in the overlap interval between the impedance ranges of the sand and shale classes. One such area corresponds to the shallow part of well 1 where there is no acoustic contrast between sands A and B and the encasing shales (See Figures 5 and 7). Note that, as expected, the differences between models also increase away from the wells.

4. CONCLUSION

We have introduced a two-step Monte Carlo procedure for modelling reservoir geometry and porosity by rigorously combining seismic impedance and log data. In the first step, the spatial arrangement of the producing and nonproducing facies, sand and shale in this case, is simulated. This step is followed by the simulation of the lateral distribution of porosity within the sand units. The simulation procedure yields a family of equally probable reservoir models which are conditioned by both the seismic and well data, and by the autocorrelation and crosscorrelation structures of the data. Compared with well-log-derived models, the seismically conditioned simulations are better constrained spatially and provide more detailed and reliable images of inter-well reservoir heterogeneities. However, in keeping with the band-limited and ambiguous nature of the impedance data, the seismically consistent simulations still differ in the connectivity of the sand units and in the details of the porosity distribution.

ACKNOWLEDGEMENTS

We are grateful to Amoco Canada Petroleum Company Ltd. for releasing the seismic and interpreted log data for this publication.

REFERENCES

[1] Journel, A.G., and Alabert, F.G., New method for reservoir mapping, Journal of Petroleum Technology, February, 1990.

[2] Doyen, P.M., Guidish, T.M., and de Buyl, M.H., Lithology prediction from seismic data, a Monte Carlo approach, Presented at the 58th Annual International SEG Meeting, Anaheim, California, October, 1988.

[3] Doyen, P.M., Porosity from seismic data: A geostatistical approach, Geophysics, Vol. 53, No. 10, October, 1988.

[4] Journel, A.G., and Huijbregts, CH.J., Mining Geostatistics, Academic Press, 1978.

[5] Pullin, N.E., Matthews, L.W., and Hirsche, W.K., Techniques applied to obtain very high resolution 3-D seismic imaging at an Athabasca Tar Sands thermal pilot, Presented at the 56th Annual International SEG Meeting, Houston, Texas, November, 1986.

[6] Lindseth, R.O., 1979, Synthetic sonic logs–a process for stratigraphic interpretation: Geophysics, Vol. 44, No. 1 , 3-26.

TRANSMISSION AND REFLECTION TOMOGRAPHIC INVERSION OF OFFSET VSP-DATA AND THE USE OF THE RESULTS FOR A TARGET ORIENTED PROCESSING OF REFLECTION SEISMIC DATA

B. Lehmann[1], D. Krollpfeifer[2], L. Dresen[2], C. Gelbke[1]

1: Deutsche Montan Technologie
Institut für Angewandte Geophysik
Herner Str. 45
D-4630 Bochum 1
Federal Republic of Germany

2: Ruhr-Universität Bochum
Institut für Geophysik
Postfach 10 21 48
D-4630 Bochum 1
Federal Republic of Germany

Summary

The development of a tomography package to evaluate and invert traveltime and amplitude data is presented. All source-receiver geometries like VSP, HSP, crosshole geometries and their combinations are possible. For the inversion not only the first traveltime or amplitude onsets but also the later arrivals and amplitudes from refracted and reflected waves are used. For the forward modelling we have favoured the Gaussian beam method among other possibilities and for the tomographic inversion we used the simultaneous iterative inversion techniques (SIRT).

The results of the tomographic inversion can be used to examine limited parts of the underground like hydrocarbon reservoirs, which are of great interest. Based on the model which is found by the inversion process a semi-empirical equation can be established to calculate the incidence angle which is necessary to send directed seismic waves to the target. These directed waves can be simulated by using linear source arrays and delay-and-sum techniques. This paper shows results of the target oriented processing for simple analogue two-dimensional models with thin gas bearing layers.

1. TOMOGRAPHIC INVERSION

Iterative inversion techniques involve the following steps: forward calculation of the synthetic data and inversion of the differences between the synthetic and the observed data. One iteration consists of performing the forward calculation of traveltimes and amplitudes and inverting them to obtain the perturbations which are used to correct the model. Iterations continue until the model traveltimes match the field data traveltimes within the data error, or until other stopping criteria are met. The result of the tomographic inversion method is a 2-D velocity distribution which yields the best agreement with the real measurement data.

Usually only the first traveltime onsets are used for the tomographic inversion. All other informations of the seismograms like later arrivals and amplitudes are not considered. In the following we describe a new technique to include traveltimes and amplitudes of later arrivals.

1.1. FORWARD MODELLING

A very simple and fast forward modelling method is to calculate rays along straight lines without any refraction or reflection, which can be done if the medium is assumed to contain only small velocity variations. This is not valid in cases where the model has stronger velocity variations with more than 10-15% velocity contrast to the surrounding bed rock. Then the rays must be traced through the velocity distribution. For the ray tracing in the latter case we use two different methods.

Firstly, the rays can be traced using the gradient method either with or without Snell's law (4),(7),(8). The ray tracing algorithm using velocity gradients, is based on the grid structure of the tomographic inversion. The gradient method without Snell's law produces smoother rays; using Snell's law the rays are refracted.

Secondly, we use the Gaussian beam method for the fast reliable computation of theoretical seismograms in complicated laterally inhomogeneous media (14). Hereby, the model is subdivided into triangles with linear density and velocity laws. This division has the advantage that the ray paths are known to be circular in each triangle and thus there is no need to numerically integrate the ray equations. The differential equations can be solved analytically and so we have a very fast kinematic ray tracing. It is possible to compute phases which are specified by single or multiple reflections, refractions and/or conversions at different discontinuities in the medium. For the computation of seismograms the wave-field of the source has to be expanded into Gaussian beams which then have to be traced through the medium and summed up at the receiver. The wave-field as the superposition of all beams is regular even at a caustic because every beam is regular everywhere.

1.2. ITERATIVE INVERSION TECHNIQUE

In geophysical tomography the medium to be imaged is discretized by most authors into a grid of rectangular elements. In each of these elements the slowness value or absorption coefficient is considered to be constant. Here the solution is iteratively improved until an optimum is reached. A lot of slightly different iterative reconstruction procedures have been developed. The iterative techniques have been proved to be the most advantageous in the geophysical image reconstruction (15). We use the simultaneous iterative reconstruction technique (SIRT) (6), which we found to be the most convenient method because of its speed and stability.

The new inversion procedure consists of the following main steps and proceeds as follows. The first step is to compute a straight-ray SIRT tomogram with few (10-20) iterations. In a second step we perform a curved-ray SIRT. Based on the two results and additional a priori informations we decide where we have some boundaries of inhomogenities. The third step is the definition of a macro model with this possible structure and the forward modelling with the ray traced Gaussian beam method. This procedure is repeated until the current guess gives a satisfactory picture of the slowness distribution, or e.g. until the root mean square of traveltime differences reaches a limit.

1.3. EXAMPLE FOR A VSP-GEOMETRY

The presented synthetic example of a reverse VSP-geometry illustrates the necessity of using more than the first arrivals of one seismogram. Figure 1 shows the shot/receiver geometry and the simple velocity distribution in greyscale representation based on the grid structure and the correct layer boundaries. Twenty sources are located in the borehole on the left side whereas twenty receivers are placed at the Earth's surface. The model consists of two layers with different interface dips over a half space. In each layer the velocity is constant. The velocity distributions resulting from any inversion are presented as greyscale plots displaying the grid structure and the velocities of the individual cells. For all following figures the same greyscale is used to display the velocities that are normalized to the synthetic model velocities (3000 m/s ≡ white and 4500 m/s ≡ black). So, all the results can be compared for the different methods. Figure 2 shows the reconstructed image after 20 iterations using straight direct rays between shot and receiver locations. Only in the nearest region of the borehole the greyscale image agrees well with the synthetic model. The dip of the layers can not be resolved. Figure 3 shows the result after 20 iterations using the true straight ray paths between sources and receivers with refraction. The region near to the borehole yields a good resolution again. The consideration of refracted rays leads to an inhomogeneous ray coverage which can be seen at the irregular greyscale distribution. Due to Snell's law the region which is covered by the refracted rays is increased to the right side compared with figure 2. Finally figure 4 displays the velocity distribution after 20 iterations using additionally the reflections from the two layer boundaries. Above the second discontinuity the ray coverage increases because the inclusion of reflected rays. Thus, the velocity distribution appears to be homogeneous in each layer.

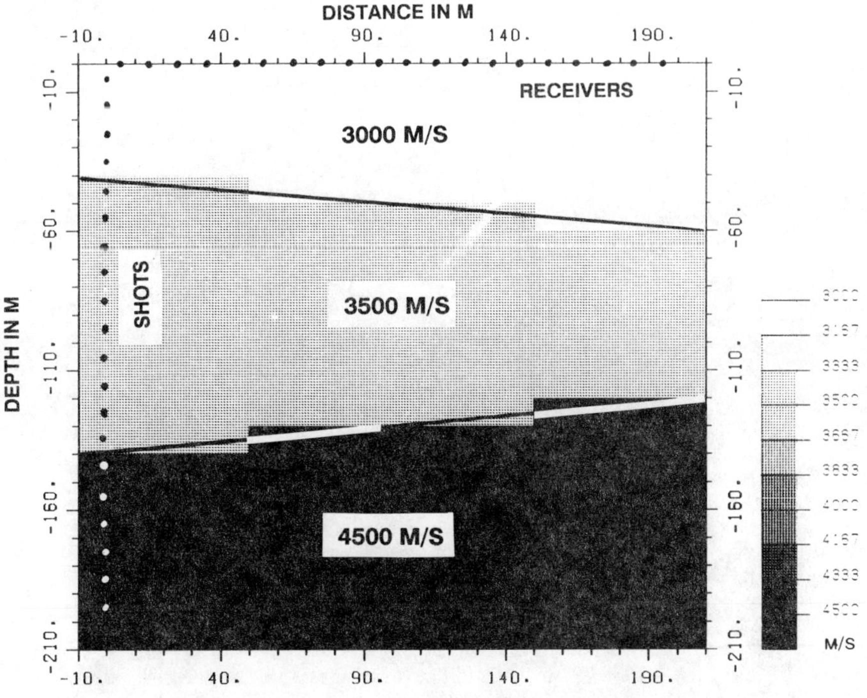

Figure 1: Synthetic model with VSP-shot/receiver geometry in greyscale representation and layer boundaries.

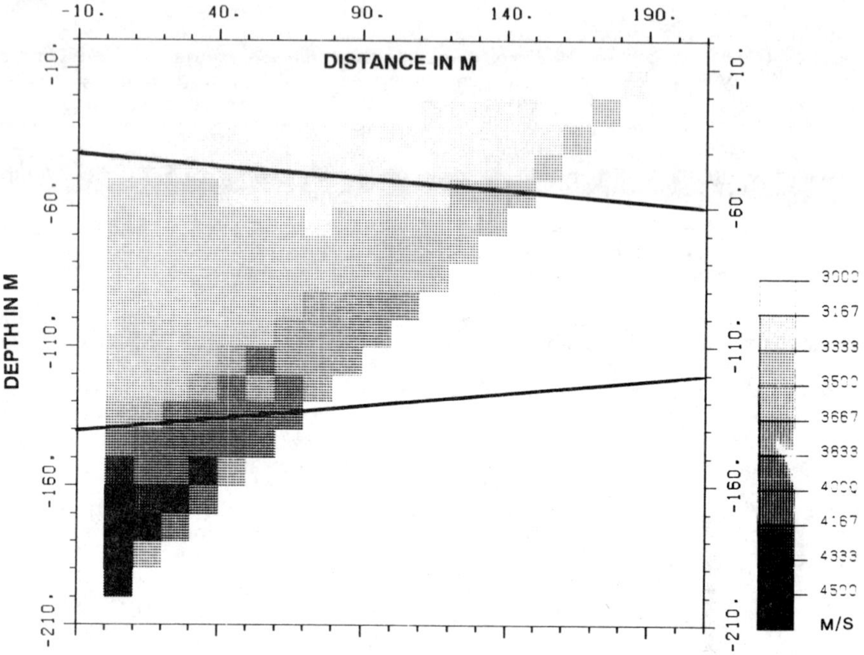

Figure 2: Tomogram with straight direct ray paths after 20 iterations.

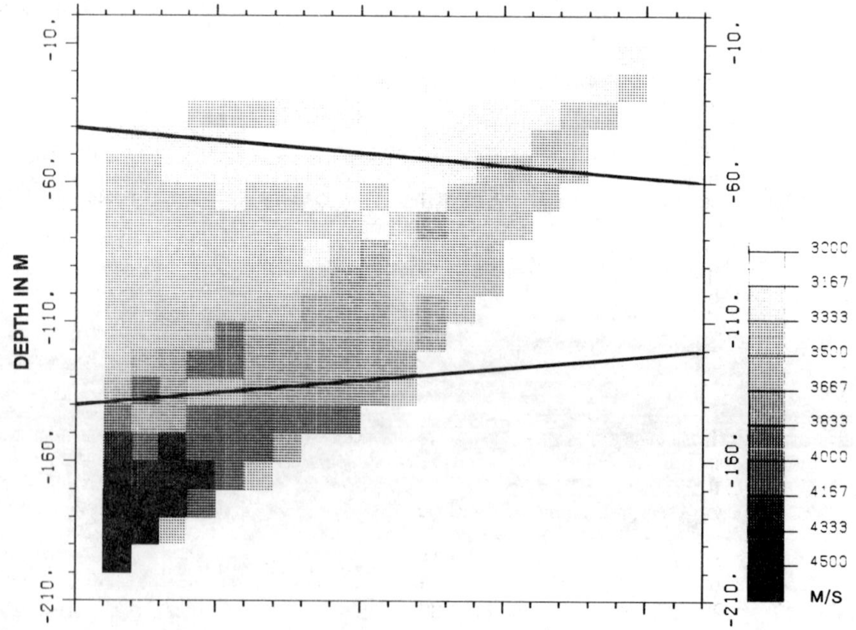

Figure 3: Tomogram with straight refracted ray paths after 20 iterations.

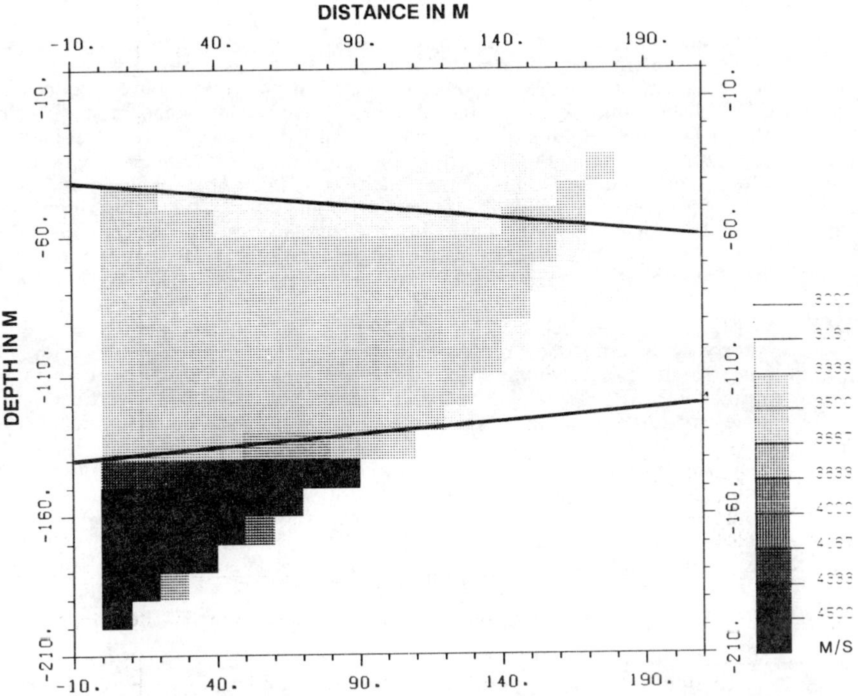

Figure 4: Tomogram using refracted and reflected arrivals after 20 iterations.

In the right region with a small ray coverage and great ray length the velocity information is smeared along the diagonal of the picture. In this figure the location and the dip of the two layers are reconstructed very well.

These results illustrate the demand for using later arrivals and the refracted ray paths. By using later arrivals the resolution of the investigated area will be greater and better.

1.4. CONCLUSIONS FOR THE TOMOGRAPHIC INVERSION

A successful adaption of the Gaussian beam method to the SIRT algorithm has been performed. For this kind of forward modelling we used traveltimes of later arrivals for the tomographic inversion. Considering later arrivals increases the ray coverage, which leads to an improvement of the uniqueness and resolution of the velocity distribution. In a later state we will show that the inclusion of exact amplitudes into the inversion process provides us with informations about the distribution of the absorption coefficient.

In the case when special parts of the area, which was investigated by the tomographic inversion, are to be examined more closely, the results of the inversion can be used for a target oriented processing.

2. TARGET ORIENTED PROCESSING

When an area is explored by means of reflection seismics usually not the whole depth section is of the same interest. In most cases special geological targets are to be examined more closely. In some field cases there is an additional problem to illuminate these targets by seismic waves. A high-impedance contrast is located above the chosen targets which reduces the transmission of seismic energy into deeper layers and lead to disturbing high-amplitude reflection signals in the recorded traces. This problem is very well known when doing offshore seismic measurements, where the seafloor represents the very high-impedance contrast. The exploration of oil and gas bearing layers in North-Germany is impeded by a very high-impedance contrast at the top and the base of the cretaceous formation above the jurassic and triassic host rock which includes the hydrocarbon reservoirs that are to be explored.

2.1. SINGLE POINT SOURCES

Nearly all land seismic measurements are carried out using the conventional single point source like dynamite or vibrators. Figure 5 shows the compressional (P) - wave radiation characteristic of a vertical single force source measured in a perspex plate by using analoque model seismic techniques (12),(3),(2). The radiation diagram of the single point source reveals three important properties: 1) The maximum energy is emitted vertically down into the model or into the earth. 2) Large amounts of energy are sent out over a wide angular range. For a radiation angle of ± 50° still 50% - 60% of the peak energy (at 0°) can be observed. 3) The directivity of this source cannot be changed. It is clearly seen that this type of source alone is not suitable to generate directed P-waves for the target oriented exploration.

2.2. GENERATION OF DIRECTED P-WAVES BY LINEAR SOURCE ARRAYS

A possibilitiy which offers an improved directed emission of seismic energy is given by a linear, homogeneous and equidistant array consisting of single point sources (9),(10),(11),(5),(1). Figure 6 shows the principle of generating a quasi-plane, directed P-wave. The circular wavefronts sent out by each single source superimpose and lead to a quasi-plane wave, which leaves the array with an angle γ to the vertical. This angle depends on the delay time function $\Delta t(n)$ that is used for triggering all sources and which is given as:

$$\Delta t(n) = (n-1) \; (a \,/\, v) \; \sin\gamma \qquad n = 1,2,3,4,5 \qquad (2.1)$$

v is the P-wave velocity, a is the distance between two sources and n is the serial number of the sources. We call the new wave 'quasi-plane' because of its limited length and due to the sampled, non-continuous line source.

The direct generation of the directed wave insitu is of course not very practical and too expensive, because a special trigger technique is needed and each line shot is merely associated with one direction. The repetition for other directions would increase the costs extensively. The indirect method, where all point source seismograms are recorded separately, shifted in time and stacked afterwards, is practical and cost reducing (13). This procedure is a numerical simulation of the directed wave.

Figure 7 presents radiation diagrams for directed P-waves of a seven sources array. The distance between two neighbouring sources is equal to $\lambda/2$, where λ is the main P-wavelength. Four different radiation

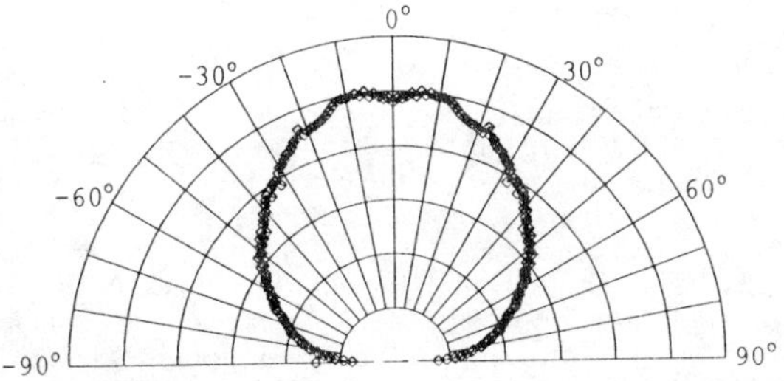

Figure 5: P-wave radiation of a vertical single force measured in a homogeneous perspex plate.

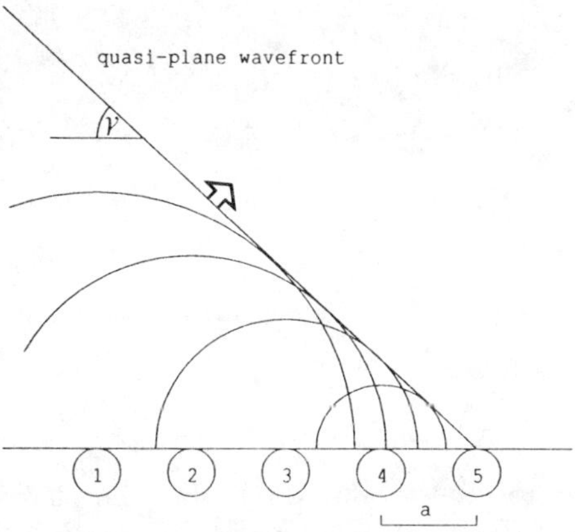

Figure 6: Generation of a directed plane wave with a linear source array
a: distance between sources γ : radiation angle.

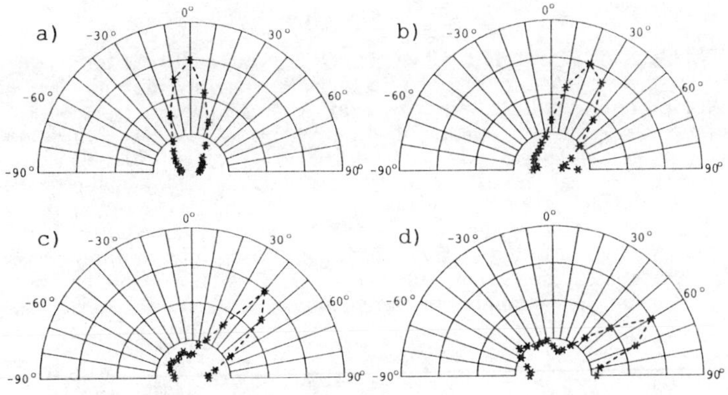

Figure 7: Normalized P-wave radiation of a seven sources array for the radiation angles 0° (a), 20° (b), 40° (c) and 60° (d).

angles for the directed wave have been chosen: 0°, 20°, 40° and 60°. The seven sources array has a very sharp directivity. Only less energy is radiated to the side of the main direction. In case of a desired radiation angle of 0° it can be seen that for angles of ± 40° there are only 20% of the maximum amplitude at 0°. The focussing effect has the same quality for all four radiation angles.

2.3. REFLECTION OF DIRECTED P-WAVES IN CASE OF A THIN GAS BEARING LAYER BENEATH A HIGH-IMPEDANCE CONTRAST

For the use of directed waves in case of a special underground target the determination of the appropriate radiation angle at the surface is most important. If a macromodel is known due to former seismic investigations the exact radiation angle can be determined by a ray tracing algorithm. But if only some main informations like the location of high-impedance contrasts in the underground are given, then an empirical formular can be used to find an approximate value α_a for the radiation angle which is:

$$\alpha_a = \begin{cases} \arctan\ (X\ /\ 2Z) & \text{for}\quad Z \leq Z_c \\ 0.5\ \arctan\ (X\ /\ 2Z) & \text{for}\quad Z > Z_c \end{cases} \tag{2.2}$$

Here X is the geophone offset, Z is the depth of interest and Z_c is the depth of the main impedance contrast.

Figure 8 shows a model of a simplified underground situation, where a thin gas bearing layer is situated below a high-impedance contrast which may represent the top of the cretaceous formation. The seismogram section of figure 9a is a single shot section where a seven sources array was used with four different geophone offsets and the same CDP-point on the thin layer. The reflection arrivals from the thin layer should appear below the arrow in figure 9a but due to the low signal to noise ratio it cannot be detected. Applying equation (2.2) the radiation angles were calculated to send the P-wave energy directly to the thin layer. The seven traces of each offset value were shifted according to the delay time function (2.1) and summed. The results are the four traces of figure 9b where the reflection signal can be seen clearly (arrows). After an additional NMO-correction and stacking the final result is illustrated in figure 9c. The thin layer reflection signal shows a very high signal to noise ratio.

2.4. CONCLUSIONS FOR THE TARGET ORIENTED PROCESSING

The conventional single point source is not appropriate for a target oriented exploration. A linear source array can be used to generate or to simulate directed P-waves which illuminate special targets of the underground. The radiation angle which must be known for this procedure, can be determined by an empirical equation. This equation is based on the depth of the main impedance contrast in the prospecting area. In case of a simple model with a thin gas bearing layer beneath a high-impedance contrast the signal to noise ratio of the reflection signal could be improved. Thus it could be shown that this method can be applied when special targets are of great interest during tomographic measurements.

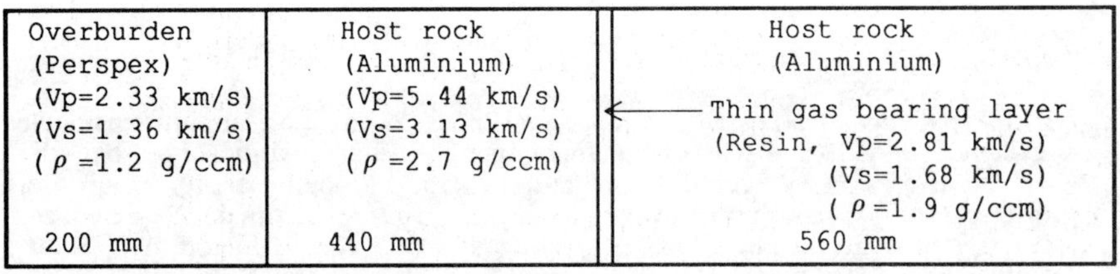

Figure 8: Two-dimensional analoque model of a thin gas bearing layer beneath a high-impedance contrast.

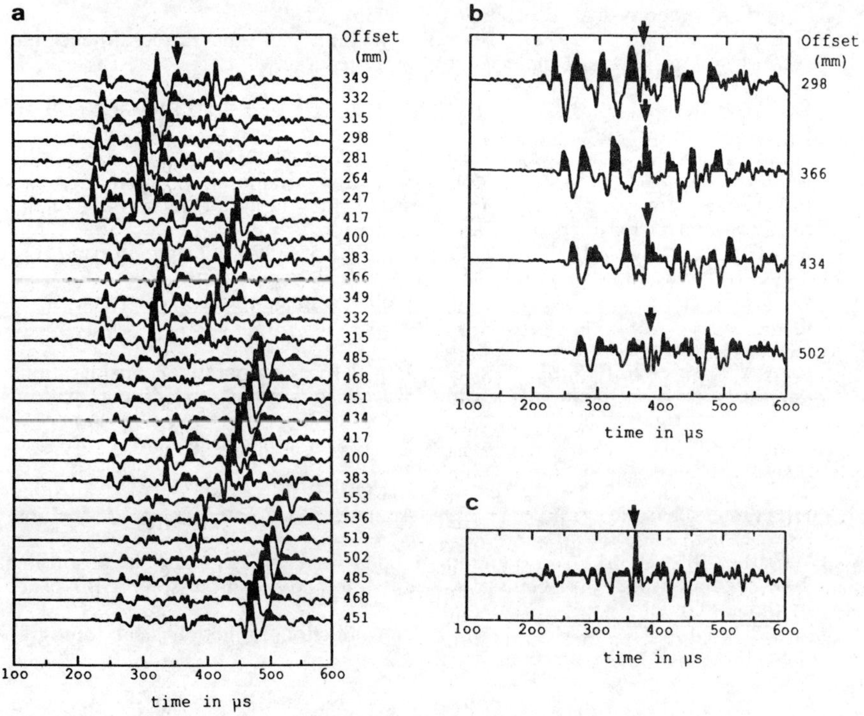

Figure 9: Single shot section for the model of figure 8 (a); P-wave reflection signal of the thin layer after the simulation of directed waves (b) and after additional NMO-correction and 4-fold stack (c).

3. REFERENCES

(1) ARNOLD, M.E. (1977). Beam forming with vibrator arrays. Geophysics 42, 1321-1338.

(2) BEHRENS, J. UND DRESEN, L. (1982). Zwei- und dreidimensionale analoge modellseismische Untersuchungen - Interpretationshilfe bei der Erkundung von Lagerstättenstrukturen. In: Modellverfahren bei der Interpretation seismischer Daten, Vol.2-2. Mintrop-Seminar, W. Budach, L.Dresen, J.Fertig und H. Rüter (eds), 295-432. Unikontakt, Ruhr-Universität Bochum.

(3) BEHRENS, J. AND WANIEK, L. (1972). Modellseismik (model seismology). Journal of Geophysics 38, 1-44.

(4) BISHOP, T.N., BUBE, K.P., CUTLER, R.T., LANGAN, R.T., LOVE, P.L., RESNICK, J.R., SHUEY, R.T., SPINDLER, D.A., WYLD, H.W. (1986). Tomographic determination of velocity and depth in laterally varying media. Geophysics 50, 903-923.

(5) BORTFELD, R., HÜRTGEN, H. AND KÖPPEL, H. (1960). Direction Shooting, Geophysical Prospecting 8, 535-562.

(6) GILBERT, P.F.C. (1972). Iterative methods for the three-dimensional reconstruction of an object from projection. J. Theor. Biol. 36, 105-117.

(7) KRAJEWSKI, C., DRESEN, L., GELBKE, C. and RÜTER, H. (1989). Iterative tomographic methods to locate seismic low-velocity anomalies: a model study. Geophysical Prospecting 37, 717-751.

(8) MIRANDA, F. (1989). Geophysical tomographic reconstruction of travel-time and amplitude anomalies. Ph.D. thesis, University College of Swansea, Department of Earth Sciences.

(9) MITCHELL, G.D. (1951). US Patent No. 2,555,806. Seismic prospecting method, including generation of a cylindrical wave front.

(10) PIERAU, H. AND MULLER, W. (1960). Improvement in the quality of deep reflections by uniformly linear shotpoint arrays. Geophysical Prospecting 8, 154-163.

(11) PIERAU, H. AND ROSENBACH, O. (1960). Comparative considerations on the energy content of seismic waves in central and linear pattern shooting. Geophysical Prospecting 8, 164-177.

(12) O'BRIEN, P.N.S. AND SYMES, M.P. (1971). Model seismology. Reports on Progress in Physics 34, 697-764.

(13) TANER, M.T., BAYSAL, E. AND KOEHLER, F. (1987). Controlled directional seismic sources. 57th Annual SEG International Meeting, New Orleans, U.S.A.

(14) WEBER, M. (1988). Computation of body-wave seismograms in absorbing 2-D media using the Gaussian beam method: comparison with exact methods. Geoph. Journ. 92, 9-24.

(15) WORTHINGTON, M.H. (1984). An introduction to geophysical tomography. First Break 8, 20-26.

NEURAL NETWORKS APPLICATIONS WITHIN IFP

Bertrand BRAUNSCHWEIG, Jean-Michel LAMBERT
Institut Français du Pétrole
Rueil-Malmaison , France
&
Patrick NAIM
Auralog
Palaiseau, France

Summary

The Institut Français du Pétrole started in 1989 an experimental activity on the potential applications of neural networks to petroleum-related problems. Since then, a few systems have been developed for various domains such as seismic processing, geochemical signal processing, experimental data analysis and thermodynamic models adjustment. We will present some results obtained from these applications, and our view of the future of neural networks applied to the Oil and Gas industry.

I. INTRODUCTION

A. The neurobiological foundations

Artificial neural networks are brainlike designed systems which mimic cognitive processes : perception, learning. Neurobiologists and mathematicians have established the foundations of the domain in the 1950s:

-a set of threshold neurons. Mc Culloch & Pitts[9] showed that the biological neuron can be simply modeled as a threshold automata, which activates when the total excitation it receives from its neighbors and the external world exceeds a given level.

-a highly structured and hierarchical architecture. Research on the visual system showed that visual information is processed in a hierarchical way from the retina to the visual cortex. Different groups of neurons are sensitive to stimuli of increasing complexity and abstraction : points, lines, directions, and specific patterns in the visual field. This structure gives a basis for the design of systems made of a very large number of very simple elements (neurons), and able to perform highly complex tasks, such as pattern recognition or obstacle avoidance.

-a system with learning abilities. According to Hebb's model[4], the learning ability of neural networks is based upon the reinforcement of connections between pairs of neurons during the interaction of the network with external stimuli.

B. Mathematical models

From these basic ideas, Rosenblatt designed the perceptron[13], the first operational neural network, made of threshold neurons gathered into layers. This very simple but plausible model of the visual system was already able to learn some pattern recognition problems. The learning principle of this model is called "supervised learning". Each time the network misclassifies a given pattern, a modification of all the connections is done. Finally, the network reaches a stable state, which gives the correct class for the whole set of patterns. Today models often share this learning scheme with the perceptron. Using a more sophisticated formalization of the network, it has been possible to use classical optimization methods, and to design efficient and universal algorithms. In this paper, we will mostly describe applications using gradient back-propagation mode[14]. This model uses a layered neural network, sigmoid neurons, and supervised learning based on squared error minimization through time. However, several other neural models have been developed, using different architectures (fully-connected, 2D-maps, ...) or learning schemes (unsupervised). Back-propagation is well-suited for classification or estimation. For other related problems such as auto-association, some other models are better, such as Kohonen's feature maps [8].

C. Characteristics and applications

Neural networks tuned with these algorithms exhibit very interesting brainlike performances:
-learning,
-high-speed processing and ease of parallelization,
-generalization abilities. Neural networks can recognize patterns they have never seen before.
-noise filtering. Incomplete or noisy patterns are also identified.

The applications of neural network technology are numerous : some major themes are :
-classification : signal or image processing and interpretation
-detection: neural networks can be used to detect defaults or special characteristics in image or signal data.
-estimation: some models of neural networks can be used in numerical model fitting. Their performance lies in non-linearity, and in the very large range of functions they can mimic.
-control : in monitoring complex processes, the use of non-linear adaptive systems like neural networks can bring a very powerful solution.

D. Some well-known applications of neural networks

Neural networks models are universal classification and estimation models, however, some applications are particularly spectacular, and well-known. We will only mention two examples.

NetTalk, of T. Sejnowski [16] is a network that learns to "read aloud", and probably the first "operational" application of neural networks. This is an alternative solution for the "text-to-speech" vocal synthesis : the input of the system is simply a text-window of seven characters, and the network learns from examples to output the phonemes corresponding to the central (4th) letter of the window. The speech produced is very acceptable, but the success of NetTalk is based on its learning process : recordings of the early stages of

NetTalk's learning showed errors very similar to children's ones...
SAIC's detection of explosives in checked airline baggage [17] is another well-known application. The basic technique is the emission of a flow of neutrons to the bags. Absorption of this neutrons causes different parts of the bag to emit different gamma energy rays. A characteristic energy spectrum is obtained. Any classification method can be then used to detect explosive bags from others. Back-propagation neural networks has proven better than discriminant analysis on this problem, and much easier to use.

E. Organization of the following chapters

In part III of this paper we will show some examples of applications, currently developed within IFP. Part II gives a short review of software and hardware tools for practical neural network applications design.

II. STATE OF THE ART

A. Hardware

Dedicated hardware for neural networks are not expected before a couple of years. Several microprocessors suppliers started developments of prototype silicon chips which possess the major characteristics expected for real-time neural networks applications. However, there is no commercially available neural network chip today. Some vendors have developed add-in boards, consisting in multiple floating-point coprocessors, generally with a parallel architecture, which simulate by software the major characteristics of a hypothetical neural network. These boards (HNC's Anza+, SAIC's DeltaII) give an user a computing power high above the average performance of micros and workstations. This solution will remain valid until the specialized chips will become available.
For all these reasons, most neural networks today are software-emulated on conventional computers. The next paragraph will look at these software simulators in detail.

B. Software

There is choice for both the novice and the expert in neural nets, ranging from public-domain software on IBM-PCs for $30 to development platforms including coprocessor and software for more than $40.000. We consider the existing software in respect with their focus, that is : development environments, i.e. basic tools for developing and integrating NN applications, usually with some programming; integrated environments, i.e. programs which allow users to build their own networks, but within a predefined framework, usually including graphical interfaces and standard algorithms; specialized tools, i.e. tools that are aimed for a specific market or application type. We do not intend to present the full range of available software but rather a few illustrative examples.

1. Development environments

a) PDP3

The third volume of Rumelhart and McClelland's book "Parallel Distributed Processing" [15] consists in two disks containing C programs and their manuals, simulating the most common paradigms of NN. Considering its very low price and the fact that source code is given, the PDP software has been and is still used as a basis for developing many NN applications on PCs and other platforms. The programs are tailored for PCs and compatibles but easily adaptable to other machines. We use enhanced versions of these programs for some developments.

b) SN/2

SN is a Lisp-based toolkit providing standard and advanced backpropagation mechanisms, together with customizable graphical interfaces and efficient computation. The software runs on Apple Macintosh and Unix workstations. The power and flexibility of Lisp and the availability of high-level functions such as matrix operations and x-y plots make it efficient for quickly building backpropagation networks and training them.

c) Conventional languages

The effort needed to simulate a neural network with a conventional language such as Fortran or C is very small. Therefore several applications embed a neural network by simply calling one of two subroutines doing all the tasks, that is, learning and recall. Neither a vector machine, nor a parallel language, are needed, although they might help.

2. Integrated environments

Several software manufacturers offer integrated environments which allow network construction, predefined models library, training and test, graphical interfaces. This seems to be very straightforward compared with the development effort spent in programming and adapting the development tools presented hereabove. However, these software often show a lower computing performance due to the amount of elements linked together, which allow a great user-friendliness and flexibility, but are less efficient for real-world problems. In this category, some popular tools are Neuralware's Explorer and Professional2, SAIC's AnSim, HNC's Axon, Neural System's Genesis. All these tools offer the most common NN architectures and algorithms, including - at least - several kinds of backpropagation, Kohonen, Hopfield, ART, etc... They are generally available on PCs, Macintosh and Unix workstations. These tools are useful for learning about NNs and for presentation purposes. Once a prototype application is built, it is usually necessary to translate the network memory into a more efficient format, in order for it to be used by conventional programs. This can be supplied by the manufacturer itself.

3. Specialized tools

a) Nestor Learning System

NLS is a classifier system based on a proprietary algorithm called RCE, developed by Leon Cooper at Brown University, which mimics some of the characteristics of neural networks. It is specially designed for classifying patterns and needs only a small amount of training in order to produce good results. Nestor offers other products derived from NLS, such as a decision support system for the finance and banking industry. All these software run on IBM PCs and compatibles.

b) Software for the coprocessor boards

Both HNC and SAIC offer a wide range of products allowing shallow to deep programming of their coprocessor boards. This includes versions of the neural network development environments running on these platforms, implementation of a conventional language, usually C, and connection to the outside world.

C. Trends for hardware and software

The major step towards a wide use of neural networks would be the availability of chips implementing at least one algorithm and capable of overriding the present limits of software simulators and boards, both in terms of speed and capacity. With software simulation, it is unrealistic to build networks of more than a few hundred processing nodes. The boards move this limit to a few thousands, which is still orders of magnitude below the number of neurons of living creatures.

Among other evolutions, we will mention : 1) the development of more realistic models of the natural neuron, which implies more complex transfer functions and learning algorithms; 2) more knowledge about the capacities of the networks, including guidelines on how to tailor a network for a specific application; and 3) facilities for making the networks capable of producing explanations when they reach a conclusion. Several current developments tackle these questions.

III. SOME APPLICATIONS

We will now briefly review some prototypes developed or under development within our Institute. Our current focus is to evaluate the potentials of neural networks for applications in the Oil and Gas industry. With this in mind, several attempts have been made on various problems, with various goals and techniques. The examples shown below illustrate our approach.

A. Identifying pinch-outs in seismic sections

Our first attempt was for a pattern recognition problem, finding geometrical patterns (pinch-outs) in seismic sections. One of the tasks of the geophysicist consists in a reconstruction of stratal terminations geometry. This is done by identifying pinch-outs which indicate horizons' endings. Today, only experts are able to identify these endings by visual recognition, this being a complex task due to 1) a great variety in shape, size, orientation; 2) uncompleted patterns; 3) noisy sections.

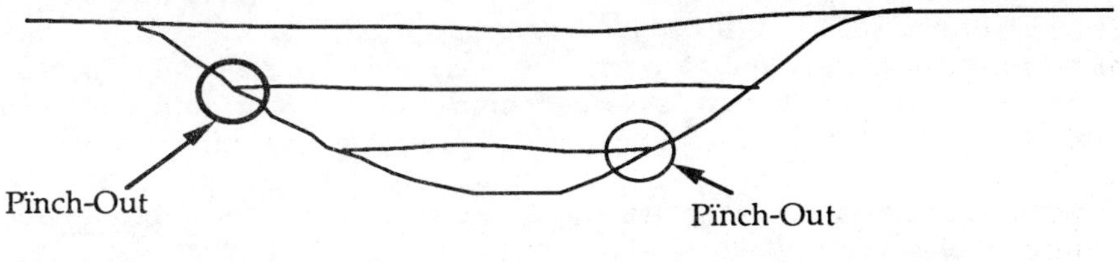

Three kinds of pinch-outs can be distinguished :

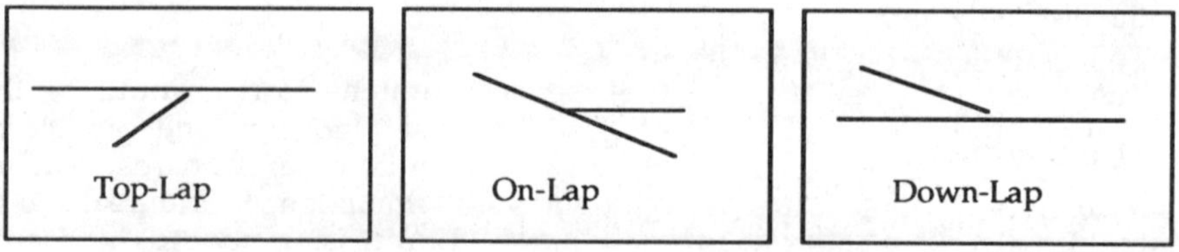

We simplified the problem by a segmentation which allowed us to work on small pieces of a section. A standard three-layers backpropagation architecture was used, the input layer being a window moving along the seismic section, the output layer consisting in four units, one for each pinch-out type, plus one for absence of pinch-out. The training was made with two hundred examples in three hours on a SUN 3 workstation.

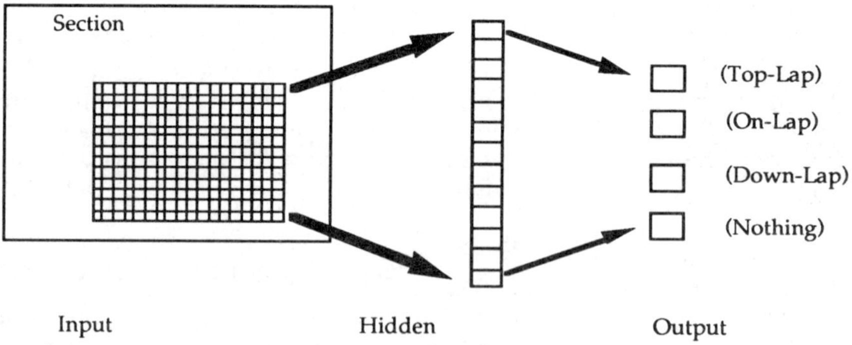

The problems encountered were the following :

- rotation : a same shape has to be presented with several orientations. It is impossible to use signal preprocessing because we need the system to be sensitive to strong rotations and insensitive to small ones.
- translation: the network only recognized centered patterns; therefore the input window has to be shifted by very small increments.
- over-classification : the system produces a class even if the window contains only noise. The solution consists in training the network with a lot more examples.

Once these problems were given an answer, the system was perfectly able to recognize more than 80% of distorted patterns.

B. <u>Estimating polymers properties with NNs.</u>

In this application, we wanted to test NN's effectiveness in estimation properties. The network has been designed to produce a continuous value representing the permeation of a polymers mixing. Considering physical parameters of a two-polymers blending, we want to train a NN, to predict its permeation. Usually, experts elaborate numerical models and adjust parameters with experimental data. The same architecture as above was used. We trained the NN with a set of sixty experimental mixings. The input parameters were the following:

- blending composition : relatives percentages of polymers, polymers characteristics.
- blending conditions: temperature, rotation speed.

All these values were normalized between 0 and 1.

As output we presented the value measured in experimental conditions. After adjusting network parameters (number of units in the hidden layer), we tested the network by considering a sample of the database and changed the different parameters. According to the small number of data presented to the NN during the training period, we obtained good results in prediction which compared well with the model-computed approach.

A neural network has been applied to the problem of estimation. It has shown good results. Comparing to conventional analysis, backpropagation takes a small amount of human supervision, considers the database as a whole and elaborates its own best estimation criteria.
Moreover, this study can be extended easily to other kinds of polymers by just adding samples in the database.

C. Noise Filtering in seismic data

In this application, our goal is to teach networks how to identify several kinds of noise in seismic traces. The presence of noise in raw seismic data has always been a problem for the downstream processing phases. Both organized and non-organized noise will generate possibilities of wrong interpretation by the geophysicist. Some conventional signal processing methods are used for certain kinds of noise, but there is still a lot of work done by human operators who visually locate the presence of noise in a seismogram and manually eliminate it.

Our first attempt in this direction is to train a network to detect the presence of spikes in the seismic signal. Spikes are very short and high-amplitude events artificially created in a seismogram, often due to connection problems between the sensors and the recording devices. We want to eliminate spikes where they create the most problems, that is, where the amplitude of the signal is high and it is not easy to distinguish them from normal behavior.

We built a two-step network : a first step for detecting the presence of a spike within a trace, which only answers yes or no; a second step will then take as input the traces identified as potentially containing a spike and try to precisely locate it within the signal. We have not worked on the next stage, that is, deciding by what the spike will be replaced in the signal for downstream processing. We tried to locate spikes in individual traces - 1D - rather than using 2D information and lateral coherence.

The amount of data taken as input for the network lead us to use advanced backpropagation architectures such as shared weights and shifting masks, which diminish the number of processing elements and therefore reduce the training time.

Our intention is to continue to apply NN techniques to other various types of noise in seismic traces.

D. Classification of pyrolyzes

The Rock-Eval method developed by IFP is based on the rapid, inert atmosphere pyrolysis of the organic matter present in small rock samples. A signal output by pyrolysis device is used to classify the rocks in three classes called I, II and III. This interpretation is based on the presence of peeks within certain temperature ranges, the peeks time and their shape.

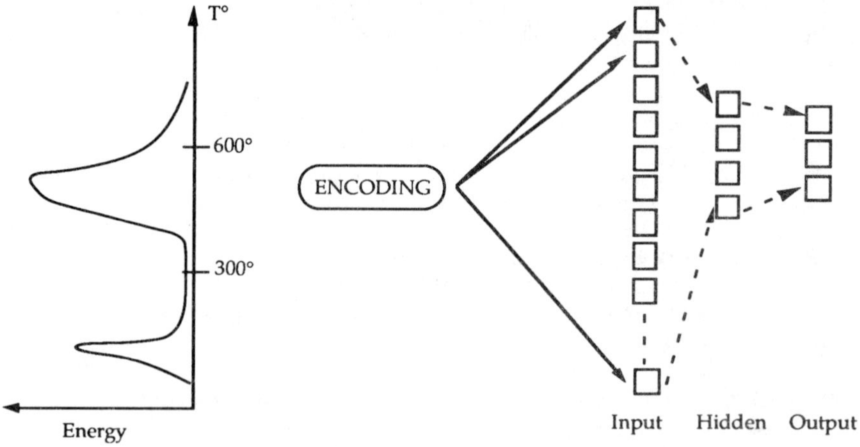

A pyrolysis signal contains three thousand values taken every 150 ms. The vector input to the network is limited to 100-300 values by extraction of local extrema over small segments of the signal. The network was trained with a set of one hundred samples. We obtained very good results both for generalization and resistance to noise. The next step will be to test the network's ability to recognize atypical signals for complex rocks. No conventional method has proven to be efficient on such matter.

E. Neural networks for process control

The features we briefly reviewed in the previous sections could make neural networks a very powerful and relatively easy-to-use tool in control problems : robotics, manufacturing process control, plant control ...

The first, and maybe simplest control problem which can be solved using neural networks is the problem of inverse dynamics. Suppose a model of the system to be controlled is given, that is, a relation $Y(t)=F(U(t),Y(t-1))$, where U is the input (command), Y the state of the system and F the model.

The problem is to find a correct approximation of the inverse F-1 of the model, so that it is possible to give the system state any given trajectory. Neural networks can provide a good numeric solution for this purpose, in the case that no theoretical inverse model is known :

-a neural network can approach a wide range of functions, including highly

non-linear ones.
-as the direct model is known, there is no difficulties to gather data for network training.

Another important problem in process control or monitoring is signal filtering and estimation. Given again the model of the system, and random perturbations on both the input and the measurements of the system internal parameters, one has to estimate in real time the internal state of the system. Traditional filtering methods usually make strong hypothesis on the perturbations. Neural networks have shown on some signal filtering problems a stronger noise resistance, and with no a priori hypothesis of the noise [5].
Such applications, or even more difficult ones, such as dynamic programming are just beginning to be investigated by the researchers in the field of neural networks. Results are promising, but it is often necessary to design a special network architecture and algorithm (Werbos' back-propagation through time [19], Jordan's recurrent network [6]).
A very appealing work is the "truck backer-upper" of Widrow[12]. A new formalization of the problem, i.e. the use of both a neural emulator (neural simulation of the truck kinematics) and a neural controller, allowed him to teach a network how to back a trailer truck to a loading dock, using standard back-propagation.
In practical cases, however, a significant part of process control and monitoring is often done by human operators :
-sensor failure diagnosis,
-"pattern recognition" for predicting evolution of the system,

These tasks are far from theoretical global modeling of the process, but are rather based upon the practical experience of the operators. In fact, expert systems are often used in order to help, or to gradually replace, such operators. But this is a domain in which knowledge acquisition has proven one of the most difficult. In these tasks, neural networks can be used like in "standard" pattern recognition tasks, such as the ones previously reviewed. Human operators will no more have to explain the rules governing their diagnosis, but only to teach it to the network for several examples.

Process control, and other control problems, such as sensori-motor control in robotics, are probably the future of neural networks : they are more difficult applications than simple classification or estimation problems, but the potentiality of neural systems is also much greater.

IV. CONCLUSION

We have shown some examples of neural networks applied to petroleum-related problems. The field is still very young, but several attempts made within several organizations [1,2,3,7,11] prove that there is a great potential for using this technology in our industry. The benefits of neural networks are numerous once good data is available, which is the major difficulty for setting up an application : ease of development, low cost, fast processing time once the network is trained, small amount of programming, noise resistance, generalization.

V. ACKNOWLEDGEMENTS

We would like to thank the Institut Français du Pétrole for permission to publish this paper, and the following persons who are major contributors to the neural networks project : Alain Bamberger, Françoise Coppens, Jean-Paul Diet, Jean Espitalié, Eric Mousset, Yann Parrod, Gaëtan Serpe.

VI. REFERENCES

[1] Arehart, R.A. (1989). Drill Bit Diagnosis Using Neural Networks. *Proceedings of the 1989 Conference on Artificial Intelligence in Petroleum Exploration and Production*. College Station, Tx

[2] Baldwin, J.L. et al (1989). Computer Emulation of Human Mental Processes : Application of Neural Network Simulators to Problems in Well Log Pattern Recognition. *Proceedings of the 1989 Conference on Artificial Intelligence in Petroleum Exploration and Production*. College Station, Tx

[3] Derek, H. et al (1990). Comparative Study of Backpropagation Neural Network and Statistical Pattern Recognition Techniques in Identifying Sandstone Lithofacies. *Proceedings of the 1990 Conference on Artificial Intelligence in Petroleum Exploration and Production*. College Station, Tx

[4] Hebb, D.O. (1949). *The organization of behavior*. New York : Wiley.

[5] Himmelblau D. (1989). Neural Networks and Chemical Engineering; *University of Texas, Austin , 1989*

[6] Jordan M.I. (1986). Attractor dynamics and parallelism in a connectionist sequential machine. *Proceedings of the Eighth Annual Meeting of the Cognitive Science Society*, Hillsdale, N.J. Erlbaum

[7] Kim,C.S. & McCauley, C.C. (1990). Hybrid Kohonen/Backpropagation Network Tolerant to Missing Inputs. *Proceedings of the 1990 Conference on Artificial Intelligence in Petroleum Exploration and Production*. College Station, Tx

[8] Kohonen, T. (1988). *Self-Organization and Associative Memory*. Springer-Verlag.

[9] McCulloch, W.S., & Pitts, W. (1943). A logical calculus of the ideas immanent in nervous activity. *Bulletin of Mathematical Biophysics*, 5, 115-133.

[10] Minsky, M., & Papert, S. (1969). *Perceptrons*. Cambridge, MA : MIT Press.

[11] Morgan, K. & Morgan, C. (1990). Hydrocarbon Deposits in the Shannon Basin. *Proceedings of the 1990 Conference on Artificial Intelligence in Petroleum Exploration and Production*. College Station, Tx

[12] Nguyen D., Widrow B. (1989).The Truck Backer-Upper : An Example of Self-Learning in Neural Networks, *Proceedings of the International Joint Conference on Neural Networks*, Washington D.C.

[13] Rosenblatt, F. (1962). *Principles of neurodynamics*. New York : Spartan.

[14] Rumelhart, D. E., Hinton, G. E., Williams, R. J. (1986). Learning representations by back-propagating errors. *Nature*, 323, 533-536.

[15] Rumelhart, D.E. & McClelland, J.L. (1986). *Parallel Distributed Processing : Explorations in the Microstructure of Cognition. Vol. 1 (Foundations), Vol.*

2 *(Psychological and Biological Models), Vol.3 (Explorations in Parallel Distributed Processing).* Cambridge, MA : MIT Press
[16] Sejnowski, T. J., Rosenberg, C. R. NETTalk : a parallel network that learns to read aloud. *The John Hopkins University Electrical Engineering and Computer Science Technical Report,* JHU/EECS-86/01, 32pp.
[17] Shea P.M., Lin V.(1989). Detection of Explosives in Checked Airline Baggage Using an Artificial Neural System. Proceedings *of the International Joint Conference on Neural Networks,* Washington D.C.
[18] Werbos P.J. (1988).Generalization of Backpropagation with Application to a Recurrent Gas Market Model, *Neural Networks Vol. 1, 1988*
[19] Werbos P.J. (1989). Backpropagation and Neurocontrol: a Review and Prospectus. *Proceedings of the International Joint Conference on Neural Networks,* Washington D.C.

LOBSTER: AN EXPERT SYSTEM FOR THE INTERPRETATION OF SEDIMENTARY ENVIRONMENT FROM CORE ANALYSIS

L.Matteini, F.Fonnesu and G.Di Dio

AGIP S.p.A.
P.O. Box : 12069
20100 MILAN (ITALY)

SUMMARY

The recognition of ancient sedimentary environments is a very important step in defining both exploration and development strategies.

LOBSTER (Lisp Object Based System Tool for Environment Recognition) is an expert system, developed and tested by Agip, which is able to identify the original depositional environment of cored clastic sediments, with good success rate.

Since it's impossible to fully describe and classify the huge variability of natural environments on the basis of lithological and textural characteristics only, LOBSTER uses the depositional processes inferred from facies descriptors as the major sedimentary environment descriminators. With this approach the system performs consistent facies analysis.

Input data are collected by sedimentologists as in normal core inspection (i.e. facies thickness, lithology, bedding, sedimentary structures, etc.). The depositional process is then inferred, and facies are grouped vertically into facies associations.

On the basis of facies associations, the system identifies the depositional environment skillfully comparing them by means of tailored matching with a reference taxonomy.

LOBSTER is a frame-based object-oriented system written in LISP. It is currently at an advanced prototype state and it has been successfully tested with both core and outcrop data on continental to deep water sediments.

LOBSTER knowledge-base includes about 70 elementary environments; facies can be decribed with a choice of almost twelve sets of descriptors.

The results achieved lead us to consider LOBSTER as a powerful support for both experts, in evaluating multiple hypotheses, and novices, in reaching high quality results.

1. INTRODUCTION

In an hydrocarbon field, knowledge of the distribution of reservoir rocks and related fluid flow barriers is of primary importance for selecting, planning and implementing recovery projects.

The distribution and continuity of sand bodies, their petrophysical properties and internal characteristics are generally strongly controlled by the original depositional environment in which the sediments were deposited. For istance, large differences in reservoir performances can in fact be expected between two fields one producing from clastics bodies representing stacked fluvial channels and another producing from an ancient coastal plain characterized by linear and well sorted beaches.

In subsurface studies the interpretation of depositional environments is carried out by a sedimentologist through a detailed inspection of the available cores. This inspection, called

"Facies analysis", is a logical process that, if well conducted, leds to infer the depositional environment.

The reliability of this interpretation is definitely related to the sedimentologist's experience. It is thus a strategic commitment for an oil company to preserve and make such experience available.

The expert system technology is a well suitable solution to achieve this goal. Some experiences are already available in this domain. Shultz et al., (1986), attempted to recognize the depositional environment directly on the basis of textural and lithological characteristics alone. This approach, being far from a real facies analysis, forces the sedimentologist to oversimplify the core description, thus resulting in a lack of operativness.

The system implemented by Shultz, based on a shallow knowledge model, uses object-attribute-value triplets which were simply through the use of rules.

Mac Donald et al., (1986), followed the same approach but implemented it using frames theory.

A different alternative is described in this paper. LOBSTER (Lisp Object Based System Tool for Environments Recognition), developped and tested by Agip in the last two years, currently at the advanced prototype stage, directly uses data collected during routine core analysis and exactly emulate the interpretation process carried out by a skilled sedimentologist.

LOBSTER is a knowledge-based system using frames theory and object oriented programming techniques, implemented in Common Lisp language.

2. THE DOMAIN: FACIES ANALYSIS

The term "facies analysis" means the logical deductive process that the sedimentologist carries out to identify the original depositional environment of a sedimentary rock.

This analysis is based on 4 fundamental concepts:

1 - Facies: this is a rocks body with well-defined characteristics which make it easy to distinguish from over- and underlying deposits;

2 - Depositional process: this is the physical, chemical or biological process that operates in a natural environment, capable of generating, transporting and depositing sediments within the environment;

3 - Facies association: this is created by the superimposition of two or more facies which are genetically interconnected reflecting the natural evolution of a simple sedimentation environment; it represents the vertical section of a

4 - Depositional element: i.e. of the same depositional environment seen as a sedimentary body extending 3- dimensionally over space and time.

Through a detailed analysis of a facies characteristics, the expert sedimentologist can identify the processes operating at the time of the facies deposition.

By identifying the main processes responsible for the deposition of each facies it is possible to establish which facies, one directly overlying the other, form an association. In order to identify a facies association the sedimentologist relies on observation of the current environments which demonstrate that only the facies currently depositing in lateral sedimentation continuity can be found associated in vertical sedimentation continuity.

The sedimentologist thus compares facies associations identified in the section observed with reference models of current environments, whose operating processes and interrelationships are known.

This comparison leads to identification of the various depositional elements present in the series.

However, while only a limited number of depositional elements exists in nature, the morphological and textural expression of each element can vary infinitely. An example is the infinite variety that a depositional element such as a beach can develop in nature as a function of the intensity of wave movement and the grain size of the sediments. Furthermore, the models in the literature do not offer summaries which are clearly and immediately usable in everyday practice, or else not all the depositional elements have been studied in equal detail.

The sedimentologist then, according to his own experience and knowledge, must define the characteristics peculiar to each depositional element in order to identify them in the sedimentary sequences through a reasoned comparison of the really significant parameters.

In developing the expert system all the typical phases of facies analysis described so far were followed. The system can interpret the facies association under examination by discriminating between 70 different reference depositional elements. These are described by comparison parameters which are truly diagnostic and sufficiently elastic to cater for great natural variability.

3. KNOWLEDGE REPRESENTATION

A sedimentary facies can be characterized and completely described with a certain number of descriptors.

There are two different kinds of descriptors: those that have to be directly observed on the core and those that are inferred from the previous ones.

The directly observed descriptors are the following:

- facies top and bottom depth,
- lithology and grain size,
- lithology alternations,
- bedding (geometry, thickness, upper and lower stratal boundary),
- sedimentary structures,
- byogenic structures,
- bathimetry from paleontological contents,
- color,
- auxiliary characteristics (i.e.: paleosoils, authigenic minerals, plant fragments, etc...)
- lower and upper facies boundary.

Most of these decriptors have to be quantitatively evaluated.

The inferred descriptors, derived from the preceding ones, are:

- facies thickness,
- sand thickness,
- facies depositional processes: inferred from sedimentary structures, auxiliary characteristics, grain size and eventually bedding type. The processes considered are as follows:
    - unidirectional tractive currents,

- tidal or bottom currents,
- wave action,
- wave swash and back-swash,
- traction plus fallout,
- fallout,
- mass flow,
- water escapes ,
- biological ,
- chemical ,
- others physical ,
- hydraulic jump;

– prevalent process and its energy: the most common facies depositional process,
– prevalent bedding thickness: the most common bedding thickness,
– prevalent grain size: the most common grain size.

When facies are groupped into associations new descriptors can be inferred:

– facies association thickness,
– sand percentage: the ratio between the comulative sand thickness and the facies association thickness,
– grain size trend: the vertical trend of the prevalent grain size of each componing facies,
– bedding thickness trend: the vertical trend of the prevalent bedding thickness of each componing facies,
– process trend: the vertical trend of the prevalent process of each componing facies,
– energy trend: the vertical trend of the energy related to the prevalent process of each componing facies,
– facies association depositional processes: all the depositional processes recognized into the association ordered according to their frequency.

All the facies association descriptors, both observed and inferred, are utilized for matching with the reference depositional elements which are ordered in a hierarchical taxonomy.

The same set of decriptors is used to define the reference depositional elements. Within each element the descriptors are subdivided into four main categories. This assignement is peculiar of each element. The categories are:

1- necessary descriptors: if absent, comparison fails,
2- non necessary descriptors: if present, comparison score increases,
3- negative descriptors: if present comparison fails,
4- insignificant descriptors: their presence or absence does not affect the comparison.

The use of this categories makes the matching very selective.

The established facies association, along with its descriptors, is compared to the depositional element taxonomy starting from the highest hierarchical level. If the comparison conditions are satisfied, lower hierarchical levels are then tested.

The final result is identification of the depositional element with a score reflecting the interpretation reliability.

If multiple interpretations are possible the lowest in the hyerarchy with the highest score is then selected as the most reliable.

4. SYSTEM ARCHITECTURE

A knowledge-based system may be developped using various paradigms. The knowledge engineer usually select the appropriate paradigm in order to achieve the best fit of the characteristics of the knowledge domain and of the reasoning process applied by the expert.

Our domain is characterized by the description of the reference depositional elements and by the classification, through a stepping process, of a facies association described.

In this case frames are the best solution as basic descriptive units to describe the entire domain.

The reference depositional elements are organized in a hierarchical taxonomy of classes. So far, based on the inheritance of attributes values, the depositional elements are classified with increasing details as shown in fig. 1.

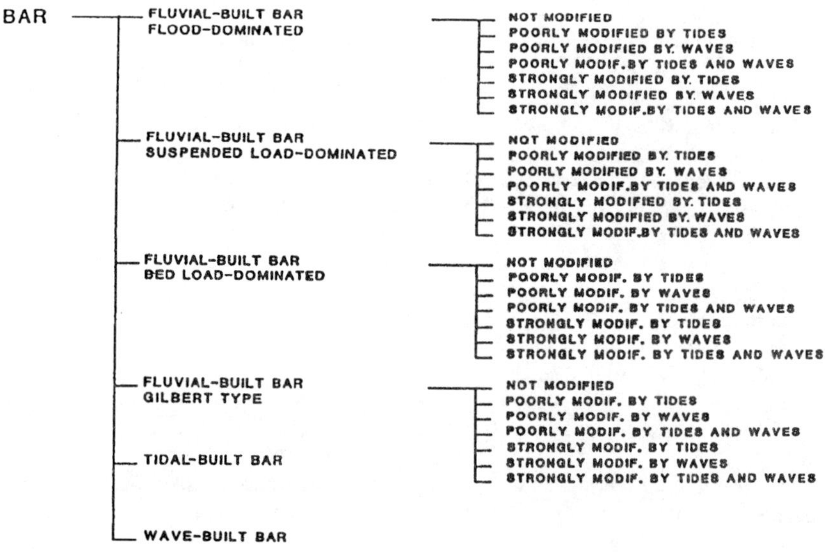

Figure 1 - Taxonomy of the bar depositional element

All the domain elements (observed descriptors, inferred descriptors, facies, facies association and depositional elements) are implemented as objects. The relationships between objects are made explicit using Object Oriented Programming techniques as message-passing, demons, active- values.

The whole classification procedure is implemented as an attribute in the classes definition. A method performing the general classification is defined at the root level of the reference taxonomy. This method is then inherited by all the subclasses. Moreover, a local more detailed method specifies, for each subclass, a peculiar matching procedure, by using different "views" of the same descriptor, according to their four categories.

In such a way the recognition of a facies association is carried out simply by sending a message to the root of the reference taxonomy. This message is then recursively propagated down to all the subclasses. Using a recursive breadth-first search algorithm, the classification process is optimized: only the subclasses of classes that succesfully match the described facies association will be tested in the next search cycle.

This technique provide good search performances also with wide and deep taxonomies and, moerover, it definitely keeps static knowledge (reference depositional elements), dynamic knowledge (described facies association) and reasoning separate. Therefore, adding new classes does not affect the matching procedure and the static knowledge base can be updated without any impact on the classification inferences.

The "what-if" capability is implemented by the definition of an explicit dependency network. The network maps the existing relationships between different objects (descriptors, facies, facies association,...). If an object attribute is modified, by means of active-values, all the effects are propagated along the network, mantaining the consistency of all the other related objects.

In such a way it is possible to verify how, for a facies association, the final interpretation changes if different values of the same descriptor are alternatively specified.

5. CONCLUSIONS

LOBSTER is at an advanced prototype stage. It can currently recognize all the silicoclastic and some mixed, silicoclastic-carbonatic, depositional elements.

In future, the carbonatic depositional elements and the whole spectrum of natural depositional systems (vertical associations of genetically related depositional elements) will be implemented.

The system has been widely tested, with a good rate of success, on a variety of facies associations spanning continental to deep-marine, derived either from subsurface cores and outcrop sections.

The experience reported in this paper has led to the following benefits.

Through the implementation in a knowledge-based system, the methodology internally used for core description and depositional environments recognition has been standardized. The integration of many different types of expertise provided an accurate description of a wide range of depositional elements. So, the accuracy of both description and interpretation is not related only to the sedimentologist's experience any more; novices are driven by the system to carry out consistent interpretations. The generation of multiple hypotheses, with related score, provides the interpreter with all the different alternatives possible with the given set of input descriptors. No reasonable hypothesis is missed thus supporting the interpretation objectivity.

Expert system technology enables preservation of the sedimentologist's know-how in this strategic field, making it broadly available inside the company.

LOBSTER emulates the kind of reasoning followed by the sedimentologist by using object oriented programming and frames.

ACKNOWLEDGEMENT

The authors wish to thank AGIP Hydrocarbon Exploration for the kind permission to publish this work.

REFERENCES

BRANCHMAN , J.G.SHMOLZE, "An Overview of the KL-ONE Knowledge Representation System", Cognitive Science 9 (2), 1985.

DECIO E., L.SPAMPINATO, "QSL 3.0", Tech.Rept.QR-87-1, Quinary, Milano 1987.

DUCORNAU, M.HABIB, "On Some Algorithms for Multiple Inheritance in Object Oriented Programming", Proc. of the European Conference on Object Oriented Programming, Paris, France, 1987.

FIKES, T.KEHLER, "The Role of Frame-based Representation in Reasoning", Communications of the ACM, Vol.28 n.9, September 1985.

HARRIS D.R., "A Hybrid Structured Object and Constraint Representation Language", Proc. 5th National Conference on Artificial Intelligence, Philadelphia, PA, 1986.

MAC DONALD A., AAMODT A., EGGEN J., OLA JOHUSEN S., "Facexp an Expert Sistem to assist interpretation of sedimentary environments from borehole sequences", Sintef Report, n. STF14A86006, The Computing Center at the University of Trondheim, 1986.

SHULTZ A.W., FANG J.H., BURSTON M.R., CHEN H.C., BEASLEY M., "An Expert System for the Determination of Clastic Depositional Environments", AAPG Bulletin, 1986, v.70, p.647

STEFIK AND D.BOBROW, "Object Oriented Programming: Theme and Variations", AI Magazine, Vol.VI n.4, Winter 1986.

EXCHANGE FORMAT FOR OPTIMAL TRANSFER OF GEOLOGICAL AND GEOPHYSICAL 3D SUBSURFACE MODEL DATA.

M.H. Mulder - S. Pen - I.L. Ritsema*

TNO Institute of Applied Geoscience
Geoenergy Department
P.O.Box 285
2600 AG Delft, The Netherlands

Summary

Data transfer between participants of the Geoscience project (which is part of the Joule European Community Research Programme) is considered of vital importance to guarantee its best success and to develop the optimal corporation between the Universities, Institutes and Oil companies involved. The most efficient data transfer demands a generally accepted standard for all the data that are relevant in the project. For some data types such standards do already exist, however, in the Geoscience project, notably the geologic and geophysical 3D subsurface model data lack a prescribed format for storage, retrieval and transfer. The storage format and the type of information deserves special attention in the design of the proposed Data Exchange Format (DEF), which will be based on relational information management techniques. The DEF will serve as a common communication platform for the Geoscience project participants, as a complement to already existing standards such as SEGY, LIS, UKOOA, etc. It will define relations recognized between the data items (relational data model), and derive extendible relational tables from that data model. Finally the physical structure for exchange media will be standardized.

1. Introduction

The data in the Geoscience project can be subdivided into three groups: well related data, seismic related data and 1, 2 and 3 dimensional subsurface model data. These data represent geological as well as geophysical information generated in the Geoscience Project. In particular the 1, 2 and 3 dimensional subsurface model data will be of common interest to both the geological and geophysical disciplines in the Geoscience Project, and existing standards (SEGY, UKOOA, LIS, etc.) do not provide common storage and exchange facilities for such data. Therefore the design of the proposed Data Exchange Format (DEF) focuses on the method to properly store and exchange these kind of data. Parallel developments to define standard formats for exchanging petroleum oriented data are closely monitored, to avoid overlapping activities, such as the current AAPG-B (American Association of Petroleum Geologists) format developments (see also Reference 1) and the POSC (Petrotechnical Open Software Corporation) activities.

The DEF format will be unique in that
a) the design is based on a <u>relational</u> datamodel;
b) the 1, 2 and 3 dimensional subsurface macromodel data will be included.
This paper describes the methodology that will be followed to define the proposed DEF format.

* denotes speaker

2. Methodology

The Data Exchange Format will include a definition of

- the data incorporated for exchange
- the physical format for storage on different media

Two important factors have to be considered in a new design. Firstly, old formats were often very inflexible, such as the SEG-Y format with its concise definition of the datatypes, and their physical formats exclusively used for its traditional storage medium (tape). These kinds of formats are not easily extendible. It is important for the design of a new DEF to incorporate extendibility. Secondly, data are nowadays more and more stored in relational databases. Databases employing a relational structure consist of a set of tables in which the data are stored, and the most effective way to import or export data is adhering to these table structures, which are described in their data dictionaries.

The proposed strategy to define DEF requirements includes both existing and anticipated needs. Such an anticipatory development strategy makes use of Conceptual Datamodeling (as opposed to Physical Datamodeling), in which the data requirements are developed from a description of the entities and relationships among entities.

With these considerations in mind the following methodology is proposed and used to arrive at the Data Exchange Format definition:

A. Data analysis

This includes the inventorization of
a) the data (Universe of Discourse)
b) commonly accepted definitions of the data
c) mutual relations or associations of data and constraints on these relations and/or data values.

This results in a conceptual model of the data that will be incorporated in the Data Exchange Format. The data can be classified as mandatory versus optional.

B. Transformation of conceptual model to relational scheme

This includes a normalization phase resulting in table definitions, field formats and referential and domain constraints. This scheme is extendible if needed.

C. Specifications of flexible Data Exchange Format

From the relational conceptual scheme the Data Exchange Format can now be derived. The mandatory information or data coming from different relational tables will be grouped into a sequential structure. Rules how to extend the Data Exchange Format with new datatypes (attributes) have to be given. In the output of extensive amounts of table information (e.g. surface grids, seismic picks, etc.) the optional information should be omitted, enabling a more condensed way of exchanging and storing data.

The described method has been applied in this EC program to 1, 2 and 3D subsurface model data, including for example 1D well markers or velocity models, 2D seismic picks or velocity models and 3D lithologic, stratigraphic or velocity models.

3. Data analysis

The data analysis is carried out according to Nijssen's Information Analysis Method (NIAM). This method makes use of a sentence-predicate calculus, whereby the data is defined by sentences having deep structure, in the sense that they can be transformed into a variety of other representations (called surface structures). The sentences may be presented as natural language statements, tables or graphs. A detailed description of this data analysis method can be found in Reference 2.

The inventarization of necessary datatypes has resulted in the following list of major data object types(with mandatory attributes)
- model (MOD) (identification name)
- model property class (MPR) (property, property unit)
- spatial domain (DOM) (XYZ/t, coordinate system, coordinate units)
- interface (INT) (identification and version name)
- layer block (LAY) (identification and version name)
- 1, 2 and 3D interface/layer block geometry (GEO) (point/segment/grid id/coordinates)
- 1,2 and 3D interface/layer block properties (PRO) (point/segment/grid/data id/data properties).

These data object types and their mutual one-to-one (1:1) or one-to-many (1:N) relationships are shown in figure 1. This scheme is a simplification of an Entity Relationship Diagram (ERD method) or Information Structure Diagram (ISD) of the NIAM method (Reference 2). Intentionally not all constraints are shown and no extensive list of attributes for these data objects is given, as this will be user definable. This means that at all levels extra attributes may be related to the major objects.

4. Transformation to conceptual scheme

Once the NIAM analysis and documentation of the data or object model has been completed, it is then possible to subsequently derive the relational table/field/definitions and associations between tables from the information structure diagram in the 5th order normalized form (5NF).These tables can directly be used to fill the data dictionary of a RDBMS and to create the tables to store 1, 2 and 3D subsurface models. It depends on the specific RDBMS if some tables can be further optimized at the internal level to improve access performance for selection or manipulation. However this is considered to be transparent for the user. Adding new attributes to the main data object types will result in adding fields to a table.

5. Specification of Data Exchange Format

The information presented in the Data Exchange Format can be compared to information retrieved from a RDBMS into an output report. Exchange information has to be stored in one logical file which can be put on (optical) disc, tape or other media. This implies that information from different tables with one to many relationships to each other has to be assembled and reorganized into one file.
The tables with general information and limited number of occurrences will be presented as header records in a header block. The original tables, objects and fields will be recognizable in the output records. An example of header records is given below

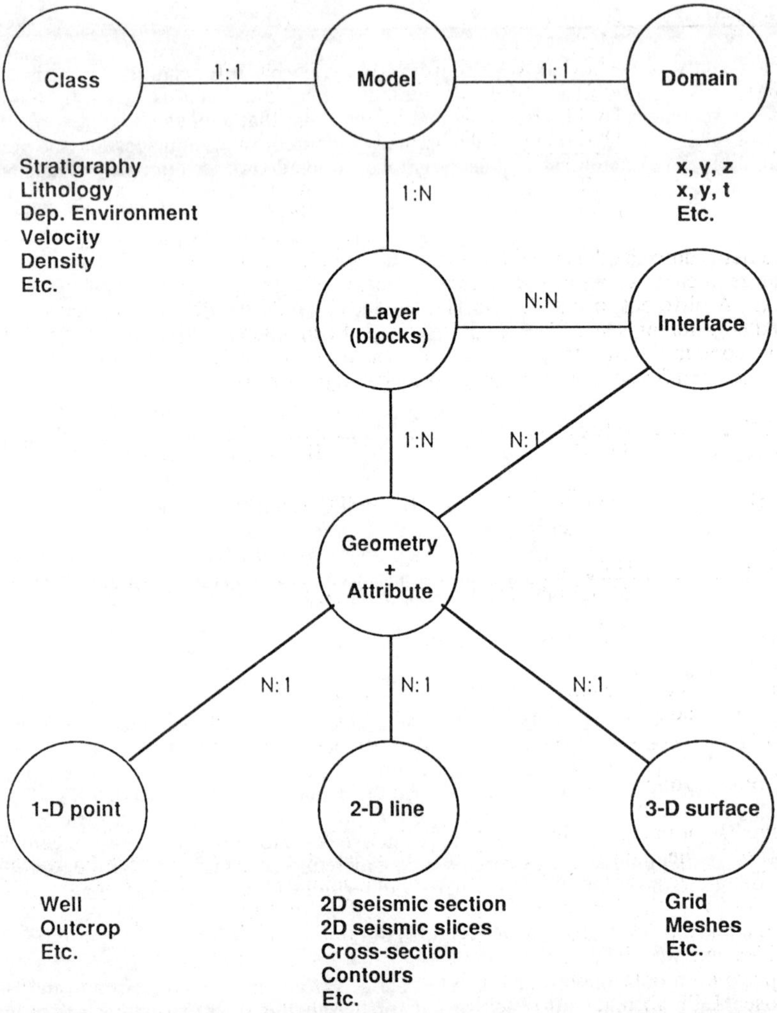

Figure 1. Simplified Information Structure Diagram showing 1D, 2D and 3D Subsurface Model Data

table.id	field id	fields
MOD	'MOD-NAME'	'model name'
MPP	'MOD-PROP, MOD-UNIT'	'velocity', 'm/s'
DOM	'DOM-1, DOM-2, DOM-3'	'x, 'y', 'z' (or 't')
DOX	'DOM-LAT, DOM-ELL'	'latitude', 'ellipsoid'
DOY	'DOM-LONG, DOM-ELL'	'longitude', 'ellipsoid'
DOT	'DOM-T'	'millisec'
INT	'INT'	'horizon1', 'horizon2'
LAY	'LAY'	'layer0','layer1','layer2'

These items describe the layer sequence for a velocity model.The bulk information, i.e. the actual geometrics of an interface and properties of a layer can be given in a condensed format. A different format for the 1D, 2D and 3D bulk data is needed, because they are related to different 'source' locations (see fig. 2).Each record of the bulk information should always contain the interface or layer identification, the model dimension and source type (well, etc.) and further geometry or attribute data. For example:

table.id	field id
GEO	'DIMENSION, INT, SOURCE ID , X COORD, Y COORD, TIME'

fields: '1D', 'horizon1', 'well', 'xcoord', 'ycoord', 'time'

table.id	field id
PRO	'DIMENSION, LAY, SOURCE ID, PROPERTY'

fields: '1D', 'layer0', 'well', 'velocity'

In some cases a more condensed format can be accepted e.g. for regular grids only GEO with TIME and PRO with PROPERTY. This can be extrapolated to a way to exchange 3D regular grids, where each vertical grid increment represents a layer.

6. Conclusions

A standard format for exchanging geologic and geophysical macromodelling data has been proposed by the management of the Joule European Community Research Programme. This Data Exchange Format (DEF) is currently developed by the TNO Institute of Applied Geoscience in the Netherlands, and is specifically designed to optimize (electronic) communication between the participants of the Joule Research Programme. Current activities on a worldwide scale to develop similar transfer formats for geologic, geophysical and petroleum data underline the importance of the application of such formats. The proposed DEF format will complement other existing or elsewhere developed format standards in that an open and flexible format for1, 2 or 3D macromodel subsurface data will be added to the geologic and geophysical data, and that the design will be based on relational information management techniques.

7. References

Reference 1. Shaw B.R. and H.O. Waller, 1990, Exchange Format for Transfer of Geologic and Petroleum Data: AAPG-B, Geobyte, April 1990.

Reference 2. Nijssen G.M. 1976, Modeling in Data Base Management Systems, North Holland , Amsterdam.

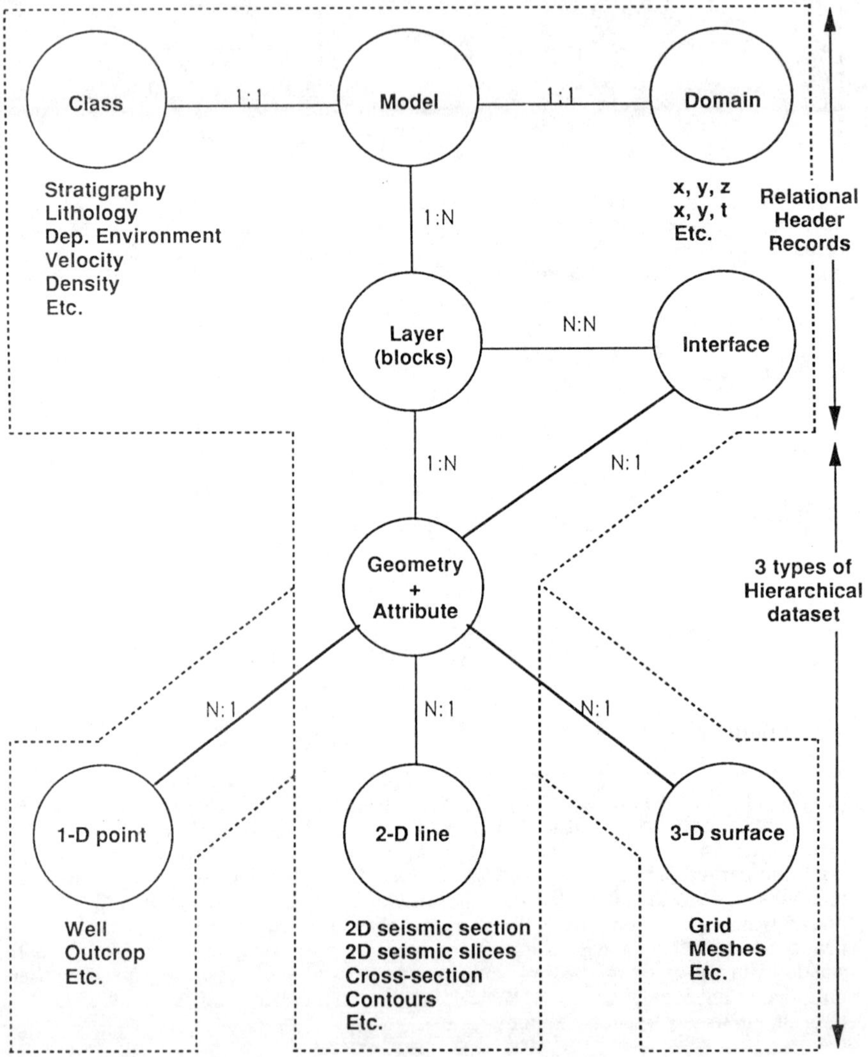

Figure 2. Classification of Data Object Types contained in header records and extensive dataset records of DEF

XPS-FROCKI
AN EXPERT SYSTEM FOR FLUID ROCK INTERACTION PROBLEMS IN OIL PRODUCTION

M. ALBERTSEN*; H. KNOKE**; A. PEKDEGER***;
D. SCHENK**; L. THOMAS***

*) DGMK Deutsche Wissenschaftliche Gesellschaft
für Erdöl, Erdgas und Kohle e.V.,
Steinstraße 7, 2000 Hamburg 1
(corresponding author)
**) Geologisch-Paläontologisches Institut
und Museum der Universität Kiel,
Olshausenstraße 40, 2300 Kiel 1
***) Institut für Angewandte Geologie,
Fachbereich Geowissenschaften der
Freien Universität Berlin,
Wichernstraße 16, 1000 Berlin 33

Summary

Scale precipitation and reservoir damage are actual technical problems in oil and gas production with a strong impact on the economy of recovery projects. Mechanical as well as chemical interactions between the solid phases like rock or well installations and the fluids desequilibrated along their production or injection path are the main damaging processes. To avoid these problems in oil and gas production the interactions must be predictable by qualitative or even quantitative model calculations.

Subject of this paper is the development of an expert system as tool for the prediction of rock-water interactions. This system called XPS-FROCKI combines both physics and geochemistry of the different mechanical and geochemical rock-water interactions with logical rules of practical experience. As main processes the mobilization and filtration of fines, the effects of clay disintegration and the thermodynamics of mineral solution and precipitation are handled.

The paper will present a view of the fundamentals of the model and gives a first outlook on its prototype application.

1. INTRODUCTION

In all field operations where fluids are penetrating the subsurface, damaging effects due to fluid-rock interactions can occur. The resulting decrease in well productivity or injectivity can often interrupt the economic base of technical projects seriously even if the change is a relatively low one.

Concerning the different fields of practical applications fluid-rock interactions can lead to serious problems in the following cases:

- o during production of drinking water from bank filtration,

- o injection of waste fluids into wells,
- o circulation of fluids in geothermal projects,
- o circulation of drilling fluids,
- o production of reservoir brines,
- o stimulation of pay zones.

Generally there are three different pathways to consider. First of all we have to look at natural fluids like reservoir brines uplifted from subsurface conditions to the surface. Due to pressure and temperature decline a large range of potential changes in the chemical character of the fluid can occur. The very best known case in this respect is the degassing of carbon dioxide which can result in precipitation of carbonate scales in the well or in production pipes. Another source of serious problems can appear during the injection of waste or operation fluids into the subsurface. In those cases the fluid is not in equilibrium with its new host. The fluid can be either agressive against minerales of the rock or it can on the other hand react with components and generate new precipitations.

In all cases where fluids are circulated, the whole range of effects can occur due the continous injection, mixing, and uplifting.

Typical problems combined with the injection of technical fluids are often obtained when the injection fluids are not free of oxygen. In this case a high potential for precipitation of metal oxides is to expect which can effectively plug all available pathways in the subsurface.

In any case, to repair formation damage of the so-called skin or to clean the well installations from scale is usually difficult and costly. Therefore the basic approach should be to prevent damage.

In oil companies this is achieved by specialized experts who have to take care of the application of formation protecting fluids only. In smaller companies, especially in the other field of fluid application in subsurface operations, a lack of experience and know how is often noticed in this area.

With that background situation it is the objective of the project to collect the published experience in this area and to combine it into an expert system. It is directed to fit as a serving tool for the prediction or problem analysis in order to prevent or overcome fluid-rock interaction problems in practice.

Especially, the system shall fit the following aspects of application:

1. As assisting tool it shall serve experts with its fundamental data base and its calculation subroutines.
2. As predicting tool it shall replace experts up to a certain level with respect to fluid-rock interaction problems in practice.
3. As education tool it shall assist in the training of employees on a subexpert level.

Concerning to fluid-rock interactions in the recovery of oil, gas and drinking water as well as in waste disposal activities the system is directed to a wide field of application.

2. PROJECT APPROACH

The general arrangement of XPS-FROCKI is outlined in Fig. 1. It contains the following elements:

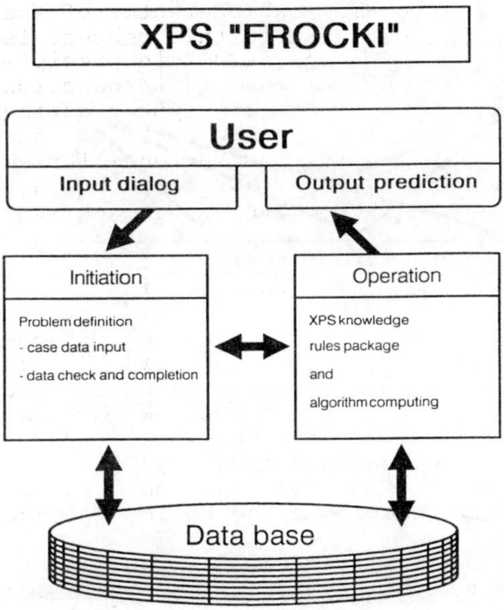

Fig. 1: General Composition of the expert System

1. An initiation unit, which handles the input dialog, the data check and the completion of missing data by correlation steps or by user guided picking of values out of the data base (Fig. 2). The input values are needed to describe both the rock of the formation destinated and the fluid of interest. Concerning the rock the mineralogical composition and the main petrophysical parameters are requested. The fluid is characterized by its ionic content of dissolved spezies and other physicochemical parameters.

2. An operation unit, where the rule-based set of operations to analyse and calculate fluid-rock interactions is organized. The whole knowledge of XPS-FROCKI is arranged in a set of rules which are combined with algorithms and subprograms as far as special calculations are involved.

3. A data base, where rock and fluid data are stored using a genetic organization structure. The data base is continously maintained by every new case handled by the system.

4. An user port, which maintains the interactive communication with the system in all input and output operations.

XPS-FROCKI is developed by use of the XPS-tool BABYLON from GMD/VW GEDAS which is based on the computer language LISP (List Processing Language). BABYLON is a development tool for expert systems with special constructs using LISP as a background language.

The system development by experienced engineers of VW GEDAS, Berlin was realized on a SYMBOLICS 3620, the first computer where BABYLON is installed.

The prototype will be transfered to a SUN SPARCstation by use of the program tool ALLEGRO COMMON-LISP. As computer of the new workstation generation the SUN SPARCstation is chosen as sophisticated host for XPS-FROCKI. The FORTRAN program PHREEQE is implemented as subprogram into XPS-FROCKI. It is compiled to the SUN system so that direct data transfer between XPS-FROCKI and PHREEQE is possible.

The general intention of XPS-FROCKI is to predict potential fluid-rock interactions during percolation contact.

The different types of interaction handled by the system are outlined in Fig. 3.

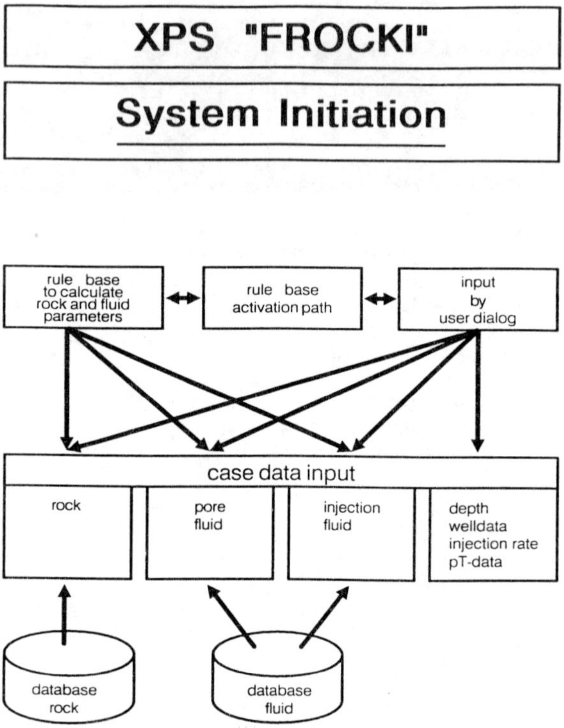

Fig. 2: Schematic configuration of the interactive initiation unit

The clay destabilization potential of a fluid is obtained by calculating the sodium adsorption ratio (SAR), which gives a first estimate of the potential clay sensitivity of a fluid.

The destabilizable content of rocks is generally viewed on the basis of the clay minerals and cation exchange capacity.

The fines content of fluids is considered as far as input values concerning the content and size distribution are available.

The fines content of the rock is obtained via the particle size distribution. Particles smaller than 10 µm are generally approximated as potentially mobilizable fines.

The filtration potential from a geometrical point of view is judged by comparing of the fines size with the hydraulic effective pore size which takes into account different petrophysical parameters like porosity, unconformity and particle size distribution. And last not least the hydraulic filtration potential is considered by calculating the suffusion potential which bases on permeability, porosity, unconformity and particle size distribution.

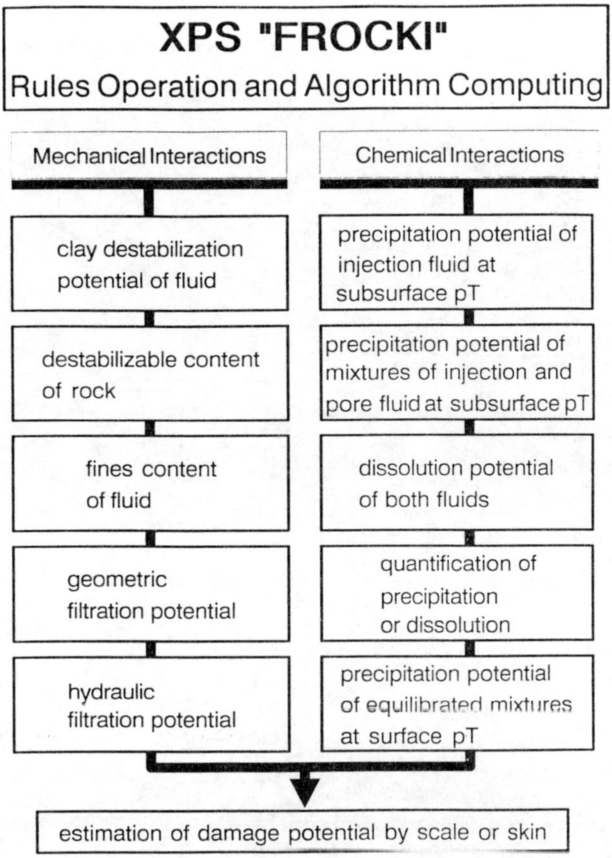

Fig. 3: Fluid-rock interaction processes considered in the expert system

A critical potential around the well bore is calculated, where fines filtration and damaging is to be expected.

The critical potential corresponds to the parameter "critical velocity" described in detail by MUECKE (1) and other authors (2, 3). This rock dependent parameter distinguishes among two types of fines deposition in porous media. At velocities below the critical velocity an adsorption-like surface deposition with only little effect on permeability is predominant. Above the critical velocity, pore throat bridging leads to a plugging-like deposition with a strong effect on permeability (2). If the critical velocity or the critical potential is exeeded by the actual flow near the wellbore, there is a point at a definite distance from the well, due to pressure gradient decline, where the fines deposition switches over from plugging-type to surface-type. This radius or mantle-type area defines the boundary of permeability effective damage. XPS-FROCKI is directed to calculate the extension and the damage ratio of this so-called "skin" zone around the well bore. As shown by KRUEGER (4) there does exist a simple interrelation between the skin and the productivity decline of a well (Fig. 4). This figure clearly outlines the strong effect of even a small skin zone on productivity. If for example permeability has been damaged to 20 % of its original value in a skin zone of only 60 cm, the well productivity is nearly decreased by 100 %.

As outlined in figure 3, different chemical interactions are handled by XPS-FROCKI by geochemical modelling of thermodynamic equilibrium between minerals and aqueous solutions.

The general intention of this part of XPS-FROCKI is to characterize the saturation status of the fluids at different positions on their pathway and to predict the potential reactions to be expected during an equilibrating process. XPS-FROCKI includes a subprogram PHREEQE which is activated on distinct positions during the rule-based XPS-calculations.

The computer code PHREEQE provides a variation of different options which allows a flexible way of problem solution (Fig. 5).

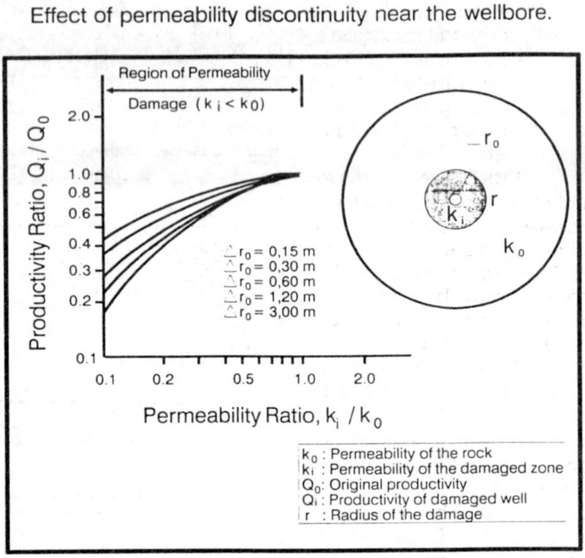

Fig. 4: Effect of formation damage on well productivity (4)

PHREEQE is developed by USGS (5) on the base of pioneer papers on the application of computer technology for the calculation of mass transfer in geochemistry (6-8).

PHREEQE simulates geochemical reactions on the base of an ion pairing aqueous model.

A given set of analytical concentrations of cations and anions including physicochemical parameters like temperature, pH and redox potential are used as input data for the calculations. Assuming different equilibrium approaches, the model calculates in a first step the concentrations of the free ions, the dissolved complexed ions and the saturation vs. possible solid phases.

The equilibrium expressions for the solution used in the computer code are:

- total masses of each element in the system. These may be real or assumed concentrations, whereas the total concentration of a given ion must be equal to the sum of all species in solution bearing this ion.
- mass action for ion pairs. The concentration of the ion pair, i.e. the complexed species in solution, are calculated by a mass action equation

$$K = (X_n Y_m)/(X)^n (Y)^m$$

with X;Y being activities of ions, n;m stoichiometric factors and $(X_n Y_m)$ the concentration of the complex in the solution. The temperature dependent equilibrium constant K may be an empirical expression or calculated from the standard enthalpies.

-Electrical neutrality. If the electrical neutrality in the chemical analysis is not given, PHREEQE offers the possibility to adjust neutrality either by changing the pH-value or by titration with a user defined ion.

PHREEQE also allows to calculate phase equilibria with respect to solid phases and redox reactions by expressions for electron conservation.

The numerical procedure involves the calculation of the ionic strength to obtain individual activity coefficients. The concentrations of the free ions are reduced by formation of ion pairs. The new concentrations are now used to modify the ionic strength and the activity coefficients. These iterations are continued until the mentioned equilibria are obtained within a given range of accuracy.

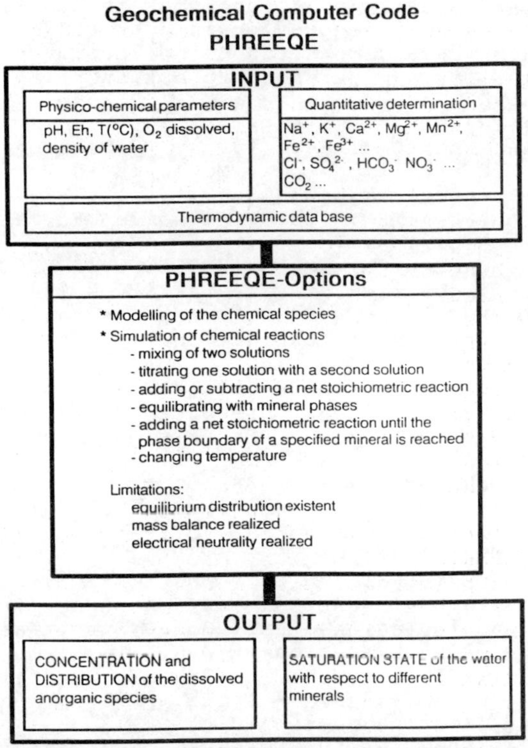

Fig. 5: Schematic representation of the geochemical computer code PHREEQE integrated as subprogram in the expert system (after DAHMKE (9))

The saturation of the water with respect to possible solid phases are then calculated with a mass action equation, describing the state of under- or supersaturation. On the base of this adjusted hydrochemical model, different reaction types may now be simulated. These are:

- adding of reactants to a given solution
- mixing of two waters
- titration of one solution into an other.

Each type of reaction can be simulated including temperature changes. The equilibrium with respect to a mineral may also be postulated by the user. This involves also CO_2-fugacity calculated as pCO_2. With mass transfer calculation reaction pathes as well as phase diagrams may be calculated.

3. STATE OF THE PROJECT

The XPS-development is planned in form of a three phase schedule with a one year duration of each phase:

- prototype development
- completion of prototype and data base
- test at case histories and further improvement.

Actually the prototype is completed, so that the first phase of the project is just finished. First test calculations have demonstrated the prototype to be a sophisticated base for the further project steps.

REFERENCES

(1) MUECKE, T.W. (1979). Formation Fines and Factors Controlling their Movement in Porous Media. Journal of Petroleum Technology. Paper SPE 7007, presented at the SPE-AIME Third Symposium on Formation Damage, Lafayette, Feb. 15-16, 1978.

(2) GRUESBECK, C. and COLLINS, R.E. (1982). Entrainment and Deposition of Fine Particles in Porous Media. PSEJ, SPE 8430, presented at the 1979 SPE Annual Conference and Exhibition, Las Vegas, Sept. 23-26.

(3) GABRIAL, G.A. and INAMDAR, G.R. (1983). An Experimental Investigation of Fines Migration in Porous Media. Paper SPE 12 168, presented at the 58th Annual Technical Conference and Exhibition, San Francisco, Oct. 5-8, 1983.

(4) KRUEGER, R.F. (1986). An Overview of Formation Damage and Well Productivity in Oilfield Operations. SPE 10029, Journal of Petroleum Technology (Feb. 1986), 131-152

(5) PARKHURST, D.L., THORSTENSON, D.C. and PLUMMER, L.N. (1980). PHREEQE - A Computer Program for Geochemical Calculations. U.S. Geol. Surv. Water Resour. Inv. 80-96, 210 S., Washington D.C.

(6) GARRELS, R.M. and CHRIST, C.L. (1965). Solutions Minerals and Equilibria. - 450 p, San Francisco (Freeman, Cooper & Co.).

(7) HELGESON, H.C. (1968). Evaluation of Irreversible Reactions in Geochemical Processes Involving Mineral and Aqueous Solutions. - I. Thermodynamic Relations. Geochim. Cosmochim. Acta, 32, p. 853-877, Oxford.

(8) HELGESON, H.C., GARRELS, R.M. & MACKENZIE, F.T. (1969). Evaluation of Irreversible Reactions in Geochemical Processes Involving Minerals and Aqueous Solutions.- II. Applications. Geochem. Cosmochim. Acta, 33, p. 455-481, Oxford.

(9) DAHMKE, A. (1986). Die geochemischen Modelle WATEQF und PHREEQE. Bericht der Bundesanstalt für Geowissenschaften und Rohstoffe, Hannover, Archiv-Nr. 99397, Tagebuch-Nr. 10901/86.

OUTCROP STUDIES AND GEOSTATISTICAL MODELLING OF A MIDDLE JURASSIC BRENT ANALOGUE

C. RAVENNE*, A. GALLI**, H. BEUCHER**, R. ESCHARD*, D. GUERILLOT* and HERESIM GROUP*/**

* Institut Français du Pétrole, 1–4 av. de Bois Préau, BP 311, 92506 Rueil–Malmaison, France
** Centre de Géostatistique, Ecole des Mines de Paris, 35 rue Saint–Honoré, 77305 Fontainebleau, France

ABSTRACT

This paper first presents the results of a 2D and 3D geological study which aims at describing reservoir heterogeneities. This study then serves as a basis for the 2D and 3D conditional simulation of the geometry of a fluviodeltaic reservoir using the Heresim softwares. These two studies should solve the difficult problem of interpolating between wells and thereby the problem of forecasting the extension of sedimentary bodies. The research is being carried out by both IFP and the Centre de Géostatistique of the Paris School of Mines, which form the HERESIM Group.

The study was carried out in the middle Jurassic in Yorkshire, which is very similar to the Brent formation in the North Sea. This paper also describes the new sedimentological interpretation of the Scalby formation (U.K.) from the core–drill data, the different correlations made between the core–drills, the quantification in terms of proportion curves and of variograms of lithofacies sets, and last, some conditional simulations.

These studies will provide new tools for understanding and describing the shape and heterogeneities of a reservoir.

1. INTRODUCTION

The problem of heterogeneities in reservoirs has often been emphasized, particularly when a reliable production forecast is required in reservoirs developed with wide spacing or where enhanced oil recovery techniques are implemented. Several types of heterogeneities may be found, either on different scales, or resulting from different causes (sedimentology, diagenesis, tectonics, etc.). The results given here concern a 2D and 3D field study and the method used to model the geometry of sedimentary bodies on a pluridecametric to plurihectometric scale. We will first present the methodology used in order to get the geological data and then the main geological results. The geostatistical method for the conditional modelling of the lithofacies of a fluvio–deltaic series will be given in Chapter III.

The aim of the work presented in this paper is to obtain 2D and 3D images through which geological hypothesis for the reservoir models can be tested. Geologists and geostatisticians are joining forces to develop a simulation methodology to model geometrical patterns that could later be transposed in terms of hydrodynamic properties.

The method used here to obtain conditional simulations of lithofacies is based on truncating a Gaussian random function. The thresholds are the Gaussian values of the lithofacies proportions. The geostatistical properties of the Gaussian random function are determined by the characteristics of the lithofacies sets. The parameters needed there to implement this method are:

- the proportion curves showing the global distribution of the lithofacies,
- the geometrical variograms of the sets showing the deposit anisotropies, and to which the parameters of the underlying Gaussian variables are fitted.

In order to model sedimentary bodies on a scale which can be handled by simulators, we have to simplify the data. This led us to seek the geological parameters that can provide the most representative geometrical definition of the sedimentary bodies on the required scale.

The prime importance of the problem of vertical and lateral shale and sand intercalation in the fluvio–deltaic series of the North Sea reservoirs led us to study the Middle Jurassic outcrops on the cliffs of Yorkshire (Great–Britain) (Fig. 1). The cliff chosen displays remarkably well a great variety of fluvial to fluvio–deltaic deposits over a length of about 10 km and a height of 30 to 150 m.

RAVENSCAR GROUP

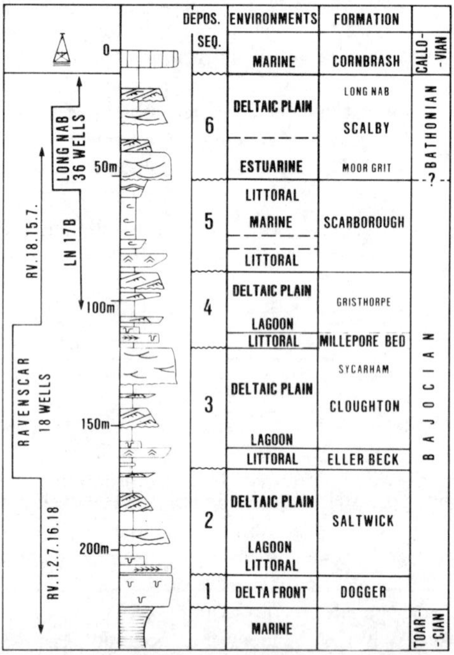

Fig. 1. Formations and main environment of the Ravenscar Group.
The studied outcrops in site 1 belong to the Scalby Formation which is the upper part of the Ravenscar Group. The Scalby Formation overlays the Scarborough marine formation. It is, roughly speaking, made of fluvial to deltaic deposits. The Scalby Fm is overlaid by the Cornbrash marine Fm. Location of the coredrills with regard to the cross–section of the formation is indicated on the left. The field geological study was carried out by the HERESIM Group and with the help of:
- J. ALEXANDER from Leeds University (now at Cardiff University),
- R. KNOX, G. LOTT and A. HOWARD from British Geological Survey,
- CAYO PUIG de FABREGAS from Servei Geologie de Catalunya,
- M. MARZO from Universidad de Barcelona.

2. GEOLOGICAL STUDY

First step : Methodology used in order to get geological data and first results.

a) Former knowledge

These outcrops have already been studied by Archer and Hancock (1) and Hancock and Fischer (2) among others and have been compared to the sandy deltaic reservoirs in north-west Europe and mainly to the sandy reservoirs in the Middle Brent. The studied outcrops belong to the Ravenscar group and the terminology used is that of Hemingway and Knox (3). The regional geological knowledge and the sedimentological data have been provided by the work of Knox, Nami, and Leeder, Livera, Alexander (4, 5, 6, 7, 8).

The study concerns two formations of the Ravenscar group. In the first one, the Cloughton formation, we have chiefly studied the lower part between the open sea events of the Ellerbeck Formation and of the Millepore bed. It includes the Sycarham member which is a massive sandy layer. In the second one, the Scalby formation, which is the focus of this paper, two members have been distinguished: the Moor Grit and the Long Nab. In 1976, Nami (5) divided the Long Nab member into its upper and lower parts. Nevertheless, as J. Alexander (8) has pointed out, these divisions are not time-stratigraphic units but are based on variations in architectural styles.

b) Methodology used for the 2D study (9)

All the Middle Jurassic outcrops of the cliff were continuously photographed from a helicopter so as to limit the distortion of the photos due to the angle and distance of shot-points and to the dip of the layers. In the first interpretation, we were able to describe the large units, in order to separate the sand bodies and to select the two sites. Detailed shots of the Scalby formation, near Long Nab (Cloughton site) and of the Cloughton formation at the Ravenscar site were taken. The photographical study of the Scalby formation made it possible to distinguish the geometry of the different depositional units, the internal structure of the sandstones, the nature of the contacts – erosional or concordant – of the sand bodies with the surrounding shales, and the continuity of the impermeable layers.

The Scalby formation was studied over a length of 600 m and over a height of 30 m. This photogeological study serves as the basis for the first 2D modelling. The petrographic nature and the distinction between the different facies and structures were determined with 16 vertical sections covering the entire height of the cliff (Fig. 2). The sections were spaced in view of the statistical computations (no preferential sampling as usually carried out for the sedimentological study..., but another set of sections were carried out in view of the sedimentological study). Topographic measurements were taken of the vertical sections and of sedimentary bodies in order to obtain the optimum correlation between the data of the cliff and of the neighboring core-drills. 500 m of sections were described. Samples were systematically taken from the sandy parts of the sections to obtain permeability and porosity profiles constituting hypothetical wells with spacing impossible to obtain in actual fields. Superficial alteration obliterates the representative of the measurements. However, trends observed remain valid. The data contained in the photo-geological study were then simplified so that they could be processed on a scale in accordance with the requirements of the reservoir description. The topographic correction and an interpretation on four lithofacies were carried out. As a result, quantified parameters were obtained which will be used with a greater confidence interval and with more precision.

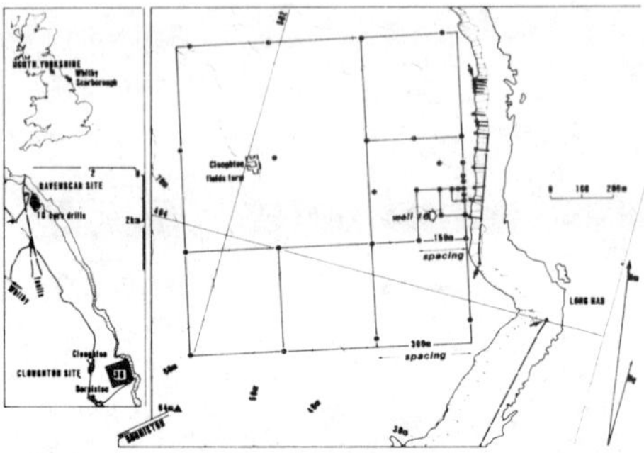

Fig. 2. Location map of the two sites studied in the middle Jurassic of Yorkshire, and detail map of the Cloughton site.
The studied formations are very similar to some of the North Sea reservoirs.
The cliff is continuous over 10 km in the horizontal direction and 30 to 200 m in the vertical direction.
On the detail map, 1) the location of the 16 vertical cross-sections carried out in the cliff (black lines in the cliff, right of the figure) and, 2) the location of the core-drills are indicated.
36 coredrills have been cored and logged in the site over an area of approximately 1 km^2.
Each coredrill is about 30 to 50 m long. Thus, more than 1000 m of cores have been recovered.
We chose a square grid with coredrills located 300 m apart. In one grid, the drillhole spacing was progressively reduced from 300 to 10 m in order to study the small scale heterogeneities and to get a better understanding of the sedimentological evolution.
The lines of the coredrills are parallel and perpendicular to the cliff. The first parallel line is located just a few meters behind the cliff to provide an unambiguous correlation with the observations of the cliff face.

Figure 3 shows part of the lithofacies interpretation (sandstone, shaly sandstone, silty clay, and clay). On this figure the beddings have been represented horizontally. The guide marks LN4 ... LN6 correspond to the projections of the core-drills, the guide mark M11 indicates the site of a large vertical section made in the cliff: the guide marks M22 ... show the location of each small vertical section (mean height : 4 m) plotted every 10 m to follow the evolution of the sedimentologic and petrophysic properties in the basal sandstone lithofacies (Unit A1, Fig. 3, defined in Fig. 5).

The precise interpretation in terms of lithology is obtained from the vertical sections and the core-drills. Figure 4 shows the lithostratigraphic representation of Figure 3. The very large lateral extension of the units and their almost constant thickness are obvious compared to the lithofacies.

On the lithofacies interpretation of the whole cliff, 66 vertical lines, 10 m apart, have been digitized. The digitization is made vertically every 20 cm, from a reference level, over a 30 m height (i.e. 150 points). The reference level (dashed line at the base of Figs. 3 and 4) is a marker level inside the marine shales. The reservoir levels we want to simulate overlay these shales after erosion.

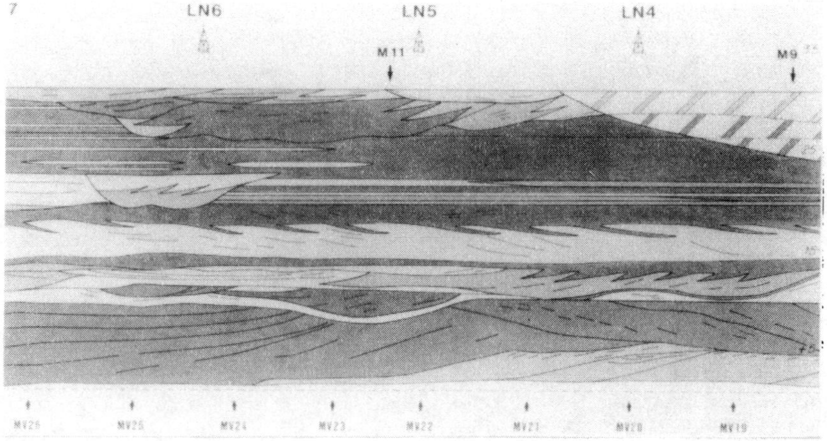

Fig. 3. Part of the lithologic interpretation of the cliff has been carried out after topologic corrections.
4 classes of lithologies are now used for the computation of quantitative parameters. There is a continuous lithologic evolution from the darker shade of grey (sandstone) to the lighter shade of grey (shales).

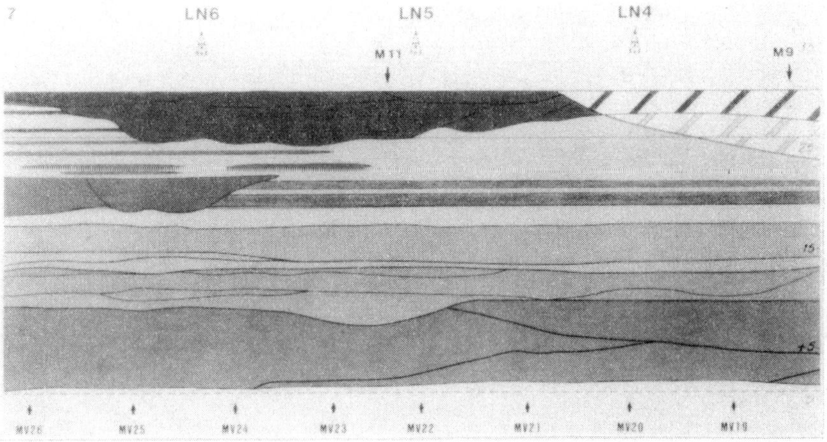

Fig. 4. Lithostratigraphic interpretation of the portion of the cliff represented on Fig. 2. Units are defined on Fig. 5. The thickness is about 30 m and the length about 100 m.
At the top, there is the projection of the location of the neighbouring coredrills.
M12, M11, M9 correspond to the entire vertical cross-sections.
MV30 to MV17 correspond to the small vertical sections.

c) Methodology used in the 3D study (10)

The ultimate goal is to provide a description of actual reservoirs, where the only available data are the well data (cores, logs) and the geophysical data. An intermediate step therefore involves applying and testing the simulations obtained on data from core–drills carried out in the vicinity of the cliff. With regard to the description of the heterogeneities and to the prediction of sedimentary body geometry, it is then possible to extend the model from 2D to 3D. Also the results obtained can be compared with those from well–logs obtained in oil reservoirs.

Core–drills were carried out on both study sites. Eighteen were cored on the Ravenscar site through the entire lower part of the Cloughton formation. Five of them were cored through the whole of the Ravenscar group in order to extend the vertical simulation on a larger scale. Thirty–six core–drills (Fig. 2) 30 to 50 meters long were carried out through the entire Scalby formation on the Cloughton site over an area of approximately 1 km^2. The size of the grid was chosen in collaboration with geostatisticians. As a result, a square grid was chosen with a 300 m spacing between the core–drills. In one pattern, the spacing was progressively reduced from 300 to 20 m (Fig. 2) to study the small–scale heterogeneities. The lines of core–drills were parallel and perpendicular to the cliff. The first line was several meters behind the cliff to provide an unambiguous correlation with the 2D observations. All the core–drills were cored with a 9.3 cm diameter (3" 5/8) through the effective thickness of the Scalby formation.

In each core–drill, the spontaneous polarisation (PS), the natural radioactivity (gamma–ray), the density (gamma–gamma LSD), the hydrogen index (thermal neutron N01), the dipmeter log (3 arms) were recorded. The sonic log was only recorded in the first two wells, because the measurements were not significant owing to the great number of fractures. The logs obtained are very similar to those form the middle part of the Brent formation. A plug, 4 cm long and 23 mm in diameter, was taken every 25 cm to study the paleo–environment (shales – silty clays) and the physical properties of the reservoirs. The values of these properties are indicated on porosity and permeability logs.

To compare the observations available on the core–drills with those made on an outcrop, the interpretations and calculations are first carried out on the vertical plane located immediately behind the cliff, against the LN1 to LN11 core–drill line (Fig. 2). This line makes possible the direct correlation between core–drill data and outcrop data, due to the small distance (less than 50 m) between the two planes. This observation can be applied in particular to the sedimentological study and makes it possible to calibrate the diagraphic interpretation.

The tridimensional extension of the sedimentary bodies can be estimated from the core–drill data. The lithofacies studied on the cliff may be analyzed more accurately with a very weak alteration (which is of interest in subsurface reservoir studies). Finally, the petrophysical measurements on the core–drills are more reliable and can be taken systematically.

d) Main sedimentological results on the Scalby formation based on core and logging data

Scalby formation

The Scalby formation overlays the silty clays of the Scarborough formation with erosional unconformity. The top of the Scalby formation is eroded by pleistocene glacial deposits. The Scalby formation has been divided into lithological units which can be correlated on a kilometric scale (Fig. 5). This division is based on a sedimentological core interpretation and on the potential reservoir properties of the deposits and can be obtained from the logs (mainly with gamma–ray and resistivity).

Two major lithologic units A and B have been distinguished (Fig. 5). In Unit A, the sandstone fraction is dominant. The sandstone bodies are approximately 10 m thick and extend laterally over several tens of meters.

The detailed study of core-drills show that sets A and B are separated by a regional discontinuity. The A1 and A2 units (a lithostratigraphic unit is a set of genetically linked sediments) show an important lateral extension. They are separated by a large erosional surface. The A1 Unit is now interpreted as being part of a system of linear distributaries, fluvial-dominated upstream of an estuary.

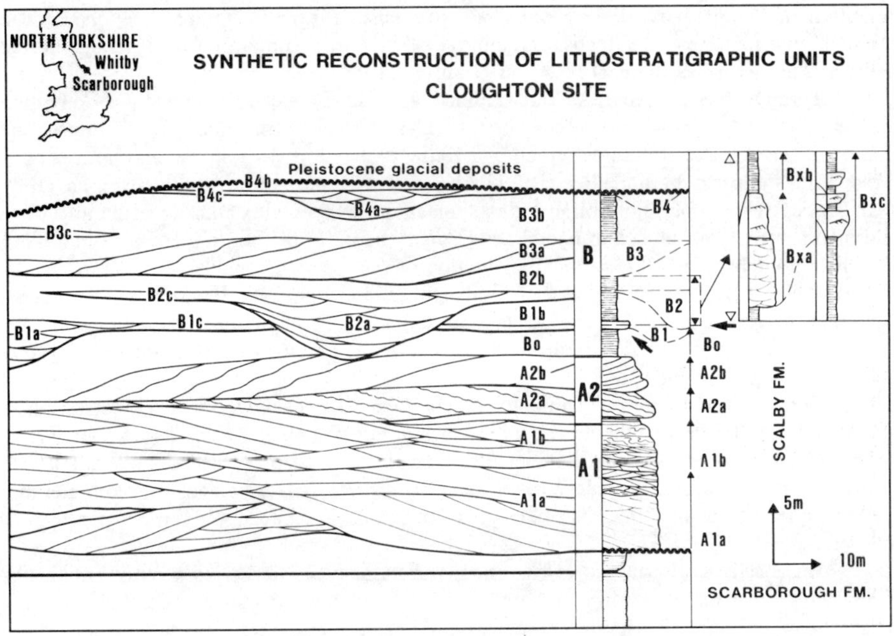

Fig. 5. Synthetic reconstruction of the lithostratigraphic units of the Scalby formation in site 1 deduced from the cores and the cliff face study.
Two major sets have already been defined in the Scalby formation and Set A corresponds to the Moor Grit Member defined by M. Leeder.
Coredrills, new vertical cross-sections along the cliff and photogeological interpretation allow us to subdivide these sets and to obtain a more precise internal description.
Set A, overlaying the Scarborough marine Formation with a regional unconformity is interpreted as a fluvial to estuarine valley fill formed by three main depositional prisms. In the cored area, we can only observe the lower and the upper ones. The vertical evolution displays an increasing upward tidal influence.
Unit A1 is made of stacked distributaries, fluvial dominated. In sub-unit A1b, we can observe the beginning of tidal influence.
Unit A2 is formed by two superposed meander belts mixed with fluvial and tidal influences.
Set B overlies unconformably Set A. It contains sandstone channels 10 to 100 m wide isolated in a deltaic plain shale.
Four sub-units B1 to B4 have been distinguished with four levels of channels. This set is interpreted as a plain deltaic environment overlaying unit B0 made of palustrine deposit.

Unit A

The two sub–units A1 and A2 have been distinguished with an upward increase in shale component in each of them. Both have a sandstone term A1a and A2a, and a shaly or shaly–silt term, A1b and A2b.

In term A1a the sandstones are fine–to–medium grained, pure and cross–stratified. The lignite fragments and the mud pebbles abound in several places. They form large–size channels which erode each other. Their filling is composed of trough cross–bedded sediment sandstones and sigmoid cross–stratified bars several meters thick. They are probably longitudinal channel bars. Bioturbated shaly silty lenses fill the topographic troughs between the main distributaries. The deposit dynamics of this term is similar to that of a large braided system. This term has the best reservoir qualities of the formation.

Term A1b is composed of fine–grained shaly sandstones and shaly silts. Convoluted fine–grained shaly sandstones and sandstones with ripple–marks and planar–oblique structures can be observed. The structures are emphasized by very thin shaly layers and mud pebbles. This term indicates both the end of the braided system by the decrease in its energy (sandstones with ripple and oblique marks accentuated by shale) and the beginning of the meander system (appearance of lateral accretionary bars). In this term, there is an increase in the shale content which leads to a deterioration of the reservoir quality.

Term A2a is formed of shaly sandstones and shaly silts with abundant plastic deformations (convolutions, slumps, water escape marks). The sigmoid stratification composed of sets several meters in size, which was observed on the outcrop cannot be seen on the cores. Its reservoir qualities are poor.

Term A2b is formed of laminated or homogeneous black shales. They correspond to the flood plain deposits of the meander system but they can also correspond, in some places, to the shaly plugs in abandoned meanders. Due to the well preserved organic matter, the varved nature of some silty shales, and the absence of well defined paleosols, we can assume the existence of shallow interdistributary swamps. This term forms a vertical permeability barrier.

The basic A1a term shows two main structural types: large festoons of about 10 m generated by distributaries filling up, and large bars with oblique planar structures interpreted as side–bars building themselves in the distributaries. The numerous internal erosion surfaces indicate an important lateral migration of the distributaries under low deposit in a fluvial–dominated environment under tidal influence. This influence is exhibited by systematic argillaceaous drapings on the structures, conjugated directions of the current and the bioturbation. The dominant structures are plane troughs filled with sigmoidal foresets.

The A2 unit is interpreted as the result of fluvial–dominated meandering distributaries, strongly influenced by tide. The overall shape is a sandsheet type composed of two superimposed meandered belts. Each meandered belt is composed of superimposed lateral elementary accretionary bars still showing a weak subsidence.

In set A, from A1 to A2 units, we observe a growing marine influence and an increasing deposit energy. This possibly results from a relative rise of the marine level, which flattens the stream equilibrium profile. Set A corresponds to an incised valley–fill.

Unit B

Unit B is formed by sandstone channels from 10 to 100 meters long and their associated overflow facies isolated in a shale matrix. Five successive channel phases (B1 to B5) overlay or erode each other (Fig. 5). In the deposits of each channel phase, three terms (a, b, c) can be observed. Term B1a corresponds to the sandstone filling of the channel, which is fine–grained and micaceous with trough cross–bedding. Term B1b is formed of homogeneous, green, silty shales deposited in the channel flood plain. Term B1c is formed of tabular layers of micaceous fine-grained sandstones with ripple marks interbedded in green silty shales, forming the lateral overflow facies of the channels (crevasse-splay and sand sheet). Terms B2 and B3 are similar.

Term B4 contains plastic deformations shaly sandstone (similar to A2a) and could indicate a local recurrence of the meander system. There are good reservoir qualities in the sandstone filling of the channels. The erosion of different channels has led to a connection between sand bodies in some places, thereby providing a link between the reservoirs.

The channels are interpreted as being the result of sinuous ribbons to meandering channels in a deltaic plain. Only four levels of channels have been observed on the core-drills.

We note the importance of the elementary erosion surfaces inside the lithostratigraphic units. These surfaces are interpreted as the result of lateral distributary migrations which erode the previous deposits. The numerous unconformities between the units seem to result from the relative variations of the sea level.

The correlations can also be defined in terms of porosity and permeability. The values are estimated from the systematic measurements made on the plugs every 25 cm and with the logs (See (9)).

Such a distribution of lithologies is comparable to the one described by Parry, Whitley and Simpson (11) in Middle-Upper Brent as described in CONOCO/BNOC/GULF well 211/19-4, and by Simpson and Whitley in the same well in the middle shaly unit (12) where one can almost superimpose lithologic descriptions and gamma-ray logs for both LN 16 and well 211/19-4.

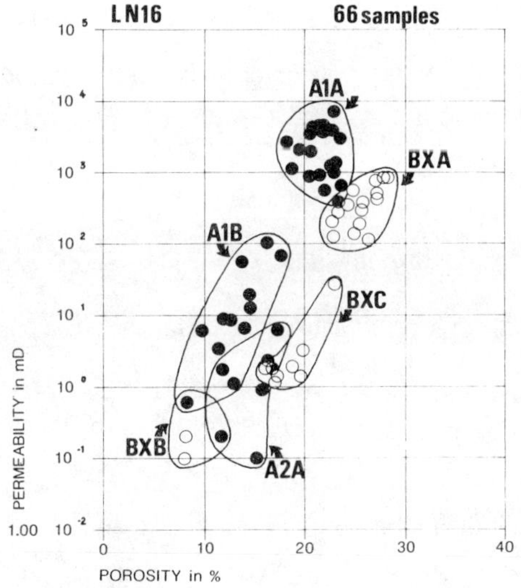

Fig. 6. Porosity-permeability cross-diagrams for some of the lithostratigraphic units.
These cross-diagrams result from the study of the 5 coredrills.
These physical properties have been measured on plugs taken every 25 cms.

Fig. 6 shows porosity–permeability measurement results on sandstone samples from core–drill L.N.16. The different points have been assembled in accordance with the terms previously defined. Term A1a (sandstone lithology) is homogeneous and closely assembled around 3 Darcy and 22% porosity. Term Bxa is very close to term A1a. Porosity and permeability both decrease in terms A1b and Bxb which correspond to the general lithology and shaly sandstone adopted for the digitization. Term A2a is very similar to terms A1b and Bxb and is homogeneous and closely assembled around 1 md and 15% porosity.

Porosity and permeability are two of the elements required for 3D simulation models, together with well–logs and core–parameters such as lithologic nature, internal configuration, etc.

As will be shown later, the outcrop study is fundamental in determining the lateral behavior of the studied variable. The density of information available here provides means of calculating quantitative parameters (variograms, proportion curves).

3. GEOSTATISTICAL STUDY

Introduction to Geostatistics

Geostatistics was created at the end of the fifties by G. Matheron (13), its main contributor. It is a part of the probability theory applied to the Earth Sciences. Its main originality consists in two basic principles, which take into account two facts: – First, the data we are dealing with in the Earth Science are located in space (or time–space): the spatial correlation between these data is of prime importance and is a constitutive part of the theory. – Secondly, these data are measured on different volumes (these are called the support of the data), but we are usually interested in quantities measured on a support which is much larger than the data themselves. (For instance, porosity data are either measured on small plugs or derived from logs which integrate a greater volume. For reservoir simulations, we need to know the porosity values on large cells).

In the sequel, we will concentrate on spatial correlations and we will introduce the principles and the tools derived from geostatistics in an intuitive way, leaving aside mathematical developments and formulae as much as possible.

The variograms and the probabilistic models

How can spatial correlation be quantified?

The basic tool is the variogram. We will show how to compute the experimental variogram, and we will then show how this experimental variogram is used to quantify the spatial correlation.

Given p data points, it is possible to compute, for each data point $z(x_i)$, all the squared differences $(z(x_i) - z(x_j))^2$ between $z(x_i)$ and the $z(x_j)$.

We now define classes of distances, generally multiples of a given lag h_o. For each class of distances, we compute the mean value of the points within this class (that is, the mean square value of the differences) and we obtain the experimental variogram.

This mean value for a given distance (i.e. the value of the experimental variogram at that distance) has the sense of a correlation. If all the point values within this class of distances are close to each other, the mean value is small (there is a high correlation for this class). Conversely, if the values are far apart, the mean value is large (the correlation for this class is low).

The way in which this correlation varies in relation to the distance constitutes in itself an interesting element of quantitative information on the data we are dealing with.

In our example, we do not take the direction into account; that is, we implicitly assume the data to be isotropic. This is generally not the case, therefore the computation has to be done for different classes of directions (i.e. we select the points $z(x_i)$ and $z(x_j)$ if the value of the angle between x_i and x_j is within a given interval). This is the way anisotropies are tested and quantified.

This can be illustrated by the following example. When working on channelized data, we would expect a better correlation in the main direction of the channels.

But we also know that for a large distributary environment even if the main direction is more or less constant, if we are concerned by a field whose dimensions are small relatively to the distributary environment expected maximum correlation in the mean direction of channels can be taken in default because of the meandering and displacements of channels.

In that case, the experimental variogram can be very useful for the interpretation, as it provides the correlation relative to the distance for some directions.

Questions :

1) What is the implicit model used and what are the assumptions made when computing the variogram?

The implicit model used is a Random Function. A random function $Z(x)$ is a collection of random variables indexed by location. That is to say, at point x_i, $Z(x_i)$ is a random variable, and we interpret the values $z(x_i)$ measured at point x_i as the outcome of this random variable.

Now the experimental variogram appears to be an approximation of the (theoretical) variogram $\gamma(h)$ defined by:

$$\gamma(h) = 1/2\ E\{(Z(x + h)-Z(x))^2\}$$

(where E stands for the mathematical expectation).

We can also define the non-centered covariance $C(h) = E(Z(x+h)\ Z\ (x))$ or the centered covariance $C(h) = E\{(Z(x + h) - m)\ (Z(x) - m))$ where $m = E(Z(x))$.

Here two assumptions are made. First the expectation $E(Z(x))$ is independent of the location of the points.

Secondly, the expectation of the squared differences $E\{(Z(x + h) - Z(x))^2\}$ does not depend on the location of the points but only on the magnitude of the distance between them. This is the hypothesis of second order stationarity (or of intrinsic stationary, which is more general). If these assumptions are not verified, the data are said to be non–stationary.

It can be shown that this probabilistic framework using random functions enables us to make clear these hypothesis, and furthermore to test them.

For example, in a stationary case, we were able to define the range of the variogram, that is the distance where the variogram begins to flatten out and we could also give a precise meaning to this parameter: it is the maximum distance for a spatial correlation, i.e. two points separated by a distance larger than the range are no longer correlated.

2) Can we do more with an experimental variogram? The answer is positive. We only have to fit to the experimental variogram a theoretical variogram $\gamma(h)$, i.e. a continuous curve representing the quantity,

$$\frac{1}{2}E\{(Z(x + h) - Z(x))^2\}$$

Not any curve may represent a variogram. It can be shown that a necessary and sufficient condition for a function $f(h)$ to be a variogram is conditional negativeness, that is, whatever n and $x_1,..., x_n$ being points in R^p, and $\lambda^1,...,\lambda^n$ being subject to the condition

$$\sum_{i=1}^{n} \lambda^i = 0, \qquad \text{we must have:} \qquad -\sum_{i=1}^{n}\sum_{j=1}^{n} \lambda^i \lambda^j f(\mid x_i - x_j \mid) \geq 0$$

where $| xi - xj |$ is the distance between points x_i and x_j.

Having fitted a theoretical variogram to the experimental one, we can use the probabilistic framework to perform a number of tasks. For the purpose of this paper, only the procedures of estimation (kriging) and simulation will be described.

Kriging (14)

Knowing a number of data points and the variogram, kriging is the "best" linear estimation taking the spatial correlation into account.

We estimate z(x) at point x by:

$$z(x) = \sum_{i=1}^{n} \lambda^i z(x_i)$$

$z(x_i)$ are the known data points, λ^i are the weighting factors. These are determined by the location of the data points in relation to each other, the location of the point to be estimated in relation to the known points, and by the variogram, simply by solving the kriging equations (14). Solving this equation also provides a reliability index: the kriging variance.

Kriging is a tool well suited to a large number of problems, for example the estimation of seismic with faults, the estimation of the top of a reservoir using the wells and the seismic data (15). It can also be used to compute the volume of oil in place, or to estimate porosities in large cells using plugs measurements and logs measurements, or to obtain point estimation of permeabilities using porosities.

To solve different types of problems, another tool is used: the simulations.

Simulations (Geostatistical simulations) (16, 17, 18)

The object of a simulation is not to provide the "best" estimation, but to produce various possible versions of a partly known reality, or to solve non-linear problems.

On the other hand, when using simulations to reproduce the variability of the reality, we overlook the notion of precision of the estimation. If we allow fluctuations to occur only between data points, and we build every possible version of the reality (simulation) so as to match the data points, the simulation is said to be conditional.

We know from experience that one of the main problems encountered when trying to obtain input parameters for reservoir simulators is the following. If we do an estimation, the continuity of the reservoir is overestimated, as is quite often the case when the correlations between wells are made by geologists. The consequences are well-known: overestimation of the percentage yield, underestimation of the optimal distance between wells, bad estimate of coning, etc.

Definintion of conditional simulations

They provide possible versions of the partly known reality:

a) Each simulation honors the data points

b) They have the same spatial variability as reality. That is, if we compute the experimental variogram of a simulation, we must be very close to the experimental variogram of the data.

c) The simulations and the data have the same histogram.

In the case of continuous variables (i.e. variables which can take any value between two extreme values), several methods are available.

In the past, this type of simulation has been widely used to obtain the probability of reserves in place, and it has occasionally been used as input for reservoir simulations.

In our case (fluvio-deltaic environment), we believe that the main factor for heterogeneities is the geometry of the different lithofacies. We know that permeability is partly controlled by the lithology, and we expect fluctuations between lithofacies. So we choose to first simulate the lithology. This also has the advantage of providing a better geological control of the results obtained.

However the lithology is not a continuous variable: it can be sandstone, shaly sandstone or shale. Sets (of sandstone, etc.) have to be simulated. For this, various methods can be used. A number of methods for simulating sets (called Random Sets) are available, but unfortunately, these are difficult if not impossible to condition to the data points. This is why we have designed a specific method for simulationg Random Sets, which is conditional and adapted to fluvio–deltaic environments: the Truncated Gaussian Random Function. But before describing this method, we shall go back to random sets, and indicator random functions.

Random sets

Basically, a random set is a set with a random aspect in the shape, the location of its elements, or the relation between the elements within the set (for a more precise definition, see Matheron (19)).

There are mainly two ways of describing random sets. One is called "object oriented". It consists of defining one or more basic shapes (rectangles, ellipses, etc.). The probability law of/for the dimensions of these shapes, for the location of their center, and also for the relation existing between these various elements, is also specified.

The most commonly used "object oriented" random set is the Boolean model (19), which was popularized by the work of H. Halderson "Stochastic Shales" (20).

In practice, the procedure goes as follows:

1) We define a basic element (a parallelepiped for instance)
2) The location of the center of this element is drawn, using a Poisson process
3) Given the center, we then draw the dimensions of the element. (To do this, we can either draw each dimension separately using one law for each dimension, or we can draw one dimension in relation to another, e.g. thickness in relation to extension).

In our example, the elements represent shale, and the background represents sandstone. It can easily be seen that such a model can include more than two facies, as well as their orientation. Another way of defining a random set is simply to use a random function: given a random function, we decide for instance to define shale as the points for which the random function is below a given value (the threshold). This given value may vary in the space. Each time we get a realization (that is a draw) of the random function, we get a realization (a draw) of the random set.

Indicator Random Functions

An indicator Random Function is a random function which can only take the value 0 or 1. This type of function is well suited to describe sets.

Let us consider the case where we have two facies (shale and sandstone for instance). To describe sandstone, we use the function:

$$1_{SAND}(x) = \begin{cases} 1 \text{ if x belongs to a sandstone facies} \\ 0 \text{ if x belongs to shale} \end{cases}$$

If we compute the mean value of this function, we get the proportion of sandstone (and in our case, the proportion of shale will be 1 minus the proportion of sandstone).

As we now know how to compute the experimental variogram of a random function, we can make the same computation for an Indicator random function.

To do this, we need to compute half the square of

$$\mathbf{1}_{SAND}(x) - \mathbf{1}_{SAND}(y)$$

for each location of x and y, we then sort out the values according to the distance between x and y, and we calculate the mean value for each class of distances. The difference above is equal to either 1 or 0. We find 1 if one of the 2 points x and y does not belong to sandstone, and 0 otherwise.

Another possibility is to compute the experimental non-centered covariance. For this, we have to calculate the product

$$\mathbf{1}_{SAND}(x) \times \mathbf{1}_{SAND}(y)$$

and again sort out the values according to the distances. As above, we compute the mean value for each class of distances. In this case, we see that each product can take a 0 value if one of the points belong to shale, or the value 1 if both points belong to sandstone. By calculating the mean value for a given distance, we obtain the probability that 2 points that far apart belong to sandstone.

1° Extreme caution should be taken concerning the interpretation: this variogram does not represent the probability that 2 points that far apart are connected within sandstone.

2° Because a non-centered covariance of an indicator is a probability, it has some strong properties. In order for a function to be the covariance of an indicator, it has to be of a positive type, whereas to be a variogram, it must be of a negative type. (This is however not sufficient).

<u>Truncated gaussian random function (21)</u>

The basic idea behind this method is to describe the n facies using one indicator function per facies. The indicator for facies i, defined by means of a Gaussian Random Function Y(x), is:

$$1_{a_{i-1} < Y(x) \leq a_i} = \begin{cases} 1 \text{ if } Y(x) \in]a_{i-1,ai}] \\ 0 \text{ otherwise} \end{cases}$$

Therefore, point x belongs to facies i if $Y(x) \in]a_{i-1,ai}]$. The a_i are called the thresholds. If these intervals are disjoint and their union covers the whole space R, the function defined by:

$$F(x) = \sum_{i=1}^{n} i \, 1_{a_{i-1} < Y(x) \leq ai}$$

takes the value i at point x if and only if x belongs to facies i, that is if

$$1_{ai-1 < Y(x) \leq a_1} = 1.$$

The a_i are called the thresholds. The main advantages of this method are:

1) Conditioning is easy to perform
2) The method requires little computation
3) It enables us to take account of the spatial correlations between points in a same lithofacies, but also between the lithofacies themselves, which is a characteristic of geological events.
4) From a methodological point of view, the theoretical variograms and cross-variograms of the indicators are consistent (that is, they satisfy all the necessary conditions)
5) Few parameters are involved, and moreover, "external" information can also be included.

We will now describe in more details this model, the parameters involved, and their physical meaning.

The threshold

The probability for the Gaussian random function Y(x) to be smaller than a_i and greater than a_{i-1} is:

$$Pr(a_{i-1} \leq Y(x) < a_i) = G(a_i) - G(a_{i-1})$$

where G is the normal cumulative distribution.

If Y(x) is between these two values, by definition point x belongs to facies i, that is, the above probability represents the exact global proportion of facies i in the field.

If we choose a_o to be $-\infty$, and a_N (N being the number of facies) to be $+\infty$, as we have:

$$G(-\infty) = 0$$

$$G(+\infty) = 1$$

we therefore find (p_i being the proportion of facies i):

$$G(a_1) - G(a_o) = G(a_1) = p_1$$

$$G(a_2) - G(a_1) = p_2$$

so that

$$G(a_2) = p_1 + p_2, \text{ and so on ...}$$

In other words, all we have to do to find a_i is to invert the cumulative normal distribution

$$a_1 = G^{-1}(p_1)$$

$$a_2 = G^{-1}(p_1 + p_2) \ldots$$

$$\ldots$$

$$a_{N-1} = G^{-1}(p_1 + p_2 + \ldots + p_{N-1})$$

Now if we look at the vertical proportion curve on Fig. 6, a realistic model would be to take the coefficients a_i which depend on z (the vertical level). For a given i, $a_i(z)$ is the proportion of facies i at level z. It follows that a point with coordinates (x,y,z) will belong to facies i if and only if $a_{i-1}(z) < Y(x,y,z) \leq a_i(z)$, and we know for sure that the proportion of points belonging to facies i at level z is the exact proportion of this facies at level z. Note: If necessary, it is possible to use a model where the a_i depend on x, and $y(a_i(x,y))$ or x,y and z $a_i(x,y,z)$.

The variogram

We know that a Gaussian random function (with a 0 mean) is entirely determined by its variogram. Therefore in order to specify Y(x), we only need to know its variogram.

In fact, there is a well known formula in geostatistics (22) relating the variograms and cross-variograms of the indicators built by truncating a Gaussian random function, to the variograms of this Gaussian random function. This comes in very handy, first because the only way we can compute the experimental variograms or cross-variograms is by using the data from the facies, and also because we want to simulate the facies distributions, respecting their variograms and cross-variograms.

To sum up, we have seen that it is possible to obtain the parameters needed to describe the truncated Gaussian function simply by using the data available to us, that is the facies. We have also shown the physical meaning of each parameter involved.

We will now explain in more details how to perform what is called the structural analysis, that is, how to compute the experimental variograms, and then fit these with the theoretical variogram.

Structural analysis

1) Proportion curves

a) Proportion curves of the lithofacies along the axis perpendicular to the bedding (on the cliff)

The cumulated proportion curves (Fig. 6) summarize the vertical distribution of each of the four lithofacies on the whole cliff. For each point of the curves, the proportions, are calculated on the 66 vertical lines. Thus, from the base to the top, we roughly observe three levels of sandstone. The most important one is located at the base, up to 14 meters, which mainly corresponds to Unit A19.

The other two sandstone levels located in the upper part, mainly correspond to the B2a and B3a units ans show very few changes. However, we note their stairway shape, bounded at 7%. The new representation also brings out more characteristic shapes of the shaly sandstone and silty clay lithofacies between 8 and 18 meters, whereas the previous interpretation showed an almost constant shape. We note the presence of a narrow band of silty clay between 13 and 18 meters, mainly corresponding to the A2b unit, not previously seen. We also note that the shaly sandstone facies and the silty clay facies are equally represented (about 40% each) between 8 and 13 meters and form the mainpart of the A1b unit. These proportion curves show a good correlation between the lithofacies and the lithostratigraphy.

b) Proportion curves of the lithofacies on the wells

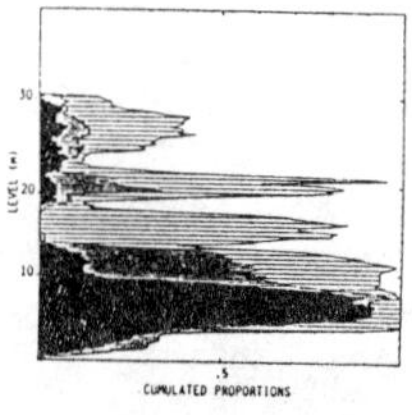

Fig. 7a. Proportion curves of the lithofacies computed from the 66 vertical lines, 10 m apart, digitized from the geological interpretation of the cliff, after topographic corrections.
Black is sandstone, dark grey is shaly sandstone, light grey is silty shale, white is shale.

The curves of the proportions cumulated according to the vertical axis (Fig. 7a), calculated on 11 coredrills (from LN1 to LN11) practically show the same aspect as those calculated on the cliff. However, the variations are less regular than on Fig. 6, based on 66 vertical lines. The use of only 11 vertical lines (or coredrills) give an important relative weight to each coredrill and a stairway shape to the curve.

Thus the overall shape of the sandstone level at the base is retained, but we observe a stronger proportion of argillaceous sandstone between 6 and 7 meters. We also note an increase of the rough fragments (sandstone and argillaceous sandstone) towards 18 meters.

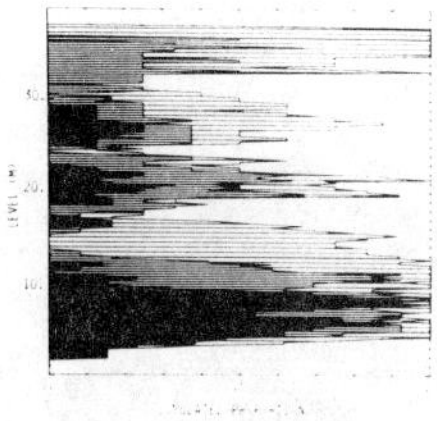

Fig. 7b. Proportion curves of the lithofacies computed from the 11 coredrills located just behind the cliff.

The curves of the proportions cumulated according to the vertical axis and calculated on 36 coredrills (Fig. 7b) are similar to those calculated on the cliff on the 11 coredrills. The differences previously described are attenuated. However, we again note that the sandstone bodies are distinct. This is due to a clear separation in space of the lithostratigraphic units and their sedimentary bodies, because the 36 coredrills give a 3–D image, while we only have a 2–D image of the cliff.

c) Proportion curves computed parallel to the bedding

They show that the stationary hypotheses for the "horizontal" direction is suitable for this environment.

2) Variograms

We computed all the variograms on the different facies, and the cross–variograms between facies. We then decided that a theoretical variogram would be accepted for the Gaussian Function if and only if the resulting fit was reasonably good on the experimental variogram and the cross variograms of the facies (see 10). As a result, we had a better control of the fit, and perhaps more importantly, this is a drastic test of the adequation to the data of the method used : Truncated Gaussian Functions.

a) Horizontal variograms

We already noted that we are in a stationary case in the horizontal direction. This was verified too for the whole field, including the cliff and the wells. We can therefore compute the standard horizontal variogram.

On the cliff

The horizontal variograms presented on Fig. 8 are mean variograms averaging 150 individual variograms. Each individual variogram is calculated on one level. The data plotted on an outcrop are crucial at this step. On account of the large amount of information available, the experimental variogram is more representative and can be fitted, which is not always obvious for a reservoir, where the core–drill data are few and scattered. So, the outcrop data give the order of magnitude of the range along the corresponding direction.

The cliff and the wells

The structural analysis of the sets is made plane by plane along 4 horizontal directions. These variograms exhibit an anisotropy, roughly in the NS and EW directions. The practical range fitted in the EW direction is 350 m, 175 m in the NS direction.

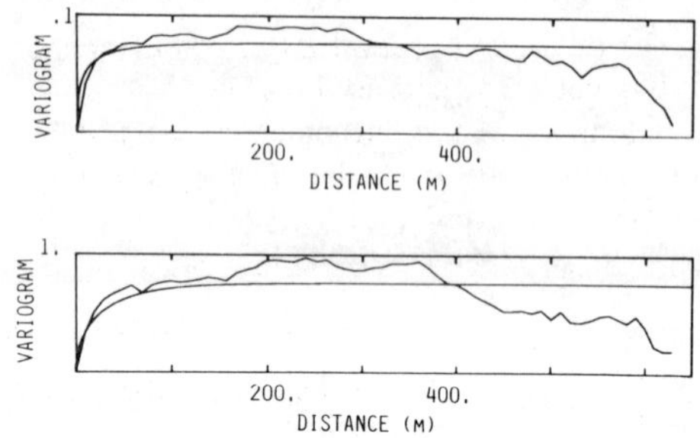

Fig. 8a. North–South horizontal mean variograms computed from 150 individual horizontal variograms. The lower one is for sandstone, the upper one is for shaly sandstone.

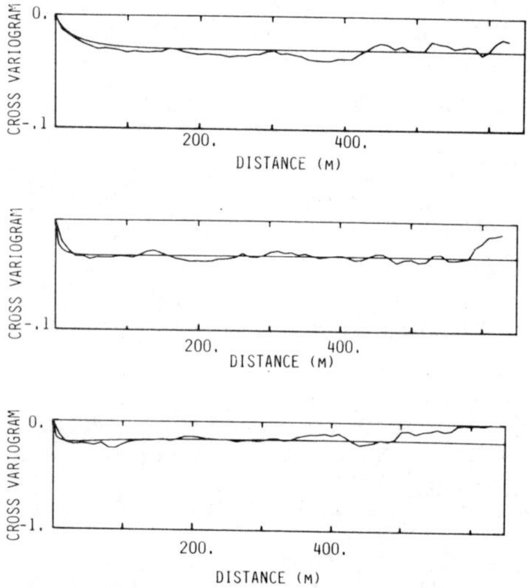

Fig. 8b. North–South mean cross–variograms computed from 150 individual cross–variograms. From top to bottom : sandstone and silty shale, shaly sandstones and silty shales, sandstone and shaly sandstone.

b) Vertical variograms

We have seen before that, as far as the vertical direction is concerned, the stationary assumption is clearly not acceptable. We then have to adopt a different approach. Within the scope of the model, this vertical non–stationarity is accounted for by the vertical variation of the threshold $a_i(z)$. We now have to compute the experimental non–stationary vertical variograms. In our particular case, we will proceed as follows.

Define: $$\gamma^i(z\ ;\ h_z) = \frac{1}{2} E\{(Ind_i(z + h_z) - Ind_i(z))2\}$$

were $Ind_i(z)$ is the indicator of facies i at level z (to simplify matters, we ignored the other two coordinates) and h_z is a vector parallel to the z (vertical) axis.

Because of the stationarity of the phenomenon in the horizontal direction, and also because we usually have several wells available, the experimental non-stationary variogram will be computed as follows:

For a given level z, we compute, for each h_z:

$$\gamma^i(z\ ;\ h_z) = \text{Mean value over all the wells of } \{Ind_i(z + h_z) - Ind_i(Z)\}^2$$

Note that, in this case, the variograms depend not only on the level, but also on the direction of the vector h (i.e. upward or downward).

<u>This method is well suited for vertical studies particularly when we have sequences</u>.

Therefore this non-stationary variogram has a 0 value for $h_z = 0$ (that is, at level z), but it is <u>not symmetrical</u>.

In what follows, we will only present the mean experimental vertical variograms and their fits, that is, the mean values over all possible z of the variograms defined in the preceding section.

It is easy to see that this mean experimental variogram is symmetrical, and we only present the part $h_z \geq 0$ (as is usually the case). The fit was nevertheless checked on several individual vertical variograms.

Note: We will also present the fits for the non-stationary vertical cross-variogram on the mean vertical cross variogram.

The mean of the 150 vertical variograms was taken on the 30 m height of the cliff. These curves present a strong analogy with the proportion curves. The fit of these variograms show a range superior to that performed on the corresponding variograms of the core-drill data. This is the reason why the practical range fitted on the coredrill variogram will be later used for the simulations. This practical range is about 10 m. (Fig. 9).

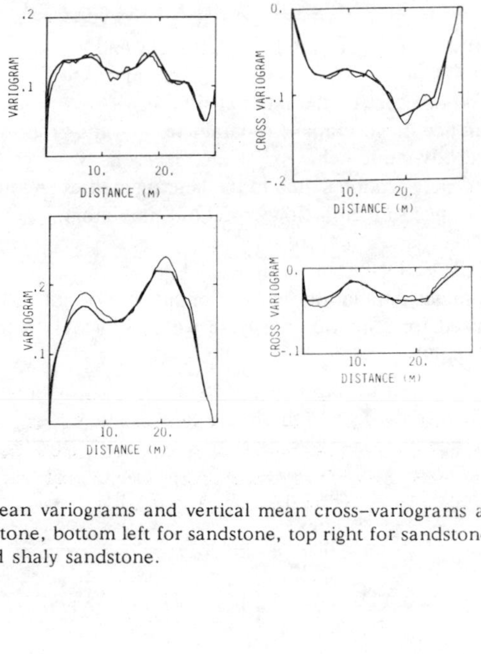

Fig. 9. Vertical mean variograms and vertical mean cross-variograms and their fit.
Top left for shaly sandstone, bottom left for sandstone, top right for sandstone and silty shale, bottom right for sandstone and shaly sandstone.

To sum up the structural analysis, we fit a model of variogram, which is, for the Gaussian Random Function, a product of exponentials

Vertical range 10 m
N–S range 175 m
E–W range 350 m

We have shown that this variogram provides a reasonably good fit for both the horizontal and vertical variograms and also the cross–variograms of the lithofacies (namely of the Indicator of lithofacies). The model takes into account this typical non–stationarity of the vertical direction.

4. THE HERESIM 3D SOFTWARE

To apply this methodology, an interactive and graphic system has been developed. This 3D software provides tools corresponding to the steps described below. It involves the following tasks:

1) Data input, choice of a simulation grid and of the correlation horizons
2) Geostatistical structural analysis
3) High resolution lithological geostatistical simulation at the decimetric scale and with the control on the sedimentary organization.

Each simulation is an image of the reservoir geology, consistent with the data. It then provides a particular geological interpretation of the reservoir. Note that a lithofacies can be defined from its petrophysical properties.

4) High resolution porosity and absolute permeability description controlled mainly by the high resolution lithology

Petrophysical data are assumed to be strongly dependent on the nature of the lithofacies. Porosity and permeability values are assigned to each facies, using different deterministic or stochastic algorithms. Note that probabilistic methods are required to generate petrophysical heterogeneities inside a given lithofacies.

5) Scaling up in order to calculate effective parameters directly suitable for fluid flow purposes.

Work is in progress in order to implement steps 4 and 5.

This software allows geologists and reservoir engineers to:

- check different hypotheses for the internal lithology
- estimate the influence of geostatistical parameters on the proportion of lithofacies in place and the hydrodynamic behavior of the reservoir
- build a reservoir descretization suited to the heterogeneities encountered and the usual size of grid blocks used in fluid flow reservoir simulators.

5. LITHOFACIES SIMULATION

The full 3D lithofacies simulation was run on an area of about 1 km x 1 km x 30 m including the 36 wells used for conditioning. Here we present only a few outputs (Fig. 10, 11). (See 10 for more details).

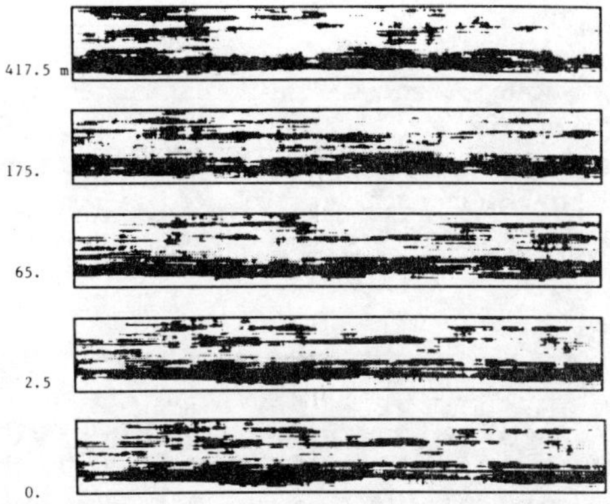

Fig. 10. Example of a 3D conditional simulation with 36 coredrills. (Ranges are 175 m in the SN direction and 300 m in the WE direction). SN successive vertical sections parallel to the cliff (the distance from the cliff face is indicated on the left). The length is 675 m and the height 20 m.

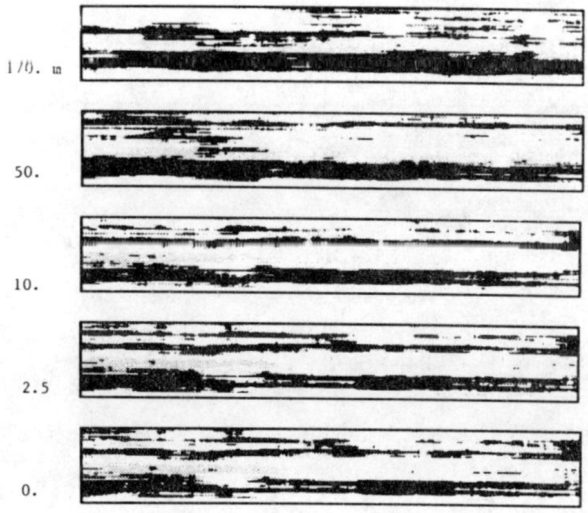

Fig. 11. Example of the same 3D conditional simulation : EW successive vertical sections perpendicular to the cliff. The length is 625 m and the height 20 m.

6. FLUID FLOW MODELLING

To illustrate the link between geostatistical imaging and fluid flow simulation, let us briefly describe work in progress in 3D (23, 24).

After the geostatistical simulations, fluid flow simulation model has been performed giving petrophysical characteristics to each facies. Simulations of waterflooding has been achieved, with water injection well in the central well and the producing wells are in the corners. Injection and production rates were balanced in reservoir conditions, and kept constant for the various simulations. We have used the multipurpose ΣCORE Reservoir Simulator.

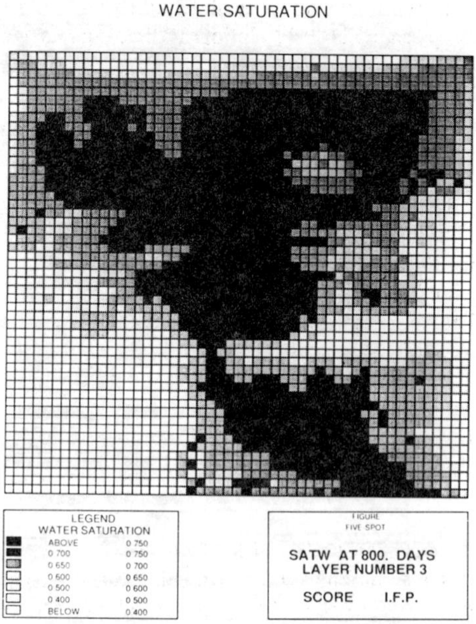

Fig. 12. Example of water saturation after 800 days in one layer of a 5 spot site showing the impact of heterogeneities on fluid flow.

One result is displayed on Fig. 12 as a water saturation map of one layer. The water saturation maps show a tortuous pattern very different from that obtained in an homogeneous reservoir. It can be observed in some simulations that some layers are not swept by the water injected into these layers, but by water coming from an upper or lower part of the reservoir.

7. CONCLUSIONS

In order to solve the problem of including general geological knowledge in a reservoir fluid flow simulator, (1) a new methodology of outcrops and geological data study, reservoir oriented, has been established, (2) a new simulator method of random sets using truncated gaussian functions has been elaborated, (3) an integrated software designed to characterize a reservoir has been developed.

This integrated software provides the advantage of being flexible and interdisciplinar. Flexible, as each parameter can be easily modified, thus different assumpions, relying on different values of parameters, may be rapidly tested. Interdisciplinar, as the joint experience of geologists and reservoir engineers is needed, in order to use it efficiently.

The main philosophy behind this software is that a correct reservoir characterization can only be achieved through a realistic geological description. In most reservoirs, permeability and porosity heterogeneity is mainly related to lithofacies heterogeneity. Furthermore, there is an important advantage in simulating first the lithofacies, then porosity and permeability, as stochastic lithofacies descriptions can be compared with conceptual sedimentological models. Their consistency with geological knowledge can be tested. Conceptual models of sedimentary environments exist and are well established, whereas only a few conceptual models of porosity and permeability distributions exists. Thus, a drawback of direct porosity and/or permeability simulation is that such a control of consistency is certainly more difficult, perhaps impossible.

As a deterministic description of a reservoir seems impossible, due to the high number of possible involved parameters, a stochastic description of reservoir is a better solution.

The stochastic techniques presented in this paper are based upon joint use of correlation functions and proportions. They make it possible to rapidly generate different scales of heterogeneity, linked to the depositional environment of each reservoir unit.

Our stochastic methods provide new quantitative tools for the reservoir geologist: correlation functions and lithofacies proportions. They also put some quantitative way of thinking into geological studies.

In reservoir studies, large data sets are usually available only when enough wells have been drilled. Nevertheless, HERESIM can be applied in the appraisal stage of a field, in order to test different geological assumptions. For example, the hydrocarbon reserves may be computed as a function of the size of sandstone bodies, modeled with various given proportions and correlation lenghts. But HERESIM can also be applied on a producing field, where the input parameters can be directly computed from the field data.

REFERENCES

(1) ARCHER, J.S. and HANCOCK, N.J.: "An Appreciation of Middle Brent Sand Reservoir Features by Analogy with Yorkshire Coast Outcrops", paper EUR 197, presented at the 1980 European Offshore Petroleum Conference and Exhibition, London, October 21–24.

(2) HANCOCK, N.J. and FISHER, M.J.: "Middle Jurassic North Sea Deltas with Particular Reference to Yorkshire", Proc. Conf. Petroleum Geology of the Continental Shelf of North West Europe, London, March 1980 (published by Heydon and Son Ltd)

(3) HEMINGWAY, J.E., and KNOX, R.W. O'B: "Lithostratigraphical Nomenclature of the Middle Jurassic Stata of the Yorkshire Basin of North–East England", Proc. Yorks. Geol. Soc. 39, pp 527–535, 1973.

(4) KNOX, R.W. O'B: "Sedimentological Studies of the Ellerbeck Bed and Lower Deltaic Series in North–East Yorkshire", Unpublished PhD thesis, University of Newcastle Upon Tyne, 1969.

(5) NAMI, M.: "An Exhumed Jurassic Meander Belt from Yorkshire, England", Geol. Mag. 113, pp 47–52, 1976.

(6) NAMI, M. and LEEDER, M.R.: "Changing Channel Morphology and Magnitude in the Scalby Formation (M. Jurassic) of Yorkshire, England", in Fluvial Sedimentology (ed. A.D. Miall), Can. Soc. Petrol. Geol. Mem. 5 pp 431–440, 1978.

(7) LIVERA, S.E.: "Sedimentology of Bajocian Rocks from the Ravenscar Group of Yorkshire", unpublished PhD Thesis, University of Leeds, 1981.

(8) ALEXANDER, J.: "Sedimentary and Tectonic Controls of Alluvial Facies Distribution: Middle Jurassic, Yorkshire and the North Sea Basins and Holocene, SW Montana, USA", Unpublished PhD Thesis, University of Leeds, 1986.

(9) RAVENNE, CH., ESCHARD, R., GALLI, A., MATHIEU, Y., MONTADERT, L., RUDKIEWIC, J.L.: "Heterogeneities and geometry of sedimentary bodies in a fluvio-deltaic reservoir", SPE Formation Evaluation, June 1989, pp. 239–246.

(10) RAVENNE, CH., BEUCHER, H.: Recent development in the description of sedimentary bodies in a fluvio–deltaic reservoir and their 3D conditional simulation, in Proc. SPE Annual Technical Conf. and Exhibition, Oct. 2–5, 1988, Houston, Texas, pp. 463–476, SPE paper 18310.

(11) PARRY, C.C., WHITLEY, P.K.J. and SIMPSON, R.D.H.: "Integration of Palynological and Sedimentological Methods of Facies Analysis of the Brent Formation", in Petroleum Geology of the Continental Shelf of North–West Europe, pp. 205–215, Institute of Petroleum, London, 1981.

(12) SIMPSON, R.D.H. and WHITLEY, P.K.J.: "Geological Input to Reservoir Simulation of the Brent Formation", in Petroleum Geology of the Continental Shelf of North–West Europe, pp 310–314, Institute of Petroleum, London, 1981.

(13) MATHERON, G.: Traité de Géostatistique Appliquée, Tome 1, Ed. Technip, Paris, 1962.

(14) MATHERON, G.: Traité de Géostatistique Appliquée, Tome 2 – Le krigeage, Ed. Technip, Paris, 1963.

(15) GALLI, A., MEUNIER, G., Study of a gas reservoir using the external drift method, in Geostatistical Case Studies, ed. by G. Matheron and M. Armstrong, D. Reidel. Publ., Dordrecht, Netherlands, 1984.

(16) MATHERON G.: The intrinsic random functions and their applications, Adv. Appl. Prob., Vol. 5, 1973, pp. 439–468.

(17) JOURNEL, A.: Simulations conditionelles – Théorie et pratique, Thèse de Doct.–Ing., Université de Nancy, 1974.

(18) CHILES, J.–P.: Géostatistique des phénomènes non stationnaires (dans le plan), Thèse de Doct.–Ing., Université de Nancy, 1977.

(19) MATHERON, G.: Random sets and integral geometry, Wiley, N.Y. 1975.

(20) HALDORSEN, H.H.. A new approach to shale management in field scale simulation models, SPE paper 10976.

(21) MATHERON, G., BEUCHER, H., DE FOUQUET, CH., GALLI, A.: Conditional simulation of the geometry of the fluvio–deltaic reservoirs, Proc. SPE 1987 Annual Technical Conference and Exhibition, Dallas, Texas 27–30 September 1987, Vol. X: Formation, Evaluation and Reserve Geology, pp. 123–131.

(22) MATHERON, G., BEUCHER, H., DE FOUQUET, CH., GALLI, A., RAVENNE, CH.,: Simulation conditionelle à 3 faciès dans une falaise de la formation du Brent, Sciences de la Terre, n°28, Juin 1988, "Etudes Géostatistiques V", pp. 213–249.

(23) GUERILLOT, D., GALLI, A., RAVENNE, CH., Heresim Group: 3–D fluid flow behaviour in heterogeneous porous media, 1990 Latin American Petroleum Engineering Conference, Rio de Janeiro, Oct. 14–19, 1990.

(24) GALLI, A., GUERILLOT, D., RAVENNE, CH. and Heresim Group: Combining geology, geostatistics and multiphase fluid flow for 3D reservoir studies. Proc. of the Second European conference on the Mathematics of Oil Recovery, Arles, France, Sept. 11–14, 1990, Technip, Paris.

ANALYTIC NON-LINEAR FORWARD AND INVERSE SEISMIC SCATTERING: A PERTURBATION APPROACH

B. KUMMER

University of Hamburg
Institute of Geophysics
Bundesstrasse 55
2000 Hamburg 13, (FRG)

SUMMARY

A concise formulation of 1D seismic scattering is presented based on the Lippmann-Schwinger equation in the wave number domain. This transformation turns out to be of particular importance for a calculation of higher order terms of the Born series. In the first stage the scattering formalism is applied to the Born approximation and elucidates the advantages of this formulation in terms of generality and conciseness of results: While confronted with four contributions for the Green's function in the space domain, the calculations in the wave number domain yield only one analytic expression. An analysis of this result shows that the poles of the Green's function implicitly encompass the shot-geophone locations whereas the analoguous calculation in the space domain corresponds to an explicit approach.

Thus pointing out the strategy for attacking a calculation of higher order approximations of the Green's function, analytic results are presented for the second and third order iteration of the Lippmann-Schwinger equation. It is shown that the second order term yields a dynamic correction of the primary reflections. The third order term additionally produces first order multiples. Numerical examples demonstrate an accurate and fast modelling of the primary reflections for relatively large velocity contrasts.

Based on the third order approximation an analytic objective function is constructed and applied to inversion of synthetic CDP-stacked seismograms of a saltdome model. The results show that for sufficiently smooth velocity distributions a reliable reconstruction of the model is obtained. For strong lateral heterogeneities, however, the numerical example illustrates the restriction of the 1D inversion approach. In spite of the lack of lithological resolution this inversion approach has the advantage of indicating a computationally efficient way for the derivation of an appropriate apriori model. This is of particular importance for non-linear single shot inversion methods which strongly rely on well chosen starting models.

PETROPHYSICAL CHARACTERIZATION OF CARBONATIC AND CLASTIC RESERVOIRS USING ACOUSTIC WAVEFORMS LOGS.

M.GONFALINI, M.PIANA AND H.ANXIONNAZ

INTRODUCTION

The Sonic log has been used for years as openhole lithological indicator, seismic calibrator and for volumetric analysis. With the introduction of the Array Sonic tool and the recording of a full waveform train, new applications have been developed for fractured zone location and mechanical analysis.

The scope of the paper is to analyse the practical results obtained in Italy through a case study of twelve wells drilled in different type of reservoirs, the objectives being different according to the lithologies.

In hard rocks the aim of the waveform analysis and interpretation are mainly the location of fractured zones and petrophysical facies recognition.

In shaly-sand reservoirs the target is the characterization of unconsolidated sands in terms of permeability estimation.

DISCUSSION

Different operating procedures have been tried to adapt the acquisition to the lithology and to the type of reservoir (low porosity, thin beds,..).

Based both on the data set available and on the objective of the study, the capabilities of the existent processings have been carried out, compared and discussed.

The acquisition procedures are differentiated by the time of listening duration, that can be set to 5 or 10 msec allowing a detailed study of the different components of the wavetrain, respectively Compressional and Shear (body) waves and the borehole guided Stoneley waves.

The different components are analysed by means of several processings both in time and frequency domains to compute their slowness and energy losses. For the Stoneley component, the reflection coefficient between the direct and the reflected energy is also computed.

Energy losses of all the components are related to presence of fractures, lithology changes and borehole diameter variations. The Stoneley reflection is mainly related to the presence of permeable events.

The reflections of Stoneley energy occur when open fractures filled by drilling fluids are interposed between transmitters and receivers. This effect is related to the sudden change of acoustic impedance. Thus the reflection coefficient magnitude is though to be strongly related to permeability.

CONCLUSIONS

The large number of analysed wells allow to define:

* the most suitable acquisition procedures for each type of reservoir.
* the limits and advantages of each type of processing for fracture location and formation characterization capabilities.
* the good potential of Stoneley energy losses to estimate the permeability in case of unconsolidated sand reservoirs.

DAMAGE DIAGNOSIS USING KNOWLEDGE PROCESSING TECHNIQUES

M.TAMBINI AND G.COSENZA
AGIP S.p.A.
P.O.BOX: 12069
20100 MILAN (ITALY)

The great impact of formation damage on decreasing well productivity and consequently on oil company economics has led in the past to a number of lab test investigations and theoretical studies. Many publications (1) have focused on various types of damage and recognized their mechanisms and locations in the productive formation. Moreover experience has shown that appropriate design of a remedial treatment (2) absolutely requires exact knowledge about the nature of the damage.

The authors have approached this problem by using expert systems technology applied to formation damage analysis and diagnosis. "DAMAGE" is a knowledge based system aimed at identifying the potential causes of damage by reviewing all the well operations from drilling to production phases: cementing, completion, stimulation, gravel packing, well servicing, workover and injection for enhanced recovery.

A lot of specific and heterogeneous experience is required to establish the potential formation damages in every field condition. It is possible to say that for every type of damage there is an associated specialist with the specific knowledge to recognize it and evaluate the diagnosis reliability. Analysis of the way this diagnosis is carried out by the expert shows that he really refines his diagnosis hypothesis using specific experiences. Therefore an expert system has been developped that applies the same knowledge structure used by the expert to approach the problem. "General Task Architecture" (3) has been chosen as the conceptual model representing problem solving activity in the diagnosis task. In this architecture the knowledge engineer defines a taxonomy where every class is a "specialist" knowing the data and test required to identify a generic malfunction. In the figure a portion of this taxonomy is shown.

The system was realized using a frame-based shell developped on the symbolic programming language Common Lisp (4).

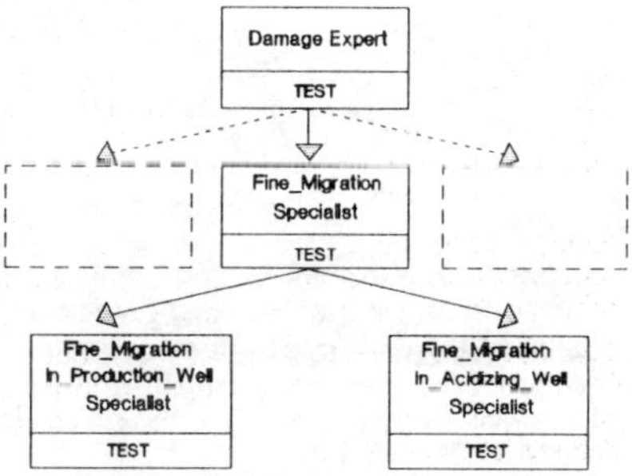

Four major features have been included in the system in order to make the analysis and the diagnosis specific to field applications:

1. Dynamic acquisition of field data: a careful examination of operation reports is necessary with a "detective mind" so that the system output data highlight significant clues of damage;
2. Search for all possible types of damage: plugging action has been considered to be caused by foreign fluids (i.e. mud, completion or workover brine, well treating mixture, injection water) or by precipitates changing the chemical physical initial state of reservoir (i.e. asphaltanes, paraffines, scales) or by movement of solids (i.e. fine migration, sand production);
3. Quantification of reliability factors: the system pinpoints and identifies the severity of each possible type of damage;
4. Suggestions for remedial action: specific optimized designs for stimulation treatments or prevention techniques able to minimize the damage are proposed.

REFERENCES

(1) Krueger, R.F.: "An Overview of Formation Damage and Well Productivity in Oilfield Operations" JPT (Feb. 1986) 131-152

(2) Williams, B.P., Gidley, J.L., and Schechter, R.S.: "Acidicing Fundamentals" Monograph Series, SPE, Dallas (1979)

(3) Chandrasekaran,B: "Generic Task in Knowledge-Based Reasoning: High-Level Building Blocks for Expert System Design" IEEE Expert, Fall 1986, 23-30

(4) Fikes, R., Kehler ,T.: "The Role of Frame-Based Representation in Reasoning" ACM Communications (Sept. 1985) Vol.28 n.9.

POLARIZATION: A NEW TOOL

Ch. CLIET and M. DUBESSET

Institut Français du Pétrole
4, avenue de Bois Préau
92506 RUEIL-MALMAISON CEDEX (France)

For some years now, three-component recordings have allowed access to the true movement of the sensor, the "particle trajectory". Over a time-window containing a seismic arrival, some polarization attributes are computed, that characterize the shape and the angular attitude of the trajectory in space. Several techniques can be used, such as the covariance matrix, for example (1).

This set of parameters is a useful tool for studying and analyzing the signal. It permits numerous applications, in particular, the following:

- spatial directional filtering is used to sort events according to their polarization or to attenuate events with undesired polarization, whether it be in quality or in direction; it also provides an enhancement of the signal-to-noise ratio;

- plotting any parameter of well seismics versus depth is a way of generating pseudo-logs; plotting it versus time permits to get polarization traces, and versus offset to get polarization horizons. Particularly, correlation of pseudo-logs between themselves or with conventional logs (as sonic, e.g.) can reveal very minute lithological variations of the layers that have been traveled through;

- the principal stress directions of a medium can be found by the study of shear-waves birefringence; the six-component "source rotation" method in the time domain, or the "propagator" method in the frequency domain, can help detect and evaluate weak variations of shear-wave polarization;

- polarization determinations also serve to localize unknown microseismic sources for well monitoring or hydraulic fractures mapping.

But these attributes more generally permit other applications such as the study of the behaviour of sources by an estimation of the nature and wealth of their emission, verification of the reorientation of offset-VSPs, reorientation of a sonde in a deviated well, etc.

Some examples of application are presented in several domains, notably, surface and well seismics, reservoir characterization, anisotropy study and hydraulic fracture mapping aid. In these domains, just as in seismology or reservoir monitoring, it is clear that polarization information is not sufficiently exploited; polarization can bring us more, in a lot or petroleum-related techniques.

REFERENCE

(1) CLIET, Ch. and DUBESSET, M, 1988, Polarization analysis in three-component seismics. Geophysical Transactions, 34, 1, 101-119.

COMPLEX : AN EXPERT SYSTEM FOR WELL COMPLETION DESIGN

J. MONNET, C. PRESLES, A. RICORDEAU, D. EHRET and J. VANDEVELDE

Societé Nationale Elf Aquitaine (Production),
C.S.T.J.F. EP/S/PRO/FIP, Avenue Larribau, F-64018 Pau Cedex (France)

1. INTRODUCTION/ABSTRACT

While it takes several days to select appropriate hole completion equipment, the COMPLEX expert system rationalizes and performs the whole task in a few hours by implementing its specific knowledge. This expert system was developed by ELF AQUITAINE for internal use in order to gain time, elaborate more standard completion designs, and avoid costly errors. It works on HP 9000 computer and is supported by the FRAMENTEC S1 inference engine.

2. DESCRIPTION OF THE SYSTEM

Knowledge

Knowledge is shared out in three tools:

- Rules, which are the computer translation of know-how. They were consolidated by several ELF specialists.
- Scientific computation routines, which help support solutions. They consist in existing programs that are automatically activated while remaining invisible to the user. the two mains are OPERA, a diphasic pressure loss calculation, and SCOLIOSE, a tubing stress analysis.
- Equipment file, which includes about 400 equipment from various suppliers, representing the most common tools used by ELF operating subsidiaries over the world. Are concerned casing, tubing, packers, packer/tubing connections, circulation devices, nipples and SCSSV's.

Consultation

An interactive exchange during which system asks questions to user about well data and development program. Through the control blocks that trigger the rules to be applied, the system progresses, from type of perforations to size of wire line nipples, including choice of packers, tubing, SCSSV's and returning backward if necessary. The user can make the decision if various solutions are possible, and in any case at each step when the system presents its recommendations.

Coverage

System can handle a maximum of 2 producing zones, with single, dual or tandem completion and associated equipment for an eruptive oil or gas well. It will be extended soon for gas lifted wells.

Hardware/Software

System runs on a HEWLETT PACKARD 9000 series 835 under unix. The inference engine is Technowledge S1 marketed in FRANCE by FRAMENTEC. Operating softwares are GKS for graphic out-puts and EMPRESS for equipment file. The OPERA and SCOLIOSE programs are in FORTRAN. Total number of rules is about 1000.

Out-puts

The work, is materialized by a report including well data sheets and a drawing of the selected completion. System gives recommendations on type of completion (single, dual...), perforations, choice and depth of packers, size and weight of tubing, SCSSV's, circulating devices and WL nipples. It calculates production parameters (well head pressure, temperature...) and tubing stress.

3. CONCLUSION

Easily accessible to any well completion engineer, this COMPLEX expert system gives safe guide lines in combining knowledge with know-how. Its systematic progression guarantees that no major aspect of the completion problem could be forgotten, but the system is flexible enough to allow the user to direct his choice according to his local contigencies.

CONVERSION OF NATURAL GAS TO LIQUIDS

Session Chairman: G. DONAT, Gaz de France (F)

NATURAL GAS EXPLOITATION: SNAMPROGETTI TECHNOLOGIES NETWORK

D.Sanfilippo and A.Paggini

Snamprogetti S.p.A.
Via Maritano, 26
20097 S.Donato Milanese
Italy

SUMMARY
Natural gas is an energetic resource of increasing interest in the world due to its huge reserves, geographic distribution and intrinsic characteristics.
Presently natural gas is mainly used as fuel. The use as chemical feedstock of its components is limited respect to the potentiality.
This paper summarizes the activities devoted by Snamprogetti to the development, in several years of experience, of a network of technologies related to the chemical conversion of natural gas.
Processes for natural gas purification by acid gas removal, for alcohols formation via syngas route, for dehydrogenation of light paraffins to olefins followed by upgrading to MTBE and other transportation fuels, are already commercialized.
New technologies, including direct methane conversion, are at different levels of R&D.

1. INTRODUCTION
Natural gas is an energetic resource that seems destined to play an increasing role in the world due to its huge reserves, geographic distribution and intrinsic characteristics.
At the beginning of 1990 the estimated proved reserves of natural gas amounted to over 113,000 billion cubic meters, corresponding to 98 billions tons of oil equivalent. That means the same order of magnitude as crude, the reserves of which at the same date were estimated as 135 billions TOE (1). The average ratio of 0.73 between the world gas and oil reserves varies widely in different countries ranging, for instance, from 0.1 in Kuwait up 2.2 in Algeria.
However, natural gas consumption represents about one half of the oil one: in 1989 1.76 billions TOE (2) as compared with 3.0 of oil (3). This means that the driving force towards the use of natural gas for oil substitution will increase, whenever possible.
Today over 90% natural gas is used as fuel.
As a chemical feedstock, the use of methane is limited to the production of hydrogen for ammonia, methanol and derivatives, some chlorinated compounds and other minor productions.
Other light natural gas paraffins have had only few exploitations apart from their direct use as fuel: ethane steam cracking, butane to maleic anhydride, to butadiene and, very recently, to isobutene are the most important examples.

Undoubtedly, the low chemical reactivity of all light paraffins, their location which is often remote, and the availability at a competitive price of alternative feedstocks have strongly limited their use in the petrochemical industry.
On the other hand, the changed scenario of the chemical commodities producer countries and the increasing tendency of developing countries to better exploit their internal resources, and not only for captive utilization, have led to the development of technologies aimed at transforming natural gas components into more valuable or transportable products.
In terms of future prospects, there is plenty of space for growth in the use of natural gas even taking into account the economic competitivity of other available resources.

Natural gas is often a complex mixture and its chemical composition varies widely from reservoir to reservoir all over the world. Main component is generally methane, but ethane, propane, butanes, inert gases (like nitrogen or helium) and acid gases (CO_2 or H_2S) often represent very important components that a strategy of a carefully designed use of natural gas has to take into consideration.
Snamprogetti's strategy for economically attractive transformation of natural gas takes into consideration the complexity of its composition and, during the last twenty years, has assembled a network of technologies aimed at making available wider possibilities of natural gas exploitation.

Snamprogetti technologies for NG

Fig. 1

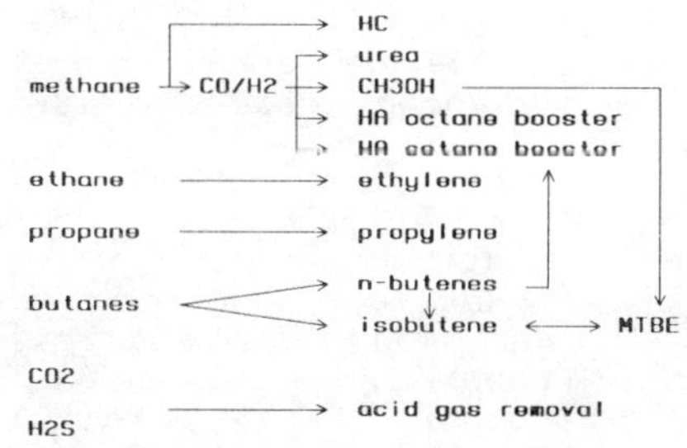

Fig. 1 shows the main lines of technological development activities pursued by the Company. Some of these are at explorative level, some are at different stages of development and some have reached the commercialization phase and the industrial implementation:

- purification of natural gas from CO_2 and/or from the toxic H_2S,
- activation of light paraffins through their conversion to olefins and technologies related to further exploitation
- indirect conversion (through synthesis gas) to chemical products like urea, methanol, and higher alcohols,
- direct conversion of methane to liquid products.

2. ACID GAS REMOVAL

Several wells produce natural gas with a considerable content of CO_2 and/or H_2S that heavily limit its use even as fuel: CO_2 lowers its heating value and H_2S is toxic, cannot be burned, is a poison for all catalysts and then it has to be removed.

Cryofrac

The most energy demanding step in CO_2 separation is the CO_2/CH_4 split. Snamprogetti has developed a technology, called CRYOFRAC (4), for its removal from:

- medium to ultrahigh (>25%) acid gas concentration,
- gaseous streams from Enhanced Oil Recovery and similar.

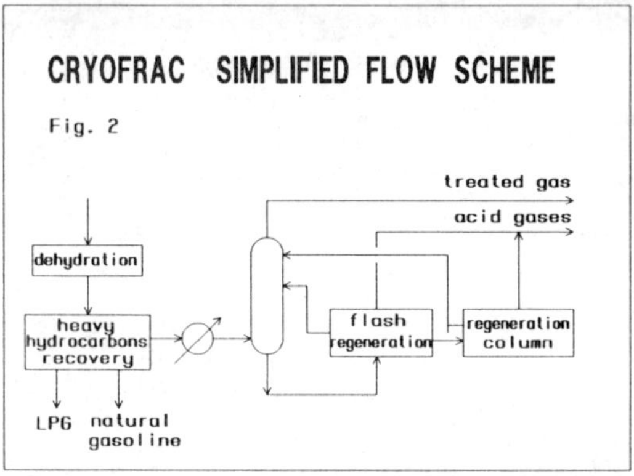

Fig. 2 shows a simplified flow scheme of the process, that is however generally tailored to specific local requirements.

A gas liquid absorption tower operating at low temperature (200 K) and high pressure (4.0 Mpa) gives lean gas at the top. Solvent assures the complete absence of solid CO_2, high CO_2 transport capacity and high selectivity in CO_2/HC separation. Solvent is regenerated in a second column giving a concentrated stream of acid gas. Solvent used can be alcohols, glycols, esters, ethers, hydrocarbons or mixtures.

The process is optimized as concerns energy recovery by utilizing the expansion of separated CO_2.

CRYOFRAC has been demonstrated at the pilot plant stage.

Selefining

For the selective removal of H_2S from gases containing also CO_2, Snamprogetti has developed a technology called SELEFINING (5).

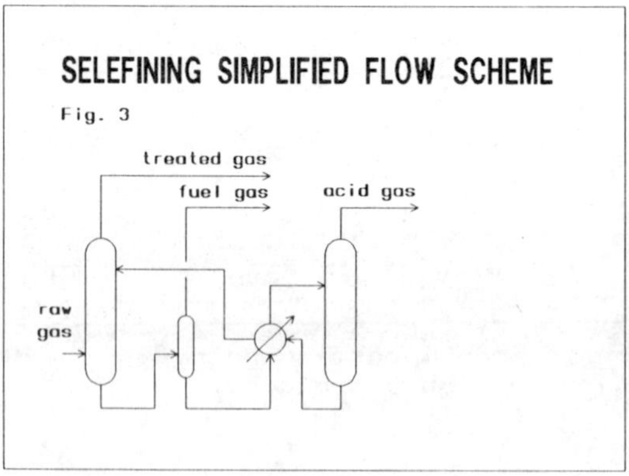

Fig. 3 shows the flow scheme of this process that is based on the traditional scheme of a gas liquid absorption tower and a regeneration column. A particular feature of this technology is that H_2S/CO_2 separation is controlled by thermodynamics and not by kinetic factors. Concentrated H_2S coming from the top of the regeneration column can be treated in a traditional Claus plant.

Solvent is a mixture of tertalkanolamines in organic solvent with the presence of some water.

SELEFINING is at present used for H_2S removal from natural gas in a plant in Southern Italy treating 1 Mm^3/d. Another plant of this type has been realized to sweat 19 Mm^3/d of associated gas from an off-shore oil field in Libya.

Total Deacidification
For the removal of H_2S and CO_2 present in natural gas (or syngas) at low-medium concentration (≤25%) Snamprogetti has developed a new technology called GTD process (Gas Total Deacidification). The process scheme is the same of SELEFINING with a gas - liquid adsorption and a regeneration column. Solvent is a water containing promoted tertalkanolamine. In comparison with competing processes, GTD offers very low solvent circulation flow rate, higher solvent capacity and then lower investment costs and lower utilities consumption. GTD has reached the pilot plant stage.

3. LIGHT PARAFFINS ACTIVATION

MTBE (and other ethers) Process
The chemistry of C_4s has always greatly interested Snamprogetti. This aspect, together with the Company's activities for lead phase down from gasoline pool, led to the development of MTBE process and to the design and the construction of the first industrial plant in the world, which was started up in Ravenna in 1973. Since then the Snamprogetti process has realized the most significative applications: the world largest plant (500,000 MTPY), first in the use of all possible feedstocks like cat cracking butilenes, butadiene-rich streams, dehydrogenated field butanes (6),(7),(8),(9).
Over 50% of the world's MTBE installed capacity is still based on the Snamprogetti technology, a simplified flow scheme of which is reported in fig.4.

MTBE PROCESS FLOW SCHEME

Fig. 4

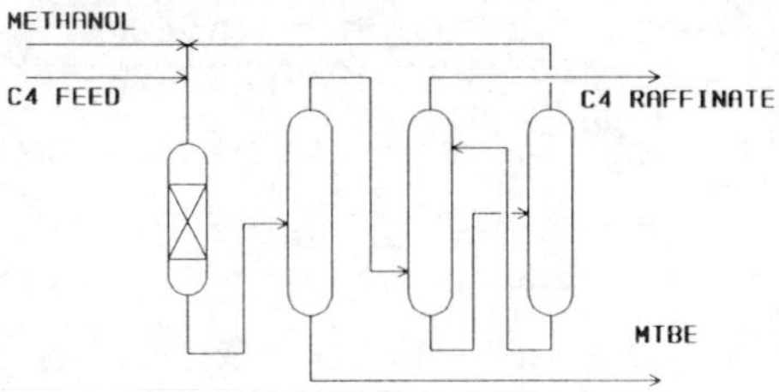

MTBE process has been easily extended to other ethers (TAME, ETBE, ...). Today MTBE represents one of the best exploitation of wet natural gas even in remote sources being fully accepted by refiners and since its very good blending characteristics lend it a higher market value respect to premium gasoline.

Hiqh purity isobutene (99.99%)

In the frame of the MTBE technology maintenance, Snamprogetti has also developed a process for high purity isobutene production via MTBE decomposition. A simplified flow scheme is reported in fig.5. The first plant of this kind with a capacity of 1,500 t/y was started up in 1987 in Hungary (10). A 50,000 MTPY plant will start up before the end of 1990 in South Korea and another in Hungary is under design.

HIGH PURITY ISOBUTENE PROCESS SCHEME

Fig. 5

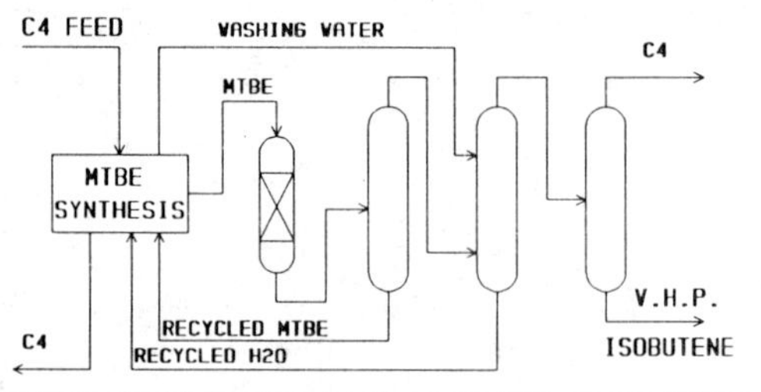

Butenes isomerization

We have studied a catalyst and a process scheme for the skeletal isomerization of n-butenes present in several upgradable streams to isobutene in order to increase its availability to produce MTBE (11).

Paraffins dehydrogenation

In this connection, the foreseen shortage in isobutene connected with a heavy utilization of refinery and steam cracker streams together with the increasing demand for MTBE and a potential increasing interest for propylene, encouraged us to look for new sources of olefins through the dehydrogenation of natural gas paraffins.

In the early 80's Snamprogetti and Yarsintez of Yaroslavl entered into an agreement for the improvement of the Soviet isobutane dehydrogenation technology, that has been applied since early 60's in 34 industrial units presently running in the USSR.

ISOBUTANE DEHYDROGENATION PROCESS

Fig. 6

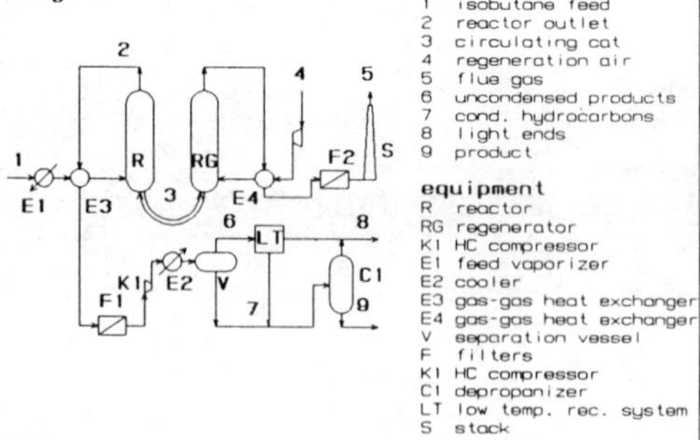

The dehydrogenation scheme, reported in fig.6, can be applied to all light paraffins.

Being the dehydrogenation reaction highly endothermic and limited by thermodynamics that imposes very high temperatures in order to reach economically acceptable conversions, very large quantities of heat at a high temperature have to be supplied, while at the same time reducing side reactions.

The conceptual design of the reactor and nature of the catalyst are therefore of paramount importance.

PARAFFINS DEHYDROGENATION CYCLE

Fig. 7

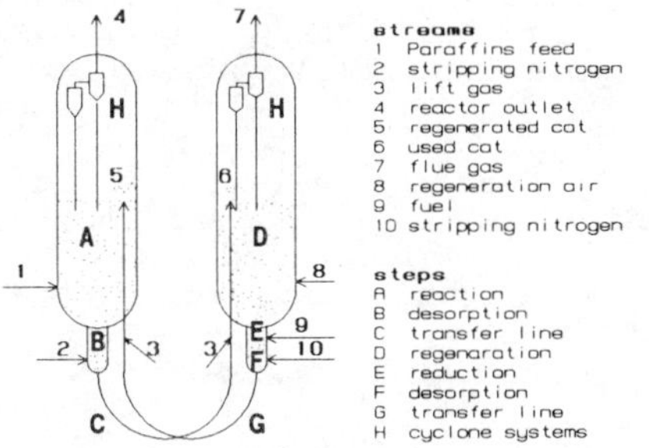

The core of the technology is the reactor - regenerator system (Fig.7), based on the fluidized bed concept.
Hydrocarbons are fed and well distributed in the lower part of reactor; they are dehydrogenated according to thermodynamic equilibrium with very high selectivity (≥90% mol.) to the corresponding olefins and leave the reactor after separation of the entrained catalyst.
Reaction heat is supplied by the heat capacity of circulating catalyst. Catalyst leaves the bottom of the reactor and is transferred to the regenerator where deposited coke is burned off and the catalyst is heated up to the temperature required by the heat balance (12).
Besides the existing plants the improved technology has been selected for a new plant to be built in Lithuania.

4. INDIRECT METHANE CONVERSION

The value of natural gas has been always significatively lower respect to the obtainable energetic products, particularly for remote sources.
The three energy price shocks of 1973, 1980 and 1986 had a significant impact on worldwide R&D strategies.
Interest in natural gas exploitation, especially in its transformation into automotive fuel, grew whenever there was a considerable gap between the valorization of natural gas heating value and that of the other sources.
Remote natural gas can reach developed countries also as LNG. Internal evaluations estimated a minimum crude break-even price of 32 $/bbl to allow this solution.
Besides LNG, there are several possibilities for the conversion of methane into a liquid product, for example through synthesis gas.
The production of synthesis gas and its further conversion into ammonia, methanol, and Fischer - Tropsch hydrocarbons represents the history of Chemistry.
Syngas chemistry, requiring strong investments and some energy consumption for synthesis gas production (about 30% in the simplest case of methanol) has known periods of increasing or decreasing interest according to the variation of oil prices.

<u>Methanol and higher alcohols (MAS) Process</u>

Since the late 60's Snamprogetti has been interested to the use of methanol in gasoline pool and, in the late 70's, together with Haldor Topsoe A/S and Enichem undertook considerable R&D effort to develop the MAS process for the

direct cosynthesis of methanol and higher alcohols, these latter identified by us (13) as the best solution to overcome the problems of methanol containing gasolines concerning water tolerance and vapour pressure. Being MAS a new product, it was necessary to join a wide range of expertise and know-how in the fuel world: evaluation, production, distribution, marketing, etc.
We chose the way of methanol synthesis catalysts modifications, since we were interested in branched higher alcohols. The catalyst identified was able to produce with a very high selectivity a 50/50 to 80/20 by weight mixture of methanol and higher alcohols, mainly isobutanol.
The product was evaluated, both as such and blended with gasoline, with regard to its chemical characteristics, toxicity, engine behaviour, impact on refinery and on the distribution network.
The classical step of a big pilot plant was bypassed and, with a scaling up factor of 500,000, we jumped directly from our laboratory once through, isothermal reactors to an industrial plant having a capacity of 15,000 MTPY.
This plant, located in Southern Italy, has been running successfully in the period 1982 - 1986. The depressed gasoline price forced the termination of experiments.
The MAS produced was sold blended with gasoline on the market in a network of service stations kept under control in order to test also potential distribution problems.
Distribution test has been fully successful.

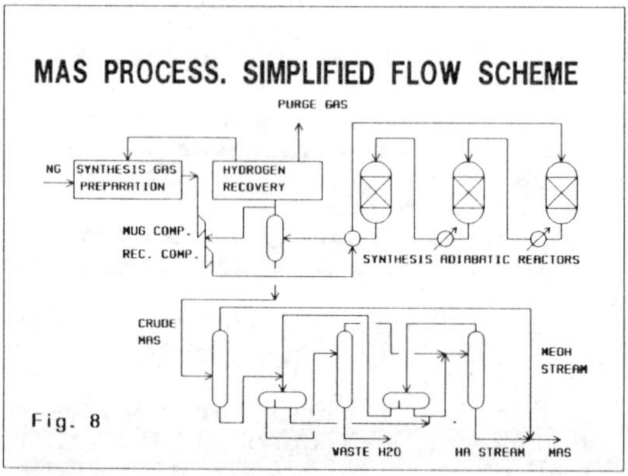

Fig. 8

Fig.8 shows a flow scheme of the MAS process (14). Although this technology is considered mature, R&D activities are moderately in progress in order to look for improved process economics.
MAS technology represents one of the most efficient ways of converting methane into liquids, but today's oil prices do not make the syngas way to liquid fuel very attractive.

<u>Higher linear alcohols</u>
Another R&D activity that we have been performing in the field of motor fuels concerns the synthesis from syngas of linear primary alcohols containing more than 4 carbon atoms for extenders in diesel fuel.
The method adopted is the modification of Fischer - Tropsch catalysts with a view of suppressing hydrocarbons formation and improve the oxygenated production. The results already obtained support this approach.

Low temperature methanol synthesis

Although highly debated between EPA and API, the Clean Air Act for Alternative Fuels, is foreseeing the use of neat methanol (M85) as alternative fuel.

This energetic use of methanol, to be competitive with existing fuels, requires a considerably lower production cost. In existing technologies the major costs (≥50%) are associated to the syngas preparation. A significative reduction in these costs can be achieved by the use of the partial oxidation with air, but this requires, because of the nitrogen presence, much higher conversion per pass respect to the actual heterogeneous catalysts based processes. Since the methanol synthesis is adversely affected by the increasing temperature, we are active respect to the study of a very low temperature methanol synthesis process.

5. DIRECT METHANE CONVERSION

The potential advantages of a direct conversion of methane to produce higher homologues have induced an increasing number of R&D centres to study its direct functionalization.

Sharing these concepts, even with low energy prices and strong worldwide competition in this line, we have set up a long-term R&D program with the purpose of finding new economic ways for the production of higher hydrocarbons or chemicals.

By developing and adopting a theoretical approach to catalyst design using new techniques based on microkinetic considerations on reaction mechanisms (15) we are trying to rationalize experimental results and to schedule the next step to be taken.

The main lines of this R&D work that is currently under implementation at bench scale concern:

- selective ethylene formation via oxidative coupling,
- non selective higher hydrocarbons formation,
- new syngas production ways.

6. CONCLUSIONS

One of the most important consequences of the existing differences in the relative economic costs of gas and petroleum products in gas-rich countries (compared to the developed countries) is that they might develop rather different, gas based productions and systems that would not be economical elsewhere.

However, it is probable that the technologies more indicated in the exploitation of natural gas will be the ones able to give products having or a substantial added value or a "premium" in terms of quality, since in industrialized countries the market price of natural gas will allow only high value chemicals and in gas-rich developing countries the transport costs will in practice hinder the production of simple "extenders" for exportation.

An optimized use of technologies and know-how would help mobilize part of the huge natural gas resources to meet a number of objectives of vital importance to both industrial

and developing countries that could boost their industrial development and alleviate chronic deficit of foreign exchange.
It may well be highly beneficial to introduce the new technologies on a integrated basis, in order to benefit from economies of scale.

REFERENCES

(1) Oil & Gas Journal 25.12.1989
(2) Cedigaz News Report, March 16, 1990
(3) Market Economy Bulletin, 75 Feb. 1990. + Planecon, Winter 1990: Long Term Energy outlook.
(4) L.GAZZI, R.D'AMBRA, R.DICINTIO, C.RESCALLI and A.VETERE: Treat high acid gases; Hydr. Proc. July 1984,99-103.
(5) L.GAZZI, and C.RESCALLI: A MEA-DEG Plant successfully retrofitted to Selefining; Energy Progress, 8, (1988), 113-117.
(6) F.ANCILLOTTI, M.MASSI MAURI and E.PESCAROLLO: Mechanism in the reaction between olefins and alcohols catalyzed by ion exchange resins; J. Mol. Cat., 4 (1978) 37-48.
(7) T.FLORIS, G.PECCI and G.ORIANI: Snamprogetti/Anic MTBE process benefits unleaded gasoline refiners; NPRA Annual Meeting March 22-25, 1979, San Antonio, Texas
(8) A.CLEMENTI, G.ORIANI, F.ANCILLOTTI and G.PECCI: Upgrade C_4s with MTBE process; Hydr. Proc. Dec. 1979, 109-113.
(9) F.ANCILLOTTI, E.PESCAROLLO, E.SZATMARI, L.LAZAR: MTBE from butadiene-rich C_4s. Hydr. Proc. Dec. 1987, 50-53.
(10) O.FORLANI, V.PICCOLI, S.ROSSINI, D.SANFILIPPO: A new catalyst for the selective MTBE decomposition to high purity isobutene. Italian Chem. Soc. Meeting, Oct. 1990
(11) US Patent 4,038,337 to Snamprogetti (1977)
(12) F.BUONOMO, G.FUSCO, D.SANFILIPPO, G.R.KOTELNIKOV and R.A.MICHAILOV: The Fluidized Bed Technology for paraffins dehydrogenation. Snamprogetti - Yarsintez Process. 1990 Petrochemical Review DeWitt & Co. Inc. Conference
(13) G.TERZONI, R.PEA and F.ANCILLOTTI: Improvement of the water tolerability of methanol gasoline blend; III Int. Symposium on Alcohol Fuels, Asilomar, Ca., May 1979
(14) A.PAGGINI and D.SANFILIPPO: Implementation of the Methanol plus higher alcohols process "MAS Technology". VII Int. Symposium on Alcohol Fuels, Paris Oct.1986
(15) O.FORLANI, M.LUPIERI, V.PICCOLI, S.ROSSINI, D.SANFILIPPO, J.A.DUMESIC, L.A.APARICIO, J.A.REKOSKE and A.A.TREVINO: General mechanism for the oxidative coupling of methane. New Dev. in Selective Oxidtn. Elsevier, Studies in Surf. Sc. and Cat., 55, 1990.

THE OXIDATIVE COUPLING OF METHANE IN A RECIRCULATING FAST FLUID BED REACTOR

S.J. Korf, J.A. Roos, P.T. Coolen,
J.G. van Ommen and J.R.H. Ross

Faculty of Chemical Technology,
University of Twente,
PO Box 217,
7500 AE Enschede,
The Netherlands

SUMMARY

The oxidative coupling of methane over Li/MgO, Na/CaO and PbO/Al_2O_3 has been studied in a Recirculating Fast Fluid Bed Reactor ("Riser"). Due to the agglomeration of catalyst particles it was not possible to operate the "Riser" in a satisfactory way using Li/MgO and Na/CaO. With a PbO/Al_2O_3 catalyst it was possible to obtain a good particle flow in the "Riser". A comparison has been made between the operation in a fixed bed and a "Riser" using the PbO/Al_2O_3 catalyst. It was concluded that both the oxidation of methane and the oxidation of the C_2 products are responsible for the CO_x formation in the case of a PbO/Al_2O_3 catalyst.

INTRODUCTION

The production of higher hydrocarbons (principally ethane and ethylene) directly from methane by catalytic oxidative coupling is a novel methane conversion process which is currently the subject of considerable research [1-5].

The overall oxidative coupling process is highly exothermic. As we have shown that ethane and ethylene are susceptible to further oxidation, optimum selectivity is reached under plug flow conditions [6]. To reach plug flow, a relatively high rate of gas flow is needed. The combination of these factors can cause the occurrence of (severe) hot spots in the catalyst bed, especially at high oxygen concentrations. To avoid the formation of hot spots, we studied the oxidative coupling of methane in a so called "Recirculating Fast Fluid Bed" Reactor ("Riser"). This type of reactor has the advantage of better temperature control under plug flow operation conditions.

We studied the behaviour of three catalyst systems in the "Riser": Li/MgO, Na/CaO and PbO/Al_2O_3. The use of Li/MgO and Na/CaO catalysts in the "Riser" gave rise to large problems; the abrasion resistance of these catalysts was not high enough and furthermore the catalyst particles were found to agglomerate. It would appear that both molten carbonate and hydroxide, mobile and volatile under the reaction conditions used were responsible for the agglomeration effect. For this reason we were not able to run the "Riser" in a satisfactory way using Li/MgO or Na/CaO. We therefore concentrated on PbO/Al_2O_3, despite the fact that this catalyst is not as effective in the production of ethane and ethylene as is Li/MgO or Na/CaO. With the PbO/Al_2O_3 catalyst we made a comparison between the operation in a fixed bed and a "Riser". We also studied the effect of oxygen partial pressure in the "Riser". From these measurements and from measurements where we added ethane and ethylene to the reaction feed in a fixed bed reactor containing PbO/Al_2O_3 we concluded that both the oxidation of methane and the oxidation of the C_2 products are responsible for the CO_x formation in the case of a PbO/Al_2O_3 catalyst.

EXPERIMENTAL

Catalysts

The PbO/Al_2O_3 catalyst was prepared by wet impregnation of γ-Al_2O_3 (Ketjen 240 m^2/g) with $Pb(NO_3)_2$. After drying at 110°C, the material was calcined at 850°C for 6h. After calcination, the catalyst contained 23 wt% Pb and the surface area of the catalyst was 43 m^2/g. XRD showed the presence of $PbAl_2O_4$, PbO and γ-Al_2O_3.

Li/MgO and Na/CaO catalysts were prepared by wet impregnation [7,8] in the presence of a stream of CO_2. After drying at 140°C, the samples were calcined in air at temperatures of 850°C and higher.

Catalytic Experiments

Set up of the Riser Figure 1 shows the arrangement of the Riser system. The construction material was quartz. The particle size range used for the catalyst was $0.2 < d_p < 0.5$ mm. The gas flow through the reactor was ca. 42 $cm^3(STP)s^{-1}$, while the total aeration flow was 1.67 $cm^3(STP)s^{-1}$. The reactor feed consisted of methane, oxygen and helium. In a typical experiment, 10 g of catalyst was used.

Operation of the Riser

Bed Porosity By stopping simultaneously the aeration gas flows and the gas flow through the Riser both the catalyst flow to and out of the reactor stops. After that the length of the catalyst bed in the Riser can be measured. The W/F value and the bed porosity can be calculated from the catalyst volume in the Riser. Using the conditions stated above the value found for W/F was ca. 0.25 $gscm^{-3}(STP)$ and the value for the bed porosity was ca. 90%.

Flow of Catalyst Circulation By stopping the aeration flow the catalyst flow to the Riser stops. However the transport of catalyst particles out the reactor will continue for some time. The catalyst flow can be determined by measurement of the initial velocity of the increase of the catalyst level in the standpipe after stopping the aeration flow. Measurement of the initial velocity is necessary because the velocity decreases with time due to the decreasing bed porosity. This method was not very well applicable for the PbO/Al_2O_3 catalyst because after stopping the aeration flow the accumulation of the catalyst particles became more dense and in that case the catalyst level was decreasing instead of increasing. Only after the decrease in catalyst level, measurement of the circulation flow was possible. However the velocity had already decreased then. The values found for PbO/Al_2O_3 were in fact minimum values. A typical circulation flow was ca. 3 g/s (catalyst flux: ca. 60 kg/m^2s). The residence time in the Riser could be calculated from the quotient of the amount of catalyst in the Riser and the circulation flow of the catalyst.

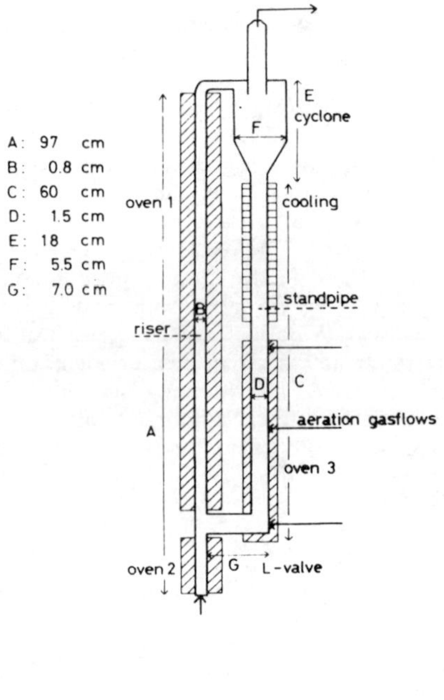

Figure 1: Schematic representation of the "Riser" reactor

Fixed bed The reactor consisted of a quartz tube (5 mm i.d.) heated in an electric oven. Two sets of reaction conditions were used to compare with the results in the Riser: (i) a low superficial gas velocity, using a catalyst weight, W, of 0.093 g and a gas flow rate, F, of 0.42 $cm^3(STP)s^{-1}$; and (ii) a high specific gas velocity with W = 0.750 g and a flow F of 3.33 $cm^3(STP)s^{-1}$. The reactor feed consisted of methane, oxygen and helium; in some experiments, ethylene (see Fig.5) or ethane (see Fig.6) was also added to the reactor feed. The experiments with addition of C_2H_4 and C_2H_6 to the gas feed were carried out with a catalyst weight of 0.075 g and a gas flow of 1.67 $cm^3(STP)s^{-1}$.
For each experiment in both types of reactor, the composition of the gas feed is given in the text. All gases were analyzed by gas chromatography [9].

RESULTS AND DISCUSSION

Preliminary Results of the Oxidative Coupling of Methane over Li/MgO, Na/CaO and PbO/Al_2O_3 in a Riser.

The catalyst which is used in the "Riser" should have a high abrasion resistance; this is because the catalyst is transported with high velocity and the frictional forces with the walls of the system are relatively high, especially in the Riser reactor itself and in the cyclone.
Li/MgO To prepare a Li/MgO catalyst with a relatively high abrasion resistance we usually calcined these materials at 850°C or 950°C for 20h. This resulted in a relatively inactive catalyst when compared to a standard Li/MgO catalyst which was calcined for 6h at 850°C [7]. However flaking which produced very fine particles was still a problem. A major problem with the use of the Li/MgO catalyst was found to be the agglomeration of catalyst particles in the "Riser" and the standpipe. It would appear that both molten, Li_2CO_3 and LiOH, mobile and volatile under the reaction conditions used, are responsible for this effect. Following various attempts to counteract these problems (varying preparation, calcination temperature et cetera) it was found that addition of quartz particles gave the best results in particle flow through the reactor (dilutions up to 70% were used). However due to the poor thermal conductivity of quartz, desired reaction temperatures (ca. 800°C) were never obtained. Such high temperatures were necessary in the case of these Li/MgO samples, due to their poorer activities relative to the standard Li/MgO samples. No reaction was observed during the experiments using quartz dilution.
Na/CaO With Na/CaO the same problems were found as with Li/MgO. In this case the addition of quartz particles enhanced the agglomeration of catalyst particles and was thus not useful. This result was also found by Andorf and Baerns [10]. It was found to be possible to run the "Riser" (with some difficulty) with a Na/CaO catalyst which was prepared by wet impregnation of limestone (Duwa-Calcaire) with Na_2CO_3. A typical result at 760°C is shown in Table I. However during the measurements hot spots of particles which sticked together were found in the upper part of the "Riser".

Table I Oxidative coupling of methane in a "Riser". Gas composition: P_{CH4} = 50 kPa, P_{O2} = 10 kPa, balance He. T_R = 760°C.

Catalyst	C_2 selectivity	C_2 yield
Li/MgO	-	-
Na/CaO	67	4.2
PbO/Al_2O_3	50	8.3

Because of the problems described above we were not able to run the "Riser" in a satisfactory way using Li/MgO or Na/CaO.

<u>PbO/Al_2O_3</u> With a PbO/Al_2O_3 catalyst it was possible to obtain a good particle flow in the "Riser". Table I gives a typical result at 760°C. We therefore concentrated on PbO/Al_2O_3, despite the fact that this catalyst is not as effective in the production of ethane and ethylene as is Li/MgO or Na/CaO [3,7,8,9].

<u>Comparison of Catalytic Results over PbO/Al_2O_3 in a Riser and a Fixed Bed Reactor</u>

We have shown previously for Li/MgO [6] and doped Sm_2O_3 [11] catalysts that both the conversion and selectivity are increased considerably if the superficial gas velocity is increased while the contact time in a fixed bed reactor is maintained constant. These improvements were thought to be due to better plug-flow behaviour and a consequent reduction of the degree of further oxidation of the C_2 products to carbon oxides.

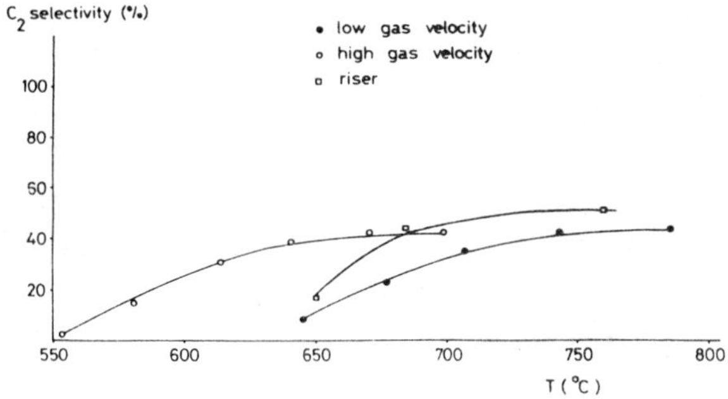

Figure 2: Variation in % C_2 Selectivity with both reaction temperature and space velocity for the PbO/Al_2O_3 system

Figure 2 and Figure 3 show the C_2 selectivities and the C_2 yields respectively as a function of the <u>oven</u> temperature for the PbO/Al_2O_3 catalyst operated in a fixed bed reactor at both high and low gas velocity conditions and in the Riser reactor. Essentially the same value of W/F was used in all three cases and the gas feed consisted of 50 kPa CH_4, 10 kPa O_2, the balance being He. Under the high gas flow conditions, the "light-off temperature" decreased by some 100°C. This effect seems to be due to the development of hot-spots in the catalyst bed; measurements in which a thermocouple attached to the outside wall of the reactor was moved along the bed showed the development of a hot-spot of up to 150°C. However, in contradiction to previously reported results for Li/MgO and the doped Sm_2O_3 catalysts there was no increase in C_2 selectivity or C_2 yield when using high gas-flow rates. On the other hand under the plug flow conditions of the Riser reactor there was a (small) increase in both C_2 selectivity and C_2 yield. The lack of an increase in C_2 selectivity and C_2 yield at the high flow conditions is surprising and seems to indicate that a large part of the CO_x observed must be formed by the direct oxidation of methane or an intermediate methyl species i.e. in a parallel route rather than by oxidation of the C_2 species in a consecutive route. The small increase in the C_2 selectivity and C_2 yield observed in the Riser reactor may be due to rather slower deactivation of the catalyst in this system than in the fixed bed reactor; Roos et al. [9] and Baerns et al. [12] have shown that the disadvantage of this catalyst is its low stability due to the loss of lead. In the presence of a hot spot present in the fixed bed reactor, the loss of lead will be accelerated compared with the Riser reactor, where the temperature profile is more uniform. Furthermore the high temperatures present in a hot spot can enhance the total oxidation.

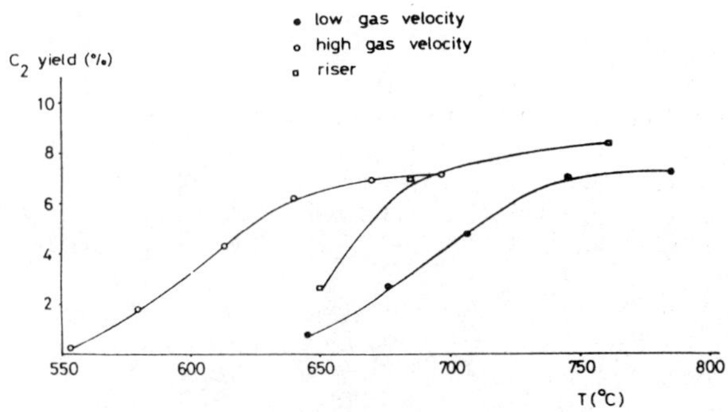

Figure 3: Variation in % C_2 Yield with both reaction temperature and space velocity for the PbO/Al_2O_3 system

The Effect of the Concentration of Oxygen in the Gas Feed of the Riser

A number of experiments were carried out to show the effect of the partial pressure of oxygen in the feed of the Riser reactor on the product composition and the results are shown in Figure 4. The partial pressure of CH_4 in the feed for these experiments was kept at 50 kPa. The following conclusions can be drawn from the results shown in Figure 4:

i) the C_2H_4/C_2H_6 ratio increases with increasing P_{O2};

ii) there is no production of CO; and

iii) the partial pressure of CO_2 is more affected by the partial pressure of oxygen in the reactor feed than are the partial pressures of the C_2 products, the sum of which remains almost constant with increasing P_{O2}.

The increase in the C_2H_4/C_2H_6 ratio can be explained by an increase in the contribution of the oxidative dehydrogenation of the initial product, C_2H_6, with increase in oxygen concentration. To clarify this point, the experiments in the following section were carried out.

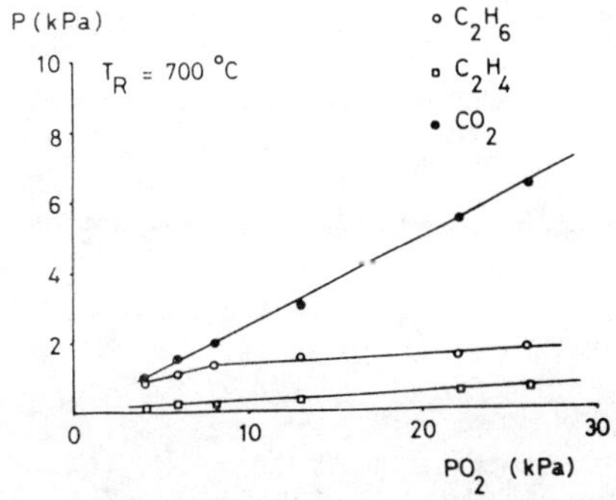

Figure 4: Product distribution as a function of O_2 partial pressure in the "Riser" reactor containing PbO/Al_2O_3

Addition of Ethylene and Ethane to a Gas Mixture Containing Methane and Oxygen

Figure 5 shows the C_2 concentration at the exit of the reactor as a function of the reaction temperature in a series of experiments in which 5.7 kPa of ethylene was added to the gas feed containing 62 kPa CH_4, 12 kPa O_2, the balance being He. Figure 6 shows equivalent results for a series of experiments in which 4.7 kPa of ethane was added to the gas feed. If the C_2 products were not oxidised and the coupling reaction of methane was not influenced by the addition of ethane or ethylene, the total C_2 concentration in the effluent would be the sum of the C_2 concentration produced from the methane and that of C_2H_4 or C_2H_6 added to the feed (curve 1, open squares, ▫). That substantial oxidation of the C_2 products also took place follows from the observed behaviour of the total concentration of C_2 products in the effluent as a function of the reaction temperature (curve 2, filled circles, •). If there were no oxidation of ethane or ethylene, curve 1 and curve 2 would be the same. The fact that this is obviously not so in the cases of either C_2H_4 or C_2H_6 confirms the conclusion that total oxidation of the C_2 products must occur.

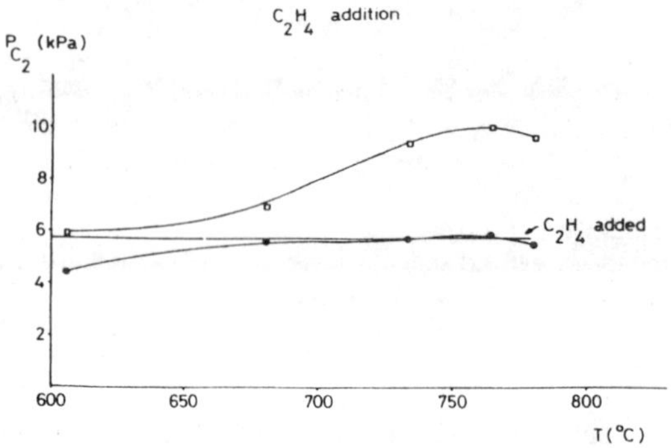

Figure 5: C_2 concentration at exit of "Riser" reactor as a function of reaction temperature (C_2H_4 added to the feedstream)

At the lower reaction temperatures shown in Figures 5 and 6 there was no conversion of methane (T < 650°C). The only reaction which occurred at this temperature was the oxidation of ethane and ethylene; in this situation the partial pressure of C_2 in the reactor feed was higher than the partial pressure of C_2 at the exit of the reactor. At higher reaction temperatures, the methane coupling reaction started but the total amount of C_2 products barely exceeded the amount of ethane or ethylene added to the gas feed. In comparison with experiments without added ethane or ethylene, the methane conversion is lowered by competition for the available oxygen between the methane and the added C_2 hydrocarbons. In agreement with measurements which we have carried out using Li/MgO and Sm_2O_3 catalysts [13], we conclude that the C_2 products are not stable under methane coupling reaction conditions.

Work carried out recently in our laboratory [14] has shown that doped Li/MgO catalysts (where the dopant may be SnO_2, CoO, etc.) offer considerable advantages over both the undoped system and other systems as they operate at considerably lower temperatures than do Li/MgO materials while maintaining the high selectivities given by Li/MgO. Such systems are likely to be applicable in the riser reactor. Further work is in progress on these systems, not only for use in riser reactors but for more conventional reactors, for which the results are equally promising.

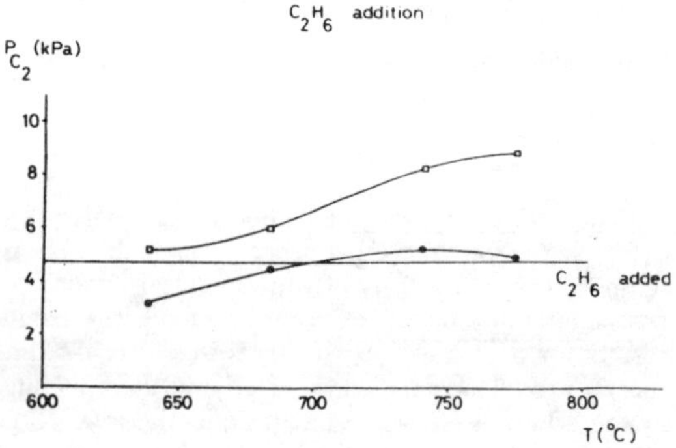

Figure 6: C_2 concentration at exit of "Riser" reactor as a function of reaction temperature (C_2H_6 added to the feedstream)

CONCLUSIONS

1) It is possible to operate a Riser reactor with a PbO/Al_2O_3 catalyst.

2) The results presented here for a PbO/Al_2O_3 catalyst show that both the oxidation of methane and the oxidation of the C_2 products are responsible for the CO_x formation.

3) Further research is necessary to prepare a good catalyst for a Riser reactor system. Results with a Li/Sn/MgO system look promising.

ACKNOWLEDGEMENTS

S.J.K. thanks the Dutch Foundation for Scientific Research for financial support. We also thank the Non-Nuclear Energy programme of the European Community for partial support of the work (Contract No. EN3C-039-NL (GDF)).

REFERENCES

(1) T. Ito, J-X Wang, C-H Lin and J.H. Lunsford, J. Amer. Chem. Soc. 107 (1985) 5062.
(2) J.A. Sofranko, J.J. Leonard and C.A. Jones, J. Catal., 103 (1987) 302.
(3) W. Bytyn and M. Baerns, Appl. Catal., 28 (1987) 199.
(4) J.M. Deboy and R.F. Hicks, J. Catal. 113 (1988) 517.
(5) C. Miradatos, A. Holmen, R. Mariscal and G.A. Martin, Catal. Today 6 (1990) 601.
(6) J.A. Roos, S.J. Korf, A.G. Bakker, N.A. de Bruijn, J.G. van Ommen and J.R.H. Ross, in "Methane Conversion", ed. D.M. Bibby, C.D. Chang, R.F. Howe and S. Yurchak, Elsevier SSSC Series, 36 (1987) 427.
(7) S.J. Korf, J.A. Roos, N.A. de Bruijn, J.G. van Ommen and J.R.H. Ross, Catal. Today, 2 (1988) 535.
(8) S.J. Korf, J.A. Roos, N.A. de Bruijn, J.G. van Ommen and J.R.H. Ross, Appl. Catal., 58 (1990) 131.
(9) J.A. Roos, A.G. Bakker, H. Bosch, J.G. van Ommen and J.R.H. Ross, Catal. Today, 1 (1987) 133.
(10) R. Andorf and M. Baerns, Catal. Today, 6 (1990) 445.
(11) S.J. Korf, J.A. Roos, J.M. Diphoorn, R.H.J. Veehof, J.G. van Ommen and J.R.H. Ross, Catal. Today, 4 (1989) 279.
(12) W. Hinsen, W. Bytyn and M. Baerns, Proc. 8th. Int. Congr. Catal., Berlin, (1984) 3 581.
(13) J.A. Roos, S.J. Korf, R.H.J. Veehof, J.G. van Ommen and J.R.H. Ross., Catal. Today 4 (1989) 441.
(14) S.J. Korf, J.A. Roos, L.J. Veltman, J.G. van Ommen and J.R.HG. Ross, Appl. Catal., 56 (1989) 119.

IFP PROCESSES FOR THE DIRECT CONVERSION OF METHANE INTO HIGHER HYDROCARBONS

J. Weill(*), C.J. Cameron and C. Raimbault

Institut Français du Pétrole
B.P 311, 92506 Rueil-Malmaison, France
(*) B.P 3, 69360 Vernaison, France

Abstract

This paper describes two original approaches developed by IFP to the direct chemical conversion of natural gas. The first one is the oxypyrolysis of natural gas, which combines, in a first step, the oxidative coupling of methane, a heterogeneously catalyzed reaction producing ethane as a primary product, and, in a second step, the cracking of ethane, a homogeneous gas-phase reaction producing ethylene and hydrogen.

The second approach is the thermal coupling of methane, mainly producing acetylene, ethylene, benzene, hydrogen and coke, for which we are looking for an original design for reactors stemming from heat-exchanger technologies using new materials such as ceramics.

Both approaches will be described (technologies, flow sheet, separation scheme, economic assessments, etc.) together with the main results and advantages of each approach.

Introduction

The chemical conversion of natural gas into higher hydrocarbons is an important challenge at present, for which the Institut Français du Pétrole is developing original approaches, both by the traditional indirect route using syngas and by the unconventional direct route. This paper describes our research on the latter approach. IFP has been led to examine two distinct routes, *i.e.* the oxidizing coupling of methane and the thermal coupling of methane via a different methodological and technological approach.

For the oxidizing coupling of methane, we were searching for the best possible catalyst and the optimum flow sheet for a new catalytic reaction that appeared for the first time in the early1980s [1]. For the thermal coupling of methane, which is a reaction that has been known for more than 50 years [2], our general idea was that the emergence of new technologies, and especially new materials, would provide important technological contributions to the development of new reactor designs for implementing this reaction.

1. Oxypyrolysis Process for Natural Gas

Principle and Main Results

The principle of the process developed by IFP consists in combining the oxidizing coupling of methane and the pyrolysis of the C_2+ fraction [3]. Natural gas is separated into two fractions. The first one, containing methane without any C_2+ mixed with oxygen, is preheated and then placed in contact with a catalyst made up of a mixture of $SrCO_3$ and $La_2O_2CO_3$. This mixture is spread out thin so as to achieve a very short residence time (minimum space velocity of 0.7 m/s). Since the oxidizing coupling reaction is exothermic, the energy generated can be used by introducing the second fraction containing C_2+ from natural gas into the hot gas flow coming from the catalytic reaction, so as to convert, in particular, ethane into ethylene. The operating pressure is slightly higher than normal pressure, and the temperature of the catalyst is 850°C to 900°C depending on the proportion of oxygen taken in.

Table 1 gives the typical operating conditions and the results obtained with a reactor (13 mm in diameter). The ethane content is 9.1% in cases 2 and 3, and the oxygen content goes from 9.9% (cases 1 and 2) to 15% (case 3) compared to the methane content. A comparison with case 4, in which ethane has not been separated from methane in the natural gas before the oxidizing coupling reaction, and where there is less methane conversion and a degradation of the selectivity by an increases CO and CO_2 production, clearly illustrates the advantages of the IFP process. A more complete study can be found in Reference [7].

Overall Process Flow Sheet (Figure 1)

CH_4/C_2+ separation can be performed upstream from the oxypyrolysis reactor as shown in Figure 1, or else in the effluent fractionating section, especially if the natural gas contains CO_2.

The hydrogen, methane and CO produced by the reaction are separated in a demethanizer, and the hydrogen is used to methanize the CO and part of the CO_2. The C_2+ fraction is decarbonated and deethanized. By draining off part of the recycling, the inert compounds (N_2, Ar) coming from the oxygen are eliminated, and this also makes up for the need for fuel. Saturated hydrocarbons are recycled with natural-gas makeup. Fuels can be produced from ethylene by concatenations of IFP processes such as *Alphabutol* and *Polynaphta*.

2. IFP Methane Pyrolysis Process

IFP's general concept in the field of hydrocarbon pyrolysis is that new technologies must lead to substantial improvements. Indeed, since pyrolysis reactions are highly endothermic, the key point in such reactions is often their capacity to provide a great deal of enthalpy at high temperature and with controlled residence time. We feel that the use of such new technologies consists in using new materials such as ceramics together with new reactor designs inspired by those of heat exchangers. This concept, moreover, has already been successfully demonstrated by IFP with regard to the steam cracking of hydrocarbons [8].

Starting from initial micropilot-plant research [9], which involved a parametric study, we have succeeded in determining the average operating conditions, and a resulting target has been reached (Table 2) that represents a compromise between the chemical need to diminish coke formation from the extensive dilution by hydrogen and the economic necessity of reducing the dilution so as not to result in the ballooning of investment costs.

Technology Retained -- Chief Results

To complete the development of this process, we decided to retain an energy supplied by electrical heating, and we did this for several reasons. First of all, at 1200°C, the thermal heating efficiency by the Joule effect is quasi quantitative. Likewise, the technological advance required seems to be somewhat less. The technology retained consists in using a design of the shell-and-tube type heat exchanger, for which the general principle is shown in Figure 2 [10]. According to this design, each tube is made up of a sheath protecting a high-temperature resistance made of silicium carbide (SiC). The mixture of gas to be cracked (natural gas + hydrogen) circulates among the tubes. Based on this principle, IFP built a pilot-plant furnace consisting of a row of 21 heating units (see the flow sheet in Figure 3 and the photo in Figure 4) with a power of 12 kW and a total throughput of 10 m^3/h, which is capable of preheating gases from 500°C to 1200°C and then of pyrolysis at 1200°C. This pilot-plant furnace started operating in early May 1989. After various technical problems had been solved and after a purely thermal analysis had been made (heat exchange, temperature profile), this furnace has been operating with natural gas since the start of January 1990, and a complete parametric study has been begun. This study is being carried out as follows. Outside of normal working hours (nights, weekends), the furnace remains at 1000°C with an inert-gas current of several m^3/h. Each run consists in first pyrolyzing the natural gas for several hours at 1200°C, and then of decoking it, currently at 1000°C, by mild oxidation with dilute air so as to obtain a complete material balance. As of 30 May 1990, the following results could be announced:

• The target result (Table 2) was achieved during pilot-plant experimenting.
• Without any noteworthy incident, we performed 57 runs, representing more than 3300 hours at 1000°C and 230 hours at 1200°C.

This experiment is continuing, and a complete study of the results will be published subsequently, but we can already conclude that we have a new and reliable technology for pyrolyzing methane, even if extensive work still remains to be done for a technological development of this amplitude.

Principle of the Process -- Flow Sheet

Dilute methane with hydrogen is preheated by heat exchange and is then fed into the furnace, which pyrolyzes it at 1200°C. This produces a mixture of hydrogen, unconverted methane, acetylene, ethylene and benzene. Since the coke formed mostly remains inside the furnace, this furnace is periodically decoked by steam, which produces a $CO + H_2$ mixture. The process flow sheet [8] is described in Figure 5. The natural gas makeup, mixed with recycled methane and hydrogen, is preheated by the hot effluent. After being cooled, this effluent is fed to the separation section. After the heavy compounds have been eliminated, the gases are compressed. On a membrane installation, the hydrogen produced during the reaction is eliminated. This is quite favorable since the hydrogen yield sought after is limited, and the methane content of the permeate is small. An initial separation of methane and hydrogen is done by absorption of the C_2+ fraction in a cooled solvent. The residual methane is separated by expansion of the liquid phase, which is then processsed in a reactor where acetylene is selectively hydrogenated to ethylene, which is separated from the other products after the solvent has been recycled. Just like the oxypyrolysis process, fuels can be produced from ethylene.

3. Comparison of Processes -- Economic Assessment -- Conclusion

Tables 3 and 4 show two examples of economic assessments, one for fuel production and the other for ethylene production. More detailed economic assessments can be found in the References [9, 10].

The following comments can be made:

• Fuel production can be considered only for oil producing countries. The location factor of 1.4 in these countries multiplies the already high investment cost. The average price of the fuels produced is comparable to that for competitive routes and is not cost effective at the current price of oil, unless jet fuels and high-quality diesel fuels (absence of sulfur and aromatics) have to be produced.
• Ethylene can be produced in industrialized countries at a cost comparable to the mean price of ethylene on the market.

Furthermore, the following important points can be emphasized:

• These two processes are being developed. The separation part of these processes should be able to be optimized to achieve a significant reduction in investments.
• The excess LPG available, added to the C_2+ fraction from natural gas, could also increase the importance of converting methane by one or the other of these two routes.
• Both processes can be considered at a wide variety of scales. This is not the case with steam cracking.

Lastly, concerning thermal coupling, consideration must be given to the future demand for hydrogen, which can be produced by this process, and further consideration must be given to the advantage of producing acetylene and the upgrading routes that this opens up.

To conclude, in a medium-term outlook, we feel that several possible applications of these two processes can be considered.

References

[1] Keller, G. E. and Bhasin, M. M., J. Catal., 73, 9-19 (1982).
[2] Billaud, F., Baronnet, F.,, Freund, E., Busson, C. and Weill, J., Rev. Inst. Franç. du Pétrole, Vol., 44 No. 6, 813-823 (1989).
[3] Cameron, C.J., Mimoun, H., Robine, A., Bonnaudet, S., Chaumette, P. and Quanq, D.V., French Patent Application 88/04588, (5 April 1988).
[4] Cameron, C.J. Quanq, D.V., Lepage, J.F. and Mimoun, H., French Patent Application 88/11312 (25 August 1988).
[5] Cameron, C.J., Mimoun, H., Robine, A., Bonnaudet, S., Chaumette, P. and Quanq, D.V., French Patent Application 89/00 188 (6 January 1989).
[6] Mimoun, H., Robine, A., Bonnaudet, S. and Cameron, C.J., Appl. Catal., 58, 269 (1990).
[7] Robine, A. and Cameron, C.J., 199th ACS Meeting in Boston, MA (22-27 April 1990).
[8] Broutin, P., Busson, C., Weill, J., Heynderickx, G. and Froment, G., AIChE Meeting in Orlando, FL (18-23 March 1990).
[9] Broutin, P., Busson, C., Weill, J. and Billaud, F., 199th ACS Meeting, Boston, MA, (22-27 April 1990).
[10] Alagy, J., Busson, C., Broutin, P. and Weill, J., European Patent Application 323287 (5 July 1989).
[11] Juguin, G., Collin, J.C., Larue, J. and Busson, C., French Patent Application 89/15767 (28 November 1989).
[12] Busson, C., Weill, J. and Raimbault, C., EUROGAS 90, Trondheim (28-30 May 1990).
[13] Raimbault, C., and Cameron, C.J., Natural Gas Conversion Symposium, Oslo (12 17 August 1990).

Table 1 : IFP oxypyrolysis process

IFP OXYPYROLYSIS PROCESS

Experiment No	**1**	**2**	**3**	**4**
Temperature (°C)	850	850	880	850
Flow-rate in (mol/h)				
CH_4	2.500	2.500	2.500	2.500
C_2H_6		0.250	0.250	0.250 (a)
O_2	0.248	0.248	0.375	0.248
Flow-rate out (mol/h)				
H_2	0.124	0.303	0.310	0.222
O_2	0.001	0.001	0.001	0.002
CO	0.017	0.018	0.020	0.039
CO_2	0.059	0.059	0.112	0.064
C_2H_4	0.105	0.267	0.297	0.204
C_2H_6	0.036	0.094	0.106	0.067
C_3H_6	0.008	0.011	0.012	0.012
C_3H_8	0.000	0.000	0.000	0.000
CH_4	2.118	2.168	2.026	2.319
H_2O	0.359	0.358	0.504	0.325
Conversion (CH_4) %				
CH_4	15.3	13.3	19.0	7.2
C_2H_4/C_2H_6 ratio	2.9	2.8	2.8	3.0

(a) C_2H_6 mixed with CH_4 and O_2

Table 2 : Target results for methane pyrolysis

Operating conditions		**Selectivity** (in % of CH_4 converted)	
Pyrolysis temperature (°C)	1200	C_2H_2	32
Residence time (ms)	200	C_2H_4	23
Diluent gas	H_2	C3-C5	4
Percent of diluent (vol %)	50	C_6H_6	15
Conversion (%)	31	C7-C12	8
		Heavy ends + coke	18

Overall chemical reaction : $CH_4 \rightarrow CH_{1.1} + 1.45\ H_2$

Table 3

NON INDUSTRIALIZED AREA (LOCALISATION FACTOR 1.4)

	IFP Therm.	IFP Oxypyr.
CAPACITY (t/year)	651000	610000
Investments US$ million Fixed capital Start-up expenses	1075 75 9	840 58 10
TOTAL INVESTMENT	1159	909
PRODUCTION COST US $ MILLION		
variable Costs		
-Natural Gas at 1.75 $/MMBTU	37.2	33.7
- others	0.9	10.0
Fixed Costs		
-depreciation (10 % TI)	115.9	90.9
-maintenance (4% FC)	43.0	33.6
-overheads (3 % FC)	32.3	25.2
Labour (200000 $/shift)	2.5	1.8
PRODUCTION COST		
US $ million	231.8	195.2
$/t	356	320
c/gallon	100	90

Table 4

INDUSTRIALIZED AREA ; NATURAL GAS PRODUCER

	IFP Therm.	IFP Oxypyr.
CAPACITY		
- ethylene equivalent (*)	550290	639000
- ethylene only	456920	639000
Investments US$ million	620	600
Fixed capital	44	42
Start-up expenses	12	8
TOTAL INVESTMENT	679	656
PRODUCTION COST US $ MILLION		
Variable Costs		
- Natural Gas at 1.75 $/MMBTU	84.3	118.0
- electricity at 3c/Kwh	100.0	
- others		10.0
By products		
- Hydrogen at 750 $/t	-103.6	
Fixed Costs		
-depreciation (10 % TI)	67.9	65.6
-maintenance (4% FC)	24.8	24.0
-overheads (3 % FC)	18.6	18.0
Labour (200000 $/shift)	2.0	2.0
Other by-products (t/t)		
- propylene	0.0119	
- benzene	0.2429	
- solvents	0.1712	
PRODUCTION COST		
US $ MILLION/Y	194.0	237.6
$ /t	352.0	372.0

(*) $C_3H_6 = 0.8 \ .C_2H_4$

Figure 1 simplified flow sheet : oxypyrolysis

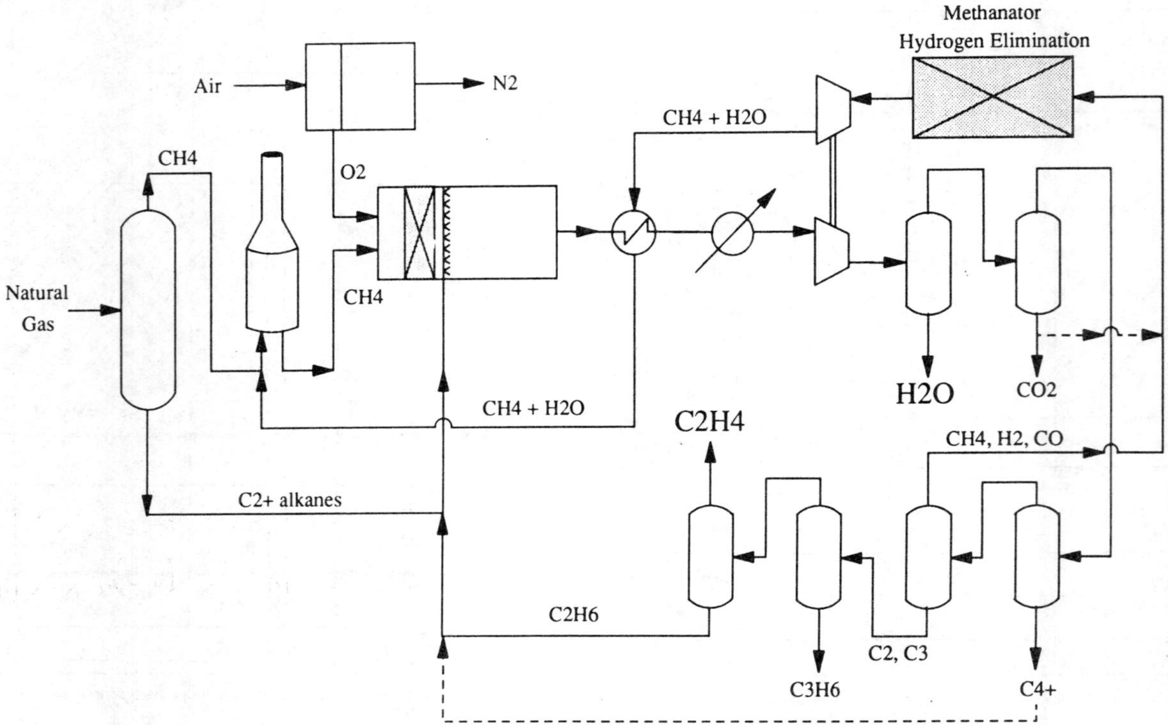

Figure 2 general principle of IFP electric furnace

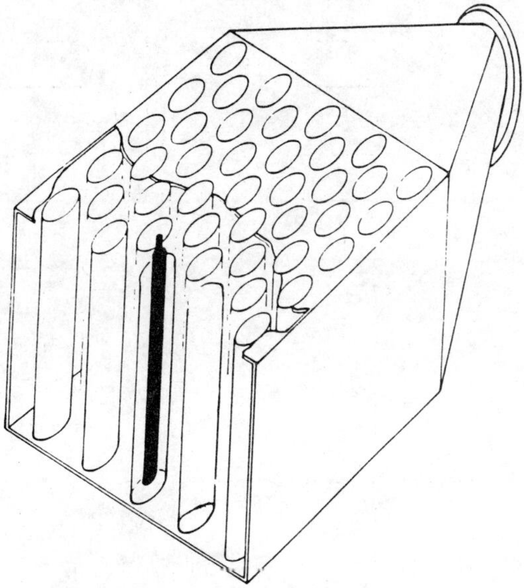

Figure 3 : IFP electric pilot furnace : simplified flow sheet

Figure 4 : photography of IFP pilot plan for methane pyrolysis

Figure 5 simplified flow sheet : thermal coupling of methane

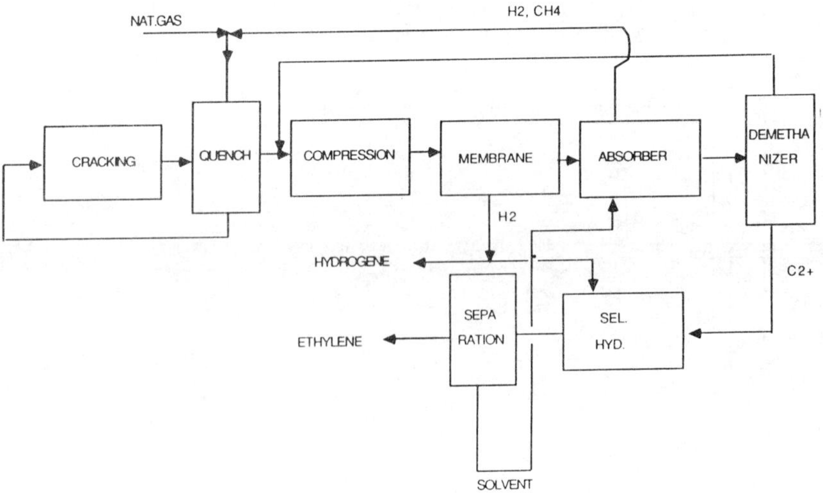

THE SHELL MIDDLE DISTILLATE SYNTHESIS PROCESS

H.M.H. van Wechem*, P.L. Zuideveld** and M.M.G. Senden***
*Shell Internationale Petroleum Maatschappij, The Hague
P.O. Box 162, 2501 AN The Hague
** Shell International Gas Limited, London
Shell Centre, London, SE1 7NA
***Koninklijke/Shell Laboratorium Amsterdam
P.O. Box 3003, 1003 AA Amsterdam

ABSTRACT

A description is given of the Shell Middle Distillate Synthesis (SMDS) process. In this process natural gas is converted into middle distillates. Naphtha, kerosine and gas oil yield ratios can be varied from 15:25:60 to 25:50:25. The products are of an unique nature in the sense that they do not contain any sulphur, nitrogen or aromatic compounds. Both the kerosine and the gas oil have excellent combustion properties: the smoke point of the kerosine is about 100 mm and the gas oil has a cetane number in excess of 70. Starting from natural gas, a thermal efficiency for a stand-alone plant of over 60% can be achieved by using Shell technology for both syngas manufacture and middle distillates synthesis.
The world's first commercial SMDS plant is being constructed in Bintulu, Sarawak, Malaysia and will come on stream in the last quarter of 1992.

1. INTRODUCTION

There are a considerable number of possibilities to produce synfuels, all of which have received attention in the past: direct conversion of heavy oils, tar sands, shale and, most challenging, coal; and the indirect way of first producing syngas from any carbonaceous source followed by synthesis.
The main reasons to start the development of a technology to produce **middle distillates** from **natural gas** have been:

- Remote and relatively small gas fields cannot support the high investments needed to develop a gas pipeline system or LNG production facilities and may well otherwise remain unutilized.
- Locally produced synfuels derived from an indigenous resource may carry a premium owing to the high cost required to import and manufacture transportation fuels.
- Energy strategy reasons in terms of balance of payment as well as strategic considerations.
- Synfuels derived from coal and natural gas via syngas and Fischer-Tropsch have an inherent high quality and, hence, could be used as quality improvers for conventional components when blended. The capital expenditure required for a coal based synthesis plant is, however, about twice that for a natural gas based plant.
- A global long term demand picture showed there was a large need and growth for middle distillates, in particular in the developing countries.

The Shell Middle Distillate Synthesis (SMDS) process has been developed with these considerations in mind.

2. PROCESS DESCRIPTION

The Shell Middle Distillate Synthesis (SMDS) process is a process for the conversion of a mixture of carbon monoxide and hydrogen (generally known as synthesis gas) into middle distillates. It consists of three major processing steps:

1. Conversion of natural gas into synthesis gas. For the production of predominantly saturated hydrocarbons, the syngas components H_2 and CO are consumed in a ratio of about 2:1, so a production in about that ratio is required.
2. Conversion of synthesis gas into normal paraffins. This step in the process is in fact a highly modernized version of the classical Fischer-Tropsch process, with emphasis on a high yield of liquid products and on a favourable catalyst performance, i.e. activity, stability and selectivity.
3. Conversion of normal paraffins into middle distillates. An efficient Fischer-Tropsch process leads to a raw product of a rather waxy nature, which is unsuitable for transportation fuels. Combined with a hydro-isomerization and hydrocracking stage, the product can be converted with minimum gas make selectivity to give a maximum yield of middle distillates.

The syngas production part and the synthesis unit are both net producers of energy. A large energy sink is the oxygen plant. Comparatively, the hydroconversion step turns over much less energy.

Managing the consequences of the stoichiometric relationships and the release and uptake of heat and energy in these process steps offers a challenging field for integration and adaptation, and success in these areas forms the basis for an optimized and economic process. In the present scheme an overall thermal efficiency of over 60% has been achieved.

The overall configuration is shown in Fig. 1.

2.1 SYNTHESIS GAS MANUFACTURE

When starting from methane, the most common conversion via steam reforming produces a synthesis gas with an H_2/CO ratio of about 5-7. This ratio is somewhat smaller if the natural gas contains higher hydrocarbons, and smaller still if it contains CO_2. By judiciously recycling the CO_2 produced in the process this ratio can be reduced to about 3, the CO_2 recovery process and compression being expensive operations.

Since the utilization ratio is about 2:1, it is clear that application of steam reforming will always result in surplus hydrogen production. An often practised solution is to burn the surplus of hydrogen in the reforming furnace, which implies that part of the expensive synthesis gas is used as fuel gas. Another drawback is that the non-reacting hydrogen has to be carried through the synthesis reactors, which adds to the various recycles involved. Though stoichiometrically water-balanced, the process will be a net consumer of water since it is impractical to recycle the water from the synthesis step.

A gas with a H_2/CO ratio of about 2:1 can be produced by non-catalytic autothermal partial oxidation.

	Approximate H_2/CO ratio
$2\ CH_4 + O_2 \rightarrow 2\ CO + 4\ H_2$	2:1

Without much correction such gas is suitable for the production of high quality middle distillates and, incidentally, also for the production of methanol. To reach a correct H_2/CO ratio, air gasification would do, but the high temperatures/pressures required and the large build-up of inert nitrogen in the synthesis loop renders such practice uneconomic. For oxygen gasification of natural gas, an eligible technology is the Shell Gasification Process (SGP). The main economic question is whether the advantages of the partial oxidation route offset the production cost of oxygen. For the case of application of SGP in combination with modern oxygen production facilities, the latter route has been shown to be the most economical choice.

For Fischer-Tropsch type of catalysis, the synthesis gas must be essentially free of sulphur. Sulphur removal upstream of the gasification unit has been selected. The high gas pressure and consequently low gas volume favour gas treating at that point. Various well-known methods for sulphur removal from natural gas are available. However, zinc oxide beds are employed in any case, to remove the last traces of sulphur and to act as an absolute safeguard.

2.2 THE HEAVY PARAFFIN SYNTHESIS STEP (HPS)

In the Heavy Paraffin Synthesis (HPS) step, the reaction mechanism follows the well-known Schultz-Flory polymerization kinetics, which is characterized by the probability of chain growth, α, vs. chain termination. There is always a regular mol. weight distribution in the total product and, as will be seen from Fig. 2, a high α corresponds with a high average mol. weight of the paraffinic product.

During the reaction

$$2\ H_2 + CO \rightarrow H_2O + -(CH_2)-$$

a considerable amount of heat is released. The classical catalyst systems suffered from a number of constraints, chief among which were:

- the temperature window of stable operation was rather narrow i.e. it limits catalyst life;
- at only moderately higher temperatures a side reaction leading to methane became more dominant at the expense of selectivity to C_5^+ production;
- the massive heat release was a challenge for the applied reactor technology.

The new, and proprietary, catalyst system establishes substantial improvement in all these areas. Its robustness allows a fixed bed pipe reactor system at a temperature level where heat recovery, via production of steam, leads to an efficient energy recovery. A high

C_5^+ selectively has been obtained by realising a high carbon chain growth probability. This inevitably results to a very waxy product of high molecular weight. This product consists predominantly of long straight chain paraffins.

2.3 HEAVY PARAFFIN CONVERSION (HPC)

In the heavy paraffin conversion unit the olefins are saturated, the oxygenates are hydrogenated and the paraffins are isomerized and cracked according to the product requirements. A commercial Shell catalyst is used in a trickle-flow reactor as is employed in refinery hydrocracking operations, but under rather mild conditions. The HPC is operated in a recycle mode, as shown in Fig. 3. Conversion per pass is optimized to improve the selectivity towards middle distillates. By adjusting cut points and conversion per pass, the composition of the product package can, within a certain margin, be adjusted to local requirements.

In practice the following product outturn flexibility has been proven:

Product split % wt	Gas oil mode	Kerosine mode
Tops/naphtha	15	25
Kerosine	25	50
Gas oil	60	25

Compared to the syngas requirements of the total complex, the hydrogen demand for the HPC step is very modest. This hydrogen can be recovered from a slipstream of the syngas via any of several methods; one modern method makes use of membranes. Steam reforming of the synthesis purge gas and/or natural gas is also a possibility.

3. SMDS PRODUCT QUALITY

As can be expected from a Fischer-Tropsch type process, products manufactured in SMDS are **predominantly paraffinic and free from nitrogen, sulphur and aromatics**. Both the kerosine and the gas oil have excellent combustion properties and a high thermal stability, and even the cold-flow characteristics meet all relevant specifications, as the following typical product data reveal:

Kerosine (boiling range 150-250°C)		
Density (15°C),	kg/l	0.75
Freezing point,	°C	-47
Smoke point,	mm	100
Gas Oil (boiling range 250-360°C)		
Density (15°C),	kg/l	0.78
Pour point,	°C	-10
Blending cetane number		75

These fractions also make excellent blending components for upgrading of low-quality stock originating from straight distillation/cracking of certain crudes.

Although the process has been developed to produce middle distillates, alternative outlets could be low molecular weight paraffins and waxes.

The naphtha/tops fraction is similar to what can be obtained from most crudes. It needs octane upgrading for use as a gasoline blending component, but as such it would have excellent value as chemical feedstock for an ethylene cracker.

The strategic decision to develop the SMDS route was based on medium- and longer-term expectations that the demand for middle distillate fuels would continue to grow while the factor "quality" could eventually place limits on the production of these transportation fuels from crude oil.

4. RESEARCH AND DEVELOPMENT

Process research has been carried out in bench scale units to develop the catalyst system. This work has continued in larger scale pilot plants to investigate the process configuration, to assist in the development of the reactor details and to prove the overall process concept. An important aid in this work was extensive modelling of the process. The result has led to a basis for all process design data required. All this work was carried out in the Shell Research laboratories in Amsterdam.

The largest of the pilot plants comprised fully integrated units including syngas manufacture and multitubular reactor with a rated capacity of 2 bbl/day product.

5. ENVIRONMENTAL ASPECTS

The production of synthetic liquid hydrocarbons from natural gas has a low environmental impact. Off-gases from the different process units are used inside the plant for utility generation: for instance, to fire the SMR. Flue gases leaving the plant are free of SO_x and low in NO_x and fulfil the most stringent European specifications. Waste water leaving the plant can be treated to allow discharge as surface water. As a general principle, process water and condensate would be re-used to minimize discharge, but the process based on partial oxidation of natural gas is a net producer of water. In areas of water scarcity, this may even be turned to advantage, e.g. for irrigation purposes.

The catalyst has a life of several years, including infrequent in-situ regeneration. Spent catalyst, the only solid waste, can conceivably be returned to the manufacturer for metals recovery.

Owing to the excellent product quality, emissions of harmful exhaust products are also low when applied in a jet or diesel engine.

6. OVERALL ENERGY EFFICIENCY

The theoretical maximum thermal efficiency for the conversion of methane into paraffins is 78% (based on lower heating values). The efficiency attainable in practice is, of course, lower.
Proper heat integration means that the SMDS plant is heat-balanced: that is to say, no extra natural gas to be burned for utility purposes. For example, the steam produced in both the SGP and the HPS processes is used in the turbines of the oxygen manufacturing plant and the different process off-gases are used to fire the SMR. As a result, the Bintulu SMDS plant will operate at an overall thermal efficiency of 63% or, in other words, 80% of the theoretical maximum.

7. ECONOMICS AND OUTLOOK

In August 1989 it was officially announced that the world's first commercial SMDS plant will be built in Bintulu, Sarawak, Malaysia and will come on stream in the last quarter of 1992. The project is a joint venture between the Malaysia state oil company Petronas (10%), the state of Sarawak (10%), Mitsubishi (20%) and Shell Malaysia (60%).

As is well known, capital and operating costs for synfuel complexes are highly dependent on location. The total investment for the first commercial project, the 12,000 bbl/d Bintulu project, is about US $660 million. Of the total process capital cost, more than 50% is required for syngas manufacture. This indicates that work aiming for improvement in this area is at least as important as work on synthesis proper. Further points worth mentioning are improvements in the catalyst, in the reactors employed, and in even better integration of the operation within the complex.

For the gas at US $0.5/MMBtu, the feedstock cost element in the product is about US $5/bbl. The operating costs are estimated at a further US $5/bbl. The total required selling price for a viable project will depend on fiscal regimes, debt/equity ratio, type of loans and corporate return requirements.
The premium to be realized for the high quality products is also a locally influenced aspect, which might become increasingly important as environmental pressure increases.

Another important factor is whether the products are for inland use or for export. For countries with sufficient gas, but which need to import oil or oil products to meet their local demand, SMDS products manufactured in that country should realize at least import parity values. In some cases these may be far above the normal world spot market values. For such countries, therefore, the national benefit of the SMDS process may be substantial.

Though the unit, as now designed for the Bintulu project, is a full scale commercial one, the matter of scaling-up to still larger units will be actively pursued. Future plants are expected to be up to 50,000 bbl/d, and significant economies of scale apply in this capacity range.

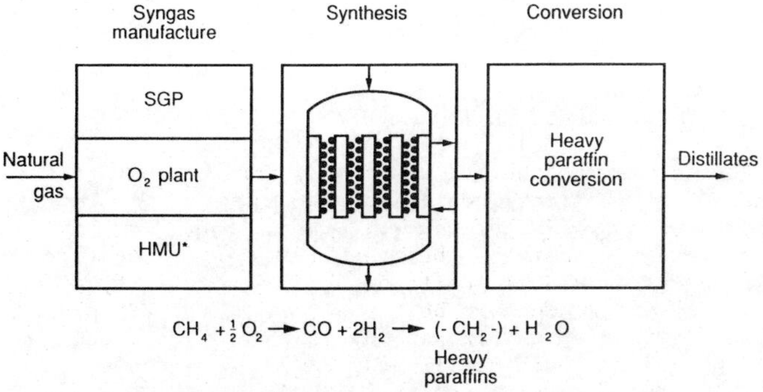

FIG. 1 :SHELL MIDDLE DISTILLATE SYNTHESIS (SMDS) : BASIC CONCEPT

* HMU = Hydrogen Manufacturing Unit

S.G.03194/1

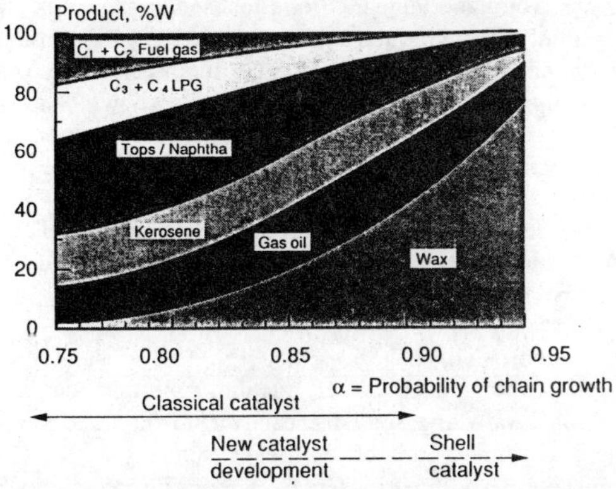

FIG. 2 :PRODUCT DISTRIBUTION IN FISCHER - TROPSCH SYNTHESIS

S.G.03194/2

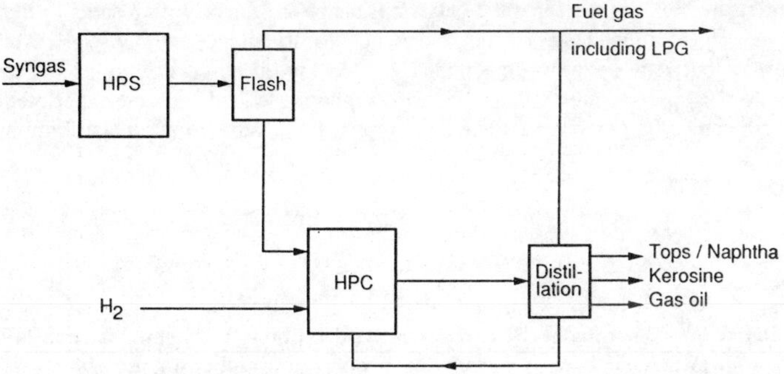

FIG. 3 :SHELL MIDDLE DISTILLATE SYNTHESIS (SMDS) SIMPLIFIED FLOW SCHEME

S.G.03194/3

COMPUTER–AIDED THERMODYNAMIC ANALYSIS OF LIGHT OLEFIN CONVERSION

M. MOLINARI and M. BRUNELLI
Eniricerche S.p.A., S. Donato Milanese (Italy)

M. LUNELLI
Dipartimento di Matematica, Università degli Studi di Milano (Italy)

Light olefins, obtained by dehydrogenation of propane and butane, can be converted to heavier olefins via acidic catalysis. The technological properties of gasoline and gasoil depend on the molecular structure of the most abundant mixture components. In spite of the large number of isomers, the Alberty's algorithm (1), based on the concept of "isomer groups", makes possible the detailed calculation of the equilibrium composition of olefin mixtures and therefore the determination of the most abundant components as functions of pressure and temperature. The "isomer group" is a pseudocomponent that strictly substitutes the isomers of a given carbon number in the calculation of equilibrium distributions with respect to the carbon number. In order to define the thermochemical properties of an "isomer group" the thermochemical properties of all the isomers must be known. The lack of thermochemical data for heavy olefins can be overcome using the Benson group contribution method (2): the number of isomers makes almost impossible to derive the thermochemical tables without computer procedures.

First of all we found a way to identify the structure of each isomer with a sequence of digits (3). On this basis we have devised a computer procedure which, when given a carbon number, generates the sequences of all the structural isomers. Then the sequences are analyzed according to the Benson method to obtain the thermodynamic properties of each isomer. We have applied this procedure up to C12.

Our approach allows simulation of shape–selectivity effects, using "restricted isomer groups". These are defined selecting the isomers that satisfy a structural condition. We compare the experimental data of Garwood (4) on ZSM–5, a shape–selective catalyst, to the distributions obtained using the unrestricted "isomer groups" and three kinds of "restricted isomer groups": linear olefins, linear and monomethylbranched olefins, olefins without quaternary carbons or non–methyl branching. There is no need to exclude the most stable isomers in the calculation of the Gibbs energy of formation of "isomer groups" to predict the data of Garwood. This result contrasts with Tabak's results (5). That author fits the data of Garwood increasing the slope of the line that gives the Gibbs energy of formation of "isomer groups" as a function of carbon number, obtained by linear extrapolation of light olefins data. The increase is justified by the argument that the catalyst is shape–selective with respect to the most stable isomers. Different extrapolations lead to different conclusions: the shape–selectivity effect can't be easily demonstrated by comparison of experimental and calculated equilibrium distributions with respect to the carbon number. We suggest to check that effect by comparing calculated and experimental isomeric distributions at a given carbon number: the method proposed gives the isomeric distributions up to C12.

ACKNOWLEDGEMENT

This paper was supported by Progetto Finalizzato Energia–2, CNR, ROMA.

REFERENCES

(1) ALBERTY, R.A., J. Phys. Chem., 87, 4999–5002 (1983)
(2) BENSON, S.W., "Thermochemical Kinetics" 2nd Edition Wiley–Interscience (1976)
(3) BELLOMO, M., GOBBI, R., "Generazione e conteggio di alberi rappresentanti molecole" tesi di laurea in Matematica (Milano 1987)
(4) GARWOOD, W.E., Symposium On Advances in Zeolite Chemistry, ACS, Las Vegas Meeting 1982
(5) TABAK, S.A., KRAMBECK, F.J., GARWOOD, W.E., AIChE Journal, 32,9 (1986)

CONVERSION OF SYNTHESIS GAS TO LIQUID OXYGENATED PRODUCTS OVER PROMOTED RHODIUM-BASED CATALYSTS

P. COMOTTI*, S. MARENGO*, S. MARTINENGO** and L. ZANDERIGHI***

*Stazione sperimentale per i Combustibili, 20097 San Donato Mil., Italy
**Dipartimento di Chimica Inorganica e Metallorganica and
***Dipartimento di Chimica Fisica ed Elettrochimica,
Università di Milano, Via G. Venezian 21, 20133 Milan, Italy

1. INTRODUCTION

The conversion, either direct or indirect, of methane to liquid products is a theme of increasing interest due to the growing availability of this fossil fuel and to the high cost of transportation. The indirect conversion via synthesis gas obtained by partial oxidation or by steam reforming of methane is the more promising perspective at present. Among the many available processes, the synthesis of alcohol mixtures (C_1-C_6) is particularly actual in connection with the good properties of these products as octane boosters.

The following catalysts are to be considered for syngas conversion to alcohols: modified methanol synthesis catalysts based on $Cu/ZnO/Al_2O_3$; IFP catalysts based on oxides of copper and cobalt; supported and unsupported MoS_2; rhodium-based catalysts (1, 2).

We report here on the properties of CO hydrogenation catalysts based on Rh highly dispersed on porous zirconia.

2. RESULTS

The catalysts, prepared from Rh and Mo carbonyl precursors, were characterized in a computer-controlled micropilot plant operating in the pressure range 1-100 bar. As the operating conditions were found to exert a strong influence on product distribution, techniques of parameter-programmed reaction were developed to define the optimal range of process variables.

The main results can be summarized as follows: ZrO_2 appears a suitable rhodium support for the production of oxygenated compounds (C_1-C_4 alcohols and esters); the catalyst functionality is markedly modified upon addition of doping agents (K, P, Y) to the support; K enhances selectivity to oxygenated compounds by suppressing hydrocarbon formation; addition of Mo to Rh-ZrO_2 promotes markedly both CO conversion and selectivity to oxygenates.

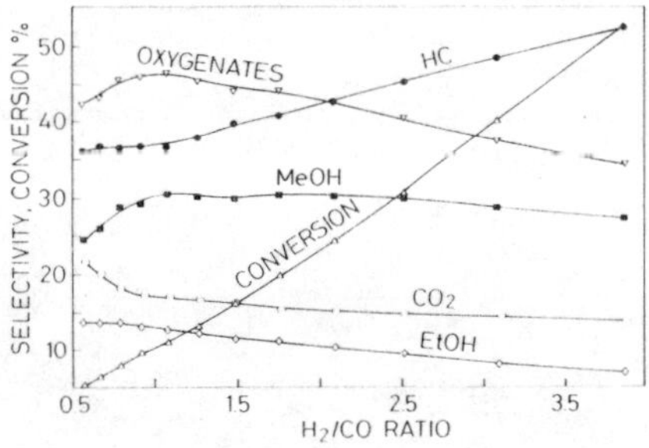

Figure 1 Activity of Rh-Mo/ZrO_2 (Mo/Rh=2) measured in composition-programmed reaction at 503 K, 20 bar, GHSV=2400 h^{-1}

Temperature-programmed measurements showed that CH_4 and CO_2 become predominant above 520 K on Rh/ZrO_2 and above 500 K on Rh-Mo/ZrO_2. With the latter catalyst, the effect of the H_2/CO ratio in the feed is strong both on conversion and on selectivity (Fig. 1).

With a low Rh content in the catalyst (1 wt%) and under relatively mild operating conditions (503 K and 30 bar), the total product yield rises from 6 g/kg.h on Rh/ZrO_2 to 100 g/kg.h on Mo-Rh/ZrO_2 (with Mo/Rh atomic ratio = 2); with the latter catalyst, a 60% selectivity to oxygenates was obtained.

REFERENCES

(1) ICHIKAWA, M., Chemtech, Nov. 1982, 674-680
(2) BENEDETTI, A., CARIMATI, A., MARENGO, S., PINNA, F., TESSARI, R., STRUKUL, G., ZERLIA, T., and ZANDERIGHI, L., J.Catal., 122, 330-345(1990)

METHANE CONVERSION BY REACTION WITH
A - O_2/H_2, B - O_2/Cl_2, C - O_2/Metal Oxide

A - I. VEDRENNE[a], J. SAINT-JUST[a], A. BENHADID[a] and G.M. CÔME[b]

B - R. LE BEC[b], P.M. MARQUAIRE[b] and G.M. CÔME[b]

C - P. BARBE[b], P.M. MARQUAIRE[b], F. BARONNET[b] and G.M. CÔME[b]

a - GAZ DE FRANCE (DETN), BP 33 - 93311 LA PLAINE SAINT DENIS CEDEX (FRANCE)

b - CNRS (URA 328), INPL (ENSIC) et Université de Nancy I, BP 451 - 54001 NANCY CEDEX (FRANCE)

1. INTRODUCTION

The conversion of methane into C_2 hydrocarbons has been studied in $O_2/H_2/CH_4$ and $O_2/Cl_2/CH_4$ premixed and diffusion flames, and in a CFSTR with an adjustable catalytic surface (cofeed reactants O_2 and CH_4).

The influence of various parameters on methane conversion, C_2 selectivity and soot formation has been experimentally studied.

2. MODELS

The homogeneous model is mainly based on a free radical gas phase mechanism. It includes subsystems, such as H_2-O_2, CH_4-O_2, CH_4-Cl_2 and soot formation mechanisms.

The heterogeneous kinetic model is written in a formal way, and includes mainly three categories of processes : production of free radicals, formation of CO_2, regeneration of the catalyst by O_2.

The flames have been modelled either as a single CFSTR or two CFSTR's in series. The catalytic CFSTR has been modelled with no resistance to material transfer.

3. SIMULATIONS

In case A, the model predicts a substantial increase of selectivity and yield, going from a one-stage to a two stage operation.
In case B, a sensitivity analysis has shown the role of C_2H_3 free radical in soot formation.

In case C, the model does not predict inherent limitations to yield.

ACKNOWLEDGEMENTS
The works B and C have been funded in part by the EEC (Contrat EN 3C/0035 F/CD) and by the ACTANE Consortium respectively.

SINGLE STEP CATALYTIC OXIDATION OF METHANE TO FORMALDEHYDE

E. Mac Giolla Coda, M. Kennedy, J.B. McMonagle, B.K. Hodnett
Dept of Materials and Industrial Chemistry, University of Limerick, Ireland.

J. van Ommen and J.R.H. Ross
Dept of Chemical Technology, University of Twente, The Netherlands.

J.W.M.H. Geerts and K. van der Wiele.
Eindhoven University of Technology, The Netherlands.

The selective oxidation of methane to methanol and formaldehyde over a series of supported molybdena and vanadia catalysts has been investigated for a wide range of experimental conditions.

Silica supported molybdena catalysts are suitable for the transformation of methane to formaldehyde at ambient and at elevated pressures in the temperature range 500-600°C but improved performance in terms of conversion of methane and selectivity to formaldehyde can be achieved by doping the catalysts with small amounts of sodium , copper or iron.

Formaldehyde is the only selective oxidation product observed at ambient pressure, but a mixture of this product and methanol forms at pressures of five bar and above. At ambient pressure conversion of methane increased as the methane partial pressure was decreased but selectivity to formaldehyde was not affected. At higher pressures selectivity to methanol and formaldehyde was less but the total amount of methane consumed was greater. Comparison is made between the performances of catalysts at ambient and at elevated pressures which show diminishing selectivity to methanol and/or formaldehyde for increasing methane consumption.

An infra-red spectroscopic investigation of supported molybdena catalysts revealed the presence of methoxy and oxymethylene species on the catalyst surface following exposure to methane in typical reaction conditions.

Comparison between methods currently in development for the transformation of methane into methanol or formaldehyde, namely the homogeneous oxidation at high pressures and the corresponding heterogeneously catalysed reaction, carried out at ambient or at elevated pressures reveal that space time yields of selective oxidation products, i.e. methanol and formaldehyde, are similar for each method, as shown in the table.

TABLE
Production of methanol and/or formaldehyde from methane

Catalyst	Temp /°C	Pressure /Atm	Oxidant	Contact Time /sec	%C	%S	%Y	Rate mole $min^{-1}(g^{-1})^{\beta}$
no catalyst(1)	452	25	O_2	8.6	13.3	55+	7.3	$6.3x10^{-5}$
2V-Si	550	15	O_2	0.08	16.6	3*	0.5	$6.1x10^{-4}$
2V-Si	550	5	N_2O	0.25	1.6	8*	0.14	$5.7x10^{-5}$
2V-Si	550	1	N_2O	1.3	0.2	60**	0.12	$9.4x10^{-6}$

(+)Methanol; (*)Methanol and Formaldehyde; (**)Formaldehyde; (β)rate per gram of catalyst where appropriate.

REFERENCES
(1) P.S.Yarlagadda, L.A.Morton, N.R.Hunter and H.D. Gesser, Ind. Eng.Chem.Res.,27 (1988) 252

FINAL DISCUSSION AND CLOSURE SESSION

by Dr. M. Frias

The last session of the Conference was intended as an open debate for discussion of the status and future evolution of the major points high-lighted during the different sessions for synthesizing the results, for drawing the first conclusions on the potential merits of the "multidisciplinary" approach of the event and for presenting the future development of the Conference.

The overviews of the content of the different sessions and of the corresponding discussions have been prepared by the respective chairmen and are reported here briefly as follows:

GEOSCIENCES

In this branch of science, in which the interplay of different disciplines is so strong, the "multidisciplinary approach" is particularly effective.

Synergy in the field of geosciences is of paramount importance as it more immediately results in improvements of knowledge. Of the many fields covered, important new aspects of the evaluation and description of rock permeability and hydrocarbons migration have been reported and the possibility of treating them with statistical and stochastic methods was particularly interesting.

Progress in reservoir characterization has been hampered by the problems of incomplete data, the problem of 2D-3D analysis and the scale problem.
Reservoir modelling requires - for improved effectiveness - more extensive data for the effective use of geostatistical models.
Basin modelling techniques have reached a certain maturity; it is however desirable that they are used in interactive processes (as for example in sequential stratigraphy) for a better basin characterization. The key to effective and economic field development and planning lies in early recognition of those reservoir characteristics that control drive mechanisms and well spacing requirements.
To identify and describe such characteristics one can make use of a large number of different data, including seismic, geological, petrophysical and well test data. Analogue data from similar well-studied fields and data from outcrops of formations similar to the reservoir are very useful at early stages of development planning. Modern 3D seismic have become a particularly powerful tool in reservoir analysis.

INFORMATION PROCESSING IN GEOSCIENCES

Improvements are still needed in seismic processing but the resolution is intrinsically limited by the related costs. To improve resolution, improved sources and better techniques such as "VSP's" crosswell tomography, should be developed.
3D seismic is nowadays of the utmost importance for thorough and efficient reservoir studies. There is still a need to improve resolution at reasonable costs.
Accurate seismic modelling and geostatistics are nowadays the methods most successfully applied to improve the limits of seismic resolution.

To further increase the resolution, other means are in the development phase:

1. Statistical approach mixed with geological information.
2. Artificial Intelligence approach to integrate methods for reservoir characterization, such as expert systems and self-learning systems.
3. Improvement of hardware (faster supercomputers) and software : advanced mathematical algorithms for seismic modelling, 3D graphical software for the visualization of seismic wave fields for data handling and integration, object - oriented data base.

We can note a very interesting evolving situation where software tools are moving from data processing to knowledge and concept processing.

OPERATIONS

Both sessions on "Operations" were dedicated almost entirely to safety and environmental issues. This is a clear indication of the importance attached to these aspects in oil and gas production activities.

With respect to safety regulations for offshore structures the fair degree of cooperation already existing in Europe could be stimulated further by the EEC.

Regarding environmental protection, there are four main areas of concern:

1. Sea water pollution,
2. Disturbance of sensitive land areas (e.g. due to subsidence),
3. Ground water pollution,
4. Removal of offshore structures.

In spite of the progress already achieved there is a clear need for more complete data bases and for improved models to describe, predict and prevent the first three phenomena.
In view of the many platforms that will have to be removed in the coming years, this may well become a service industry for which new technology is required.

Finally, it seems that research and development in the spheres of safety and environmental protection are areas where oil companies can only gain by working together. Here also the EEC may see opportunities to further such cooperation.

MECHANICS AND NEW MATERIALS

In offshore developments the general tendency of operators is to go to unmanned lighter and automated platforms, the objective being to reduce costs and increase safety.

Weight reduction is a constant, whether it is looked for by using new materials or techniques: composite fibres for risers, forged nodes for jackets, standardized design for topside structures.

Simplification of topsides by using multiphase production systems looks very promising.

Conception and implementation of unmanned platforms and the management of these platforms organized into networks are becoming a reality.

Many new tools were presented in technical sessions as well as in the poster sessions. They all have in common a high level of integration of electronics and informatics.

To summarize it appears that the industry is making a considerable effort to make all types of device or equipment simple, reliable and safe.

CONVERSION OF NATURAL GAS

All experts agree on the increasingly important role of natural gas on the energy scene.

The only limitations to its development are due to the difficulties inherent in transport of a gaseous fluid: high investment for the overall gas networks which indeed are very rigid. There is an evident interest in converting natural gas into other high added value products or liquid fuels. This leads to the necessity of further developing the chemistry of natural gas.

In the short term, the break-even oil price for the indirect conversion of natural gas into liquid fuels lies between 25 and 35 $/bbl for an installation at the gas well head. With the gas market price prevailing in Europe, the break-even oil price rises to 40 to 50 $/bbl. The future of natural gas conversion into liquid fuels is therefore uncertain.

In the medium term, the direct conversion of natural gas by oxidative coupling or thermal coupling might be viable if researchers can find economically adequate catalysts and processes, including an economical separation of the valuable products from undesired by-products with unreacted feedstock being recycled.

GENERAL DISCUSSION

The very lively discussion which followed can be summarized as follows:

A number of substitutes for oil products and, in particular, fuels are technically available at costs which can already be competitive with the oil prices at the date of the

Conference. The conclusions of a workshop on "Advanced Transport Fuels" held a few days before were summarized by the Chairman. In synthesis that meeting showed that advanced fuels produced from coal could very probably be produced at 30-35 $/bbl and that the technology is nearly fully developed: a full scale plant will need, however, several years to be built. As we can read in the conversion of natural gas session, transport fuels can be produced from natural gas at costs of about 25-35 $/bbl and the technology is fully available. The general feeling was that the whole sector should be critically reviewed from the technical and techno-economic points of view but also from the supply strategy side since, even if break-even conditions are met by any one of these alternative fuels, delays in demonstration and in plant construction will render this alternative late. Cooperative research is still necessary for the successful conversion of natural gas into liquid fuels, but a word of warning was raised on the competitivity of the processes because of the oil-gas price link. Eventually the factory should be placed near the gas wells to profit from gas cost advantages.

Enhanced oil recovery was cited as the most effective "replacement technology" permitting replacement of oil by <u>secondary oil</u> : it could become competitive even under present conditions but it demands much more accurate and complete knowledge of the basin structure and behaviour. This in turn demands the development of the advanced multidisciplinary technologies and methods proposed during this Conference.

The potential usefulness of the new "horizontal wells" in this field has been mentioned, but again specific and precise characterization of their behaviour (still insufficiently explored) is needed. This type of well is a good example of the benefits resulting from R&D as it is a real breakthrough by comparison with the standard technology.

Multidisciplinary research has been confirmed to be the only real way to progress in the very broad range of sciences covered by petroleum activities. In this respect it has been said that even large companies do not have sufficient funds and specialized staff to cover the big multidisciplinary projects needed to make sufficient progress. Probably they will continue to provide for their proprietary R&D in fields of their direct interest, but major projects should be handled in a cooperative way. Cooperation between different and competing companies has been defined as possible (though difficult to start) and desirable.

In fact most of the R&D can be defined as non-proprietary: for example the development of methods, tools, computer codes, etc. On the other hand the results, their interpretation and evaluation are the really proprietary matters, which can be easily avoided during the cooperative part of the research.

This point might also be one of the themes of the next Conference: what is proprietary and what is not? The necessity of long term planning for research was discussed, in particular in relation to the fluctuations of oil price (leading indirectly to varying size of R&D staffs). It was proposed that R&D should be considered as a long term investment and R&D staff as company assets, to keep a steady planned programme of ongoing research. Multinational R&D is another possibility. Irreversible allocation of funds to research on reasonable terms (e.g. five years) could also be an alternative. This problem raised much attention and comment from the participants and appears to be an essential point. Most of them identified R&D as the only way out of uncertainty, the only way to escape the strategic impasse due to a limited number of oil supply sources.

In fact, the development of new advanced technologies will permit utilization of minor basins as well as environmentally difficult ones, besides improving the production of existing wells, thus easing the supply side and favourably influencing prices. In this context it was recalled that in USSR large untapped reserves still exist which might become economically available if the technology is improved.

Further to this it was stressed that transfer from R&D to technological applications and to industrial practice is often a difficult step because the information is transferred slowly. The Commission's role and the means utilized in this field was recalled and it was the general opinion that this action deserves extension of the means available. It should receive serious attention and care, as its catalyzing and unifying effect was felt essential, particularly for obtaining further advances.

FUTURE OF THIS CONFERENCE

At the end of the Final Session, the results of the last meeting of the Organizing Committee of the Conference were reported as follows:

The Organizing Committee in its last meeting on Monday 8th October 1990, fully discussed the future development of this Conference in the light of the present situation in the specific fields.

The outcome of this discussion was the following:
It appears that there is an excessive number of scientific and specialized conferences in the world and in particular in Europe. It seems therefore useful and desirable to attempt to group together conferences of complementary, though similar nature, not only to limit their number (and therefore simplify the task of participating in them) but also to rationalize their task, which is to favour techno-scientific contacts and exchanges.

This could be achieved under the "multidisciplinary" approach used for this Conference but the Organizing Committee is fully aware of the corresponding difficulties.
Timing problems, previous firm engagements of the other organisations for their next conferences and similar difficulties should be solved.

It is however felt that this task is worth the effort needed and should be started as soon as possible to minimise the problems mentioned.

The Organizing Committee solicits the assistance of the Commission of the European Communities as a "supra partes" catalyst in this task of gathering together in an effective way a number of European Conferences in the field of Petroleum (and related matters) under the common denomination of Multidisciplinary Approach in Petroleum Research and Engineering, which could be held, as suggested, every second year, leaving ample space for more specialized conferences in between.

On these lines therefore this type of Conference could take place in a wider context, with wider scope and with the indispensable collaboration and contribution of a number of professional organisations, in two years time. It will include (hoping that the negotiations succeed) one or two other events, presently running separately.

ACKNOWLEDGEMENTS

Before closing the session and the Conference the Chairman, in the name of the Organizing Committee, gratefully thanked all the Chairmen of the Conference sessions for their invaluable contribution, the Plenary Sessions speakers whose reports raised a very high interest, in particular stressing multidisciplinarity in the specific field covered. All the authors and speakers, who, by presenting their work, have contributed to the high scientific level of the Conference were also thanked.

Thanks were also addressed to all the participants for their active role in the discussions throughout the Conference inside and outside the meeting rooms.

Particular thanks were expressed to all those who often unnoticed, contributed to the organization and running of the Conference by their active work.

LIST OF PARTICIPANTS

AANESTAD P.
STATOIL
P.O. Box 300
4000 STAVANGER
NORWAY
Tel. 47-4-808080
Tlx.
Fax 47-4-807042

ABOUD J.
CORPOVEN S.A.
La Campina
Avenida Libertador
CARACAS
VENEZUELA
Tel. 58-2-7081531
Tlx. 21363
Fax

ABRAHAMSEN K.A.
STATOIL
P.O. Box 300
4001 STAVANGER
NORWAY
Tel. 47-4-805130
Tlx.
Fax 47-4-806650

ALBERTSEN M.
Deutsche Gesellschaft für
Mineralölwissenschaft
und Kohlechemie E.V.
Steinstrasse 7
2000 HAMBURG 1
GERMANY
Tel. 49-40-326479
Tlx. 211446 DGMK D
Fax 49-40-326398

AMURSKY G.
NPO SOYUZGAZTEKNOLOGIVA - ALL
UNION SCI.
Research Institute of Natural Gas
Leninskiy Rayon P. Razuika
142717 MOSCOW
USSR
Tel.
Tlx.
Fax

ANDERSEN C.
GEOLOGICAL SURVEY OF DENMARK
Thoravej 8
2400 COPENHAGEN NV
DENMARK
Tel. 45-31-106600
Tlx. 19999 DANGEO DK
Fax 45-31-196868

APPLEFORD D.
ALPHA THAMES
Suite 2 - Essex House
Station Road
UPMINSTER ESSEX
UNITED KINGDOM
Tel. 44-4022-29229
Tlx.
Fax 44-4022-51273

ARAGONA L.
AGIP SpA.
Viale Europa 44
20122 MILANO
ITALY
Tel. 39-2-52027402
Tlx. 310246 ENI I
Fax 39-2-52027607

ARENDT H.
COMMISSION OF THE EUROPEAN
COMMUNITIES
DG XII/E.6
200 rue de la Loi
1049 BRUXELLES
BELGIUM
Tel. 32-2-2355648
Tlx. 21877 COMEU B
Fax 32-2-2363024

ARGYRIS P.A.
COMMISSION OF THE EUROPEAN
COMMUNITIES
DG XVII
200 rue de la Loi
1049 BRUXELLES
BELGIUM
Tel. 32-2-2361551/2361111
Tlx. 21877 COMEU B
Fax 32-2-2350150

ARMENIS D.
ALFAPI S.A.
Messoghion 304 Ave.
155 62 ATHENS
GREECE
Tel. 30-1-6532579
Tlx. 223296
Fax 30-1-6533924

ATTAYA C.
SITEP
92 Rue de Palestine
TUNIS
TUNISIA
Tel. 216-1-785244
Tlx.
Fax

AUGUSTSON J.H.
NORSK HYDRO RESEARCH CENTER
A.S.
Storaakeren 11
9400 HARSTAD
NORWAY
Tel. 47-82-13500
Tlx.
Fax 47-82-66077

AZOUZ M.
SITEP
92 Rue de Palestine
TUNIS
TUNISIA
Tel. 216-1-785244
Tlx.
Fax

BAKIR K.
WAHA OIL COMPANY
P.O. Box 395
TRIPOLI
LIBYA
Tel. 31116-4312
Tlx.
Fax

BALOSSINO P.
AGIP SMES
P.O. Box 12069
20120 MILANO
ITALY
Tel. 39-2-52024734
Tlx. 310246 ENI I
Fax 39-2-52022065

BARBERO G.
AGIP/MODG
Strada 2 - Ingresso 3
Milanofiori
ASAGO (MI)
ITALY
Tel. 39-2-8242751
Tlx. 325813 AGIPNA
Fax 39-2-8242928

BARNABA P.F.
UNIVERSITA DI MILANO
Dipartimento Scienze della Terra
Via Mangiagalli 34
20133 MILANO
ITALY
Tel. 39-2-236981
Tlx.
Fax

BECKERS H.L.
SHELL Research Intern. B.V.
Carel van Bylandtlaan, P.O. Box 162
2501 DEN HAAG
THE NETHERLANDS
Tel. 31-70773522
Tlx.
Fax 31-70774848

BELHAOUAS A.
R&D CENTER SONATRAC
Av. 1er Novembre
35000 BOUMERDES
ALGERIA
Tel. 213-2-414343
Tlx. 63465 CRD DZ
Fax 213-2-415302

BEMTGEN J.M.
COMMISSION OF THE EUROPEAN COMMUNITIES
DG XII/E.6
200 rue de la Loi
1049 BRUXELLES
BELGIUM
Tel. 32-2-2362071
Tlx. 21877 COMEU B
Fax 32-2-2363024

BENISA M.
WAHA OIL COMPANY
P.O.Box 395 - Gielo Field 59
TRIPOLI
LIBYA
Tel. 31116 EXT.3311/3312
Tlx.
Fax

BERNABEI P.
DCD SOCIETA' DELLE FUCINE
Viale B. Brin 218
05100 TERNI
ITALY
Tel. 39-744-488209
Tlx. 860008
Fax 39-744-401759/401973

BERTOCCO E.
AGIP SpA.
INIZIATIVE PROMOZIONALI E REDAZIONALI
Via del Serafico, 89/91
00142 ROMA
ITALY
Tel. 39-6-50392529
Tlx. 613525-614421
Fax 39-6-50392241

BILARDO U.
UNIVERSITA DI ROMA
Facoltà di Ingegneria
Via Eudossiana 18
00184 ROMA
ITALY
Tel. 39-6-4687334/44585625
Tlx.
Fax 39-6-44585618

BISEO S.
Consultant to ILVA (UK) LTD
31 Pattison Road
LONDON NW2 2HL
UNITED KINGDOM
Tel. 44-71-7241444
Tlx. 28697
Fax 44-71-7241754

BLASCHKE W.
POLISH ACADEMY OF SCIENCES
MAEERC
Ul. Wadowicka 12
KRAKOW
POLAND
Tel. 48-12-669029
Tlx. 0322377
Fax 48-12-669029

BOLONDI G.
AGIP / Geof. Dept.
Via Fabiani 1
20097 SAN DONATO MILANESE MI
ITALY
Tel. 39-2-5205187
Tlx. 310246 ENI I
Fax 39-2-52038609

BONNET Ph.
GULF PUBLISHING COMPANY OF HOUSTON (TEXAS, U.S.A.)
rue de la Fédération 65
75015 PARIS
FRANCE
Tel. 33-1-42732114
Tlx.
Fax 33-1-40560332

BORRIELLO G.
AGIP SpA.
Piazza Vanoni 1
20097 SAN DONATO MILANESE MI
ITALY
Tel. 39-2-5209356
Tlx.
Fax 39-2-52036349

BOSIO J.
ELF AQUITAINE
R&D et Relations Industrielles
Tour Elf Cédex 45
92078 PARIS LA DEFENSE
FRANCE
Tel. 33-1-47444477
Tlx. ELFA 615400 F
Fax 33-1-47447373

BOURDEAU Ph.
COMMISSION OF THE EUROPEAN COMMUNITIES
DG XII/E
200 rue de la Loi
1049 BRUXELLES
BELGIUM
Tel. 32-2-2354070
Tlx. 21877 COMEU B
Fax 32-2-2363024

BOUTECA M.J.
INSTITUT FRANCAIS DU PETROLE
1 et 4 Av. de Bois Préau P.O. Box 311
92506 RUEIL-MALMAISON - CEDEX
FRANCE
Tel. 33-1-47526242
Tlx. 203050 F
Fax 33-1-47527000

BOZZO G.M.
TECNOMARE SPA
San Marco 2091
30124 VENEZIA
ITALY
Tel. 39-41-796711
Tlx. 410484
Fax 39-41-5230363

BRAUCKMANN F.
BEB Erdgas und Erdöl GmbH.
Riethorst 12
3000 HANNOVER 51
GERMANY
Tel. 49-511-6412893
Tlx. 921421
Fax 49-511-6412845

BRAUNSCHWEIG B.
INSTITUT FRANCAIS DU PETROLE
P.O. Box 311
92506 RUEIL-MALMAISON
FRANCE
Tel. 33-1-47526648
Tlx. A 203050 F
Fax 33-1-47527022

BRET N.
MINISTERE DE L'INDUSTRIE ET DE L'AMENAGEMENT DU TERRITOIRE
3 et 5 rue Barbet de Jouy
75007 PARIS
FRANCE
Tel. 33-1-45563919
Tlx.
Fax 33-1-45564867

BRUNELLI M.
ENI Ricerche
Via Maritano 26
20097 SAN DONATO MILANESE - MI
ITALY
Tel. 39-2-52023435
Tlx. 310246 ENI
Fax 39-2-5204422

BUDDING M.Ch.
SHELL RESEARCH (KSEPL)
Expl. en Produktie Lab.
Volmerlaan 6, P.O. Box 60
2200 AB RYSWYK
THE NETHERLANDS
Tel. 31-70-3112758
Tlx.
Fax

BULLER D.C.
BP RESEARCH
Exploration & Production
Chertsey Road
SUNBURY-ON-THAMES TW16 7LN
UNITED KINGDOM
Tel. 44-932-762551
Tlx. 296041
Fax 44-932-763160

BURRUS J.
INSTITUT FRANCAIS DU PETROLE
Géologie, Géochimie
P.O. Box 311
92506 RUEIL-MALMAISON
FRANCE
Tel. 33-1-47526936
Tlx. IFP A 203050 F
Fax 33-1-47490411

CALVARESE L.
MINISTERO dell'INDUSTRIA
Uff. Naz. Minerario per gli Idrocarburi e la Geotermia
Via Molise 2
00187 ROMA
ITALY
Tel.
Tlx.
Fax

CAMPOBASSO S.
AGIP SpA.
Centro Studi
Via Fabiani 1
20097 SAN DONATO MILANESE MI
ITALY
Tel. 39-2-52037851
Tlx.
Fax 39-2-5207309

CARBONE I.
AGIP SPA.
Via del Serafico, 89/91
00142 ROMA
ITALY
Tel. 39-6-50392-1
Tlx. 613525 AGIPDA
Fax 39-6-50392 320

CAZZULO R.
REGISTRO ITALIANO NAVALE
Via Corsica 12
18128 GENOVA
ITALY
Tel. 39-10-5385332
Tlx. 270022 RINAV I
Fax 39-10-591877

CHADWICK D.
IMPERIAL COLLEGE OF SCIENCE
Dept. Chemical Engineering
Prince Consort Road
LONDON SW7 2BY
UNITED KINGDOM
Tel. 44-71-2258306
Tlx. 929484
Fax 44-71-5841170

CHEBBI M.
SITEP
92 Rue de Palestine
TUNIS
TUNISIA
Tel. 216-1-785244
Tlx. 15439 TN
Fax

CHIERICI G. L.
Via Triulziana 36/a
20097 SAN DONATO MILANESE - MI
ITALY
Tel. 39-2-515248 Uff. 39-2-5204086
Tlx.
Fax 39-2-52022371

CIKES M.
INA-NAFTAPLIN
Project Department
Subiceva 29
41000 ZAGREB
YUGOSLAVIA
Tel. 38-41-458011
Tlx. 21430 YU INA NP
Fax 38-41-418200

COLAMASI C.
CEOM
Via G. Di Marzo 2/F
90144 PALERMO
ITALY
Tel.
Tlx.
Fax

COLOMBO A.
SNAM SpA.
Piazza S. Barbara
20097 SAN DONATO MILANESE - MI
ITALY
Tel. 39-2-5204722
Tlx.
Fax 39-2-52024435

COMBARNOUS M.
UNIVERSITY OF BORDEAUX
L.E.P.T. ENSAM
Esplanade des Arts et Métiers
33405 TALENCE
FRANCE
Tel. 33-56375959
Tlx. 550140
Fax 33-56043889

COME G.
UNIVERSITY OF NANCY (INPL)
Dépt. de Chimie Physique des Réactions
1 rue Grandville, B.P 451
54001 NANCY CEDEX
FRANCE
Tel. 33-83352121
Tlx. ENSIC 961316
Fax 33-83378120

COMOTTI P.
Stazione Sperimentale per i Combustibili
Via A. de Gasperi 3
20097 S. DONATO MILANESE - MI
ITALY
Tel. 39-2-510031
Tlx.
Fax

CORBELLA M.
SCHLUMBERGER ITALIANA
Milanofiori - Pal.T1
20089 ROZZANO MI
ITALY
Tel. 39-2-8242651
Tlx. 310118
Fax 39-2-8243581

CORNINI S.
AGIP S.p.A.
SGEO Dept.
20097 SAN DONATO MILANESE - MI
ITALY
Tel. 39-2-52033694
Tlx. 310246 ENI I
Fax 39-2-52038609

COSTA E SILVA A.J.
PARTEX-Companhia Portuguesa de
Serviços, S.A.
Av. 5 de Outubro 160
1000 LISBOA
PORTUGAL
Tel. 351-1-735290
Tlx. 14708 PARSER
Fax 351-1-779516

COSTE J.F.
ELF AQUITAINE
CSTJF Z/
14 Avenue Larribau
64018 PAU CEDEX
FRANCE
Tel. 33-59835006
Tlx. 560804 PETRA
Fax 33-59836511

COURTEILLE J.M.
ELF AQUITAINE
CSTJF
Avenue Larribau
64018 PAU CEDEX
FRANCE
Tel. 33-59834000
Tlx. 560804 AGFP
Fax 33-5983666565

D'AGATA R.
AGIP SpA.
IMPU
Via del Serafico 89/91
00142 ROMA
ITALY
Tel. 39-6-50392 312
Tlx. 613525
Fax 30-6-50392 241

D'HEUR M.
Head Development Geology
PETROFINA
Rue de l'Industrie 52
1040 BRUXELLES
BELGIUM
Tel. 32-2-2339842
Tlx. 21556 PFINA EXPLOZ
Fax 32-2-2333337

DAKSHI A.M.
AGIP LIBYAN BRANCH
TRIPOLI
LIBYA
Tel.
Tlx. 20666
Fax

DAOUADJI M.
SONATRACH DIVISION EXPLORATION
2, rue Capt Azzoug Cote Rouge Hussein
ALGER
ALGERIE
Tel. 213-2-771140
Tlx. 65405
Fax 213-2-595758

DE BAUW R.
COMMISSION OF THE EUROPEAN COMMUNITIES
DG XVII
200 rue de la Loi
1049 BRUXELLES
BELGIUM
Tel. 32-2-2351662
Tlx. 21877 COMEU B
Fax 32-2-2350150

DE HAAN H.J.
Doornweg 7
2243 GS WASSENAAR
THE NETHERLANDS
Tel. 31-15-781328
Tlx. 38151 BUTUD
Fax 31-15-784891

DE LUCHI L.
SEG / EAEG
AGIP / GESO
20097 S. DONATO MILANESE - MI
ITALY
Tel. 39-2-5205189
Tlx.
Fax 39-2-52022065

DEAN T.L.
TEXACO
1 Knightsbridge Green
LONDON SW1X 7Q5
UNITED KINGDOM
Tel. 44-71-5845000
Tlx.
Fax

DEGOUY D.
INSTITUT FRANCAIS DU PETROLE
Avenue du Bois Préau 1-4
92506 RUEIL-MALMAISON
FRANCE
Tel. 33-1-47526766
Tlx. IFP A 203050 F
Fax 33-1-47490411

DELAS C.
TOTAL CFP
Cédéx 47
92069 PARIS LA DEFENSE
FRANCE
Tel. 33-1-42913704
Tlx. TCFP 615700 F
Fax 33-1-42914052

DEWIJN B.
WINTERSHALL NOORDZEE B.V.
Eisenhowerlaan 142
2517 KN THE HAGUE
THE NETHERLANDS
Tel. 31-70-3583100
Tlx.
Fax 31-70-3583333

DI LUISE G.
AGIP S.p.A.
P.O.Box 12069
20120 MILANO
ITALY
Tel. 39-2-5207149
Tlx. 310246 ENI I
Fax 39-2-52037904

DIALUCE G.
MINISTERO dell'INDUSTRIA
Uff. Naz. Minerario per gli Idrocarburi e la Geotermia
Via Molise 2
00187 ROMA
ITALY
Tel.
Tlx.
Fax

DJEBARA A.
SONATRACH DIVISION EXPLORATION
2, rue Capt. Azzoug Cote Rouge Hussein
ALGER
ALGERIE
Tel. 213-2-771140
Tlx. 65405
Fax 213-2-595758

DOBAY P.
HUNPETRO HUNGARIAN PETROLEUM CO.LTD.
Schönherz Z.U. 18
1117 BUDAPEST
HUNGARY
Tel. 36-30-1-1850501
Tlx. 225123-
Fax 36-1-1667760

DONAT G.
GAZ DE FRANCE
2 rue Curnonsky
75017 PARIS
FRANCE
33-1-47543644
Tel. 611319 GDFRI
Tlx. 33-1-42703344
Fax

DONSELAAR M.E.
DELFT UNIVERSITY OF TECHNOLOGY
Faculty of Mining and Petroleum Eng.
P.O. Box 5028 120
2600 GA DELFT
THE NETHERLANDS
Tel. 31-15-781328
Tlx. 38151 BUTUD NL
Fax 31-15-784891

DOREL M.
INSTITUT FRANCAIS DU PETROLE
1 et 4 avenue de Bois Préau
92506 RUEIL-MALMAISON
FRANCE
Tel. 33-1-47526826
Tlx. IFP A 203050 F
Fax 33-1-47527003

DOWTY E.
TEXACO E&P TECHNOLOGY DIVISION
5901 S. Rice Ave.
BELLAIRE TEXAS
USA
Tel. 713-432-3360
Tlx. 166752
Fax 713-432-6929

DOYEN P.M.
WESTERN GEOPHYSICAL
Reservoir Geophysics
455 London Road, P.O. Box 18
ISLE WORTH, MIDDLESEX TW7 5AB
UNITED KINGDOM
Tel. 44-81-5603160
Tlx. 24970 WESGEO G
Fax 44-81-8473131

DRONKERT H.
DELFT UNIVERSITY OF TECHNOLOGY
P.O. Box 5028
2600 GA DELFT
THE NETHERLANDS
Tel. 31-15-785108
Tlx. 38151 BUTUD NL
Fax 31-15-784891

DUBESSET M.
IFP - G.E.R.T.H.
Exploration-Production
Géophysique-Instrumentation
1&4 Av. de Bois Préau, B.P. 311
92506 RUEIL-MALMAISON
FRANCE
Tel. 33-1-47490214
Tlx. IFP A 203050 F
Fax 33-1-47490411

DUCATE D.L.
SPE
Palisades Greek Drive, 222
P.O. Box 833 836
RICHARDSON - TEXAS 75083
USA
Tel. 1-214-6693377
Tlx. 730989 SPE DAL
Fax 1-214-6690135

DUGSTAD O.
INSTITUTT FOR ENERGITEKINIK
P.O. Box 40
2007 KJELLER
NORWAY
Tel. 39-6-806000
Tlx.
Fax

EL FEGHI O.
JAWABY OIL SERVICE
33 Cavendish Square
LONDON W1M 9HF
UNITED KINGDOM
Tel. 44-71-4990855
Tlx. 261443/262602
Fax 44-71-4991771

ELSHRAYEF S.
AGIP LIBYAN BRANCH
TRIPOLI
LIBYA
Tel.
Tlx. 20666
Fax

ESCHARD R.
INSTITUT FRANCAIS DU PETROLE
1&4 Av. de Bois Préau, B.P.311
92506 RUEIL-MALMAISON
FRANCE
Tel. 33-1-47526629
Tlx. IFP A 203050 F
Fax 33-1-47490411

FALLEUR C.B.
ACKERMANN & VAN HAAREN N.V.
Begijnenvest 113
2000 ANTWERPEN
BELGIUM
Tel. 32-3-2317087
Tlx. 34129
Fax 32-3-2252533

FARRIMOND P.
UNIV. OF NEWCASTLE UPON TYNE
NRG - Drummond Building
NEWCASTLE UPON TYNE
UNITED KINGDOM
Tel. 44-91-2226000 (6513)
Tlx.
Fax 44-91-2611182

FENEYROU G.
SNEA (P) ELF AQUITAINE
26 Avenue des Lilas
64018 PAU
FRANCE
Tel. 33-59835124
Tlx. PETRA 560804 F
Fax 33-59834476

FONNESU F.
AGIP SpA.
Nuovi Laboratori Bolgiani
Via Maritano 26
20097 SAN DONATO MILANESE - MI
ITALY
Tel. 39-2-5209906
Tlx.
Fax 39-2-52022371

FRIAS M.
COMMISSION OF THE EUROPEAN COMMUNITIES
DG XII/E.6
200 rue de la Loi
1049 BRUXELLES
BELGIUM
Tel. 32-2-2361597
Tlx. 21877 COMEU B
Fax 32-2-2363024

GAAYA M.
SITEP
92 Rue de Palestine
TUNIS
TUNISIA
Tel. 216-1-785244
Tlx.
Fax

GANZ S.N.
UNIVERSITY OF BERLIN
T.U. Berlin - SFB 69
Ackerstr. 71-76
1000 BERLIN 65
GERMANY
Tel. 49-30-31472696
Tlx.
Fax

GARCIA-SIÑERIZ B.
REPSOL Exploracion S.A.
Pez Valador 2
28007 MADRID
SPAIN
Tel. 34-91-5749700/4092892
Tlx. 49544
Fax 34-91-5745784

GASPARINI M.
AGIP S.p.A.
New Production Technologies
Piazza Vanoni 1, P.O. Box 12069
20097 SAN DONATO MILANESE - MI
ITALY
Tel. 39-2-5204087
Tlx. 300246
Fax 39-2-52037908

GENOVESI F.
AGIP AFRICA SUCCURSALE DE TUNISIE
87 Ave. T. Mehiri
TUNIS
TUNISIA
Tel. 216-1-7800064
Tlx. 13649 AGIP TU-TN
Fax 216-1-782136

GHADDAB F.
SITEP
92 Rue de Palestine
TUNIS
TUNISIA
Tel. 216-1-785244
Tlx.
Fax

GIANNESINI J.P.
INSTITUT FRANCAIS DU PETROLE
Dir. des Ensembles Ind. de Production
1 & 4 Av. du Bois Préau
92506 RUEIL MALMAISON
FRANCE
Tel. 33-1-47526095
Tlx. IFP A 203050 F
Fax 33-1-47490411

GIUGLIANO A.
AGIP SpA.
Centro Studi
Via Fabiani 1
20097 SAN DONATO MILANESE - MI
ITALY
Tel. 39-2-52035647
Tlx.
Fax 39-2-5207309

GORDON A.
CALTEC LTD
34A Albyn Place
ABERDEEN AB1 IYN
UNITED KINGDOM
Tel. 44-224-210836
Tlx.
Fax 44-224-210837

GORICNIK B.
INA-NAFTAPLIN SLI
Lovinciceva 1
41000 ZAGREB
YUGOSLAVIA
Tel. 38-41-231122
Tlx. 22556 YU INA NPS
Fax 38-41-220117

GOTTSCHALK U.
BALSLEV A/S
Roedovre Centrum 155
ROEDOVRE
DENMARK
Tel. 45-31-410022
Tlx. 15579 BALSEL
Fax 45-31-708866

GOWEN M.
COMMISSION OF THE EUROPEAN COMMUNITIES
DG XVII
200 rue de la Loi
1049 BRUXELLES
BELGIUM
Tel. 32-2-2360436
Tlx. 21877 COMEU B
Fax 32-2-2350150

GRAU G.
INSTITUT FRANCAIS DU PETROLE
Direction de Recherche
Av. de Bois Préau 1&4 - B.P. 311
92506 RUEIL MALMAISON CEDEX
FRANCE
Tel. 33-1-47526538
Tlx. A203050 F
Fax 33-1-47470411

GRAVELLE P.
INSTITUT DE RECHERCHE SUR LA CATALYSE
CNRS - PIRSEM
4 rue las Cases
75007 PARIS
FRANCE
Tel. 33-1-47531639
Tlx.
Fax 33-1-47058914

GRAY C.
LASMO INT. LIMITED
Via Vittorio Veneto 116
00187 ROMA
ITALY
Tel. 39-6-4817835
Tlx.
Fax 39-6-4958398

GRIBAA R.
SITEP
Rue de Palestine 92
1002 TUNIS
TUNISIA
Tel. 216-1-785244
Tlx.
Fax

GRISI M.
SELM PETROLEUM
Via Rosellini 15/17
20124 MILANO
ITALY
Tel. 39-2-63334956
Tlx. 310679 MONTED I
Fax 39-2-62708151

GRIST D.M.
BP EXPLORATION
Reservoir Engineering - WA14
Britannic House - Moor Lane
LONDON EC2Y 9BU
UNITED KINGDOM
Tel. 44-71-9202710
Tlx. 888811
Fax 44-71-9208680

GROPPI G.
AGIP SpA.
20097 SAN DONATO MILANESE - MI
ITALY
Tel. 39-2-5392691
Tlx.
Fax

HAUGERUD O.
NORSK HYDRO
P.O.Box 200
1321 STABEKK (Oslo)
NORWAY
Tel. 47-2-738100
Tlx. 72948 HYDRO N
Fax 47-2-738861

HODNETT B.K.
UNIVERSITY OF LIMERICK (N.I.H.E.)
Dept. of Materials Eng. and Industrial Chem.
Plassey Technol. Park
LIMERICK
IRELAND
Tel. 353-61-333644
Tlx. 500-70609
Fax 353-61-330316

HOFFMANN U.
UNIVERSITY OF BRAUNSCHWEIG
Institut für Technische Chemie
Hans-Sommer-Str. 10
3300 BRAUNSCHWEIG
GERMANY
Tel. 49-531-3915360
Tlx. 952526 TU BSW
Fax 49-531-3915357

HOLM L.
DANISH ENERGY AGENCY
Landemaerket 11
1119 COPENHAGEN K
DENMARK
Tel. 45 - 33926700
Tlx.
Fax 45 - 33114743

HVOSLEF S.
AMERADA HESS NORGE A/S
Langkaten 1
0150 OSLO 1
NORWAY
Tel. 47-2-426028
Tlx. 74707 AH NOR N
Fax 47-2-426327

IDIL S.
NORSK HYDRO A/S
P.O. Box 200
1321 STABEKK (Oslo)
NORWAY
Tel. 47-2-738866
Tlx. 72848 HYDRO N
Fax 47-2-739876

IMARISIO G.
COMMISSION OF THE EUROPEAN COMMUNITIES
DG XII/E.6
200 rue de la Loi
1049 BRUXELLES
BELGIUM
Tel. 32-2-2356919
Tlx. 21877 COMEU B
Fax 32-2-2361094

INGRAIN D.
GAZ DE FRANCE (DETN) (CERSTA)
361 Avenue du Président Wilson, B.P. 33
93211 LA PLAINE ST-DENIS
FRANCE
Tel. 33-1-49225280
Tlx. 236 735 F
Fax 33-1-49225653

IOANNIDIS C.
NORTH AEGEAN PETROLEUM COMPANY
Filellinon 2 - P.O. Box 1077
65110 KAVALA
GREECE
Tel. 30-51-835521
Tlx. 0452120
Fax 30-51-832390

ROSS J.R.H.
TWENTE UNIVERSITY OF TECHNOLOGY
Faculty of Chemical Technology
P.O. Box 217
7500 AE ENSCHEDE
THE NETHERLANDS
Tel. 31-53-892858
Tlx. 44200
Fax 31-53-356024

JACOBSEN A.C.
HALDOR TOPSOE A/S
R&D Division
Nymollevej 55
2800 LYNGBY
DENMARK
Tel. 45-42878100
Tlx. 37444
Fax 45-42878494

JACQUOT D.
GAZ DE FRANCE
Research and Development Division
361 Avenue du President Wilson, B.P. 33
93211 LA PLAINE SAINT DENIS
FRANCE
Tel. 33-1-49224943
Tlx. 233 889
Fax 33-1-49224999

JERBI K.
WAHA OIL COMPANY
P.O. Box 395
TRIPOLI
LIBYA
Tel. 31116 EXT.2083
Tlx. 20158
Fax

JING X.D.
IMPERIAL COLLEGE OF LONDON
Dept. of M.R.E.
Prince Consort Road
LONDON SW7 2BP
UNITED KINGDOM
Tel. 44-71-5895111 ext.6431
Tlx.
Fax 44-71-589 6806

JONES K.
SASKOIL INTERNATIONAL RESEARCH
2500, 140 - 4 Avenue S.W.
CALGARY - ALBERTA, Alta T2P 3S3
CANADA
Tel. 403-260 4918
Tlx. 03-824750
Fax 403-262 5524

JOSLIN O.P.
SVENSKA PETROLEUM EXPLORATION
AS
P.O. Box 2318 Solli
0201 OSLO 2
NORWAY
Tel. 47-2-556660
Tlx. 74578
Fax 47-2-558184

JOULIA J.P.
COMMISSION OF THE EUROPEAN
COMMUNITIES
DG XVII
200 rue de la Loi
1049 BRUXELLES
BELGIUM
Tel. 32-2-2357210
Tlx. 21877 COMEU B
Fax 32-2-2350150

KADDOUR K.
SITEP
92 Rue de Palestine
TUNIS
TUNISIA
Tel. 216-1-785244
Tlx.
Fax

KAPPEL J.J.
J.J. KAPPEL MARINE CONCEPT
Hoymoseyes 4
3400 HILLERφD
DENMARK
Tel. 45-42260711
Tlx.
Fax 45-48242294

KESSEL D.
INSTITUT FOR PETROLEUM RESEARCH
Walter-Nernst-Strasse 7
3392 CLAUSTHAL-ZELLERFELD
GERMANY
Tel. 49-5323-711100
Tlx.
Fax 49-5323-711200

KHELIFA Ch.
SITEP
92 Rue de Palestine
TUNIS
TUNISIA
Tel. 216-1-785244
Tlx. 14539 TN
Fax

KING G.
HALLIBURTON GEOPHYSICAL SERVICES
G. Horne Lane
BEDFORD
UNITED KINGDOM
Tel. 44-234-273388
Tlx. 827737 HGS
Fax 44-234-273323

KIRKHUS ϕ.
STATOIL
P.O. Box 1212
5001 BERGEN
NORWAY
Tel. 47-5-9929011
Tlx.
Fax

KOHLER N.
INSTITUT FRANCAIS DU PETROLE
Direction Gisements
1&4 Av. de Bois Préau, B.P. 311
92506 RUEIL-MALMAISON - CEDEX
FRANCE
Tel. 33-1-47526632
Tlx. A 203050 F
Fax 33-1-47490411

KOHRTZ J.W.
DANISH ENERGY AGENCY
Landemaerket 11
1119 COPENHAGEN
DENMARK
Tel. 45-33-926700
Tlx. 22450 DENERG
Fax 45-33-114743

KOMATITSCH J.M.
Société Nationale ELF AQUITAINE
Production - CSTJF
Ave. Larribeau
64000 PAU
FRANCE
Tel. 33-59834792
Tlx. PETRA 560804 F
Fax 33-59834369

KOSMAS G.
NORTH AEGEAN PETROLEUM COMPANY
2, Kapodistria Street and Kifisias Ave.
15210 CHALANDRI
GREECE
Tel. 30-1-6830350
Tlx. 218315 NAPC GR
Fax 30-1-6842091

KOUCHIT A.K.
SONATRACH DIVISION EXPLORATION
2 rue Capt. Azzoug Cote Rouge Hussein
ALGER
ALGERIE
Tel. 213-2-771140
Tlx. 65405
Fax 213-2-595758

KRILOV Z.
INA - NAFTAPLIN SLI
Lovinciceva 1
41100 ZAGREB
YUGOSLAVIA
Tel. 38-41-2218849
Tlx. 2255 L
Fax 38-41-220117

KUNZ U.
UNIVERSITY OF BRAUNSCHWEIG
Institut für Technische Chemie
Hans-Sommer-Str. 10
3300 BRAUNSCHWEIG
GERMANY
Tel. 49-531-3915360
Tlx. 952526 TU BSW
Fax 49-531-3915357

LøKEN T.A.
N.T.N.F.
Sognsveien 72
OSLO 8
NORWAY
Tel. 47-2-237685
Tlx. 76951 NTNF N
Fax 47-2-181139/184137

LACAZE J.
SNEA-P
8 Bd des Pyrénées
64000 PAU
FRANCE
Tel. 33-59834511
Tlx.
Fax 33-59834858

LAFAILLE A.
TOTAL - Cie. Française des Pétroles
Tour TOTAL, 24 Cours Michelet Puteaux
92069 PARIS LA DEFENSE - CEDEX 47
FRANCE
Tel. 33-1-42914247
Tlx. TCFP 615700 F
Fax 33-1-42913924

LAGOURETTE B.
LAB. DE PHYSIQUE DES MATERIAUX INDUSTRIELS
Faculté des Sciences
Avenue de L'Université
64000 PAU
FRANCE
Tel. 33-59923039
Tlx.
Fax 33-59841696

LAHIANI M.
Immeuble BNDT
Ave. Mohamed V
TUNIS
TUNISIA
Tel. 216-1-284674
Tlx. 14652 TN
Fax

LALLI D.
TECNOMARE SpA.
San Marco 3584
30100 VENEZIA
ITALY
Tel. 39-41-796711
Tlx. 410484 MAREVE
Fax 39-41-5230363

LAMMERS J.N.J.J.
SHELL Amsterdam
Shell Research B.V.
Badhuisweg 3 - P.O.Box 3003
1003 AMSTERDAM
THE NETHERLANDS
Tel. 31-20-303539
Tlx. 11224 KSLANL
Fax 31-20-308025

LARSEN V.B.
STATOIL
P.O. Box 1212
5001 BERGEN
NORWAY
Tel. 47-5-9929011
Tlx.
Fax

LAURITZEN O.
NOPEC A/S
P.O. Box 88
NAERSNES
NORWAY
Tel. 47-3-280001
Tlx. 72390
Fax 47-3-281667

LEBLANC M.
GAZ DE FRANCE - D.E.T.N.
Dét. Res. Souterrains Serv. Géologie
Ave du Président Wilson 361, B.P. 33
93211 LA PLAINE ST DENIS
FRANCE
Tel. 33-1-49224909
Tlx. GDFSOUT 233863 F
Fax 33-1-49224954

LEBLOND M.
GERTH
4 avenue de Bois Préau
92506 RUEIL MALMAISON CEDEX
FRANCE
Tel. 33-1-47526139
Tlx.
Fax 33-1-47526927

LEDUC J.
INSTITUT FRANCAIS DU PETROLE
120 Ave. Félix Faure
75015 PARIS
FRANCE
Tel. 33-1-47526586
Tlx.
Fax 33-1-47526754

LEFEVRE P.
ELF Aquitaine
SNEA (P)
26 Av. des Lilas
64018 PAU Cédex
FRANCE
Tel. 33-59834000
Tlx. 560804
Fax 33-59834476

LEGALLAIS L.
DUCO - DULOP COFLEXIP
UMBILICALS LTD.
1 Rue d'Artois
78630 ORGEVAL
FRANCE
Tel. 33-39757344
Tlx.
Fax

LEHMAN B.
DEUTSCHE MONTAN TECHNOLOGIE
Institut für Angewandte Geophysik
Herner Str. 45
4630 BOCHUM 1
GERMANY
Tel. 49-234-625613
Tlx.
Fax 49-234-625607

LEHNER F.K.
SHELL - Expl. en Produktie Lab.
P.O. Box 60
2280 AB RIJSWIJK Z.H.
THE NETHERLANDS
Tel. 31-70-3112558
Tlx.
Fax 31-70-3113110

LENORMAND R.
INSTITUT FRANCAIS DU PETROLE
Av. de Bois Préau 1&4, P.O. Box 311
92506 RUEIL-MALMAISON CEDEX
FRANCE
Tel. 33 1 - 47526198
Tlx. IFP A 203050 F
Fax 33-1-47490411

LEPRINCE P.
INSTITUT FRANCAIS DU PETROLE
Av. de Bois Préau 1 & 4
92506 RUEIL-MALMAISON - CEDEX
FRANCE
Tel. 33-1-47490214
Tlx. IFP A 203050 F
Fax 33-1-47323092

LEWIS H.
THE PETROLEUM SCIENCE AND
TECHNOLOGY INSTITUTE
Research Park Riccarton
EDINBURGH EH14 4AS
UNITED KINGDOM
Tel. 44-31-451 5231
Tlx. 931 211 0965
Fax 44-31-451 3127

LONGBOTTOM P.L.
THE PETROLEUM SCIENCE AND
TECHNOLOGY INSTITUTE
Research Park Riccarton
EDINBURGH EH14 4AS
UNITED KINGDOM
Tel. 44-31-4515231
Tlx. 9312110965
Fax 44-31--4513127

LORENZ U.
POLISH ACADEMY OF SCIENCES
MAEERC
Ul. Wadowicka 12
KRAKOW
POLAND
Tel. 48-12-669029
Tlx. 0322377
Fax 48-12-669029

LOUNES A.
SONATRACH DIVISION EXPLORATION
2, rue Capt. Azzoug Cote Rouge Hussein
ALGER
ALGERIE
Tel. 213-2-771140
Tlx. 65405
Fax 213-2-595758

LOUWRIER K.L.
COMMISSION OF THE EUROPEAN COMMUNITIES
DG XII/E.4
200 Rue de la Loi
1049 BRUXELLES
BELGIUM
Tel. 32-2-2356962
Tlx. 21877 COMEU B
Fax 32-2-2363024

MøLLER J.J.
DANISH GEOLOGICAL SURVEY
Thoravej 8
2400 COPENHAGEN NV
DENMARK
Tel. 45-106600
Tlx. 19999 DANGED DK
Fax 45-196868

MACAULEY C.
THE PETROLEUM SCIENCE AND TECHNOLOGY INSTITUTE
Research Park Riccarton
EDINBURGH EH14 4AS
UNITED KINGDOM
Tel. 44-31-4515231
Tlx. 9312110965
Fax 44-31-4513127

MADDALENA D.
TECNOMARE SpA.
San Marco 3584
30124 VENEZIA
ITALY
Tel. 39-41-796711
Tlx. 410484
Fax 39-41-5230363

MAHMUD MOHAMED B.
WAHA OIL COMPANY
P.O. Box 375
TRIPOLI
LIBYA
Tel. 31116 EXT.2457
Tlx.
Fax

MAKRIS J.
UNIVERSITY OF HAMBURG
Institute of Geophysics
Bundestr. 55
2000 HAMBURG 13
GERMANY
Tel. 49-40-41233969
Tlx. 214732 UNIHHD
Fax 49-40-41232449

MALLEN J.
COFLEXIP
Rue Jean Hure - B.P.7
76580 LE TRAIT
FRANCE
Tel. 33-35055000
Tlx. 180024
Fax 33-35574960

MANN U.
KFA JÜLICH - Nuclear Research Center - ICH-5
P.O. Box 1913
5170 JÜLICH
GERMANY
Tel. 49-2461-614187
Tlx. 883556 OKFD
Fax 49-2461-612484

MARCHAND D.
BERTIN
B.P. 3
78373 PLAISIR CEDEX
FRANCE
Tel. 33-1-34818500
Tlx.
Fax

MARIANI E.
AGIP SpA. RIST/BREL
Viale Europa 38/44
20093 COLOGNO MONZESE MI
ITALY
Tel. 39-2-52027828
Tlx. 310246 ENI
Fax 39-2-52027515

MARTIN BOURGON J.M.
REPSOL Exploracion S.A.
Pez Valador 2
28007 MADRID
SPAIN
Tel. 34-1-5747200
Tlx. 49544
Fax

MARTINO D.
MINISTERO INDUSTRIA
Capo del Corpo Delle Miniere - Ufficio Idrocarburi
Via Molise
00187 ROMA
ITALY
Tel.
Tlx.
Fax

MATTEINI L.
AGIP SpA.
APSE Dept.
Via Fabiani 1
20097 SAN DONATO MILANESE - MI
ITALY
Tel. 39-2-5209362
Tlx. 310246
Fax 39-2- 52038609

MAURY V.
Société Nationale ELF Aquitaine Production
Avenue Larribau
64018 PAU CEDEX
FRANCE
Tel. 33-59834024
Tlx.
Fax 33-59835302

MAURY J.
C.F.T.P.
116, Av. de la Liberté
1002 TUNIS
TUNISIA
Tel. 213-1-287 114
Tlx. 14192 CF TP
Fax 213-1-894 400

MAVER K.G.
DANISH GEOLOGICAL SURVEY
Thoravej 8
2400 COPENHAGEN NV
DENMARK
Tel. 31-106600
Tlx. 19999 DANGED DK
Fax 31-196868

McMULLAN J.T.
UNIVERSITY OF ULSTER
Centre for Energy Research
Coleraine Co.
LONDONDERRY BTS2 15A
UNITED KINGDOM
Tel. 44-265-4441
Tlx. 51-747597
Fax 44-265-40900

McNAMARA S.
MCS INTERNATIONAL
3 Buttermilk Walk
GALWAY
IRELAND
Tel. 353-91-66455
Tlx.
Fax 353-91-66457

MELVIN A.
BRITISH GAS Plc.
London Research Station
Michael Road
LONDON SW6 2AD
UNITED KINGDOM
Tel. 44-71-736 33 44
Tlx. 24670
Fax 44-71-7365296

MEYN V.
INSTITUT FÜR ERDÖLFORSCHUNG
Physik der Erdöl- & Erdgasgewinnung
Walther-Nernst-Str. 7
3392 CLAUSTHAL-ZELLERFELD
GERMANY
Tel. 49-5323-711173
Tlx.
Fax 49-5323-711200

MEZINI A.
O.P.G.N.
PATOS
ALBANIA
Tel.
Tlx.
Fax

MILLER M.
THE PETROLEUM SCIENCE AND TECHNOLOGY INSTITUTE
Research Park Riccarton
EDINBURGH EH14 4AS
UNITED KINGDOM
Tel. 44-31-4515231
Tlx. 9312110965
Fax 44-31-4513127

MOLINARI M.
ENIRICERCHE S.p.A.
Via Maritano 26
20097 SAN DONATO MILANESE - MI
ITALY
Tel. 39-2-52038223
Tlx.
Fax

MONDT J.C.
SHELL - K.S.E.P.L.
Volmerlaan 6
2288 GD RIJSWIJK
THE NETHERLANDS
Tel. 31-70-3112668/3113911
Tlx. 31527 KSEPL NL
Fax 31-70-3113110

MONDY J.-F.
TOTAL CFP
Cédéx 47
92069 PARIS LA DEFENSE
FRANCE
Tel. 33-1-42913706
Tlx. TCFP 615700 F
Fax 33-1-42914052

MONNET J.
(SNEA) ELF AQUITAINE
Drilling Completion & Eng. Dept.
Avenue Larribau
64018 PAU CEDEX
FRANCE
Tel. 33-59836942
Tlx. 560840
Fax 33-59836565

MONTADERT L.
INSTITUT FRANCAIS DU PETROLE
Exploration - Gisements
Av. de Bois Préau, 1 & 4
92506 RUEIL-MALMAISON Cédex
FRANCE
Tel. 33-1-47490214
Tlx. IFP A 203050 F
Fax 33-1-47490411

MORVIK N.
STATOIL
P.O. Box 1212
5001 BERGEN
NORWAY
Tel. 47-5-9929011
Tlx.
Fax

MOSDITCHIAN G.
MINISTERE DE L'INDUSTRIE ET DE L'AMENAG. DU TERRITOIRE
Direction des Hydrocarbures
3-5 Rue Barbet
75116 PARIS
FRANCE
Tel. 33-1-45563821
Tlx.
Fax 33-1-45562274

MOUGENOT D.
COMPAGNIE GENERALE DE GEOPHYSIQUE
1 Rue Léon Migaux
MASSY CEDEX
FRANCE
Tel. 33-1-64473384
Tlx.
Fax 33-1-64473970

MULLER G.
FACULTE SCIENCES ROUEN
URA 500 CNRS
MONT SAINT AIGNAN CEDEX
FRANCE
Tel. 33-35146693
Tlx. 770127 F
Fax 33-35146349

MYSKO A.
POLISH ACADEMY OF SCIENCES
MAEERC
Ul. Wadowicka 12
KRAKOW
POLAND
Tel. 48-12-669029
Tlx. 0322377
Fax 48-12-669029

NEGGACHE A.
TOTAL CFP
Tour TOTAL
24 Cours Michelet, La Défense 10
PUTEAUX
FRANCE
Tel. 33-1-42913416
Tlx. 615700
Fax 33-1-42914291

NEIGHBOUR J.
ALPHA THAMES
Suite 2 - Essex House
Station Road
UPMINSTER ESSEX
UNITED KINGDOM
Tel. 44-4022-29229
Tlx.
Fax 44-4022-51273

NERGER M.
Technische Universität Berlin
SFB 69, SEHR. ACK. 9
Ackerstr. 71 - 76
1000 BERLIN 65
GERMANY
Tel. 49-30-31472696
Tlx.
Fax 49-30-31472837

NICOLAY D.
COMMISSION OF THE EUROPEAN COMMUNITIES
DG XVII
Rue Alcide de Gasperi
2920 LUXEMBOURG
LUXEMBOURG
Tel. 352-43012946
Tlx. 2752 EURODOC LU
Fax 352-43014129

NICOLUSSI F.
TECNOMARE SpA.
San Marco 3584
30124 VENEZIA
ITALY
Tel. 39-41-796711
Tlx. 410484
Fax 39-41-5230363

NIELSEN J.B.
DANISH HYDRAULIC INSTITUTE
Hydro Informatics Centre
Agern Allé 5
2970 HORSHOLM
DENMARK
Tel. 45-42-868033
Tlx. 37402 DHICPH
Fax 45-42-867951

NIO S.D.
INTERNATIONAL GEOSERVICES BV
Reaal 5 Q
2353 TK LEIDERDORP
THE NETHERLANDS
Tel. 31-71-417003
Tlx. 39337
Fax 31-71-416939

NOIK C.
INSTITUT FRANCAIS DU PETROLE
1-4 Avenue de Bois Préau
92506 RUEIL-MALMAISON
FRANCE
Tel. 33-1-47526269
Tlx. IFP A 203050F
Fax 33-1-47490411

NOLA L.
AGIP SpA.
Via del Serafico 89/91
00142 ROMA
ITALY
Tel. 39-6-50392497
Tlx. 613525
Fax 39-6-50392241/50392320

STEEN A.
NORSK HYDRO RESEARCH CENTER
A.S.
P.O. Box 4313
5028 BERGEN
NORWAY
Tel. 47-5-996805
Tlx. 40920 HYDRO N
Fax 47-5-996196

ORTOLANO L.
SELM PETROLEUM
V.le Teracati 102
96100 SIRACUSA
ITALY
Tel. 39-931-32699
Tlx. 971367
Fax 39-931-442610

OVERLI J.M.
STATOIL
Postuttak
7004 TRONDHEIM
NORWAY
Tel. 47-7-584011
Tlx. 55278 STATD N
Fax 47-7-584532

PANENI M.
AGIP SpA.
Viale Europa 38/44
20093 COLOGNO MONZESE MI
ITALY
Tel. 39-2-52027867
Tlx. 310246 ENI
Fax 39-2-52027856

PAOLETTI
AGIP SpA.
Via del Serafico 89-91
00100 ROMA
ITALY
Tel. 39-6-50392443
Tlx. 613525
Fax 39-6-50392320

PAOLUCCI G.
AGIP SpA.
Senior Vice President Exploration
C.P. 12069
20120 MILANO
ITALY
Tel. 39-2-5204092
Tlx. 310426 ENI
Fax 39-2-52037651

PAVONI S.
EURON - DIPRA
Via F. Maritano 26
20096 SAN DONATO MILANESE - MI
ITALY
Tel. 39-2-52033209
Tlx. 310246
Fax 39-2-52037171

PECCI G.
ECOFUEL
Via Brento 15
20139 MILANO
ITALY
Tel. 39-2-52021932
Tlx.
Fax 39-2-52021960

PERSOGLIA S.
O.G.S. DI TRIESTE
Seismic Department
Borgo Grotta Gigante, 095 C.P. 2011
34016 OPICINA TS
ITALY
Tel. 39-40-21401
Tlx. 460329 OGS I
Fax 39-40-327307

PIANA M.
AGIP SpA.
Servizio CELE
20097 SAN DONATO MILANESE - MI
ITALY
Tel. 39-2-52037321
Tlx. 310246 ENI I
Fax

PILARA R.
CORPO REGIONALE MINIERE
Via C. Camilliani, 86
90145 PALERMO
ITALY
Tel.
Tlx.
Fax

PINTO A.D.
PARTEX - Companhia Portuguesa de Serviços
Divisao Explor. e Prod.
Av. 5 de Outubro 160
1000 LISBOA
PORTUGAL
Tel. 351-1-735290
Tlx. 14708 PARSER
Fax 351-1-779516

PIRAS P.
AGIP SpA.
Via Fabiani 1
20097 SAN DONATO MILANESE - MI
ITALY
Tel. 39-2-52037937
Tlx. 310249
Fax 39-2-5207309

POLETTO F.
O.G.S. DI TRIESTE
P.O. Box 2011
34016 TRIESTE
ITALY
Tel. 39-40-21401
Tlx. 460329 OGS I
Fax 39-40-327307

PRAZZOLI M.
AGIP SpA.
Piazza Vanoni 1
20097 SAN DONATO MILANESE - MI
ITALY
Tel. 39-2-52037534
Tlx.
Fax 39-2-52036349

PRIISHOLM S.
MAERSK OLIE OG GAS AS
50 Esplanaden
1263 COPENHAGEN
DENMARK
Tel. 45-33114676
Tlx.
Fax 45-33322325

QUIGLEY Th. M.
BRITISH PETROLEUM RESEARCH
Sunbury Research Centre
Chertsey Road
SUNBURY-ON-THAMES MIDDX TW16 4LN
UNITED KINGDOM
Tel. 44-932-763318
Tlx. 296041
Fax 44-932-763747

RAMPOLLI M.
AGIP S.p.A.
C.P. 12069
20097 SAN DONATO MILANESE - MI
ITALY
Tel. 39-2-52027485
Tlx. 310246 ENI I PER AGIP FB 006
Fax 39-2-5204382

RAUBOLD H.
I.F.G.
Torfstrasse 1
D - 2085 QUICKBORN
GERMANY
Tel. 49-4106-3470
Tlx.
Fax 49-4106-2228

RESULI S.
INSTITUTE OF OIL AND GAS
TECHNOLOGY
PATOS
ALBANIA
Tel. 396-49668
Tlx.
Fax

REY-GRANGE A.
SEAMET INTERNATIONAL
61 Rue de la Garenne
92310 SEVRES
FRANCE
Tel. 33-1-45348523
Tlx.
Fax 33-1-45078476

RITSEMA I.
TNO
Schoemakerstraat 97
2600 AG DELFT
THE NETHERLANDS
Tel.
Tlx.
Fax

ROCCA F.
POLITECNICO DI MILANO
Dipt. Elettronica
Piazza Leonardo da Vinci 32
20133 MILANO
ITALY
Tel. 39-2-23993573
Tlx. 333467 POLIMI I
Fax 39-2-23993587

ROETHEL Th.
RWTH AACHEN
Keltenstr. 50
5160 DUEREN
GERMANY
Tel. 49-2421-72812
Tlx.
Fax 49-241-806493

ROSSANI A.
SITEP
B.P. 1000
TUNIS
TUNISIA
Tel.
Tlx.
Fax

ROSSINI C.
AGIP S.p.A.
Centro Studi
Via Fabiani 1 1
20097 SAN DONATO MILANESE - MI
ITALY
Tel. 39-2-52033501
Tlx.
Fax 39-2-5207309

ROUGEAUX M.
G.E.R.T.H.
4 Av. de Bois Préau
92502 RUEIL MALMAISON
FRANCE
Tel. 33-1-47526588
Tlx.
Fax 33-1-47526927

ROUGIER R.
COMEX SERVICES
36, Bld. des Océans
13275 CEDEX 9 MARSEILLE
FRANCE
Tel. 33-91235000
Tlx. 41098 SF
Fax 33-91401280

ROVENSKAYA A.S.
IGIRGI
50 Fersman Str.
117312 MOSCOW
U.S.S.R.
Tel.
Tlx.
Fax

RUMPLER J.
GEOPHYSICAL EXPLORATION COMPANY
Gorkij Fasor 42
1068 BUDAPEST
HUNGARY
Tel. 36-1-1423939
Tlx. 225123
Fax 36-1-1667760

RUSSOMANNO F.
CORPOVEN S.A.
La Campina
Avenida Libertador
CARACAS
VENEZUELA
Tel. 58-2-7081228
Tlx. 21363
Fax

SAGOT A.
ELF AQUITAINE
B.P. 65
64000 PAU
FRANCE
Tel. 33-59835123
Tlx. 560804 APUI
Fax 33-59836829

SAID F.H.
MEDITERRANEAN OILFIELD SERVICES COMPANY LTD
Manoel Island
GZIRA
MALTA
Tel. 356-314666/318949
Tlx. 1485 MW
Fax 356-339511

SAINT-GUIRONS H.
LAB DE PHYSIQUE DES MATERIAUX INDUSTRIELS
Faculté des Sciences
Avenue de L'Université
64000 PAU
FRANCE
Tel. 33-59923126
Tlx.
Fax 33-59841696

SANDVIK K.O.
IKU
Haakon Magnussonsgatan 1B,
7002 TRONDHEIM
NORWAY
Tel. 47-7-591100
Tlx. 55434 IKU N
Fax 47-7-591102

SANFILIPPO D.
SNAMPROGETTI SpA.
Via Maritano 26
20097 SAN DONATO MILANESE - MI
ITALY
Tel. 39-2-5205520
Tlx. 310246 ENI
Fax 39-2-52022757

SARANTINOS V.
NORTH AEGEAN PETROLEUM COMPANY
EPE
Filellinon 2 - P.O. Box 1307
65302 KAVALA
GREECE
Tel. 30-51-832442
Tlx.
Fax 30-51-229287

SARDA J.-P.
INSTITUT FRANCAIS DU PETROLE
G.E.R.T.H.
Exploration-Production
4 Av. de Bois Préau B.P. 311
92506 RUEIL-MALMAISON Cédex
FRANCE
Tel. 33-1-47526313
Tlx. 203050 F
Fax 33-1-47527000

SCHUPPERS J.D.
DELFT UNIVERSITY OF TECHNOLOGY
Faculty of Mining & Petroleum Eng.
P.O. Box 5028
2600 GA DELFT
THE NETHERLANDS
Tel. 31-15-785108
Tlx.
Fax 31-15-784891

SCHWOCHAU K.
KFA Jülich- Institute of Petroleum
and Organic Geochemistry
P.O. Box 1913
5170 JÜLICH
GERMANY
Tel. 49-2461-613264
Tlx. 833556 KFA D
Fax 49-2461-612484/2

SCIUTO F.
MINISTERO dell'INDUSTRIA
Uff. Naz. Minerario per gli Idrocarburi e la
Geotermia
Via Molise 2
00187 ROMA
ITALY
Tel.
Tlx.
Fax

SEBASTIANI G.
CEOM S.C.P.A.
Via G. di Marzo 2/F
90144 PALERMO
ITALY
Tel.
Tlx.
Fax

SFLIGIOTTI G.
AGIP SpA.
Via del Serafico, 89-91
00142 ROMA
ITALY
Tel. 39-6-50392-1
Tlx. 613525 AGIPDA
Fax 39-6-50392320

SGUBINI L.
AGIP SpA.
P.O. Box 12069
20100 MILANO
ITALY
Tel. 39-2-5204096
Tlx. 310426 ENI
Fax 39-2-52037651

SHAWESH M.
WAHA OIL COMPANY
Damra Field Ext. 5555
023 ZAWIA
LIBYA
Tel. 20990
Tlx.
Fax

SHEBANI A.
NATIONAL OIL COMPANY
P.O. Box 2655
TRIPOLI
LIBYA
Tel. 4681
Tlx.
Fax

SIKLOS J.
HUNGARIAN OIL AND GAS TRUST
Schönherz Z.U. 18
1117 BUDAPEST
HUNGARY
Tel. 36-1-1869 363/664 000
Tlx. 22-5123
Fax 36-1-1868 856/667 760

SIMANDOUX P.
IINSTITUT FRANCAIS DU PETROLE
Valorisation, Exploration-Production
Av. de Bois Préau, 1 & 4
92506 RUEIL-MALMAISON CEDEX
FRANCE
Tel. 33-1-47.52.66.71
Tlx. A 203050 F
Fax 33-1-47490411

SOLBAKKEN A.
STATOIL
R&D
Postuttak
7004 TRONDHEIM
NORWAY
Tel. 47-7-584011
Tlx. 55278 STATD N
Fax 47-7-967286

SPARKS Ch.
INSTITUT FRANCAIS DU PETROLE
1 & 4, Avenue de Bois Préau
92506 RUEIL-MALMAISON
FRANCE
Tel. 33-1-47526395
Tlx. 203050
Fax 33-1-47527002

SPREUX A.
ELF Aquitaine
Centre Jean Feger
Avenue des Lilas 26
64018 PAU CEDEX
FRANCE
Tel. 33-59-835682
Tlx.
Fax 33-59-836829

STEENKEN W.F.
(E.A.P.G.) European Association of
Petroleum Geoscientists
Utrechtseweg, P.O. Box 298 62
3704 HE ZEIST
THE NETHERLANDS
Tel. 31-3404-62655
Tlx. 33480 CALL NL
Fax 31-3404-62640

STRANDE J.
INDUSTRIAL LIAISON COUNCIL
Nat. Agency of Industry and Trade
Tagensvej 137
2200 KOBENHAVN N
DENMARK
Tel. 45-31-851066
Tlx. 15768 INDIRA DK
Fax 45-31-817068

SUGIER A.
INSTITUT FRANCAIS DU PETROLE
Matériaux et Chimie Appliquée
1&4 Av. de Bois Préau
92506 RUEIL-MALMAISON
FRANCE
Tel. 33-1-47 32 92 06
Tlx. IFP A 203050 F
Fax 33-1-47526429

TALLEC E.
GAZ DE FRANCE
D.E.T.N.
361 Ave. du Président Wilson
93211 LA PLAINE ST.DENIS
FRANCE
Tel. 33-1-49224932
Tlx.
Fax

TAMBINI M.
AGIP SpA./TEOP
P.O. Box 12069
20120 MILANO
ITALY
Tel. 39-2-52027460
Tlx.
Fax 39-2-52027483

TARANTOLA A.
UNIVERSITE DE PARIS VI
Institut de Physique du Globe de Paris
4 Place Jussieu - Tour 14
75252 PARIS CEDEX 05
FRANCE
Tel. 33-1-43267898
Tlx. VOLSISM 202 BIO F
Fax 33-1-43264029

TERDICH P.
AGIP SpA.
Via Fabiani 1
20097 SAN DONATO MILANESE - MI
ITALY
Tel. 39-2-52037939
Tlx. 310249
Fax 39-2-5207309

THONON C.
40B Rue St. Christophe
84000 AVIGNON
FRANCE
Tel. 33-90851843
Tlx.
Fax 33-90851358

TISSOT B.P.
INSTITUT FRANCAIS DU PETROLE
Recherche Scientifique
1 et 4 av. de Bois Préau
92506 RUEIL-MALMAISON CEDEX
FRANCE
Tel. 33-1-47526665
Tlx. 203050 F
Fax 33-1-47490411

TOLLEFSEN B.
NORSK HYDRO A/S
P.O. Box 200
1321 STABEKK
NORWAY
Tel. 47-2-738100
Tlx. 72948 HYDRO N
Fax 47-2-734876

TONELLI S.
AGIP SpA.
PROI/MAIT
P.O. Box 12069
20097 SAN DONATO MILANESE - MI
ITALY
Tel. 39-2-52033670
Tlx.
Fax 39-2-52033828

TORP T.A.
STATOIL
R&D Dept.
Postuttak
7004 TRONDHEIM
NORWAY
Tel. 47-7-584011
Tlx. 55278
Fax 47-7-967286

TOSCHEK P.H.
WINTERSHALL NOORDZEE
CORPORATION
P.O. Box 82301
2508 EH THE HAGUE
THE NETHERLANDS
Tel. 31-70-3583100
Tlx.
Fax 31-70-3583333

TRABELSI K.
INSTITUT FRANCAIS DU PETROLE
1&4 Av. de Bois Préau
92506 RUEIL-MALMAISON
FRANCE
Tel. 33-1-47490214
Tlx. IFP A 203050F
Fax 33-1-47490411

UNGERER Ph.
INSTITUT FRANCAIS DU PETROLE
1&4, Av. de Bois Préau
92506 RUEIL MALMAISON
FRANCE
33-1-47526624
Tel. IFP A 203050F
Tlx. 33-1-47490411
Fax

URSIN J.R.
ROGALAND UNIVERSITY CENTRE
P.O. Box 2557
4004 STAVANGER
NORWAY
Tel. 47-4-874283
Tlx.
Fax 47-4-874300

VAN ASSELT D.
COMMISSION OF THE EUROPEAN COMMUNITIES
DG XVII
200 rue de la Loi
1049 BRUXELLES
BELGIUM
Tel. 32-2-2353472
Tlx. 21877 COMEU B
Fax 32-2-2350150

VAN DAM J.
Domaine du Cap Benat
81 Route du Sémaphore
83230 BORMES LES MIMOSAS
FRANCE
Tel. 33-94715615
Tlx.
Fax 33-94718333

VAN DER BOK J.
WINTERSHALL NORDZEE B.V.
P.O. Box 82301
2508 EH THE HAGUE
THE NETHERLANDS
Tel. 31-70-3583169
Tlx. 31708
Fax 31-70-3583321

VAN WECHEM H.M.H.
SHELL Intern. Petroleum Maatschappij
P.O. Box 162
2501 AM THE HAGUE
THE NETHERLANDS
Tel. 31-70-3772423
Tlx. 36000
Fax 31-70-3772779

VARGA F.
HUNGARIAN HYDROCARBON INSTITUTE
P.O. Box 32
SZAZH ALOMBATTA
HUNGARY
Tel. 36-1-1800122
Tlx.
Fax 36-1-1802145

VERLY G.
BP RESEARCH
SUNBURY ON THAMES TW16 7LN
UNITED KINGDOM
Tel. 44-932-763552
Tlx.
Fax 44-932-763352

VESTAVIK O.M.
ROGALAND RESEARCH INSTITUTE
P.O. Box 2503
4004 STAVANGER
NORWAY
Tel. 47-4-875289
Tlx.
Fax 47-4-875200

VINCIGUERRA C.
CEOM
Via G. di Marzo, 2/F
90144 PALERMO
ITALY
Tel.
Tlx.
Fax

VINCKEN L.M.J.
DIETSMANN INTERNATIONAL N.V.
Noorderlaan 133 - Box 33
2030 ANTWERPEN
BELGIUM
Tel. 32-3-5417233
Tlx. 31849
Fax 32-3-5412051

VLIERBOOM F.W.
OCCIDENTAL INT. EXPLORATION AND PRODUCTION COMPANY
1200 Discovery Way
BAKENSFIELD, CALIFORNIA 93309
USA
Tel. 805-231-6993
Tlx. 188337 OXYBAK UT
Fax 805-322-7457

VOLPI B.
AGIP SpA.
Centro Studi
Via Fabiani 1
20097 SAN DONATO MILANESE - MI
ITALY
Tel. 39-2-5206365
Tlx.
Fax 39-2-5207309

WEBER K.J.
SHELL International Petroleum Mij.
Postbus 162
2501 AN DEN HAAG
THE NETHERLANDS
Tel. 31-70-3774476
Tlx. 36000
Fax 31-70774848

WEIDENBACH G.
COMMISSION OF THE EUROPEAN COMMUNITIES
DG XII/E
200 rue de la Loi
1049 BRUXELLES
BELGIUM
Tel. 32-2-2354393
Tlx. 21877 COMEU B
Fax 32-2-2363024

WEILL J.
INSTITUT FRANCAIS DU PETROLE
Direction de R&D - CEDI
B.P. 3
69390 VERNAISON
FRANCE
Tel. 33-78022020
Tlx.
Fax 33-78022008

WELTE D.
Kernforschungsanlage Jülich GmbH.
Institut für Chemie 5
Postfach 1913
5170 JÜLICH
GERMANY
Tel. 49-2461-614701
Tlx. 83355680
Fax 49-2461-615370

WEROVSZKY PIPICZ V.
OIL AND GAS PROD. COMPANY OF NAGYALFOLD
P.O. Box 86, ADY E.N. 20
5000 SZOLNOK
HUNGARY
Tel. 36-56-31611
Tlx. 23320/23372
Fax 36-56-30103

WILCOCKSON A.
AEA TECHNOLOGY - PETROLEUM SERVICES
Building 329
HARWELL OXFORDSHIRE OX11 ORA
UNITED KINGDOM
Tel. 44-235-432353
Tlx. 88135
Fax 44-235-436660

WOCHENFUSS A.
RWE DEA
Ueberseerding 40
2000 HAMBURG 60
GERMANY
Tel. 49-40-63752083
Tlx.
Fax

WRIGHT I.W.
BP PETROLEUM DEVELOPMENT
Dimlington Terminal
HUUL HU12 OSU
UNITED KINGDOM
Tel. 44-964-652135
Tlx.
Fax 44-964-650078

XERRI S.
OFFICE OF THE PRIME MINISTER
Oil Division
Auberge de Castille
VALLETTA
MALTA
Tel. 356-246065
Tlx. 1485 CEDOPM MW
Fax 356-248015

ZAHO S.
INSTITUTE OF OIL AND GAS
TECHNOLOGY
PATOS
ALBANIA
Tel. 396-49668
Tlx.
Fax

ZEMMOURI E.K.
SONATRACH DIVISION EXPLORATION
2, rue Capt. Azzoug Cote Rouge Hussein
ALGER
ALGERIE
Tel. 213-2-771140
Tlx. 65405
Fax 213-2-595758

ZAPPARDINO R.
CORPO REGIONALE MINIERE
Via Ugo Bassi 7
40100 BOLOGNA
ITALY
Tel.
Tlx.
Fax

INDEX OF AUTHORS

AAS, B., 406
ALBERTSEN, M., 489
ALLIEZ, J., 155
ANGELERI, G.P., 422
ANXIONNAZ, H., 522
ARCHER, J.S., 208
ARMENIS, D., 317
ATTAYA, C., 80
AZOUZ, M., 80

BACH, H., 228
BALOSSINO, P., 204
BARBE, P., 562
BARONNET, F., 562
BENETTI, M., 54
BENHADID, A., 562
BEUCHER, H., 497
BJØRNSTAD, T., 212
BLIGHT, J., 404
BORRIELLO, G., 407
BOURDEAU, PH., 4
BOURGON, J.M.M., 282
BOUTECA, M.J., 260, 330
BOUTROLLE, P., 370
BOZZO, G.M., 54
BRAUNSCHWEIG, B., 465
BREKHUNTSOV, A., 116
BRUNELLI, M., 560
BULLER, D.C., 182
BURRUS, J., 106

CAMERON, C.J., 544
CARLINI, A., 412
CARVALHO, J., 192
CAZZULO, R., 250
CEFFA, L., 216
CHARLEZ, P., 330
CLIET, CH., 524
CÔME, G.M., 562
COMOTTI, P., 561
COOLEN, P.T., 537
CORNINI, S., 412
COSENZA, G., 523
COSTA E SILVA, A., 192
COSTE, J.F., 398
COURTEILLE, J.-M., 370
CSERMELY, Z., 400

DALTABAN, T.S., 208
DE BAUW, R., 8
DEAN, T.L., 359
DEGOUY, D., 401
DESCALZI, C., 80
DESERT, G., 397
DETIENNE, J.L., 330
DI DIO, G., 476
DI LUISE, G., 216
DONCHE, A., 206
DONSELAAR, M.E., 90
DOSSENA, G., 216
DOWTY, E.L., 359
DOYEN, P.M., 445
DRESEN, L., 455
DRONKERT, H., 211
DUBESSET, M., 524
DUGSTAD, Ø., 212

EBDON, R.W., 237
EHRET, D., 525
ESCHARD, R., 497

FENDEL, A., 210
FOAKES, A.P., 272
FONNESU, F., 476
FOURMAINTRAUX, D., 260
FRIAS, M., 564

GAAYA, H., 80
GALLI, A., 497
GANZ, SH. N., 207
GEEL, C.R., 90
GEERTS, J.W.M.H., 563
GELBKE, C., 455
GENOVESI, F., 80
GIBBONS, K., 121
GONFALINI, M., 522
GORIČNIK, B., 205
GRIBAA, R., 80
GUDMUNDSSON, T., 228
GUERILLOT, D., 497

GUIDA, M., 296
GUIDISH, T.M., 445

HALSEY, G.W., 406
HARPER, T.R., 182
HEDGE, T., 349
HELLEM, T., 121
HODNETT, B.K., 563

INGRAIN, D., 397

JACK, R.L., 317
JING, X.D., 208
JISKOOT, R.J.J., 359
JOURDAN, A., 14

KADDOUR, K., 80
KALKREUTH, W., 207
KENNEDY, M., 563
KJEMPERUD, A., 121
KNOKE, H., 489
KOHLER, N., 398
KORF, S.J., 537
KRILOV, Z., 205
KROLLPFEIFER, D., 455
KULAKHMETOV, N., 116
KUMMER, B., 521

LAFAILLE, A., 286
LAGOURETTE, B., 155
LAMBERT, J.-M., 465
LAMMERS, J.N.J.J., 387
LANE, M., 405
LE BEC, R., 562
LEBLANC, M., 70
LECOURTIER, J., 206, 401
LEFEVRE, P., 307
LEHMANN, B., 455
LEHNER, F.K., 131
LENORMAND, R., 166
LEPRINCE, P., 44
LOMBARDINI, M., 204
LUNELLI, M., 560

MAC GIOLLA CODA, E., 563
MADDALENA, D., 349, 410
MALLEN, J., 376
MANN, U., 142
MARCHAND, P., 376
MARENGO, S., 561
MARION, A., 376
MAROTTI, M., 205
MARQUAIRE, P.M., 562
MARTINENGO, S., 561
MATTEINI, L., 476
McMONAGLE, J.B., 563
McNAMARA, J.F., 405
MEIMON, Y., 260
MINARDI, P., 237, 410
MOLINARI, M., 560
MONDEIL, L., 397
MONDT, J.C., 432
MONNET, J., 525
MONTEL, F., 155
MOREU, M., 282
MORON, A., 282
MULDER, M.H., 483
MURGIA, C., 250

NAIM, P., 465
NEMCHENKO, N., 176
NESTEROV, I., 116
NICOLUSSI, F., 340
NIELSEN, J.B., 228
NIO, S.D., 121
NOÏK, C., 206

ODRU, P., 403
ÖNER, F., 207

PAGGINI, A., 528
PEARSON, M.J., 207
PEKDEGER, A., 489
PELLIZZARI, L., 340
PEN, S., 483
PERREAU, P.J., 330
PERSOGLIA, S., 422
PIANA, M., 522
PIRAS, P.P., 80
POLETTO, F., 422
PRAZZOLI, M., 407
PRENDIN, W., 349
PRESLES, C., 525

QUIGLEY, T.M., 272

RAIMBAULT, C., 544
RAMPOLLI, M., 409
RAVENNE, C., 497
REY-GRANGE, A., 317
RICORDEAU, A., 525
RITSEMA, I.L., 483
RIVENQ, R., 206
ROOS, J.A., 537
ROSS, J.R.H., 537, 563
ROSSINI, C., 80
ROUGE, J., 70
ROUGIER, R., 404
ROVENSKAYA, A., 176
RYLKOV, A., 116

SAINT-GUIRONS, H., 155
SAINT-JUST, J., 562
SANFILIPPO, D., 528
SARDA, J.P., 330
SASSI, W., 106
SCHENK, D., 489
SCHNEIDER, F., 106
SCHUPPERS, J.D., 90
SCHWOCHAU, K., 210
SEBASTIANI, G., 399
SENDEN, M.M.G., 553
SFLIGIOTTI, G.M., 7
SHPILMAN, V., 116
SIKLOS, J., 400
SIMMONDS, S., 272
SIMON, F., 400
SOLBAKKEN, Å, 34
SPARKS, C.P., 403
SPREUX, A., 14
STORESUND, W., 340
SUGIER, A., 376
SURLE, R., 237

TALLEC, E., 70
TAM, V.H.Y., 272
TAMBINI, M., 523
TERDICH, P., 80
TERRIBILE, A., 349
THOMAS, L., 489
TONELLI, S., 296

URSIN, J.-R., 209

VALAIS, M., 44
VALDISTURLO, A., 204
VAN WECHEM, H.M.H., 553
VAN DER WIELE, K., 563
VAN OMMEN, J.G., 537, 563
VANDEVELDE, J., 525
VEBENSTAD, K., 121
VEDRENNE, I., 562
VESTAVIK, O.M., 406
VINCKEN, L.M.J., 23
VOLPI, B., 80

WEHNER, H., 207
WEILL, J., 544
WOLF, S., 106
WRIGHT, I.W., 97

XANS, P., 155

YANG, C.S., 121
YE, S., 155

ZAITOUN, A., 398
ZANDERIGHI, L., 561
ZILIOTTO, F., 250
ZOETEMEIJER, R., 106
ZORIO, J.H., 397
ZUIDEVELD, P.L., 553
ZUNDEL, J.P., 398